Analysis
of
Vertebrate
Structure

Analysis of Vertebrate Structure

SECOND EDITION

MILTON HILDEBRAND
Professor of Zoology
University of California, Davis

Illustrated by
Viola and Milton Hildebrand

175 YEARS OF PUBLISHING
1807 1982

JOHN WILEY & SONS
New York • Chichester • Brisbane • Toronto • Singapore

Library of Congress Cataloging in Publication Data

Hildebrand, Milton, 1918-
   Analysis of vertebrate structure.

   Bibliography: p.
   Includes index.
     1. Vertebrates—Anatomy. 2. Morphology (Animals)
I. Title. [DNLM: 1. Vertebrates—Anatomy and
histology. 2. Anatomy, Comparative. QL 805 H642a]

QL805.H64 1982	596'.04	81-13049
ISBN 0-471-09058-1		AACR2

Printed in the United States of America

10 9 8 7

Preface

A first characteristic of this book is breadth. Description of structure is included, as it always is, because knowledge must precede interpretation. Interpretation of structure in terms of phylogeny is included, as is usual, because organic evolution is one of the greatest stories biology has to tell, and the lineages of vertebrate animals illustrate the story with more continuity and persuasion than do the known lineages of other animals or plants. Also included, however, are interpretations of structure that relate to function, development, and some other factors.

A second characteristic of this book is its placement of emphasis. Phylogeny is stressed, yet is less dominant than in some other texts. More than usual emphasis is given to interpretation of structure on the basis of function. This is currently the emphasis of much research, and thus provides the opportunity to add recent advances to classical knowledge. It brings to attention the variety and perfection of vertebrate structure, and it lends itself to analytical treatment to which, in my experience, students respond with interest. Moderate attention is given to the evident and engaging relationship between development and adult structure. Developmental biologists now place less emphasis on the broader aspects of evolution than they did several generations ago, yet much excellent work is being done on relative growth and evolutionary morphogenesis. Interpretations of structure based on body size, age, sex, and individual variation round out the narrative.

A third characteristic of the book is its style and coverage. It is uniform, integrated, and easy to use. All illustrations are cited in text by figure number if in the same chapter and by page number if elsewhere. Figures are fully labeled (not keyed to complex legends) and legends distinguish (by capital and lowercase letters) between the point illustrated and subordinate material. I intend that the presentation be sound and solid, yet not overly technical. The book does offer a lot of material. However, description as an end in itself is minimized—particularly in Part III. Details that do not serve interpretations are omitted. The qualifying words "usually" and "sometimes" are used frequently for the sake of accuracy, but specific exceptions to usual structure are "usually" omitted. A reasonably full vocabulary of the basic terms of morphology is provided without introducing unusual terminology or saying in Latin what can better be said in English. An effort has been made to sort concepts from illustrative material, and free use has been made of parenthetical statements to subordinate the examples and qualifications.

Fourth, the book is profusely illustrated with original artwork of high quality and uniform style. My wife and I worked together closely on the illustrations. It is principally her talent that makes the artwork distinctive; she made all the carbon pencil drawings and some of the pen and ink drawings (about 44% of the total). I selected materials (largely from my teaching collection), and made most of the pen and ink drawings (about 43% of the total). The publisher's staff artists completed the more diagrammatic illustrations (about 13% of the total) from my sketches and did all labeling. For each subject we sought an appropriate compromise between illustration that is so pictorial as to introduce extraneous detail, and illustration that is so simplistic as to reduce the living body to mechanical analogs. In preparing the second edition, 27 illustrations were deleted, 12 modified, and 47 added.

Fifth, this book presents vertebrate morphology as a living discipline. The transformations from fin to limb, from jawbone to ear ossicle, from branchial artery to carotid circulation, and many more, are classic stories that should be retold to new generations of students, yet note is also made of recent studies, unsolved problems, tentative explanations, active areas of research, and current trends in the discipline.

Finally, it is hoped that this book will be found interesting. No field of study as large and complex as this one is likely to interest every student in all its aspects. Application will be needed: The book is not intended as light entertainment. Nevertheless, it is a tragedy that the teaching of comparative anatomy has sometimes been a parade of dull facts, dusty skeletons, and much-preserved specimens. Nothing else in nature has more exquisite structure than the vertebrate body. I will be pleased if the reader occasionally forgets the forthcoming examination and reads on thinking "Wow, that's really something!"

Part I of the book is a survey of the vertebrates. Students must be able to recognize and relate the major taxa in order to follow Parts II and III. Brief descriptions that stress typical features and recognition characters serve as preparation for these parts. Extinct groups are included or not, according to their relation to what will follow.

In Part II, the customary organ system approach is used to present the general structure of the classes and subclasses of vertebrates and to review the structural evidence for their evolutionary relationships. Features that do not characterize major taxa or do not show progressive change between successive categories are deemphasized or omitted. The treatment is less detailed than in several other texts, yet includes, I believe, as much "meat" as can be learned in one course of study. Many students think that teachers of anatomy tend to pack into their courses more detail than will be useful. The proportionately few students who will become professional morphologists can easily find supplementary information elsewhere.

Part III presents knowledge and analysis of the major functional groups of vertebrates. Following two chapters on bone-muscle mechanics, successive chapters consider the major locomotor and feeding adaptations in order. Unrelated, often convergent, groups of animals are taken together to see how evolution has provided for their common requirements. Extreme modifications (which would be distractions in Part II) are included.

About 327 selected and partly annotated references are listed at the end of the book to give students a start on assignments and seminars and to indulge the curiosity of anyone having unanswered questions. Some 85 titles were deleted, and 133 added in updating for the second edition. The meanings of more than 150 word roots are given parenthetically where first used in the text. These are intended as aids to understanding, and hence memory, not as etymology lessons. A glossary of 565 terms is added to the second edition.

A preparatory course in general biology or zoology is assumed; most of the requisite fundamentals and terms are reviewed here, but would come as a big dose if none were already familiar. A prior foundation in embryology, physiology, or evolution is desirable to give the student the benefit of additional familiar ground, but is not assumed. Similarly, recollection of algebra and geometry, and a course in physics would make Part III easier, but more for the security provided than for formulas remembered.

Some students complain that textbooks present more than can be learned. Well, the gourmet is not able to eat all that is on the menu, but does not therefore desire few selections; others have other tastes and he will himself choose differently another time. I urge instructors not to serve up more of this book than can be assimilated in the time available. In a two-term course, most of the chapters can be presented. In a one-term course, selection is necessary. A balanced diet can be obtained by combining portions of both Parts II and III, thus illustrating more than one of the major approaches to the analysis of structure and abandoning the myth that any one approach would be "covered" merely by assigning all relevant chapters in this, or any other, text. Unassigned chapters can become the bases for special reports. Furthermore, I hope that students and instructors will permit the terms and the illustrative and parenthetical material in the book to support but not substitute for the ideas.

Of the nearly 1000 separate drawings comprising the illustrations, about 64% are completely original, about 20% are largely original, and the remainder are all redrawn with some modification. We acknowledge our debt to more than 150 authors whose illustrations have been used to a greater or lesser extent as a guide. Twenty-eight drawing are based on photographs in *Mammals of the World,* by E. P. Walker, 18 drawings are taken at least in part from one or another of the books by A. S. Romer, and 15 drawings are patterned on illustrations in *Traité de Zoologie.*

Prepublication reviewers are never responsible for the deficiencies of a book, yet they are responsible, if competent and conscientious, for great improvement in the manuscript. I have been fortunate in having such reviewers. Carl Gans and David B. Wake each read the entire text and made hundreds of very valuable suggestions. Five chapters were reviewed by Karel F. Liem, two chapters each by R. Glenn Northcutt and Irving H. Wagman, and one chapter each by Paul F. A. Maderson, Paul S. Moller, Terry A. Vaughan, and Marvalee H. Wake. I also thank reviewers of the first edition and the following reviewers of the manuscript for the second edition: Margery C. Coombs, University of Massachusetts; James Edwards, Michigan State University; Abbot S. Gaunt, Ohio State University; William B. Muchmore, University of Rochester; Bedford M. Vestal, University of Oklahoma; and Ralph M. Wetzel, University of Connecticut. I gratefully accepted most of the suggestions of most of them.

I shall be thankful to instructors and students who send me corrections, comments, and sources of material for subsequent revisions.

Milton Hildebrand
Davis, California

Contents

INTRODUCTION   **1**

1   The Nature of Vertebrate Morphology   3

Definition, Scope, and Relation to Other Disciplines   3
Why Study Vertebrate Morphology?   4
Some Principles and Considerations   4
Some Trends in the Study and Teaching of Vertebrate Morphology   22

PART I

SURVEY OF VERTEBRATE ANIMALS:
The Principal Structural Patterns   25

2   Nature, Origin, and Classification of Vertebrates   27

Relation of Vertebrates to Nonchordates   27
Relation of Vertebrates to Other Chordates   28
Origin of Vertebrates   33
The Construction and Interpretation of Classifications   34
A Classification of the Vertebrates   38

3   Fishes   41

Class Agnatha   41
Jaw-Bearing Fishes   45
Class Placodermi   45

Class Chondrichthyes   *47*
Class Acanthodii   *51*
Class Osteichthyes   *51*

4   Tetrapods   *57*

Transition to Terrestrial Life   *57*
Class Amphibia   *58*
Class Reptilia   *60*
Class Aves   *66*
Class Mammalia   *68*

PART II

THE PHYLOGENY OF STRUCTURE:
Evolution in Relation to Time and Major Taxa   *73*

5   Early Development   *75*

Gametes and fertilization   *75*
Cleavage   *77*
Gastrulation and mesoderm formation   *78*
Neurulation and neural crests   *82*
Establishment of the body plan   *82*
Fetal membranes and placentation   *83*

6   Integument and Its Derivatives   *87*

Functions of the System   *87*
Development and General Structure of Skin   *88*
General Development of Skin Derivatives   *89*
Integument of Fishes: Emphasis on Dermal Derivatives   *90*
Integument of Tetrapods: Emphasis on Epidermal Derivatives   *96*
Phylogeny?   *106*

7   Teeth   *107*

Origin and Structure   *107*
Development   *108*
Attachment and Replacement   *109*
Evolution of Teeth   *111*

8   Head Skeleton   *117*

Importance of the Skeleton to Morphology   *117*
More About Hard Tissues   *118*
Origin, Function, and Segmental Nature of the Head   *119*
Components of the Head Skeleton   *120*
Relations of the Cranial Components   *125*
Evolution of the Head Skeleton   *127*
Features of the Skull   *144*

9 Body Skeleton   147

   Structure and Development of Vertebrae   *147*
   Evolution of the Spine   *150*
   Ribs   *157*
   Sternum   *158*
   Origin of Appendages   *159*
   Median Fins   *160*
   Structure and Evolution of Girdles   *162*
   Structure and Evolution of Paired Appendages   *166*
   Some Miscellaneous Bones   *176*

10 Muscles and Electric Organs   177

   Function and Gross Structure of Muscles   *177*
   Fine Structure, Contraction, and Physiology   *179*
   Force, Work, and Torque   *183*
   Categories of Muscles   *185*
   Evolution of Muscles   *187*
   Electric Organs   *201*

11 Coelom and Mesenteries   203

   Nature and Function   *203*
   Development, Evolution, and Recapitulation   *204*

12 Digestive System   209

   Function and General Structure   *209*
   Development of the Gut   *210*
   Structure and Evolution   *211*

13 Respiratory System and Gas Bladder   227

   General Function and Requirements   *227*
   Aquatic Gas Exchangers   *228*
   Ventilation of Internal Gills   *235*
   Aerial Gas Exchangers   *235*
   Ventilation of Lungs   *244*

14 Circulatory System   247

   General Function and Nature of the System   *247*
   The Heart   *248*
   Blood and Blood-Forming Tissues   *256*
   Blood Vessels   *258*
   Some Functional Considerations   *269*
   Evolution of the Lymphatic System   *277*
   The Circulatory System and Thermoregulation   *280*

15   Excretory System and Osmoregulation   281

General Nature and Development of Kidneys   *281*
Evolution of Kidneys   *283*
Kidney Structure in Relation to Osmoregulation   *287*
Urinary Bladders   *291*

16   Reproductive System and Urogenital Ducts   293

Early Development and Ancestry of Gonads   *293*
Structure of Gonads   *294*
Urogenital Ducts and Accessory Organs   *300*
Cloaca and Derivatives   *305*
Copulatory Organs   *305*
Reproductive Strategies   *308*

17   Nervous System: General, Spinal Cord, and Peripheral Nerves   311

Elements of the Nervous System   *311*
Development of Spinal Cord and Peripheral Nerves   *315*
Spinal Cord   *316*
Evolution of Spinal Nerves   *319*
Cranial Nerves   *321*
Autonomic System   *325*

18   Nervous System: Brain   329

How the Brain is Studied   *329*
Development of the Brain   *330*
More About the Organization of the Brain   *332*
Posterior Brainstem: Medulla through Midbrain   *333*
Anterior Brainstem: Diencephalon   *338*
Cerebellum and Pons   *339*
Cerebrum   *341*
Evolution of the Brain   *349*
Support and Nourishment of the Central Nervous System   *353*

19   Sense Organs   355

Some Relatively Simple Receptors   *355*
Organs of Olfaction and Taste   *357*
Lateral Line System   *359*
Ear   *361*
Eyes   *367*

20   Endocrine Glands   377

General Nature of Endocrine Glands   *377*
Structure, Function, and Evolution of the Glands   *378*
Neurosecretion   *385*

PART III

STRUCTURAL ADAPTATION:
Evolution in Relation to Habit and Habitat   387

21 Structural Elements of the Body   389

    Animals as Specialists   389
    Properties of Supportive Materials   390
    Stress and Stress Lines   392
    Use and Design of Structural Elements   396
    Unions of Structural Elements   403

22 Mechanics of Support and Movement   411

    Magnitude and Direction of Forces   411
    Bone-Muscle Systems as Machines   417
    Mechanics of Body Support   421
    Mechanics of Motion   428
    Free-Body Diagrams   432
    Effort, Energy, and Efficiency   433

23 Running and Jumping   437

    Advantages of Speed and Endurance   437
    Cursors, Saltators, and Their Skills   438
    General Requirements of Cursors   440
    Length of Stride   440
    Rate of Stride   445
    Mass, Endurance, and Design for Economy of Effort   449
    Stability and Maneuverability   453
    Gaits   454
    Saltation and Bipedial Running   457

24 Digging; and Crawling without Appendages   463

    Advantages of Digging   463
    Several Ways of Digging   464
    Fossorial Vertebrates   466
    General Requirements of Diggers   468
    Keeping Dirt Out of Mouth, Sense Organs, and Lungs   469
    Maneuvering and Working Underground   469
    Breaking the Soil   470
    Transporting and Disposing of Soil   479
    Crawling without Appendages   480

25 Climbing   485

    Advantages of Climbing   485
    Scansorial Vertebrates   486

Requirements and Basic Mechanisms of Climbers   490
Adaptations for Propulsion   493
Adaptations for Maintaining Contact with the Substrate   497

26   Swimming and Diving   505

Advantages of Swimming and Diving   505
The Skills of Swimmers and Divers   506
General Requirements of Swimmers and Divers   510
Drag   510
Propulsion   514
Control of Vertical Position   522
Stability, Braking, and Steering   523
Other Adaptations of Secondary Swimmers   526

27   Flying and Gliding   531

Origin and Advantages of Flying and Gliding   531
Vertebrates That Parachute and Glide   532
Flyers and Their Skills   535
General Requirements of Flyers   538
Mostly About Lift   538
Drag   541
Gliding, Soaring, and Formation Flying   543
Flapping Flight   545
Flight Control   548
Wing Structure   551
Further Structural Adaptations   554

28   Feeding   559

Jaw Mechanics   559
Locating and Securing Food   566
Processing Food in Mouth and Pharnyx   575

CONCLUSION   **585**

29   Vertebrate Morphology: Diversity, Unity, and Challenge   587

Glossary   591

References   605

Index   621

Analysis
of
Vertebrate
Structure

INTRODUCTION

1

The Nature of
Vertebrate Morphology

DEFINITION, SCOPE, AND RELATION TO OTHER DISCIPLINES

Anatomy is the science of the observation and description of structure. Gross structure, microstructure, and ultrastructure are included, though this book stresses the first and presents little of the last. Morphology is the broader science of *relating* and *interpreting* observed structure.

Since structure is influenced in many ways, information is needed from many sources if interpretation is to be reasonably complete: (*1*) Much interpretation follows from knowledge of the historical origins, or phylogeny, of form. Thus, morphology relates closely to paleontology, taxonomy, and the principles of evolution, and draws upon other fields such as serum chemistry and analysis of chromosome structure that provide evidence of evolutionary affinity. (*2*) Also, much interpretation is dependent upon knowledge of functional adaptation (i.e., suitability to fill a need effectively). Morphology is therefore related to ecology, ethology, physiology, biophysics, and biochemistry. Some adult structure is interpreted through (*3*) knowledge of embryology, and some through (*4*) analysis of the relation between form and size. Other structure is (*5*) age-dependent or (*6*) sex-dependent. Finally, (*7*) individual variation of structure may have genetic, nutritional, pathologic, or other environmental origins.

Thus vertebrate morphology relates to many other sciences. It is desirable for the morphologist to supplement training in the principles of biology at all levels and grounding in vertebrate structure by gaining familiarity with the concepts and methodology of one, or preferably more, related disciplines.

This book describes the anatomy of the major structural and behavioral groups of vertebrate animals, and interprets their morphological differences primarily in terms of ancestry and function, employing related fields as needed to further this objective.

WHY STUDY VERTEBRATE MORPHOLOGY?

In order to generate the interest and application needed to derive full value from any course of study, the student must be convinced that the returns will justify the effort. Different students derive different benefits from studying vertebrate morphology, but the following advantages, in varying proportions, accrue to most students:

1.   Knowledge of anatomy has direct application to many specializations within biology. The surgeon and veterinarian, experimental embryologist and neural physiologist, paleontologist and pathologist, all need to be familiar with the structure of their materials.

2.   Knowledge of animal structure has made important contributions to human health and technology. The selection of experimental animals, the conduct of innumerable studies in basic and applied physiology and medicine, and the design of prosthetic devices are examples. Some engineers are studying animals for clues to improved design of bearings, ships, and aircraft.

3.   Study of morphology increases the biologist's understanding of the materials at hand. Through interpretation one becomes less of a practitioner and more of a professional, less a technician and more a scholar, less a catalog of facts and more an expert. This benefit is difficult to measure, yet can be of great value to the individual.

4.   Analysis of structure may increase the biologist's interest in, and even fascination with, animal form. This subtle benefit can be very rewarding.

5.   Vertebrate morphology provides particularly favorable evidence for the process and product of organic evolution. It contributes to the answering of questions that have long been important to man: What forces govern the stream of life? How can one gain perspective in time and space? How can one account for the perfection of the animal body?

Study of vertebrate morphology is unlikely to bring wealth, fame, or influence, but it *is* likely to increase the biologist's competence and pleasure in his or her work.

SOME PRINCIPLES AND CONSIDERATIONS

HOMOLOGY AND ANALOGY   Features of two or more organisms are **homologous** if they share common ancestry (see examples 1 to 4, Figure 1-1). Homology is established if the features can be clearly linked through time by continuity in the fossil record, and is reasonably sure if they can be shown to develop similarly in the embryo from identical primordia. Homology may be difficult to establish in specific instances, and the acceptable remoteness of the common ancestor may be disputed. Nevertheless, the concept is clear and very important.

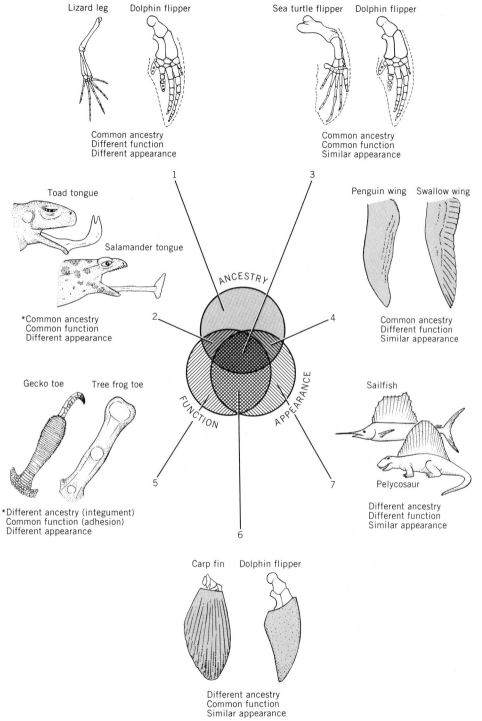

Lizard leg Dolphin flipper

Common ancestry
Different function
Different appearance

Sea turtle flipper Dolphin flipper

Common ancestry
Common function
Similar appearance

5

Toad tongue

Salamander tongue

*Common ancestry
Common function
Different appearance

ANCESTRY

1   3

2   4

FUNCTION   APPEARANCE

5   7

6

Penguin wing Swallow wing

Common ancestry
Different function
Similar appearance

Gecko toe Tree frog toe

*Different ancestry (integument)
Common function (adhesion)
Different appearance

Sailfish

Pelycosaur

Different ancestry
Different function
Similar appearance

Carp fin Dolphin flipper

Different ancestry
Common function
Similar appearance

Figure 1-1 THE DISTINCTIONS AND RELATIONS AMONG COMMON ANCESTRY (HOMOLOGY), COMMON FUNCTION (ANALOGY), AND COMMON APPEARANCE. (* In 2, the tongues are homologous, but the projection mechanisms are not; in 5, the critical parts of the integument are not homologous, but the toes are. Circle diagram modified from Hailman, 1976.)

Features of two or more organisms are **analogous** if they share common function (examples 2,3,5, and 6, Figure 1-1). Analogy may be inferred from structure, or from comparison with other similar features, but is surely established only by behavioral and biomechanical analysis. Analogous features may or may not also be homologous. Analogy can also be difficult to prove in specific instances. Acceptable levels of equivalence of function are debatable, and even the concept and its definition remain somewhat controversial. In this book the frame of reference is usually the class, subclass, or order where the concept is exceedingly useful and seldom ambiguous.

Features of two or more organisms may also be related by similarity of **appearance** (examples 3,4,6, and 7, Figure 1-1). Such features are usually also analogous, and frequently are also both analogous and homologous. Features related by appearance only are unusual unless the chance resemblance is superficial. If two features are related by homology and appearance but not by function, one of them has usually undergone an evolutionary shift of function (see mention of *preadaptation* on p. 12).

Here are further examples of various relationships among homology, analogy, and appearance; how would you place each on Figure 1-1? (*1*) The quadrate bone of reptiles, which supports the jaw, and the incus of mammals, which is an ear ossicle (see the figure on page 135). (*2*) The penis of mammals and the claspers on the pelvic fins of sharks, both of which convey sperm to the female (see the figure on p. 163). (*3*) The ear ossicles of mammals (derived from cranial bones) and the ossicles of certain fishes (derived from vertebrae) both of which transmit sound waves to the ear. (*4*) The large cannon bone in the lower leg of cloven-footed mammals (e.g., cow, vicuna) and a similar bone in the leg of the ostrich (see the figure on p. 453); each adapts the foot for running.

Serial homology is a rather independent concept, which is mentioned here because of the similarity of terms. Structures are serially homologous if they occupy different spatial positions in a series of like structures. The separate vertebrae are serially homologous, as are the different teeth of a tooth row, the several gill arches of the series, the successive muscle segments along the back of a fish, and the many tubules in the kidney of a lower vertebrate. It is usual for the structures of such series to form gradients: Vertebrae become larger toward the pelvis, teeth often become more complex toward the back of the mouth, gills may become smaller toward the posterior end of the series. Serial homologs have similar potential for change: Any embryonic vertebra behind the ribs will form an articulation with the pelvis if experimentally placed adjacent to it. Also, change usually affects more than one element of a series: If one tooth becomes larger or more complex, its neighbors tend to do likewise.

Structures have **sexual homology** if they develop from equivalent embryonic primordia, yet are sexually dimorphic. The ovary is the sexual homolog of the testis, and the clitoris is the sexual homolog of the penis.

FORM AND FUNCTION   It is clearly evident that form and function are closely correlated; appropriate structures enable an animal to perform specific tasks. Such correlations are the foundation of Part III of this book. However, the basis for the correlation is less evident and has been the subject of much debate. It seems to follow from Darwinian evolutionary theory that function precedes form and provides the selective advantage that guides change of form (i.e., mutants that are best suited to fill an existing biological need tend to survive). However, form can precede function both in embryology (lungs develop before they are used) and evolution (see mention of *preadaptation* on p. 12), and simultaneous change in form and function must be considered. Sometimes the choice of precedence can be made, if at all, only on a theoretical basis.

Although the correlation of a specific form with a specific function can usually be made with confidence, errors have too often been made by *assuming* that a particular structure is adapted to a particular purpose because that seemed obvious, or plausible, or the only probable interpretation. Also, it is tempting to generalize from one study animal to a large group of animals without adequate attention to variation within the group. Correlations of form and function should be considered tentative until it is shown by principles of functional design, coupled if possible with experiment or field observation, that the observed form does, indeed, fulfill the postulated needed interaction with the environment.

SOME APPROACHES IN MORPHOLOGY The following are some of the considerations that guide the morphologist in recognizing and distinguishing homology and analogy, and in correlating form with function.

1. It is often useful to compare form in numerous animals that are known to be unrelated, yet share the same function, such as fast swimming, or climbing by adhesion, or feeding on ants. Similarly, it is frequently helpful to study form in a large group of related animals that have developed different ways of feeding, moving, reproducing, or other functioning.

2. One must not infer that analogy indicates homology. Thus, horses and cattle share large size, hoofs, and similar molar teeth because each runs well and eats grass. The common structures evolved in response to common habits, and were not retained from common ancestry. Hedgehogs, porcupines, and one of the egg-laying mammals (echidna), though unrelated, all have quills because quills provide a satisfactory defense for animals that cannot run or fight.

3. When structures are studied in combination, it is found that correspondence of many parts bespeaks evolutionary relationship, whereas correspondence of few parts may result from other causes. For instance, numerous cartilaginous and bony fishes have electric organs. This suggests common origin, but the fishes are so different in regard to so many other characteristics that it is virtually certain their electric organs evolved independently. Likewise, horses and cattle each have complicated enamel patterns in their cheek teeth, but their stomachs, dental formulas, and skull structures argue against close affinity.

4. The study of complicated structures, although more difficult, may be more rewarding than the study of simple ones. Ribs are just too simple to hold many secrets. The skull, on the other hand, has so many functions and so many bones, foramina, and contours that it reveals much of the habits and history of its former

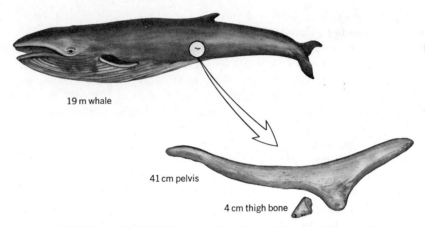

19 m whale

41 cm pelvis

4 cm thigh bone

Figure 1-2   VESTIGIAL ORGAN illustrated by the pelvis of a finback whale.

owner. Furthermore, corresponding structures of unrelated animals are unlikely to be similar merely by chance if they are also complicated.

5.   The primitive, or ancestral condition of a structure is more likely to be found in an animal that retains the ancestral condition of *other* parts. One must avoid circular reasoning. One cannot *prove* that a feature is primitive because it is found in an animal that is *assumed* to be primitive. Nor can one assume that a feature is primitive just because it is found in an animal that is *known* to be relatively primitive—even long surviving species have some highly specialized structures. Nevertheless, an animal that is *known* to be primitive in the expression of several characters is of particular importance. Thus, if one is working on the phylogeny of the types of placentas found among rodents, it would be well to see what sort of placenta the mountain beaver has. It could be quite ordinary, but if it were unusual, it would be the more interesting because this animal (not really a beaver) is known by evidence of its skeleton, muscles, and fossil record to be the most primitive living rodent. Part II of this text will pay special attention to several animals that retain numerous primitive features: *Polypterus* among fishes, *Sphenodon* among reptiles, the platypus among mammals, and others.

6.   It is also important to note the presence of vestigial or degenerate organs having no function at all. It is safe to assume that such structures were functional in the ancestor, and this tells us something about the evolution of the descendant. The blind eyes of burrowing moles and cave fishes indicate that their ancestors could see, and tiny bones in pythons and whales tell us that the remote ancestors of these animals had legs (Figure 1-2).

RECOGNITION AND ASSESSMENT OF VARIATION   The morphologist rarely describes an individual animal *as* an individual; single specimens are studied to learn about *kinds* of animals. Individuals of a kind vary so much among themselves that a single specimen is not adequately representative.

Most characters vary independently of one another—a specimen may be average for one yet extreme for another. However, some characters are related in such a way that variation in one can be correlated with variation in another. For instance, among certain mammals, an individual with extra-high tooth crowns is likely also to have extra-large premolars and an extra-long jaw, because these characters are functionally related.

Structures that have relatively great variation among individuals of a kind may be of limited value for interpretation of form. Molar teeth and ankle bones must be just right to function well, so they do not vary much. Breastbones, by contrast, can be flatter, or longer, or more segmented than usual without much impairment of function. Accordingly, there is more variation in breastbones and less can be interpreted from their configurations.

It may be important to study **sexual dimorphism** either so that the gender of a particular vertebrate can be established or so that sexual characteristics will not be mistaken for species characteristics or for adaptive features of another nature. Of course, sexual dimorphism relates most to the gonads, genital ducts, external genitalia, and accessory sex organs. These may be conspicuously different in the two sexes or closely similar. Sex differences of the genitalia may be much more evident at one stage of the breeding cycle than at another because of alterations in size, position, coloration, secretions, or vascular congestion.

Other sexual dimorphism does not relate directly to the genital system, yet correlates with distinctions between the sexes in regard to sex role and behavior. These include differences in the coloration, amount, and distribution of scales, feathers, or hair, in clasping organs and pelvic architecture, and in the presence or development of crests, spurs, antlers, tusks, and scent glands.

Adult body size and rate of growth are frequently sexually dimorphic, and gender differences in body proportions and configuration are common. Thus, males commonly have coarser and relatively heavier skeletons. There may be sexual differences in the muscular and vascular systems, and in the brain and pituitary gland.

The morphologist also needs to recognize **age variation** and to avoid confusing it with variation of another kind. The accretion of successive rings or layers on hard tissues often accurately reflects alternating periods of slow and rapid growth. Like tree rings, these relate to times of favorable climate or nutrition, which usually are seasonal. In favorable circumstances, the scales and otoliths (bones within the inner ear) of bony fishes, the scutes and long bones of turtles, the baleen and ear plugs of whales, and the dentine and cement of the teeth of mammals all show marked growth rings.

Hard parts also correlate with age in other ways: The centers where developing bones first ossify appear in regular sequence, the various teeth of mammals erupt and are replaced at different but regular ages, sutures between bones close on schedule, and epiphyses at the ends of the long bones join the shafts at given times. The ages of young humans and of some other animals (for which the sequences have been worked out) can be determined within fairly narrow limits by x-ray analysis of the skull, wrists, or other joints.

The following may also be age-related, though usually with less precision: size and structure of the baculum (a bone in the penis of some mammals), structure of the cranial wall (birds), relative size of orbits, size and configuration of horns and antlers, porosity of bones, development of crests and tuberosities on the skeleton, position of basicranial axis, and relative development of rostrum or facial area.

The structure of soft tissues tends to be less precisely age-related, yet the weight of the lens of the eye (measured after careful removal and desiccation) is sometimes a reliable index of age. Changes in the skin, in the progression of molt, in the composition of muscle, and in the circulatory system, are also ascribed to advancing age.

If an animal is extinct, rare, or otherwise difficult to obtain, it may be necessary to work with few specimens, but the morphologist tries to obtain random samples of adequate size. He may first sort the individuals by sex, age, geographic location, or other criterion, keeping enough specimens in each subsample (about 25 is usually adequate) to assess these variables. The individuals of such subsamples still vary among themselves, however, and this **individual variation** must be analyzed if the true nature of each characteristic is to be identified. The nature of typical structure, the range of variation, and the variability, or tendency to vary, are all population characteristics; they cannot be learned from single specimens.

Statistical analysis of population characteristics is beyond the scope of this book, yet should be studied by aspiring morphologists. Even several relatively simple parameters can be very helpful: The *arithmetic mean* expresses the "average" condition. The *standard deviation* shows the degree to which the separate values tend to cluster around the mean. The *standard error of the mean* enables one to judge if means derived from different populations (or samples from different populations) are significantly different. The *coefficient of variation* makes it possible to say if animals of different absolute size are comparably variable, that is, if a shrew varies as much for a shrew as an elephant does for an elephant. Morphologists commonly use these and many other parameters, and also employ various graphical techniques.

Statistics, however important, do not substitute for attention to accuracy, reliability, and significance. Results are *accurate* if free of error, *reliable* if repeatable, and *significant* if meaningful. The grade given a student on an examination is accurate if additions and recording were done with care. It is reliable if another reader would have assigned the same grade, and if the student, taking the exam again, would be assigned a closely similar score. It is significant if the grade assigned indicates the student's progress toward desirable course objectives. Research, like student grades, is best when accurate, *and* reliable, *and* significant.

CONTRIBUTIONS FROM PALEONTOLOGY   Paleontology is the study of prehistoric life as revealed by fossils. Some vertebrate fossils are mere trackways or body imprints made millions of years ago in mud that was later converted to stone. Rarely is an entire animal preserved: Insects trapped in ancient pitch have retained their delicate structures as the pitch gradually turned to amber, and great mammoths that stumbled into glacial crevasses have been preserved for 50,000 years in a natural deep freeze. Usually only teeth and bones are preserved, and even for these the slow process of petrification has replaced most of the original organic tissue by an exact copy in hard minerals.

To be preserved, the skeleton must be protected from the destructive forces of weathering, and this means that it must be covered up soon after it is released by decay from the tissues it supported in life. Landslides, wind-driven sand, and volcanic ash cover some skeletons, but most that become fossils are covered by silt and sand deposited in lakes or streams, or on the flood plains of wide valleys. In time, the mud is converted by pressure to shale and the sand to sandstone. Upheavals of the earth's crust, and subsequent erosion, expose the contained fossils. Sedimentary rocks are thus a gigantic filing cabinet for the fossils they contain: The oldest fossils are in the oldest strata and the recent fossils are in recent strata. Evolutionary changes in the structure of animals progress step by step with the accumulation of earth sediments.

As an interpreter of animal structure, the vertebrate paleontologist is confronted with many difficulties. He or she usually must work with broken materials and fragments, seldom finds fossils of animals that lived in arid uplands where skeletons disintegrate quickly in dry air and hot sun, and with rare exceptions, must work with a single organ system. Nevertheless, paleontologists have described about three times as many extinct vertebrates as there are surviving vertebrates. The morphologist, embryologist, zoogeographer, serologist, and other specialists have provided part of the story of evolution, but the paleontologist has contributed the most (but see opinion of cladistic systematists on page 35).

EVOLUTION AND HABITAT   Interpretation of structure is made more meaningful by familiarity with the major features of the evolutionary process. Supplementary reading on that subject is highly recommended (see the References at the back of this book). There is space to present only the barest outlines here, and it must be assumed that the reader has general knowledge of the origins of heritable variation, natural selection, and earth change.

A major consideration is the relation between evolutionary change and the stability of the environment. With minor exceptions, evolution results from the interplay between changing environments and adapting organisms. Each kind of animal becomes adapted to, and dependent upon, a particular kind of life (predation, seed-eating, grazing) in a particular kind of habitat (marsh, stream, meadow). If the habitat is large and constant (oceans, tropical forests, coniferous forests), the animal inhabitants have time to become well-adjusted. It is most advantageous for each species to remain about as it is, so natural selection tends to prevent change. Large habitats do move slowly in earth history (e.g., forests advance and recede), but most of the animals move with them, remain adapted, and change relatively little.

Although the general habitat may shift slowly in space, the restricted habitat of a population of animals may instead slowly change. Thus, a shoreline may gradually become more rocky, or a pine-fir forest that is isolated between mountain ranges may become more alpine in character as timberline shifts southward. The average expression of the characters of a kind of animal is then no longer quite optimum. Natural selection therefore tends to cause the animal to become a somewhat different kind of animal. If the habitat alters as a unit, the evolutionary change is in a more or less straight line and is said to be **linear.** If the old habitat subdivides into different units, different parts of the original animal population become isolated from one another, adapt independently, and evolution is **branching.**

When a habitat alters too fast for a resident species to adapt to its new character, then, so far as that species is concerned, the habitat does not change, but disappears: As the earth slowly tilts, an inland sea drains away leaving only shoreline terraces on dry hills; meadows are invaded year after year by pine seedlings, and finally succumb to the forest. As a habitat becomes less and less satisfactory, the force of natural selection, that is, **selection pressure,** for the old way of life weakens. Finally, extreme variants in the population may be better suited for life in a new habitat than in the old, provided that one or several new habitats are physically available and not already occupied by effective competitors. Selection pressure then becomes strong and shifts away from the old life style toward the new.

Several factors may contribute to success in this hazardous kind of evolution. **Specialized structures** are those that have become modified to perform restricted functions with great effectiveness, whereas **unspecialized structures** (or generalized structures) are suited to perform adequately a less restricted function or a variety of functions. Unspecialized structures have more capacity for evolutionary change, and hence favor survival when change is essential. Thus, the ancestral five-toed foot has been converted to a springing support, wing, paddle, grasping organ, and so on. Specialized organs are satisfactory as long as their restricted functions are needed (it is probable that there will long be ants for anteaters and krill for baleen whales), but if a new function is needed, such organs rarely can adapt.

When a species must adapt quickly to altered conditions, it is not granted time to evolve an entirely new complement of structural attributes. It must rely for a time on the intensified or altered use of attributes it already has. Natural selection may "discover" that a structure that was useful in one way before can now be useful for another purpose. Such structures are said to have **preadaptation.** For instance, it was a long and major task for evolution to convert the walking legs of proavian reptiles into the wings of birds. It was relatively quick and simple for several groups of birds to use their wings as waterfoils for swimming instead of as airfoils for flying—a shortcut of tens of millions of years in the evolution of effective paddles (example 4, Figure 1-1). Unspecialized form and preadaptations are good cards to hold in the game of find-a-new-way-of-life-or-become-extinct.

Just as habitats can disappear, so they can appear. A new marsh, formed close to existing marshes, will be populated from nearby, and no evolution will occur. A new inland sea, however, if extensive and isolated, is a place of evolutionary opportunity for such animals as can get a start there. Similarly, the habitat may be old, but its availability may be new: Land first became available to vertebrates when reptiles and certain amphibians evolved. There were sufficient land plants and arthropods to provide food and shelter. There, inviting colonization, were millions of square miles of diverse new habitats having no competition from established animals. In this infrequent circumstance, evolution rapidly creates various new kinds of animals. It is said to be **radiating.**

Animals that are successful in evolution's stern game of adapt or perish become extinct by progressive change: Ancestors disappear as descendants evolve. Many more kinds of animals leave no daughter species. They become extinct because they cannot meet the competition, because disbalance occurs in predator-prey relations, or because there is failure to adapt to changes in the noncompetitive parts of the environment.

EVOLUTION AND TIME   Figure 2-5 presents the time scale in relation to vertebrate evolution. The numbers are easy to read but difficult to comprehend. It helps to relate geologic time to something relevant and finite. One example equates all the time since life began on earth with one year: Then, vertebrates appeared about October 20, mammals evolved on December 7, and *Homo sapiens* made his debut at 11:48 PM on December 31.

This example teaches proportion but fails to show the magnitude of the full-scale phenomenon. Perhaps that is not possible, but consider this: Our galaxy is an assemblage of 10 billion stars distributed in the shape of a wheel so large that light requires nearly 100 thousand years to traverse its diameter. Yet, the entire galaxy has revolved once around in space since the coal bogs dried out and the dinosaur radiation began. From time to time, galaxies pass through one another. Such a collision is completed in about one-seventh of the time it took the horse to evolve from one genus to the next. Since vertebrates evolved, the far edges of our expanding universe have receded a distance 40 times the diameter of our galaxy!

The fossil record presents many relatively complete and continuous lineages: Elephants were once numerous and diversified, and camels came in all sizes and were abundant on four continents. Horses, turtles, crocodiles, the various kinds of dinosaurs, and many other groups are all assemblages of clearly related genera. Such a lineage is called a **phyletic line,** and is usually represented by genera that are related in time by linear and branching evolution, and through extinction by progressive change. Systematically, a phyletic line is often a single family but can also include all or part of other taxa. For reasons explained below, many phyletic lines appear suddenly in the fossil record and, if not still surviving, seem to disappear from the record rather than merge into one or more other lines.

Different phyletic lines evolve at different rates, each line evolves at different rates at different times, and different characters of one line evolve at different rates at the same time. With these qualifications, however, some examples can be given of the survival of lineages and of slow and fast rates of evolutionary change. The group characters of rabbits, armadillos, and turtles were established more than 65 million years ago. Opossums have been opossums for 100 million years, and the same is true for crocodiles. Some groups of fishes (coelacanths, dipnoans, sharks) have survived for 300 million years.

Peak rates of evolution for vertebrate classes in terms of numbers of new genera evolved per million years have been about as follows: placoderms, 3; cartilaginous fishes, 2; bony fishes, 6; amphibians, 3; reptiles, 12; and mammals, 30. The rate at which genus replaced genus in the horse lineage was 1 per 7.5 million years, which is high for vertebrates generally, but not particularly high for mammals. The highest rates occurred among land vertebrates, probably because they have been the most adaptable and their habitats have come and gone the fastest. Social systems that restrict the size of effective breeding populations may also contribute.

Even though there are gaps in the fossil record between phyletic lines, it is agreed by many evolutionists that the evolutionary process is the same between phyletic lines as within phyletic lines. New lines appear under the conditions noted above that favor rapid evolutionary change. Two factors contribute to the relatively incomplete nature of the fossil record at the times that new lines are established: First, fewer animals are fossilized because, when evolutionary change is maximum, kinds of animals do not survive long, and there are relatively few individuals in the affected populations. Second, such fossils as are formed are less often found because the animals tend to be small, and the same extreme conditions of earth change that increase selection pressures also tend to erode and destroy the earth's crust, thus destroying the contained fossils. Other evolutionists, however, believe that the process of evolution may be accelerated between lines. This is a subject of current interest and controversy.

EVOLUTIONARY TRENDS   It is a striking fact that *within* phyletic lines, each adaptive change tends to progress in more or less the same direction without stopping, zigzagging, or reversing. Such gradual changes are called **evolutionary trends** and, though not universal, they have been usual for large populations evolving at moderate rates. Evolutionary trends are oriented and prolonged by selection pressure. The characteristic continues to develop because it continues to be advantageous.

A common, though by no means universal, trend has been to large body size. Repeatedly, and in unrelated lineages, there has been gradual increase in size from modest ancestor to gigantic descendant. Examples of the advantages of large size, and adaptations for supporting great weight are included in Chapter 22, and some other structural responses to large body size are noted later in this chapter. A common trend for the reduction in number of serial parts is exemplified by the teeth of many lineages, lateral digits of hoofed mammals (Figure 1-3), gill arches of lower fishes, and cranial bones from fishes to mammals. Conversely, teeth and muscles have increased in number in some lineages. Trends leading to change in relative size of parts of the body have been common. The gradual development of tusks, antlers, and unusual beaks are examples. D'Arcy Thomp-

| Ancestral condylarth | Eocene horse *Hyracotherium* | Miocene horse *Miohippus* | Recent horse *Equus* |

Figure 1-3 EVOLUTIONARY TRENDS IN PROPORTIONS AND NUMBER OF SE-RIAL PARTS of the left forefoot of the horse. (Not drawn to scale.)

son demonstrated in 1917 that trends in body proportions are admirably de-scribed by progressively distorting a grid based on Cartesian coordinates as shown in Figure 1-4. Other trends involve increase in specialization of parts, such as the formation of elaborate enamel patterns in the grinding teeth of horses.

The recognition of trends is important as a means of following the path of evolution. Many examples are identified in this book. Rarely can one identify a single, isolated trend, because initial or primary changes usually necessitate de-pendent or secondary trends in response to functional necessity. Thus, increase in body size ultimately requires postural, skeletal, and muscular alterations; in-crease in the efficiency for grinding of premolar teeth requires lengthening of the face and altered jaw mechanics.

A degree of rectilinearity is inherent in the nature of most trends. Body size, body proportions, and number of parts in a series can only increase, remain the same, or decrease. Such trends seldom progress at a constant rate, and often have temporary arrests, yet the direction of change is usually constant because the direction of advantage to survival is usually constant. Trends in form and degree of complexity are less rectilinear.

Even trends of the stop-or-go variety usually come to an end if, and when, the condition of maximum advantage is reached: The sabers of saber-toothed tigers increased until the Oligocene epoch and then remained about constant for 40 million years. Furthermore, many trends have theoretical end points: Change toward longer-wearing teeth ends with ever-growing roots; change toward loss of lateral digits ends with one toe; change toward loss of vision ends with rudi-mentary eyes.

Modern horse *Equus*

Hypothetical intermediate stage
resembling the Miocene horse *Merychippus*

Ancestral horse *Hyracotherium*

Figure 1-4 EVOLUTIONARY TRENDS IN PROPORTIONS AND SIZE of the horse skull shown by the progressive distortion of a grid.

Still, some trends do *seem* to stop short of maximum advantage whereas others *seem* to go too far. For instance, some extinct elephants had tusks so large they curved and crossed and could no longer effectively thrust, pry, or dig. There are several ways in which this can come about. First, natural selection makes no provision for nature's senior citizens. It improves the fitness of breeders and potential breeders, but neglects individuals that are no longer productive.

Also, if the continuance of a trend is advantageous in one respect but disadvantageous in another, then the trend will stop when the advantages and disadvantages of further change are in balance. This is common, though often difficult to identify in specific instances. Increased dermal armor might give a fish added protection against predation but make it less maneuverable; increased slenderness of limb might give an antelope more speed but make it less able to avoid injury; increased curvature of the beak might make it easier for a bird to secure one kind of food but more difficult to secure another. The result is optimum compromise, not optimum structure in regard to any one function.

PARALLELISM AND CONVERGENCE   Descendant lineages may resemble one another closely in characters that are not present in the common ancestor. **Parallelism** is evolutionary change in two or more lineages such that corresponding features undergo equivalent alterations without becoming markedly more or less similar. Descendants are about as much alike as were their ancestors, and at each stage the correspondence may be close.

The kangaroo rats of western North America and jerboas of Africa and Asia show striking parallelism: Each has long hind limbs, short forelimbs, loss of lateral toes, lax fur of tan color, long tail with white tip, large eyes, inflated bony ear capsules, and compacted neck vertebrae (Figure 1-5). It is unlikely that the common ancestor had any of these characters; on the basis of other features, these rodents are placed in different suborders. Similarly, the golden moles of south

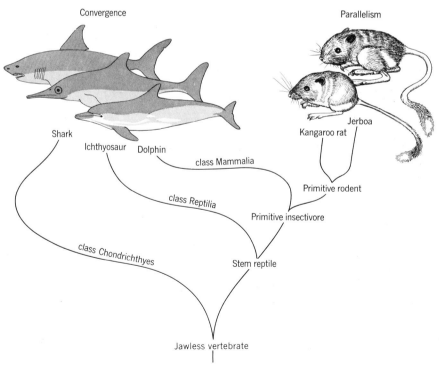

Figure 1-5   PARALLELISM AND CONVERGENCE both establish correspondence of structure in response to similar habits. They differ in the degree to which common general plan extends to similarity of detailed structure.

Africa and the marsupial moles of Australia belong to different orders, yet have in common short robust forelimbs with strong claws, rudimentary eyes, loss of external ears, tough nose pads, and other characters. Parallel lineages each remain adapted to nearly identical ways and places of living: kangaroo rats and jerboas now to hot, sandy deserts; golden moles and marsupial moles to life in the soil. It is the effectiveness of natural selection that causes basically similar animals having nearly identical functional requirements to evolve similar structural adaptations.

Convergence is evolutionary change in two or more lineages such that corresponding features that were formerly dissimilar become similar. Descendants are more alike than were their ancestors, though the similarity seldom extends to correspondence of detail. Closely similar functional requirements are met in somewhat different ways because the common ancestor is often (but not always) more remote than for parallelism.

The remarkable similarity of the shark, ichthyosaur (an extinct reptile), and dolphin is a classical example (Figure 1-5). The common ancestor was a primitive armored fish unlike any of them. One has no terrestrial ancestor; the others have dissimilar terrestrial ancestors. Nevertheless, the tail of each is a waterfoil (though one turns up, one turns down, and one is straight with lateral flukes). Each has numerous simple teeth (but they are rootless for one, rooted in grooves in another, and rooted in sockets in the third). The same sort of general resemblance with variation of detail holds for the spine, eyes, and paired and median appendages. Another example is the similarity of birds and flying reptiles: In spite of remote common origin, each has large eyes, proximal nostrils, long rostrum, long neck, short back, large breastbone firmly attached to the pectoral girdle, many vertebrae articulating with the pelvic girdle, pneumatic bones, and other common features.

The concepts of parallelism and convergence merge into one another. Each process causes taxa to become more related by function and appearance than by phylogeny. Regardless of emphasis (function or phylogeny) the morphologist will sometimes misinterpret evidence if parallelism and convergence are not recognized and interpreted.

DEVELOPMENT AND ANCESTRY   It is evident to the discerning student that the development of the individual from egg to adult (*ontogeny*) and the ancestry of the species (*phylogeny*) are closely related. However, it has been less evident, even to specialists, just what the relationship is in all its ramifications. Three authors stand out from the many who have contributed to the long, and sometimes spirited debate on this important subject.

Following the publication of Darwin's *Origin of Species* in 1859, embryologists were quick to endorse its new concept. It was thought that in the successive stages of its embryonic development, each kind of animal necessarily repeats all the stages of its evolutionary history. The former was the newly discovered key to the latter. The author and foremost champion of this doctrine was Ernst Haeckel (1834-1919), and his three-word summary of the theory has been spoken by most biologists since that day: "Ontogeny Recapitulates Phylogeny."

—Shellbreaker

Figure 1-6 THE AVIAN SHELLBREAKER illustrates an embryonic structure not present in the adult or the ancestor.

The second major contributor was Gavin de Beer, whose book, *Embryology and Evolution,* was first published in 1930, and subsequently was enlarged in two editions titled *Embryos and Ancestors.* His emphasis was on the product (not the process) of the connection between ontogeny and phylogeny. He showed that in addition to the embryonic recapitulation of ancestral stages, as recognized by Haeckel, other relationships can be identified. For instance: (*1*) The embryo may have a structure that is not present in the adult of either the ancestor or the descendant. The shellbreaker used by the chick at hatching is an example (Figure 1-6). (*2*) A structure that was present in both embryo and adult of the ancestor may become vestigial (Figure 1-3) or even be lost. (*3*) Structures that were formerly present only in the larva or embryo may come to be retained in the adult of the descendant. The gills of certain salamanders are examples (Figure 1-7). (*4*) The embryo of the descendant may repeat early developmental stages of the ancestor, but not late developmental stages. Thus, the gill apparatus of the embryo mammal resembles that of the larval ancestral fish, yet never recapitulates the functional gills of an adult fish. (*5*) The developmental sequence of the ancestor may be altered by the descendant. For example, the muscular and digestive systems of larval amphibians become functional before the appendages form, whereas in birds and mammals the appendages develop first.

The third principal contributor is Stephen Gould, whose book, *Ontogeny and Phylogeny,* was published in 1977. His emphasis is on process. He states that in the course of development, structures can be *added* or *subtracted* from those of ancestors, or they can be *accelerated* or *retarded* in the relative time of their formation. Four combinations result: (*1*) If the development of the somatic features of the descendant (but not of the reproductive organs) is accelerated, the result is **recapitulation** in the classical sense; that is, adult characters of the ancestor come to be juvenile characters of the descendant. Thus, the asymmetrical tails of advanced fishes pass through a symmetrical developmental stage which resembles the adult tail of the ancestor. (*2*) If development is accelerated

Figure 1-7   THE GILLS OF THE AMPHIBIAN *Necturus* illustrate an embryonic structure retained by the adult. Compare with the salamander shown on page 59.

only for the reproductive system (a process Gould calls **progenesis**), the resulting condition is **paedomorphosis,** or the retention of ancestral juvenile characters by adult descendants. The best examples are certain insects that breed when otherwise juvenile in form. However, the condition is also cited in theories on the ancestry of vertebrates (see page 34). (*3*) If the development of a somatic feature is instead retarded (a process called **neoteny**), then the juvenile character of that feature is retained by the adult, and paedomorphosis has been attained by a different process. The example of larval gills in adult salamanders falls in this category. Neoteny is common. Neotenous features of man include naked skin, large brain, short jaw, absence of brow ridges, and late sexual maturity. (*4*) Finally, if the maturation of the reproductive organs is retarded, then ontogeny is extended beyond its ancestral limits, usually resulting in an animal having increased body size and complexity of certain parts. The enormous growth and branching of the antlers of a large extinct elk is a classic example.

   The relation between development and ancestry is so important that all vertebrate morphologists benefit from the study of embryology. Chapter 5 presents fundamentals of embryology, and organogenesis is reviewed in the chapters on the organ systems.

FORM AND SIZE **Allometry** is the important study of the correlation between body form and size. If two structures grow at different rates, then, as size changes, body proportions also change. Allometry relates both to ontogeny (colts have relatively longer legs than adult horses) and to phylogeny (modern horses have relatively longer legs than their smaller remote ancestors). Either the altered size, or changed form, or both, are considered to be adaptive, so allometry helps the morphologist interpret structure.

If the length of structure x (e.g., the legs) is plotted against the length of structure y (e.g., the spine) for many animals of a kind, the individual plots distribute themselves on the graph in such a way that a regression line, or line of best fit, can be drawn to represent their relationship. The formula of the line might be a linear equation, or any of various polynomial equations, but the power equation $y = bx^\alpha$ gives an adequate fit, is simple, and provides a straight regression line on a logarithmic grid. Change in the value of b shifts the line parallel to itself. Change in the value of α alters the slope of the line. When x and y are both linear measures, the slope is 45° if $\alpha = 1$, more than 45° if $\alpha > 1$, and less than 45° if $\alpha < 1$. (Other values pertain if x or y is a surface or a volume.)

On Figure 1-8, the six regression lines represent allometric equations for six populations that are distinct in age or ancestry. Populations **A** and **C** have about the same body size and relative leg length, but in **C,** legs and spine are growing at the same rate, whereas in **A** the legs are growing faster. The analysis shows that the populations are less alike than it appears. The same can be said of

Figure 1-8 RELATION BETWEEN BODY SIZE AND BODY PROPORTIONS shown by graphing leg length against body length on a logarithmic grid. Each regression line represents many plotted points (not included) of one population. Sketches show body size and proportions of animals near the centers of the respective distributions.

populations **D** and **E**. These populations are larger than **A** and **C**, yet have the same proportions. However, the analysis shows that **C** differs more fundamentally from **E** than from **D**, because **C** and **E** differ also in the relative growth rate of legs and spine. How would you interpret the relationship between **A** and **B**, or **B** and **C**, or **C** and **F**?

Disproportionate growth rates among the parts of the body commonly evolve as a consequence of the size dependence of surface-to-volume ratios.

SURFACE-TO-VOLUME RATIO   Many bodily functions depend on the relation between surface and volume: Oxygen is absorbed through a surface, but is used by the entire volume of the body; rough food is ground by the surfaces of teeth and absorbed through the surface of the gut, yet nourishes the entire body; the strength of muscles is approximately proportional to their cross-sectional areas, whereas their loads are proportional to the volumes of the structures moved.

Most adaptations required to sustain large body size derive from the fact that the surface-to-volume ratio of an object is related to its size. It has been common for vertebrate giants, (examples are given in Chapter 22) to be 5–10 times as long as the ancestors of their respective phyletic lines. If such enlargement were not accompanied by alteration of form, their surface-to-volume ratios would be reduced 5–10-fold. Stated more generally, objects of similar form have surface areas that are proportional to the two-thirds power of their volumes. Clearly, as animals become larger, they fall behind in all requirements related to surface-to-volume ratios unless adaptations are made.

One way to adapt is to reduce the need for surface-related functions. The rate of gaseous exchange in the lungs, of absorption from the gut, and of excretion in the kidney all depend on the rate of metabolism. It is not surprising, therefore, that the basal metabolic rate of vertebrates is roughly proportional to the two-thirds power of body weight.

Another way large animals adapt is by making structural modifications that increase surface areas so they can "keep up" with related volumes as overall size increases. Removal of wastes from the bloodstream occurs in the surface layer, or cortex, of the mammalian kidney. In order to provide enough cortex to remove enough wastes, the kidneys of large mammals are compound, each resembling a large cluster of small, smooth kidneys. Similarly, the grey matter of the mammalian forebrain is in the surface of that organ. In order to have enough grey matter, the large mammal has a much convoluted forebrain. The occlusal surfaces of the cheek teeth of large herbivores are disproportionately large and they have particularly intricate infolding of the enamel.

(Some adaptations of the skeleton for supporting great weight·are given in Chapter 22.)

SOME TRENDS IN THE STUDY AND TEACHING OF VERTEBRATE MORPHOLOGY   It is useful to contrast the former nature of a scientific discipline with its present nature. Useful, not because it shows those who are unsure which way to jump in order to be "modern," but because it enables participants and observers to assess its progress and vitality and to make short-range projections in regard to its materials, methods, and emphasis. Vertebrate morphology now differs from the antecedent comparative anatomy of several generations ago in various ways, though it is not implied here that the changes do or should characterize all work of all researchers and teachers.

Formerly, there was more speculation than there is today. Contrasting theories about, for example, the origin of paired appendages and of multicuspid teeth were elaborated and debated on meager evidence. Now, most morphologists prefer to work with tested, or at least testable principles. Many unsolved problems (e.g., the detailed homologies of the parts of various chondrocrania, and the origin of the vertebrate type of kidney) have largely been put aside because they are doubtfully answerable with present resources.

There is now less emphasis on phylogeny and more on function, less on general adaptations and more on specific adaptations, less on homology and more on analogy. Morphologists now stress differences instead of similarities. Descriptive studies remain of the utmost importance, yet seem less satisfactory if they do not support interpretation of form or function. Major taxa are studied less and minor taxa more. There is less emphasis on isolated structures and more on functional units. Examination of preserved materials is more often supplemented by study of living animals.

Morphology is now less narrow and more eclectic in regard to materials, methods, and the identification of problems. It merges into other disciplines. Variation is considered more important, so larger samples are used—when possible. Sampling techniques are of concern. Correlation and other quantitative methods are used for analysis. Superficial studies and generalizations from insufficient data are no longer acceptable.

Dissection, including micromanipulation, remains a very important, though too often neglected, technique. In recent years, however, morphologists have used much new instrumentation and many new techniques. These are examples: microscopy (light and electron), histochemistry, cinephotography (including high-speed recording, radiography, fluoroscopy), analytical film projection (variable direction and speed, remote control), artistic and graphic techniques, stereotaxic methods, telemetry, computer programming, analysis of television tape, electrophysiological techniques (including electromyography and sensors of nervous activity), and physical, electrical, and engineering data recording techniques (including osteometry, photoelastic and photostress analysis, use of force transducers and strain gauges, use of oscilloscopes and multigraph recorders).

Also important, particularly in teaching, are techniques for fixing, preserving, and embalming animals; preparing and demonstrating bony and cartilaginous skeletons by wet and dry methods; making dry bone-muscle demonstrations; injecting vessels, ducts, and cavities; dissecting and staining nervous tissue; air-drying hollow viscera; freeze-drying; embedding; molding; modeling; and casting. I have described and illustrated these techniques elsewhere (see References at the back of this book).

Anatomists formerly asked "What is it like?" Morphologists will long continue to ask the same question, but now also ask "How does it work?" and "How did it come to be the way it is?"

PART **I**

Survey of Vertebrate Animals:
The Principal Structural Patterns

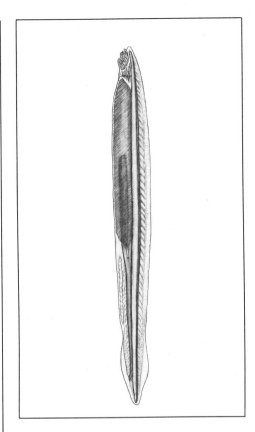

2

Nature, Origin, and Classification of Vertebrates

What is a vertebrate, and how did vertebrates originate? Some animal groups can be distinguished by one or two diagnostic characters: Any animal with feathers is a bird; any animal with mammary glands is a mammal. Similarly, any animal with a cranium is a vertebrate. However, since our objective is to understand and interpret the structure of the entire body, it is more helpful to use many characters in combination to describe vertebrates, selecting not only features that are unique to the group but also those that place vertebrates among related groups.

Remembering that VERTEBRATA is a subphylum of the phylum CHORDATA, let us start by describing vertebrates in general terms that relate their phylum to others. Vertebrates are multicellular animals derived from embryos having three tissue (or germ) layers: ectoderm outside, mesoderm, and endoderm lining the gut tube. (The few embryonic terms used in this chapter are defined further in Chapter 5 and in the Glossary.) The body has bilateral symmetry (right and left sides, anterior and posterior ends, dorsal and ventral surfaces). A body cavity, or coelom, is present and is lined by mesoderm. The gut is complete, which means that there are separate openings for mouth and anus. The anus is derived from an opening in the surface of the early embryo called the blastopore (or a point near the blastopore). There is an internal skeleton derived from mesoderm, and the mesoderm is formed at least in part from tissue derived from the embryonic gut.

These characters go far to describe vertebrates, yet they are shared by other chordates and by the phyla ECHINODERMATA (starfishes, sea urchins, sea cucumbers, etc.), HEMICHORDATA (wormlike, burrowing animals), and two lesser phyla (POGONOPHORA, CHAETOGNATHA). There are some departures: One group has no internal skeleton, another has no digestive tract, another lacks a coelom, and only the larvae of echinoderms have bilateral symmetry. In general, however, these features in combination set these animals apart from all others.

It is usually concluded that chordates are more closely related to echinoderms, hemichordates, and members of similar phyla than to animals belonging to other groups. Because these animals do not form the mouth from the blastopore, they are collectively called DEUTEROSTOMIA (= second + mouth).

RELATION OF VERTEBRATES TO OTHER CHORDATES

There are three subphyla in the phylum Chordata: UROCHORDATA (called tunicates and sea squirts), CEPHALOCHORDATA (called amphioxus), and VERTEBRATA. These animals have various characteristics in common. Most distinctive is the **notochord** (= back + cord). This is a longitudinal rod of supportive tissue generally derived from the dorsal wall of the embryonic gut. Its turgid, vacuolated cells are unique. The cord is surrounded by sheaths of connective tissue. All chordates (as the name implies) have a notochord during early development. Cephalochordates and many vertebrates (but not urochordates) retain the notochord, or remnants thereof, as adults.

Hemichordates have a muscular proboscis in which there is a short organ considered by some to be a notochord or notochord homolog. For this reason hemichordates were formerly classified as a fourth subphylum of Chordata. However, the structure is not supportive, has a cavity opening into the pharynx, and is questionably of endodermal origin. It is now usually called a stomochord instead of a notochord, and Hemichordata (also called Stomochordata) is usually considered a separate phylum.

A second chordate character is the **dorsal hollow nerve cord** derived from ectoderm by a folding process called neurulation. This feature is shared by hemichordates, but in them the central cavity, or neurocoel, is not continuous.

A third feature of importance is the presence, at least in early developmental stages, and usually also in adults, of a **pharynx** (expanded anterior portion of the gut) which is perforated by numerous slits that permit water taken into the mouth to be passed out of the body. Among nonchordates, pharyngeal clefts are found only in hemichordates.

Another chordate character is a circulatory system having a **ventral heart** (or ventral pulsating vessel in cephalochordates) which drives blood up through the bars of the pharynx in vessels called aortic arches (these are secondarily modified in higher vertebrates) and thence caudad in a dorsal vessel. (Urochordates are exceptions in that the heart drives the blood in one direction for a time and then reverses to pump in the other direction.) The blood-vascular system of chordates is closed, that is, blood remains in vessels and does not enter the tissue spaces.

Chordates tend to have their principal sense organs concentrated in a head, a condition spoken of as **cephalization.** This characteristic goes with the combination of bilateral symmetry and motility; urochordates, which are sessile, have no head. Two other chordate features that adult urochordates fail to exhibit are a **tail** extending posterior to the anus and **metamerism,** or segmentation of some features of the body.

The distinctive features of each subphylum should also be considered because differences as well as similarities are important to the establishment of relationships. UROCHORDATA inhabit coastal areas of all oceans. As already noted, only the free-swimming larva has a coelom, hollow nerve cord, and notochord. The notochord is prominent, but only in the tail (''urochordata'' = tail + cord), where it is surrounded by muscle cells (Figure 2-1). Adults are saclike or stalked, sessile, and often colonial. Colonies may be extensive, but individuals are small. The entire body is enveloped by a tough tunic (Figure 2-2). Between

Figure 2-1   Stylized LARVAL UROCHORDATE drawn to illustrate internal structure. Characters of the phylum Chordata are shown by boldface labels; characters of the subphylum are shown by standard labels.

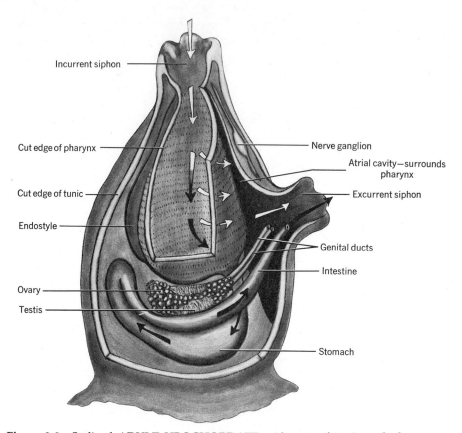

Incurrent siphon

Cut edge of pharynx

Cut edge of tunic

Endostyle

Ovary

Testis

Nerve ganglion

Atrial cavity—surrounds
pharynx

Excurrent siphon

Genital ducts

Intestine

Stomach

Figure 2-2 Stylized ADULT UROCHORDATE with part of tunic and pharynx removed to show internal structure. Light arrows show path of respiratory current;
dark arrows show path of food particles trapped on pharynx.

the tunic and the pharynx is a space called the atrium. Adult urochordates are
filter-feeders. Water enters the pharynx by an incurrent siphon, seeps through
complicated pores (one to thousands, depending on the species) in the pharynx
to reach the atrium, and escapes by the excurrent siphon. Food particles are
trapped in sticky mucus that moves from the endostyle to the pharyngeal bars
and thence to the esophagus. The simple gut loops within the body, and discharges into the atrium. Larvae do not feed, and have a short gut. A heart pumps
acellular blood first in one direction and then in the other. Individuals are hermaphroditic. Excretory cells may be grouped, but do not form organs. Sex ducts
are present. There is a nerve ganglion and a nerve net. Urochordates are a large
and diverse subphylum. Three classes (Ascidiacea, Thaliacea, Appendicularia)
are distinguished on the basis of number of pharyngeal slits, colonial versus
solitary habit, complete versus incomplete metamorphosis, nature of tunic, etc.

    The subphylum CEPHALOCHORDATA has only two living genera, but is
known by good fossils that are 530 million years old. The surviving forms are
distributed over the world, especially in coastal areas having warm shallow water.

Oral cirri

Wheel organ

Velum

Velar tentacles surrounding
mouth

Cephalization present
though not marked

SYMMETRY IS BILATERAL

Perforated pharynx
specialized for
filter–feeding

Nerve cord dorsal
and hollow

Heart absent — ventral
vessel pulsates

Notochord

COELOM PRESENT

Segmented muscles

Hepatic diverticulum

Segmented gonads
(larger in adults)

Metapleural folds
surround atrium

**TISSUES DERIVED FROM
THREE GERM LAYERS**

GUT IS COMPLETE

Fin supports

**ANUS DEVELOPS FROM
EMBRYONIC BLASTOPORE**

Tail extends posterior
to anus

Figure 2-3 AMPHIOXUS, drawn from a cleared 37 mm specimen to illustrate internal structure. Characters of the Deuterostomia are shown by boldface capital labels; characters of the phylum Chordata are shown by boldface lowercase labels; characters of the subphylum Cephalochordata are shown by standard labels.

The lance-shaped body measures 50–75 mm in length. There is a low continuous dorsal and tail fin, and paired lateral body folds (Figure 2-3). The persistent notochord extends anterior to the nerve cord. Other skeletal elements support the fins, pharynx, and oral structures. There are light-sensitive cells on the floor of the nerve cord. Like urochordates, these animals are filter-feeders with a complicated pharynx surrounded by an atrial cavity, although details of their structure and development are unlike those of urochordates. Water enters the pharynx from the mouth and passes through about 150 slitlike oblique clefts to reach the atrium. The ciliated bars of the pharynx move a mucus strand into the straight digestive tract. Unlike a "true" liver, the hepatic diverticulum is hollow. The circulatory system lacks a heart, but is otherwise closely similar to the vertebrate plan. Cephalochordates are unique and really quite peculiar in the asymmetry of various organs, which results from unusual develomental processes: The segmental muscles, pharyngeal slits, gonads, and spinal nerves alternate on the two sides of the body instead of lying in successive pairs. One other important feature departs from the vertebrate plan: The segmental excretory organs are derived from ectoderm (instead of mesoderm) and have clusters of flagellated cells such as are found in annelids.

What structures of VERTEBRATA distinguish these animals from all others? The nerve cord of larval urochordates enlarges anteriorly to form a vesicle, but only vertebrates have a true brain divided into several vesicles which serves to control and coordinate the nervous responses of the body. Furthermore, only vertebrates have a skeletal structure, the cranium, which supports and protects the brain. Eyes, ears, and olfactory organs are present. Many other animals have light receptors and chemoreceptors, but the principal sense organs of vertebrates are not derived from others and rarely resemble them closely in structure. The concentration of brain, cranium, and sense organs in the head gives vertebrates a degree of cephalization that is approached only by some arthropods.

One would expect all vertebrates to have vertebrae. Most do, but some have only a few imperfectly formed vertebrae, and there is reason to doubt that the first vertebrates had them at all. The term Vertebrata became established before this was known. Most animals have accessory digestive glands and various of these are called "livers," but the solid liver of vertebrates is not homologous and only partly analogous to others. The related hepatic portal system and gall bladder are equally unique, as are the pancreas and spleen. A heart is present and chambered. The gonads are not segmental, and the kidneys are of mesodermal origin.

Vertebrates resemble the other chordate subphyla and hemichordates in having a perforated pharynx, but relatively simple structure and predominately respiratory function of this organ are vertebrate characters. Complexity of the muscular system and attendant locomotor activities are also characteristic. Not all vertebrates have paired appendages, and some nonvertebrates have them, yet these are typical. Finally, although vertebrates include minnows, hummingbirds, and shrews, as a group they are relatively large animals.

This discussion started with a short definition of vertebrates, "any animal with a cranium is a vertebrate," and then expanded the concept for several pages. The student might find it instructive to phrase a definition in one paragraph.

The preceding discussion of structure indicates that vertebrates are more closely related to cephalochordates than to urochordates, and that the phylum chordata is more closely related to echinoderms and, particularly, to hemichordates than to other phyla.

These conclusions are accepted by most present-day zoologists. They are by no means self-evident, however, and in passing it should be said that many alternative theories have been proposed. As early as 1818, when the concept of evolution was but dimly perceived, the French naturalist St. Hilaire proposed that vertebrates were derived from insects. Near the end of the nineteenth century the American biologist Patten, among others, proposed that vertebrates evolved from the king crab. Other arthropods were also suggested as ancestors. About the same time, many zoologists, including such prominent men as the Scottish embryologist Balfour (who proposed the phylum name "Chordata") and the German anatomist Gegenbaur, considered vertebrates to have evolved from one or another annelid worm. In order to get the nerve cord in the right place and the blood flow in the right direction, it was necessary to postulate that the annelid progenitor became inverted. Still other lineages were proposed, and currently Løvtrup (see References at back of the book) favors vertebrate ancestry from arthropods and mullusks. Comparative embryology led Garstang to the currently favored theory in 1894.

The current theory is no more than a theory, for the evidence is sufficient only to guide our speculations. Nevertheless, several tentative conclusions are worthy of notice.

No known echinoderm could be the chordate ancestor. Radial symmetry, water vascular system, nerve ring, modified coelom, and other features rule them out. It is the less specialized and bilaterally symmetrical larva which resembles the larvae of simple chordates. Furthermore, no known hemichordate could be the chordate ancestor. Although hemichordates are probably closer to chordates than are echinoderms, the structure of their circulatory system, proboscis, and collar could hardly be converted to the vertebrate body plan.

Similarly, adult urochordates are far too specialized to include the sought-for ancestor. One need only recall the loss of notochord, nerve cord, and coelom, and the presence of such nonchordate features as tunic, atrium, and siphons. Amphioxus is at first glance a possible ancestor, but what of its asymmetry, atrium, unusual pattern of nerves, and annelidlike excretory organs?

One must conclude that vertebrates did not evolve from any known animal, living or extinct. It follows that echinoderms, hemichordates, and chordates must have diverged from a common lineage not later than early Cambrian times some 600 million years ago. Since those remote days, each group has gone its own way, retaining some of the common features over the ages, but altering or deleting others and adding new structures according to their separate needs. Likewise, the subphyla of chordates must have distinguished themselves soon thereafter. These tentative relationships are summarized by Figure 2-4.

There is an alternative theory in regard to cephalochordates. Some investigators think that amphioxus is a degenerate and specialized offshoot from a vertebrate ancestor. However, if this were so, one would expect some trace of

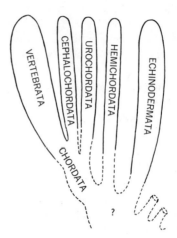

Figure 2-4 TENTATIVE PHYLOGENY of the subphyla of Chordata and of their closest nonchordate relatives.

brain, cranium, or eyes to be recapitulated in their embryonic development, and none has been found.

Embryonic and larval structure must be stressed in any discussion of vertebrate ancestry. Cleavage pattern, fate of blastopore, and origin of mesoderm and coelom relate the deuterostomes to one another. Hemichordates are linked to echinoderms primarily by the similarity of their larvae, which are called tornaria. Only larval urochordates resemble other chordates, and cephalochordates are to be compared not with adult vertebrates but with embryos and with ammocoetes, which is the larva of the eel-like lamprey.

It seems that each successive group evolved from the larva, not the adult, of the preceding group. Such an evolutionary relationship is called paedomorphosis (see p. 20). Sexual development is accelerated and the development of other organ systems is arrested so the nonreproductive larva of the ancestor becomes the reproductive adult of the descendant.

THE CONSTRUCTION AND INTERPRETATION OF CLASSIFICATIONS

When sorting many items it is desirable to establish categories of different rank arranged in a hierarchy. Thus, the postal worker can sort 6,250,000 addresses by arranging only 50 house numbers on each of 50 streets in each of 50 cities in each of 50 states. Similarly, the classifier of animals uses a number of categories arranged one within another. The general name used for a group of animals is **taxon** (plural, taxa). The hierarchy of taxa assigned to every animal, in order from the most inclusive to the least inclusive, is **phylum, class, order, family, genus, species.** Either the "higher" (more inclusive) or "lower" (less inclusive) categories may be omitted from a particular classification if the scope of the study at hand does not require them. However, every animal is assignable to each of these categories. Every animal *has* a class, order, family, etc., whether it is stated or not. The classification of large assemblages of animals, such as the bony fishes, may require additional categories. These are established either by coining new words (e.g., cohort, branch, division, tribe), or by adding a prefix such as *sub-* (below or under), *infra-* (below or inferior), or *super-* (over) to the usual categories. In this book we will most often refer to the class and subclass. Several superorders and infraclasses, and various orders will also be noted.

Animals in nature are not arranged into categories that man, by his industry, tries to discover; they differ from one another in an almost infinite number of characteristics, both of kind and degree, which man, in his wisdom tries to understand.

It may be difficult to decide the degree of relationship required to associate animals at a particular level of the hierarchy. Do these new anatomical data justify the placing of these species in different genera? Does this new fossil belong in a previously named family, or should a new family be created?

The species is a group of actually or potentially interbreeding natural populations that is reproductively isolated from other such groups. A species is not a specimen that is assigned to this rank on morphological criteria; it is a group of animals, having morphological features in common, but assigned on the basis of the relationship of its breeding members.

Although this definition emphasizes distinctions among contemporary animals, species (and also the other taxa) are regarded as having a dimension in time. The traditional view has been that over the ages species slowly change to become new species, and that gaps in the fossil record are conveniently used to designate what would otherwise be arbitrary boundaries between parent species and descendant species. Some investigators now believe that although species tend to be long surviving, change from one species to another may be relatively rapid.

A genus is one or more lineages of related species. Similarly, each higher taxon is an assemblage of related lower taxa. Animals within one species, one genus, or one family tend to have a common adaptation such as climbing ability, fish eating, or resistance to cold; animals within a taxon above the family level tend to share basic structural patterns such as four legs, feathers, amnion, or gnawing teeth.

The theoretical study of the principles, procedures, and rules of classification is called **taxonomy.** The application of names to the groups recognized is called **nomenclature.** There has been international acceptance of rules that regulate the kinds of names that may be assigned, the way new names must be announced, how priorities are established, and how mistakes can be corrected.

Nevertheless, vigorous debate continues among systematists (persons who classify animals) over both theory and procedure. One group favors **cladistic,** or **phylogenetic classification,** which pays little attention to the fossil record and follows set deductive methods to identify dichotomously branching sister categories. Another group of systematists favors **evolutionary classification,** which stresses the fossil record and uses judgment to make evolutionary trees that are not restricted to dichotomous branching. (For discussion and examples of these methods see Chapter 3 of Pielou, and the book by Løvtrup in the References listed at the end of this book.)

Since competent authorities do not agree on procedures, cannot be specialists on all kinds of vertebrates, usually cannot examine all relevant materials, and, like other people, sometimes make mistakes, it is not surprising that even though we agree that evolution is to be our guide, still there are differences of opinion. Why then, the student may ask, should I bother to learn any classification? There are three good answers: First, the general pattern of vertebrate phylogeny is evident to all specialists; therefore, they are in agreement about the major features of their classifications. Second, any reasonably sound classification serves as the powerful tool we require to create order and express meaningful relationships among animals. Finally, since a classification is an abstraction that is designed to be useful, it is to be expected that classifications intended for different uses should be somewhat different.

Since evolution has been a branching process, it is not possible to arrange animals in a linear sequence on a piece of paper and at the same time to accurately show all their phylogenetic relationships. Nevertheless, a classification that is based on evolution is consistent with our knowledge of phylogeny and tells us much about it. The points below apply to most vertebrate classifications, including the one described in the next section (which it may be well to consult as the points are considered).

1.   Animals grouped together in one of the higher, or more inclusive categories (e.g., a class or subclass) have relatively few characters in common; these are considered to delineate a structural pattern of broad evolutionary importance. The higher categories may be regarded as comprising the main channels in the stream of life. Conversely, animals grouped together in one of the lower, or less inclusive categories (e.g., genus or species) have relatively many characters in common.

2.   All the smaller categories within a common larger category share a common ancestry. The common ancestor may lie in one of these same smaller taxa or, alternatively, the common ancestor may lie in a taxon of equal or still smaller rank that is included in a different larger category.

3.   Taxa (and also characters of taxa) are said to be **primitive** if they resemble the ancestral condition of the group in question (e.g., persistent notochord among fishes). They are said to be **derived** if they depart from the ancestral condition (e.g., notochord replaced by vertebral elements), and **advanced** if they resemble the condition of a descendant group (e.g., a certain configuration of vertebral elements, but not others). The terms are comparative. When these concepts are applied to classification, it is seen that any large taxon is older in origin than the average age of the next smaller taxa that it includes. It follows that the characters shared in common by all members of a large taxon are more primitive than the characters shared in common by the members of most of the included smaller taxa.

4.   Taxa that are considered to be relatively primitive are listed before others of equal rank that are regarded as being more derived or advanced. Each animal group is usually placed adjacent to its closest relatives; proximity of listing is an indication of proximity of relationship. (Thus, amphibians are more closely related to bony fishes, which immediately precede them on the classification, than to either acanthodians or cartilaginous fishes from which they are separated by other classes.) However, it was necessary to qualify the above statement because a written classification is of necessity linear. Thus, mammals did not evolve from birds, as we might infer from the relative positions of these classes; mammals and birds each evolved from the reptiles.

Figure 2-5 shows the evolutionary relationships of the major taxa of vertebrates, their distribution in time, and the approximate relative abundance of the species in each.

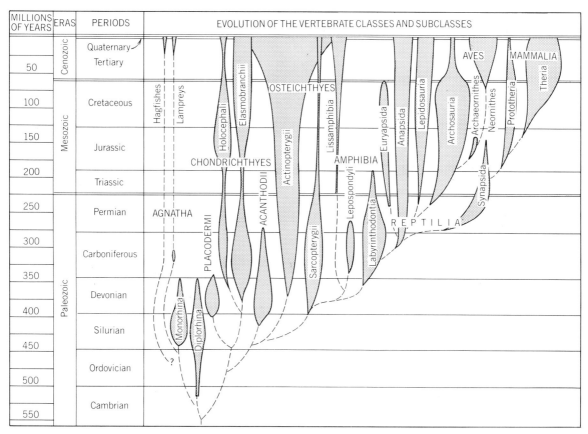

Figure 2-5 OUTLINE OF VERTEBRATE EVOLUTION IN RELATION TO THE GEOLOGIC TIME SCALE. Classes are shown in capital letters; subclasses are shown in lowercase letters, except that "hagfishes" and "lampreys" are common names for orders.

A
**CLASSIFICATION
OF THE
VERTEBRATES**

The following classification has been adapted to the needs of the remaining chapters. All classes and subclasses, and some infraclasses and superorders are included. Technical names are listed first, followed by common names and familiar examples, if such are available. Taxa preceded by an asterisk are extinct.

Class **Agnatha:** jawless vertebrates

  *Subclass **Diplorhina,** including pteraspids (Heterostraci)
  Subclass **Monorhina:** *cephalaspids (Osteostraci), *anaspids, and lampreys and hagfishes

*Class **Placodermi:** placoderms, including arthrodires, antiarchs, and sharklike forms

Class **Chondrichthyes:** cartilaginous fishes

  Subclass **Elasmobranchii:** elasmobranchs, including *pleuracanths, *cladoselachians, and sharks and rays
  Subclass **Holocephali:** chimaeras

*Class **Acanthodii:** acanthodians

Class **Osteichthyes:** bony fishes

  Subclass **Actinopterygii:** ray-finned fishes

    Superorder **Chondrostei:** chondrosteans, including *palaeoniscoids, sturgeons, paddlefishes, and bichirs
    Superorder **Holostei:** holosteans, including gars and bowfins
    Superorder **Teleostei:** teleosts, including most fishes

  Subclass **Sarcopterygii:** lobe-finned fishes

    Superorder **Dipnoi:** lungfishes
    Superorder **Crossopterygii:** crossopterygians, including *rhipidistians and coelacanths

Class **Amphibia:** amphibians

  *Subclass **Labyrinthodontia:** labyrinthodonts including diverse, often stout-bodied forms
  *Subclass **Lepospondyli:** lepospondyles, including slender aquatic forms
  Subclass **Lissamphibia:** modern amphibians, including salamanders, frogs and toads, and caecilians

Class **Reptilia:** reptiles

  Subclass **Anapsida:** *stem reptiles and turtles
  Subclass **Lepidosauria:** scaly reptiles, including tuatara, amphisbaenians, lizards, and snakes
  Subclass **Archosauria:** archosaurs, including crocodilians, *dinosaurs, and *flying reptiles
  *Subclass **Euryapsida:** euryapsids, including plesiosaurs and ichthyosaurs
  *Subclass **Synapsida:** mammal-like reptiles, including pelycosaurs and therapsids

Class **Aves:** birds

   *Subclass **Archaeornithes:** ancestral birds
   Subclass **Neornithes:** familiar birds

    *Superorder **Odontognathae:** toothed birds
    Superorder **Palaeognathae:** ratite birds
    Superorder **Neognathae:** other modern birds

Class **Mammalia:** mammals

   Subclass **Prototheria:** monotremes
   Subclass **Theria:** advanced mammals

    *Infraclass **Trituberculata:** trituberulates
    Infraclass **Metatheria:** marsupials
    Infraclass **Eutheria:** higher or "placental" mammals, including (in separate surviving orders) insectivores, colugos, bats, primates, edentates, pangolins, rabbits, rodents, cetaceans, carnivores, aardvarks, elephants, hyraxes, sirenians, perissodactyls, and artiodactyls

3

Fishes

In presenting the patterns of structure that characterize vertebrate animals, it will be necessary to refer, again and again, to the major groups of vertebrates. The classification at the end of Chapter 2 presented group names; now it is desirable to relate those names to animals. A brief introduction to the more obvious characters serves this initial purpose. More will be learned about the various taxa as Parts II and III of the text unfold, but it is desirable now that the student be able to visualize a representative of each group, recall several distinguishing group characteristics, and know the place of each group among the other vertebrates. If extinct and unusual animals seem to be stressed in this and the following chapter, it is not to give them special emphasis, but because one needs little introduction to animals that are already familiar.

CLASS AGNATHA All animals that have the general characters of the subphylum vertebrata (as reviewed in Chapter 2), but do *not* have jaws, belong to the class Agnatha. These animals are little known to laymen because all except the lampreys and hagfishes became extinct nearly 350 million years ago. Even lampreys and hagfishes are rarely seen by the public; they sometimes destroy fishes of commercial value, but otherwise are obscure.

Known Agnatha seem not to be direct ancestors of jawed vertebrates. The morphologist, however, should become acquainted with the animals in this class because they were the first vertebrates to evolve, and because the most primitive body plan of a living vertebrate is that of ammocoetes, the larva of the lamprey.

The jawless vertebrates have advanced over their progenitors among the protochordates in the possession of several important characters: They have heads with a cranium, a brain, and paired organs of sight. Although incomplete by the standards of other vertebrates, vertebrae are at least represented by cartilaginous elements on the dorsal surface of the large persistent notochord. The internal skeletons of jawless vertebrates are largely cartilaginous, but in the heavy scales and armor worn by most extinct Agnatha, that important material, bone, is present for the first time in earth history. Accordingly, extinct agnathans may be called **ostracoderms,** which means shell skins.

The Agnatha are also noteworthy for the lack of certain characters that became typical of vertebrates standing higher on the evolutionary ladder: They do not possess jaws, true teeth, girdles, or typical appendages. Pectoral spikes, folds, or lobes are often present; pelvic fins are never found. The gills are located in pouches.

The extinct agnathans were all small, fishlike animals. Most lived in freshwater streams, but they are first known from fossils in marine deposits. More than a dozen orders are recognized by some specialists; the treatment here is conservative. There are two subclasses.

Later chapters will refer to jawless vertebrates in relation to the evolution of lungs, tail shapes, some special senses, scales, vertebrae, and fins.

SUBCLASS DIPLORHINA   This diverse group of agnaths appears in the fossil record more than 500 million years ago, and 75 million years earlier than any other group of vertebrates. One poorly known order (Coelolepida) has small

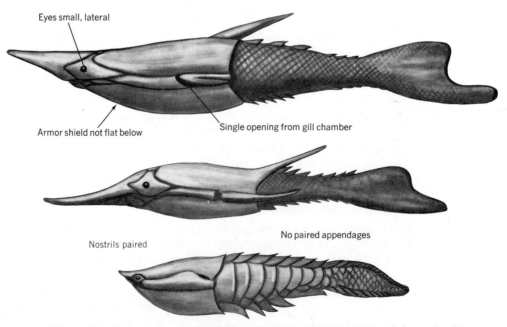

Eyes small, lateral

Armor shield not flat below

Single opening from gill chamber

Nostrils paired

No paired appendages

Figure 3-1   Restoration of REPRESENTATIVE DIPLORHINA of the order Heterostraci.

scales; it is not considered further in this book. The remainder, here called PTER-ASPIDS (order Heterostraci), have a heavy shield of armor covering the head and anterior part of the body (Figure 3-1). They usually have a rostrum projecting over the mouth, and often there are bizarre spines on the shield ("pteraspid" = wing + shield). The eyes are lateral and, unlike the other subclass, there are two nostrils ("diplorhina" = two + nose) and a common exit from all of the gill pouches. The head is not markedly flattened dorsoventrally. One infers that pteraspids were slow swimmers but not bottom feeders.

SUBCLASS MONORHINA This subclass subdivides into three groups, the cephalaspids, anaspids, and surviving lampreys and hagfishes. All have only one nostril. CEPHALASPIDS (order Osteostraci) are relatively well known for animals so long extinct; the patterns of nerves, blood vessels, and sensory structures of the head are preserved in considerable detail for some genera. These are the only jawless vertebrates to have depressed heads and dorsal eyes—characters that indicate these queer creatures moved slowly along stream bottoms to nibble food found in mud or sand (Figure 3-2). Cephalaspids are also unique among the Agnatha for their lobelike pectoral projections. Paired lateral, and median dorsal sculptured areas on the head shield are probably sense organs. The mouth is small and ventral. The upper lobe of the tail is larger than the lower lobe.

ANASPIDS (order Anaspida), which are long extinct, are the only jawless vertebrates having streamlined form. They have small, platelike scales and no head armor ("anaspid" = without + shield). Their eyes are lateral. These features indicate that they were relatively active swimmers. They have pectoral spikes, which may relate to fin folds of unknown nature. Like cephalaspids they have many separate gill openings, but unlike cephalaspids the openings are lateral instead of ventral.

The only surviving agnaths are the LAMPREYS (order Petromyzontiformes) and HAGFISHES (order Myxiniformes). These eel-like animals may conveniently be grouped together as *Cyclostomata* (Figure 3-3), but their relationship seems very remote. Lampreys are more clearly linked to other Monorhina than are the enigmatic hagfishes. Cyclostomata range in length from 100 to 900 mm, and they inhabit oceans and also freshwaters of several continents. The lamprey is commonly dissected in the comparative anatomy laboratory. We are fortunate to have this animal to show us something of the soft tissues of the class, even though we cannot tell how representative they may be. This book will refer to cyclostomes for the seemingly primitive nature of their axial musculature, circulatory system, release of sex cells, eye structure, and other parts.

The larva of the lamprey, ammocoetes, is even more primitive than the adult (Figure 3-4). This is particularly true in regard to mouth parts, pharynx, gonads, and some digestive organs. Ammocoetes becomes adult by metamorphosis.

Cyclostomes are semiparasitic upon bony fishes. Their mouths and "tongues" are adapted for holding onto their hosts and for rasping or cutting away their flesh ("cyclostome" = round + mouth). Sense organs, body form, and digestive structures are also modified in response to this way of life.

These are the only jawless vertebrates to have eel-like form, slippery naked skin, and a suctorial mouth. They are important because they retain such prim-

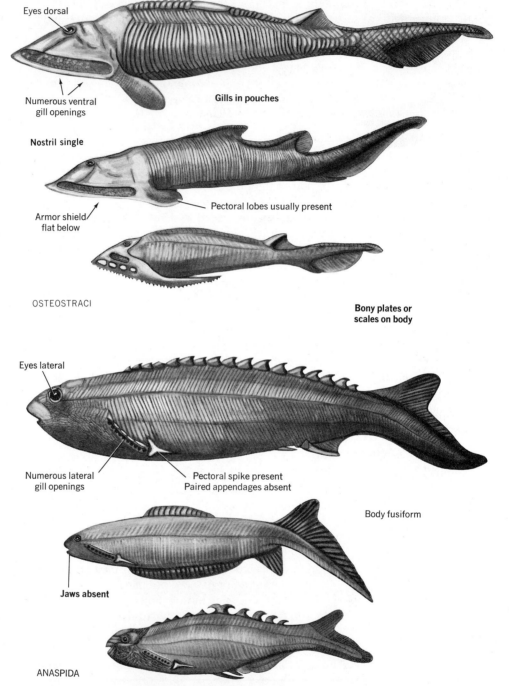

Eyes dorsal

Numerous ventral
gill openings

Gills in pouches

Nostril single

Pectoral lobes usually present

Armor shield
flat below

OSTEOSTRACI

**Bony plates or
scales on body**

Eyes lateral

Numerous lateral
gill openings

Pectoral spike present
Paired appendages absent

Body fusiform

Jaws absent

ANASPIDA

Figure 3-2 Restoration of REPRESENTATIVE AGNATHA OF THE SUBCLASS
MONORHINA illustrating some characteristics of the subclass (boldface labels) and
of the orders Osteostraci and Anaspida (standard labels).

Lamprey

Hagfish

Figure 3-3 REPRESENTATIVE CYCLOSTOMATA showing attenuated form, rasping mouthparts, and absence of jaws, scales, and paired appendages.

itive characters as a persistent notochord and absence of complete vertebrae, paired appendages, and jaws. They also display specialized features (e.g., mouth parts) which the anatomist must avoid mistaking for ancestral characters. The two orders differ markedly in the nature of their head skeletons, pouched gills, sense organs, and some other characters. Even though there are no hard parts, one remarkable fossil impression shows that a lampreylike cyclostome existed about 300 million years ago.

The most important evolutionary advance common to all the remaining fishes was the enlargement and adaptation of the first gill arch to function as jaws instead of gill supports. Chordates that lacked jaws could only filter microorganisms from water, grovel in mud, or rasp and cut algae or soft flesh. The evolution of jaws permitted fishes and their descendants to utilize larger and harder food, and thus enabled them to become adapted to many new and diverse ways of living. This advance was of sufficient importance so that fishes and tetrapods are together called *gnathostomes* (= jaw + mouth) to set them apart from the agnatha (= without + jaw).

JAW-BEARING FISHES

A second important advance common to all jaw-bearing fishes is the presence of paired appendages. True, some agnaths had pectoral folds (anaspids) or lobes (cephalaspids) which suggested incipient appendages, but it remained for jaw-bearing fishes to experiment with various types of paired appendages and perfect several models.

Most of the world's vertebrates—past and present—are jawbearing fishes. The diversity of their habits, form, and structure is enormous. They are important to the morphologist for the advancements and specializations found in nearly every organ system. Four classes are recognized here.

Most placoderms have either bony scales or armor plates, particularly on the forward part of the body—the word "placoderm" means "plate skin." The head

CLASS PLACODERMI

Figure 3-4 AMMOCOETES, larva of the lamprey, drawn from a cleared, 12 mm specimen to illustrate internal structure. Primitive and unspecialized vertebrate characters are shown by boldface labels; other characteristics are shown by standard labels.

is joined to the body by a hinge in the armor. These fishes, like jawless verte-brates, have persistent notochords. The internal skeleton contains some bone, and the gills are below, rather than behind, the braincase. Placoderms swam first in rivers and later in the oceans of the world for 50 million years. They reached maximum abundance about the time amphibians evolved, and became extinct some 345 million years ago, before reptiles, birds, and mammals had appeared.

These remote creatures are not well-known to the general biologist, and few laymen have heard of them. Nevertheless, the comparative anatomist should not entirely neglect the placoderms: They were among the first fishes to evolve two of the most successful of all vertebrate structures—jaws and paired appendages, and they pioneered gas bladders that were ultimately to become lungs. Features of their scales and gill structure are noted in later chapters.

Placoderms are diverse in structure; the evolutionary relationships within the class, and between this class and others, have not been worked out. Five to nine orders are recognized. The two largest and best-known groups of placoderms are the arthrodires and antiarchs. Each is worldwide in distribution. The class, as known, is not regarded as being ancestral to other classes. It seems more closely related to cartilaginous fishes than to bony fishes.

Typical ARTHRODIRA are the most spectacular of placoderms because most have large, predaceous jaws with serrated margins, and some, being 6–9 m long, were the largest vertebrates that had yet evolved when they roamed the rivers and oceans of the world (Figure 3-5). Most arthrodires have blunt heads and lateral eyes; they have heavy cephalic and thoracic shields that are hinged to-gether ("arthrodire" = joint + neck), and the gills are usually hidden by the cephalic armor.

ANTIARCHI are the most bizarre of placoderms because of their peculiar pectoral appendages ("antiarch" = opposite + arm). These long structures are hinged to the body and covered with bony plates. It is thought that they func-tioned as holdfasts against the currents of streams, or were used for creeping along stream bottoms. A bottom-living habit is also suggested by the dorsal po-sition of eyes and olfactory organs, and by the arched dorsal contour and flat ventral surface of the body. The small head and bulky thorax are encased in units of heavy armor. The jaws of antiarchs are small and weak. Overall body size is small.

CLASS CHON-DRICHITHYES

This class includes the familiar sharks and rays, the less familiar but striking chimaeras, and extinct relatives of these fishes. Like the arthrodires before them, the Chondrichthyes are predominantly marine and are of medium to large size. They differ from placoderms and also from most other fishes in having no bone whatsoever internally, and rarely any in their scales ("Chondrichthyes" = car-tilage + fishes). Cartilaginous fishes are also distinctive for their solid braincase, fin structure, branching pattern of blood vessels associated with the gills, and small toothlike scales (or none at all). Their teeth, unlike those of other fishes, are anchored to the integument and occur only at the margins of the jaws. Most of these fishes have a series of external gill openings and lack a gas bladder.

Until the late 1930s it was thought that a cartilaginous skeleton is more

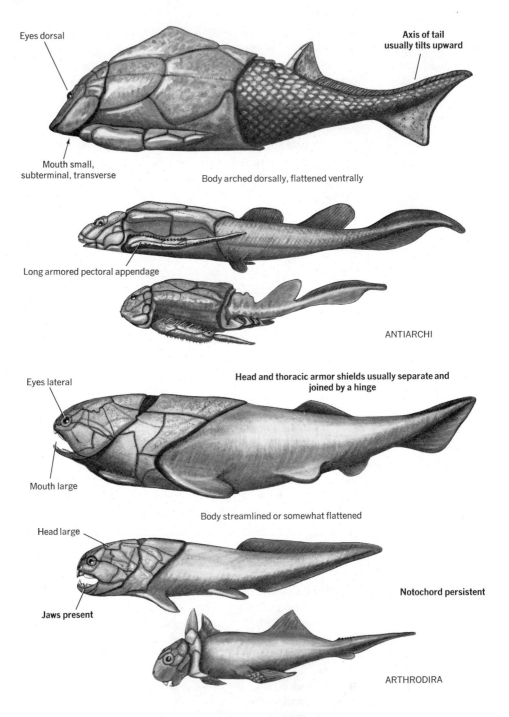

Eyes dorsal

**Axis of tail
usually tilts upward**

Mouth small,
subterminal, transverse

Body arched dorsally, flattened ventrally

Long armored pectoral appendage

ANTIARCHI

Eyes lateral

**Head and thoracic armor shields usually separate and
joined by a hinge**

Mouth large

Body streamlined or somewhat flattened

Head large

Notochord persistent

Jaws present

ARTHRODIRA

Figure 3-5 Restorations of REPRESENTATIVE PLACODERMI illustrating some characteristics of the class (boldface labels) and of the two principal orders (standard labels).

primitive than a bony skeleton. Consequently, the Chondrichthyes were accorded a key evolutionary position as the ancestors of other fishes. As the fossil record grew, another interpretation became more tenable: Cartilaginous fishes appear in the record some 30 million years later than bony fishes and have not been ancestral to any other class. Evidence for this conclusion will unfold in later chapters.

Nevertheless, cartilaginous fishes do retain from their ancestors some features that are more primitive or unspecialized than the corresponding features of bony fishes. Examples are the structure of the heart, brain, and musculature of the pharynx.

SUBCLASS ELASMOBRANCHII Cartilaginous fishes are commonly divided into two subclasses. Members of the first subclass have slitlike external gill openings and hence are called elasmobranchs (= plate + gills). Three orders of elasmobranchs—pleuracanths, cladoselachians, and selachians—will be recognized here; one or more additional orders are included in some classifications (Figure 3-6).

PLEURACANTHODII (also called Xenacanthodii) are freshwater fishes, nearly a meter in length, that became extinct about the time mammals evolved. A large spine borne on the back of the head is a handy recognition feature. The persistent notochord in this order and the next is mounted by somewhat calcified vertebral arches. Pleuracanths will be mentioned in later chapters for two other characteristic features that are unusual among fishes: Their paired fins have a jointed central axis, and the axis of the tail is straight all the way to the tip.

CLADOSELACHII were abundant in the Carboniferous period when the conquest of the land by reptiles was getting under way. These predaceous marine fishes look like large sharks except that the mouth is nearly terminal, the tail is nearly symmetrical externally, and the large pectoral fins have broad bases.

The sharks and rays comprise the order SELACHII. All surviving cartilaginous fishes having a series of external gill slits and small abrasive scales are selachians. The first ancestral gill slit is reduced to a roundish opening called a **spiracle.** These are mostly marine fishes of medium to large size. The vertebrae have centra and arches that touch one another all around the spinal cord. Sharks and rays (commonly placed in about four suborders) differ markedly from one another in their habits. Many sharks are active, predaceous fishes with fusiform bodies and large, strong tails. Their paired fins have restricted bases and their gill slits are lateral in position. Most rays, in contrast, spend much time resting on the bottom or swimming sluggishly along in search of shellfishes or other relatively inactive food. Their bodies are flattened dorsoventrally and the pectoral fins merge into the head and body. The mouth, being ventral, cannot take in water when the fish is at rest on sand or mud, so large spiracles perform this function instead.

SUBCLASS HOLOCEPHALI Surviving members of this subclass are called chimaeras. Found only at sea, they rarely come to the attention of laymen and will be mentioned infrequently in the chapters that follow. These are the only cartilaginous fishes with a fleshy **operculum** covering the gills (Figure 3-7). There are few or no scales. The notochord is persistent, and there is no spiracle. Males

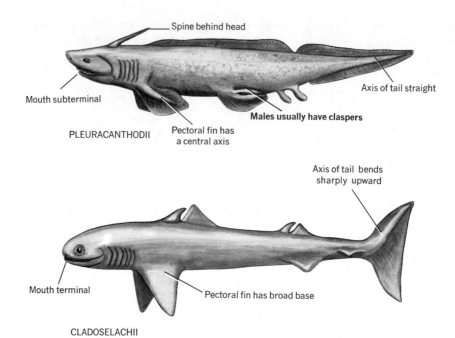

Spine behind head

Axis of tail straight

Mouth subterminal

Males usually have claspers

PLEURACANTHODII

Pectoral fin has
a central axis

Axis of tail bends
sharply upward

Mouth terminal

Pectoral fin has broad base

CLADOSELACHII

Skeleton entirely cartilaginous

Spiracle

LEOPARD SHARK

**Scales, if present,
small and abrasive**

A series of external gill slits

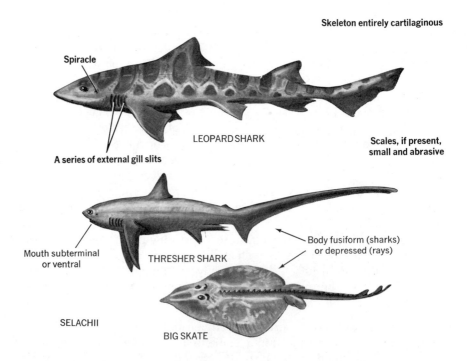

Mouth subterminal
or ventral

THRESHER SHARK

Body fusiform (sharks)
or depressed (rays)

SELACHII

BIG SKATE

Figure 3-6 REPRESENTATIVE ELASMOBRANCHII illustrating some characteristics
of the subclass (boldface labels) and of the three orders (standard labels).

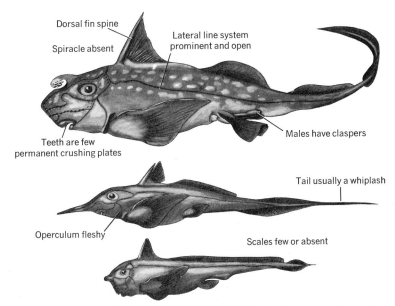

Dorsal fin spine

Spiracle absent

Lateral line system prominent and open

Males have claspers

Teeth are few permanent crushing plates

Tail usually a whiplash

Operculum fleshy

Scales few or absent

Figure 3-7 REPRESENTATIVE HOLOCEPHALI illustrating some characteristics of chimaeras.

have a unique club-shaped clasping organ on the top of the head. Large head and eyes, huge pectoral fins, whiplash tail, and striking colors give chimaeras a bizarre appearance. We will pay attention to the structure of their solid jaws, crushing teeth, and certain sense organs.

CLASS ACANTHODII

Acanthodians evolved before known placoderms, cartilaginous fishes, or bony fishes. Over the years they have been included in each of those three classes, but now, although usually considered closest to bony fishes, they are elevated to a class of their own (Figure 3-8).

Acanthodians have streamlined bodies, large lateral eyes, and wide mouths with numerous teeth. Their heads are bony and their small scales thick and hard, but they do not have the armor of most of their jawless and placoderm contemporaries. We infer from these clues that these small, mostly freshwater fishes were active swimmers. The numerous fins of acanthodians are unique in that each has a thin membrane supported at its leading edge by a long stout spine. This feature gives the subclass its name; "acanthodian" means "spine form."

CLASS OSTEICHTHYES

Bony fishes evolved from obscure ancestors at least 400 million years ago. For some 150 million years thereafter they were outnumbered, first by placoderms and then by cartilaginous fishes, but since the close of the Permian period, bony fishes have dominated the waters of the world. Since the end of the Mesozoic era, they have been the most abundant of all vertebrates. Their habits and structure are seemingly as diverse as common adaptation to an aquatic life will permit.

Most fishes of this class have bone in their skulls, vertebrae, girdles, fin

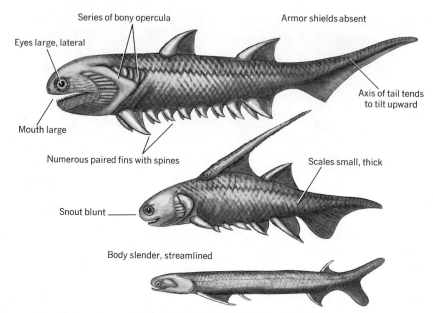

Eyes large, lateral

Series of bony opercula

Armor shields absent

Axis of tail tends to tilt upward

Mouth large

Numerous paired fins with spines

Scales small, thick

Snout blunt

Body slender, streamlined

Figure 3-8   Restoration of REPRESENTATIVE ACANTHODII illustrating some characteristics of the class.

supports, and scales. Some have secondarily substituted cartilage for much of the ancestral bone, yet even such fishes retain more bone in their internal skeletons than is present in other classes of fishes. Bony fishes are the only vertebrates having the gills of each side of the body in a common chamber covered by a movable, bony operculum. They have various sorts of fins, scales, and vertebrae, yet these structures nearly always differ from those of other classes. The pectoral girdle is joined to the skull by a chain of bones. A lung or gas bladder is usually present.

SUBCLASS ACTINOPTERYGII   Most bony fishes belong to the subclass Actinopterygii, or ray-finned fishes. The membranes of the paired fins are supported by bony rays which radiate from the fin base. Consequently, the fins do not have fleshy stalks, as do the fins of fishes in the next subclass. The pattern of cranial bones and the nature of the venous system and reproductive ducts are also distinctive and clearly show that these fishes do not include the ancestors of land vertebrates.

Ray-finned fishes have long been classified into three groups (here conservatively called superorders) termed CHONDROSTEI, HOLOSTEI, and TELEOSTEI. These have been regarded as forming an evolutionary sequence in the order named. Some specialists are now dissatisfied with this arrangement, but it will serve our purposes pending a contrary consensus. The groups differ from one another in regard to degree of ossification of the skeleton; presence or absence, mobility, or pairing of certain bones of the skull and pectoral girdle; shape of tail fin; presence or absence of spiracle; and certain features of the digestive system.

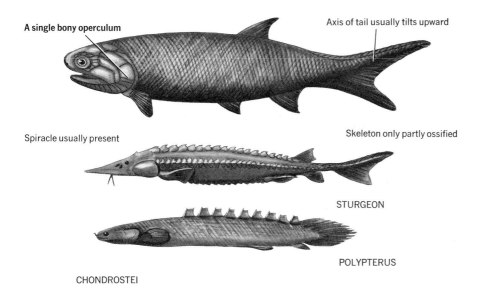

A single bony operculum

Axis of tail usually tilts upward

Spiracle usually present

Skeleton only partly ossified

STURGEON

POLYPTERUS

CHONDROSTEI

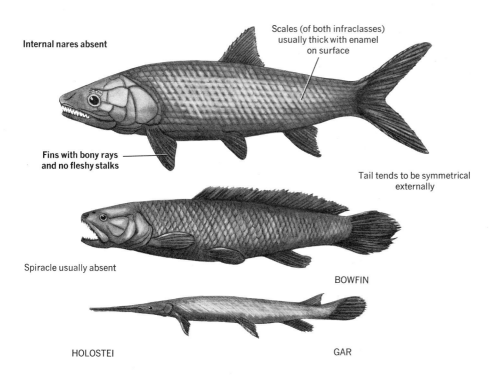

Internal nares absent

Scales (of both infraclasses) usually thick with enamel on surface

Fins with bony rays and no fleshy stalks

Tail tends to be symmetrical externally

Spiracle usually absent

BOWFIN

HOLOSTEI

GAR

Figure 3-9 REPRESENTATIVE CHONDROSTEI AND HOLOSTEI illustrating some characteristics of the subclass Actinopterygii (boldface labels) and of these two superorders (standard labels). Larger drawings are of extinct fishes; others are of surviving fishes.

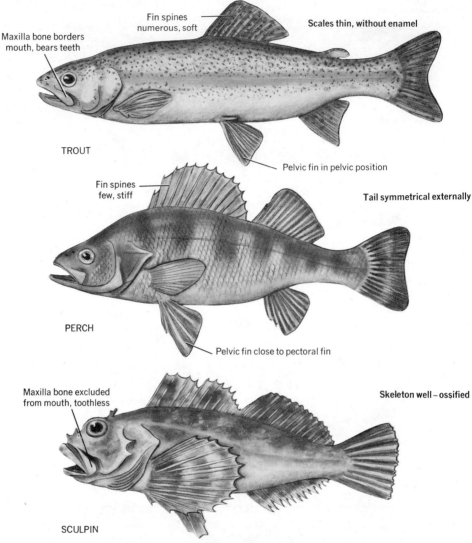

Fin spines
numerous, soft

Scales thin, without enamel

Maxilla bone borders
mouth, bears teeth

TROUT

Pelvic fin in pelvic position

Fin spines
few, stiff

Tail symmetrical externally

PERCH

Pelvic fin close to pectoral fin

Maxilla bone excluded
from mouth, toothless

Skeleton well – ossified

SCULPIN

Figure 3-10   REPRESENTATIVE TELEOSTEI illustrating some characteristics of the infraclass (boldface labels) and differences between more primitive (above) and more advanced species (center and below).

Chondrosteans include the long extinct but significant ancestral order of palaeoniscoids. The superorder is now represented by sturgeons and paddle-fishes, which are, in some respects, not representative, and by the distantly re-lated African bichirs, of which the genus *Polypterus* is the best known (see Figure 3-9). Bichirs are of special interest because of the primitive nature of their scales and gas bladder. Holosteans were most abundant in the Jurassic and Cretaceous periods. The gars and bowfin are among the few survivors. Aquarium, sport, and

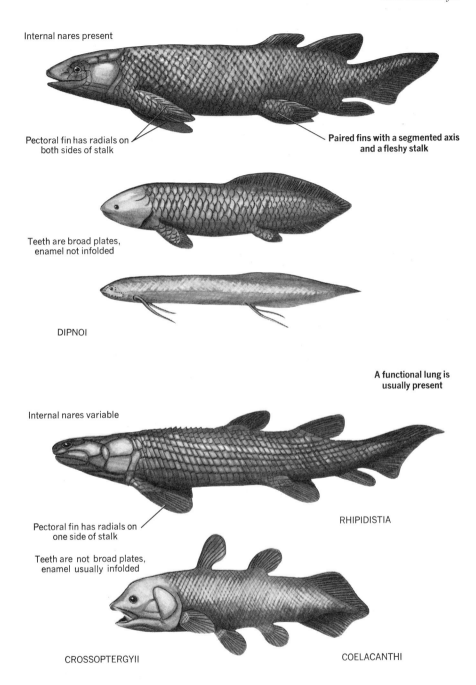

Internal nares present

Pectoral fin has radials on
both sides of stalk

**Paired fins with a segmented axis
and a fleshy stalk**

Teeth are broad plates,
enamel not infolded

DIPNOI

**A functional lung is
usually present**

Internal nares variable

Pectoral fin has radials on
one side of stalk

RHIPIDISTIA

Teeth are not broad plates,
enamel usually infolded

CROSSOPTERGYII

COELACANTHI

Figure 3-11 REPRESENTATIVE SARCOPTERYGII illustrating some characteristics
of the subclass (boldface labels) and of the two superorders (standard labels). The
most slender fish is the African lungfish; the other examples shown here are extinct.

food fishes belong to the large superorder Teleostei. About 30,000 species and subspecies are arranged in 400–600 families and are grouped variously in 6 (conservative) to 30 or 40 (usual) or more orders. Teleosts are recognized as being either primitive or advanced on the basis of mouth structure, position of pelvic fins, number and stiffness of dorsal fin rays, and other characters (Figure 3-10).

SUBCLASS SARCOPTERYGII   These fishes have fins with fleshy stalks (the term means flesh + fin). They thrived in the Devonian period, 200 million years before teleosts evolved, and became scarce after the Triassic period. Several somewhat degenerate representatives of the subclass survive today.

DIPNOI, or lungfish, comprise the first of the two superorders (Figure 3-11). These are mostly freshwater fishes of moderate size and either "normal" or elongate form. They have atypical **internal nares** (nostrils opening into the mouth), functional lungs, and advanced circulatory systems. For these reasons they were once regarded as being ancestral to tetrapods. This view is no longer tenable because the other superorder also possesses (or doubtless formerly possessed) these characters and is more primitive in regard to other features. The tendency of dipnoans to reduce internal ossification may be neotenous.

The paired fins of typical (extinct) dipnoans have a fleshy stalk and a jointed skeletal axis resembling that of pleuracanths. The few teeth are peculiar fan-shaped plates adapted for crushing. Braincase and vertebral column are poorly ossified. The scales of surviving dipnoans are simplified in structure and sometimes degenerate.

The respiratory and circulatory systems of lungfishes are at the same time advanced for fishes and primitive for tetrapods. Unfortunately, modern lungfishes are rarely dissected in the classroom because of lack of availability and the aberrant nature of much of their anatomy.

The other superorder, CROSSOPTERYGII, is of particular importance. Many ancient crossopterygians had internal nares, and it is probable that they also resembled dipnoans in having functional lungs and advanced circulatory systems. They differed from dipnoans, however, and resembled early amphibians in the pattern of their cranial bones and in having numerous conical teeth with complicated internal structure. Their fin structure is also regarded as being closer to that of the tetrapod limb than is that of dipnoans. Crossopterygians fall into two groups. Only one of these, the extinct order Rhipidistia, is ancestral to amphibians. The other group, Coelacanthi, lacks internal nares and has tail fin and scales that are more modified from the primitive condition. A relict coelacanth survives in deep water near Madagascar. Some of its features (scales, teeth, gas bladder) have lost their ancestral characteristics, whereas others are apparently primitive.

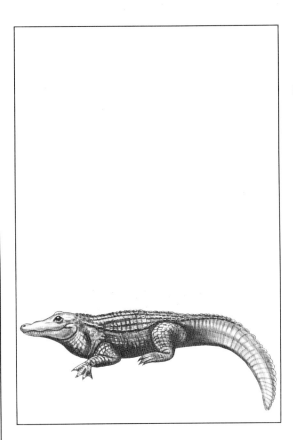

4

Tetrapods

TRANSITION TO TERRESTRIAL LIFE

Tetrapods are simply vertebrates having four legs (or at least four legs in their ancestry). It is instructive, however, to avoid this tautology by defining tetrapods instead as vertebrates that dwell on land (or that had land-dwelling ancestors), and then to consider the changes that enable descendants of fishes to live terrestrial lives.

Out of water, the body usually no longer benefits from being streamlined. A neck becomes advantageous because the head now can turn to facilitate feeding and vision without affecting the mechanics of locomotion. Median fins are no longer useful, and paired fins are converted to limbs. Deprived of the buoyancy of water, the body must be supported by the limbs, and this necessitates having appendages that are strong, girdles that are more firmly related to the axial skeleton, and a vertebral spine that can better resist bending.

Lungs and a pulmonary circulation, pioneered by air-breathing fishes, are usually retained to replace gills, which would be damaged by exposure to dry air. Gill covers can therefore be dispensed with. The superficial layer of the skin becomes sufficiently cornified to resist abrasion and drying. The eye, ear, and nose also must be modified to function in air instead of water. Oral glands are needed to moisten food that is now dry. Eggs and delicate larvae formerly were supported by water and could pass the waste products of their metabolism directly into the environment. Before becoming able to reproduce completely away from either water or moist environments, tetrapods had to accomplish the seeming miracle of evolving eggs with shells and fetal membranes to protect their embryos from desiccation and mechanical harm and to receive metabolic wastes. Some other structural changes were also necessary to make possible the shift to terrestrial life, and physiological and behavioral changes were also needed. It is not surprising, therefore, that the first tetrapods, the amphibians, did not fully accomplish the change.

CLASS AMPHIBIA Amphibians evolved from crossopterygian ancestors (of the order Rhipidistia) some 50 million years after bony fishes evolved. The class reached its greatest expansion after another 75 million years, in the Upper Carboniferous period, but continued to abound until the end of the Triassic period. Relatively few kinds of amphibians have survived to the present, yet they are distributed in tropical and temperate areas of all the world, and their habits and habitats are diverse.

The skin of modern amphibians is unable to withstand long exposure to dry air, and since fetal membranes are lacking, eggs must be laid either in water or in damp places. In this sense the class is not completely terrestrial ("amphibian" = both kinds + life), though many species do not utilize open water, and several live in remarkably arid places.

Adult amphibians have large mouths and a fleshy tongue which is attached near the front of the lower jaw. One of the bones that supported the jaws of the piscine ancestor is converted to an ear ossicle, which usually contributes to hearing in air. Lungs are usually present (they are secondarily lost by one large group of salamanders), and some respiration occurs also through the skin and lining of the mouth and throat. Eyelids and glands to moisten the eyes have evolved.

Three subclasses are commonly recognized. One of these (Lepospondyli) comprises slender aquatic forms, some of which retain characters of skull, gills, and girdles that resemble those of their fish ancestors. They are all extinct and will not be discussed further in this book. The other two subclasses are of concern here.

SUBCLASS LABYRINTHODONTIA Labyrinthodonts, extinct for 175 million years, comprise most of the amphibians that have ever lived, including the ancestors of reptiles. Different authorities recognize two or three orders. Some labyrinthodonts were entirely aquatic whereas others appear to have been true land animals with strong limbs, robust bodies, and probably dry skins (Figure 4-1). Several kinds were as large as alligators, and many had rather large, flat heads. The complicated structure of their teeth (from which the name of the subclass is derived), and the varied structure taken by their vertebrae are among the subjects to which we shall return in subsequent chapters.

SUBCLASS LISSAMPHIBIA All surviving amphibians are lissamphibia (Figure 4-2). Most are less than 30 cm in length. Their moist skin has abundant mucous glands and only rarely supports scales ("liss" = smooth). The outer cornified layer of the skin is shed periodically. Parts of the skeleton, particularly of the feet, are commonly cartilaginous, and several ancestral bones have been lost from the braincase. There are only four toes in the hand. Teeth are never complicated as in labyrinthodonts and are absent from some groups.

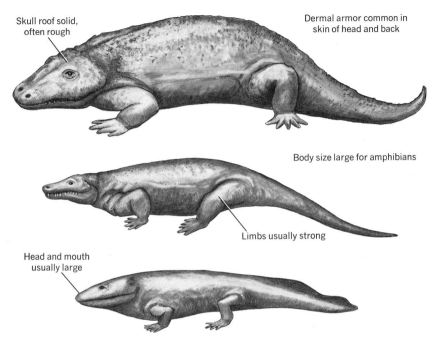

Figure 4-1 Restorations of REPRESENTATIVE LABYRINTHODONTIA illustrating some characteristics of the subclass.

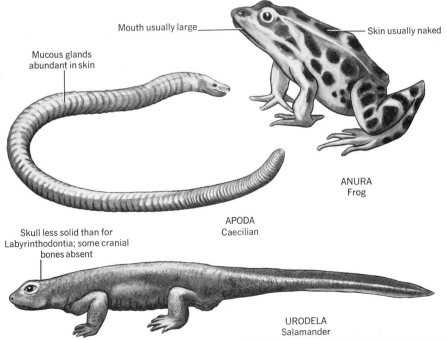

Figure 4-2 REPRESENTATIVES OF THE THREE ORDERS OF LISSAMPHIBIA showing some characteristics of the subclass.

There are three orders of lissamphibia. ANURA (= without + tail) includes frogs and toads. Salamanders belong to the URODELA (= having a tail). Finally, the APODA, commonly called caecilians, are, as the name tells us, legless. They are obscure animals that will merit mention in Part II of this book because of their primitive excretory organs and their scales (unusual in surviving amphibians), and in Part III because of their burrowing habits.

CLASS REPTILIA   Reptiles evolved from labyrinthodont amphibians some 50 million years after amphibians evolved. From the Permian through the Cretaceous periods they were the most abundant of vertebrates. This was the first class of tetrapods to have all the structures noted at the beginning of this chapter as requisite to fully terrestrial life, including fetal membranes and an integument that is resistant to drying. During the "age of reptiles," the different genera ranged from small to gigantic, from herbivorous to carnivorous, and from sluggish to swift. Several groups independently reverted to aquatic habitats, becoming highly skillful swimmers, and one group even invaded the air. No class of vertebrates had theretofore been so diverse in habits, and only the mammals have matched them since. Today, reptiles remain an important part of the faunas in tropical and temperate regions but are less numerous than bony fishes, birds, or mammals.

Reptiles are covered with horny scales. Excepting such specialized forms as snakes, most of them have claws, ribs that are used in drawing air into the lungs (amphibians instead use the mouth and throat as a force pump), and a vertebral column that is more differentiated into regions and more firmly attached to the pelvic girdle than in their amphibian ancestors. However, none of these characteristics is unique to the class. There are features of the heart and related blood vessels that *are* unique to reptiles, yet most other single structures that are typical of the class are not sufficiently distinctive to separate it from other vertebrates.

Why, one may ask, does this class have so few distinctive characters? First, it includes all the animals that made the first true invasion of land and radiated out into its varied habitats. Hence, it is a large and diverse class. Such a group can have less in common than one, such as birds, whose members have a common adaptation. Second, reptiles were ancestral to members of two other classes, birds and mammals. They have correspondingly many relatives from which to be distinguished. For these very reasons, Reptilia is a key class in vertebrate evolution and one to which the morphologist must often refer.

The class is variously grouped in 17 to about 23 orders commonly arranged in 5 or 6 subclasses. Only 4 orders survive to the present.

SUBCLASS ANAPSIDA   The term "anapsida" (not to be confused with "anaspida") refers to the absence of openings in the bones that roof over the temporal region of the skull, a feature which distinguishes this subclass from the others (see the figure on page 139). The group is divided into two principal orders.

COTYLOSAURIA (or Captorhinomorpha) are small, lizardlike creatures that were abundant in the Permian period but have long been extinct (Figure 4-3). We will rarely refer to them in later chapters, yet they are important in that they include the most primitive of reptiles and because most, and possibly all, other

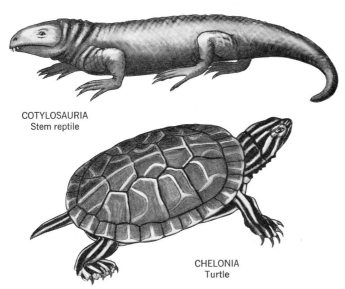

COTYLOSAURIA
Stem reptile

CHELONIA
Turtle

Figure 4-3 REPRESENTATIVES OF TWO OF THE ORDERS OF ANAPSIDA. The principal characteristic of the subclass is the absence of openings in the temporal area of the skull.

subclasses of reptiles evolved from them. Accordingly they are called **stem reptiles.**

The CHELONIA comprises the turtles and tortoises, which the layman easily recognizes by their broad, armored bodies. A turtle is often dissected in the laboratory as a representative of its class because of its availability and suitable size. Students should realize that the turtle's "shell," ribs, spine, toothless mouth, and pectoral girdle are highly specialized, and therefore not typical of the class. The skull, limbs, and (so far as we can judge) soft parts remain primitive.

SUBCLASS LEPIDOSAURIA   This was one of the first subclasses to evolve after the anapsids, but it did not reach maximum abundance until after the passing of the dinosaurs. The taxon is defined by cranial characters. There are two openings in the roofing bones of the temporal area (unless secondarily modified, see again the figure on p. 139), openings among the bones of the palate, and usually teeth on the roof of the mouth as well as at the margins. Three or more orders are recognized, all but one of which are extinct and not useful to this course of study.

Surviving lepidosaurs (Figure 4-4 and the figure on p. 487) are placed in the order SQUAMATA. Most of the 3000 species of lizards have legs and a tail. Amphisbaenians have small forelimbs only or (usually) no limbs at all. They live underground. Snakes probably evolved from burrowing, lizardlike ancestors. They have lost almost all traces of limbs and generally have modified the skull to allow them to swallow prey of diameter equal to or greater than that of their bodies. The tuatara looks like a large robust lizard. It is confined to islands off the coast of New Zealand, and is rigidly protected. The single genus, *Sphenodon*, may have survived longer (135 million years) than any other among vertebrates.

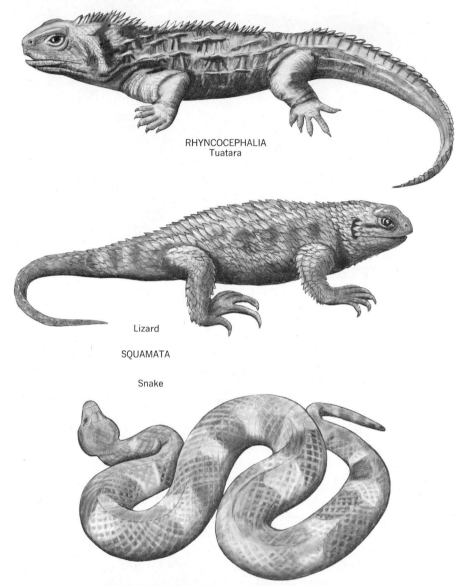

RHYNCOCEPHALIA
Tuatara

Lizard

SQUAMATA

Snake

Figure 4-4 REPRESENTATIVES OF THE ORDER SQUAMATA OF THE SUBCLASS LEPIDOSAURIA. The principal characteristics of the subclass are cranial features noted in Chapter 8.

It is of particular interest for the primitive nature of its skeletal and circulatory systems. (Some classifications assign this animal to its own order, Rynchocephalia.)

SUBCLASS ARCHOSAURIA This largest and most spectacular of reptilian subclasses is also characterized by two temporal openings and usually by an open

Figure 4-5 The alligator is a REPRESENTATIVE OF THE CROCODILIA, the only surviving order of the subclass Archosauria.

bony palate. Often there are openings in the skull in front of the orbit and in the side of the lower jaw. All teeth are marginal.

Five or more orders are recognized. The only survivors of the subclass, crocodiles and their relatives, are in the order CROCODILIA (Figure 4-5). Two orders comprise the beasts popularly known as dinosaurs: Some SAURISCHIA were carnivorous; most ORNITHISCHIA were herbivorous (Figure 4-6). These animals, which tended to gigantic size and sometimes to bipedalism, give the subclass its name: "archosauria" = ruler + lizard. Flying reptiles are in the order PTEROSAURIA (= wing + lizard) (see figure on p. 535). This leaves only the short-lived order THECODONTIA, which was ancestral to other archosaurs and to birds.

SUBCLASS EURYAPSIDA The three or four orders grouped in this subclass all have a single temporal opening and only marginal teeth. The subclass is diverse, however, and may well prove to be unnatural. We will note only two of the orders, which included some highly specialized marine animals mentioned in Chapter 26 (Figure 4-7). Aquatic members of the order SAUROPTERYGIA are called plesiosaurs. They had broad bulky bodies, tapering tails without lobes, paddlelike limbs, blunt heads, and necks that were sometimes very long. ICH-THYOSAURIA, by contrast, had dolphinlike body contours, fishlike tails, large eyes, and a large rostrum.

SUBCLASS SYNAPSIDA Members of this extinct subclass were usually terrestrial carnivores of moderate size (Figure 4-8). There is one temporal opening. Unlike archosaurs, they never tended to bipedalism. Passing over the more primitive of the two orders, PELYCOSAURIA, brings us to the THERAPSIDA, which are important as the ancestors of mammals. Commonly called **mammal-like reptiles,** they tended to have robust legs placed relatively close to the center line of the body and rooted teeth that were specialized, according to position in the mouth, for biting, tearing, or chewing. The architecture of the deep skull, palate, ear, and jaw came to resemble the corresponding features of mammals.

Roof of skull has two vacuities

SAURISCHIA
Carnosauria

Hindquarters and tail
usually relatively large

SAURISCHIA
Sauropoda

Bones common
within the skin

ORNITHISCHIA
Stegosauria

Body size moderate
to gigantic

ORNITHISCHIA
Ceratopsia

Figure 4-6 Restorations of REPRESENTATIVE ARCHOSAURIA of the orders Saurischia and Ornithischia and of the suborders indicated. Some characteristics of the
terrestrial members of the subclass are shown.

ICHTHYOSAURIA

PLESIOSAURIA

Figure 4-7 Restorations of REPRESENTATIVE EURYAPSIDA of two dissimilar aquatic groups. The principal characteristics of the subclass are cranial features noted in Chapter 8.

PELYCOSAURIA

THERAPSIDA

Figure 4-8 Restorations of REPRESENTATIVE SYNAPSIDA of the two orders. Principal characteristics of the subclass are slender form, strong limbs, and features of the cranium noted in Chapter 8.

CLASS AVES   No other locomotor adaptation requires so much structural specialization as that of flight, and all birds fly or are descendants of flyers. In striking contrast to reptiles, birds are, therefore, the most homogeneous and distinctive of all tetrapod classes. However, for all their unique characteristics among the living fauna, birds are not very different from the particular reptiles from which they evolved. Those small Mesozoic archosaurs tended to be bipedal, and therefore to have robust hind limbs with elongate feet. Like birds, they had long necks, and their pelvic bones and skulls approached avian structure. We may infer that their urogenital systems, fetal membranes, and sense organs were similar to those of birds.

Feathers are of particular importance, for birds are the most expert of flyers and the only vertebrates ever to achieve the highly successful combination of flight with bipedalism. Such flight is mechanically dependent on feathers. Furthermore, sustained flight probably requires the high metabolic rate made possible by an elevated body temperature. Contrary to former belief, many reptiles do have considerable control of body temperature, but lack of feathers or other insulation puts them at a disadvantage in some respects. We must not be overly sure that the flying reptiles could not stay aloft for long periods, and there are indications that some had hairlike insulation, but it is unlikely that they could match the birds that replaced them during the Cretaceous period.

SUBCLASS ARCHAEORNITHES   The small, light bones of birds are not easily fossilized; the ancestry of no other class of comparable size is so poorly documented. However, fossils of Jurassic birds are known from Germany, and they are "missing links" of utmost value. The Archaeornithes (= ancient + birds) were fully feathered, but unlike present-day birds had the tail feathers arranged in a row along each side of a long lizardlike tail (Figure 4-9). These arboreal climbers could already fly or at least glide but various volant adaptations of the wing skeleton, spine, and breastbone were absent or incomplete by Cenozoic standards. The skull had the large orbits and beaklike rostrum of a bird but was reptilian in other respects, including having teeth in its jaws.

SUBCLASS NEORNITHES   All remaining birds are placed in the subclass Neornithes (= recent + birds). In contrast to the preceding subclass, this one is characterized by tail feathers arranged like a fan at the end of a tail having a short bony axis, and by fusions of bones in the spine, braincase, lower leg, and "hand." A system of air sacs is usually present, and air spaces are found within most of the bones. Unless the power of flight has been secondarily lost, as in the ostrich, the breastbone has a large keel from which the flight muscles take their origins.

Although the subclass is relatively homogeneous, it is also large, and it has been convenient to establish 32 orders—as many as for the more diverse mammals and half again as many as for all reptiles, living and extinct. It follows that the characteristics that distinguish the orders and three superorders are relatively minor.

The superorder ODONTOGNATHAE has long been extinct. These toothed birds were specialized for swimming. The superorder PALEOGNATHAE includes

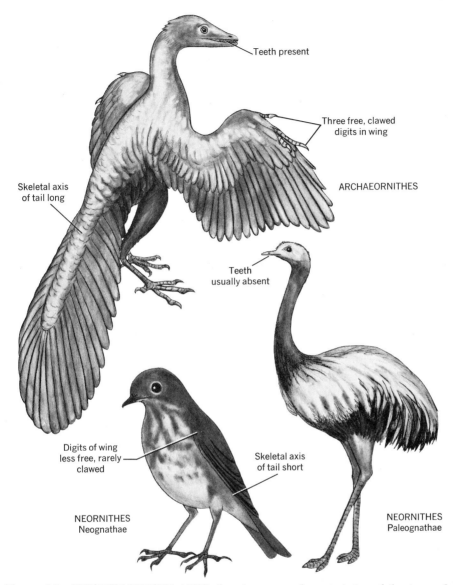

Teeth present

Three free, clawed
digits in wing

ARCHAEORNITHES

Skeletal axis
of tail long

Teeth
usually absent

Digits of wing
less free, rarely
clawed

Skeletal axis
of tail short

NEORNITHES
Neognathae

NEORNITHES
Paleognathae

Figure 4-9 REPRESENTATIVE AVES showing some characteristics of the two sub-classes and examples of the two surviving superorders of Neornithes.

the ostrich, emu, cassowary, and their relatives. Features of the pelvis and palate unite the group. However, all but one member (the tinamou) have secondarily lost the power of flight, and many of the characters they have in common (including large size, strong legs, reduced pectoral girdle) are apparently due to convergent evolution. Nearly all surviving birds belong to the remaining superorder, NEOGNATHAE.

CLASS MAMMALIA Mammals, like birds, are familiar and distinctive. Children learn that only mammals have hair and mammary glands which give the class its name. Succeeding chapters will present unique features of the cranium, jaw, teeth, ear, pectoral girdle, pelvis, muscles, brain, and other structures. Mammals are also numerous. Some 3000 genera are known (2000 are extinct).

Unlike the transition from reptiles to birds, which is documented by only several fossils, the transition from reptiles to mammals is well-recorded—so well, in fact, that the conventional boundary based on the structure of the jaw, ear, and cranium is somewhat arbitrary.

SUBCLASS PROTOTHERIA   Several orders of Mesozoic prototheres are recognized. The only living members of the subclass comprise the order MONOTREMATA, which includes the aquatic platypus and the insect-eating echidnas (Figure 4-10). These odd creatures are rare both in zoos and in their native Australia and New Guinea. If known only by the pectoral girdle, they surely would be classified as reptiles. Furthermore, they are oviparous (egg-laying), which is equally unique among mammals. The young are nourished by milk, however, and the presence of hair and a single bone in each half of the lower jaw qualify monotremes as mammals.

SUBCLASS THERIA   All familiar mammals belong to the subclass Theria. They are viviparous (give birth to live young), which sets them apart from Prototheria. One infraclass, the extinct TRITUBERCULATA, may be passed over here. Two infraclasses survive to the present. METATHERIA includes the single and long-enduring order MARSUPIALIA. Opossums, bandicoots, phalangers, wombats, and kangaroos are marsupials. They give birth to tiny embryonic young which are nourished in the pouch ("marsupium" = pouch) of the mother until they are able to walk about.

The other surviving infraclass, EUTHERIA (= true + beasts), comprises the animals commonly known as "placental mammals." The term is misleading: A placenta is an organ that accomplishes physiological exchange between mother and fetus. Some reptiles and even several fishes and amphibians have placentas, and so do all marsupials. The placenta of marsupials is always vascularized on the fetal side by a membrane called the yolk sac; that of eutherian mammals is usually vascularized by the allantoic membrane. However, some marsupials have both allantoic and yolk sac circulations, and some eutherian mammals have no allantoic circulation. Numerous features of the skeletal, reproductive, and nervous systems provide more technical but more exact ways to distinguish the infraclass Eutheria.

There are 16 surviving and 10 extinct orders of eutherian mammals. Diagnosis of ordinal characters must be left for textbooks of mammalogy. However, in Part II, and particularly in Part III, of this book, reference will be made to representatives of all surviving orders. It will be useful, therefore, to identify them here (Figure 4-11).

INSECTIVORA includes shrews, moles, and hedgehogs. All are small animals with numerous sharp teeth. This is the oldest and most primitive order of the infraclass. DERMOPTERA is represented only by the gliding colugo (see figure

Figure 4-10   REPRESENTATIVES OF THE SUBCLASS PROTOTHERIA AND THE INFRACLASS METATHERIA. These taxa comprise, respectively, the orders Monotremata and Marsupialia.

on p. 533). The only mammals capable of sustained flight are the bats, order CHIROPTERA. PRIMATES include lemurs, monkeys, apes, and man. As the name indicates, EDENTATA have simple teeth or none at all. They are the anteaters, sloths, and armadillos. The scaly pangolins also eat insects but comprise the PHOLIDOTA (see figure on p. 467). Rabbits and their smaller relatives, the pikas, are in the order LAGOMORPHA. The largest order is the RODENTIA, which includes squirrels, beavers, rats, mice, porcupines, and a host of other small mammals with gnawing incisors.

Whales and dolphins of the order CETACEA are among the most modified of mammals. Bears, dogs, weasels, raccoons, civets, cats, hyenas, seals, and most other flesh-eating mammals are in the order CARNIVORA. Only the aard-

Tetrapods

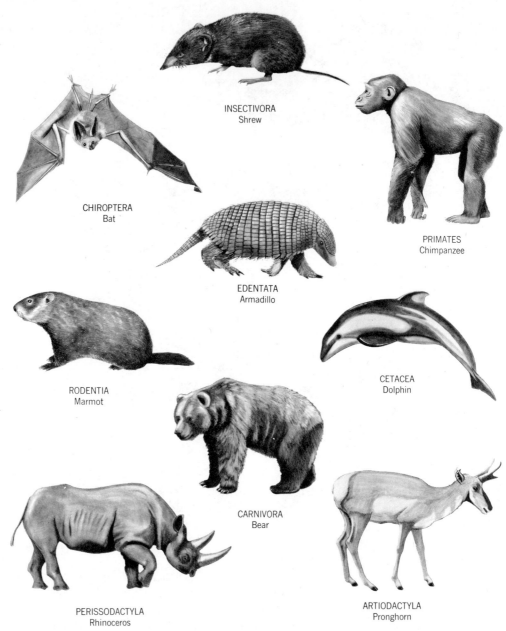

INSECTIVORA
Shrew

CHIROPTERA
Bat

PRIMATES
Chimpanzee

EDENTATA
Armadillo

RODENTIA
Marmot

CETACEA
Dolphin

CARNIVORA
Bear

PERISSODACTYLA
Rhinoceros

ARTIODACTYLA
Pronghorn

Figure 4-11 REPRESENTATIVES OF NINE ORDERS OF THE INFRACLASS EUTHERIA.

vark is in the TUBULIDENTATA. Elephants, order PROBOSCIDEA, were once more numerous both in numbers and kinds. The stocky hyraxes of Africa and Asia are in the order HYRACOIDEA. The large aquatic grazers called dugongs and manatees are SIRENIA (see figure on p. 509). Horses, tapirs, and rhinoceroses comprise the order PERISSODACTYLA. The term means odd-toed and distinguishes them from the even-toed ARTIODACTYLA such as pigs, camels, deer, antelopes, and cattle. Perissodactyls and artiodactyls include most hoofed mammals and are collectively known as *ungulates* ("unguis" = hoof), a useful term of no systematic rank.

The Phylogeny of Structure:
Evolution in Relation to
Time and Major Taxa

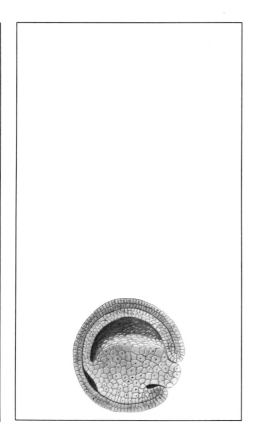

5

Early Development

The complexities of the significant relationship between embryology and evolutionary morphology were outlined in Chapter 1 (p. 18 to 20) where principles of our subject were set forth. Embryology is often studied prior to, or concurrently with, morphology. This short chapter provides an introduction to relevant aspects of early development for students who have not yet had embryology, and a refresher for those who have. Later development is included in subsequent chapters.

GAMETES AND FERTILIZATION

The mature sex cells, or **gametes,** are the male **sperm** and female **ovum** or egg. Recall that each carries a haploid, or half set, of chromosomes. The sperm cells of vertebrates are highly varied in appearance (Figure 5-1), but always have a **head,** which contains the nucleus, a **middle piece** containing the mitochondria needed to provide energy, and a **flagellum,** which propels the cell. The head may be spherical, spatulate, hooked, lance-shaped, or spiraled. It is capped by the **acrosome.**

The small eggs of Amphioxus contain little yolk and hence are said to be **microlecithal** (= small + yolk). This may be the ancestral chordate condition, but most vertebrate eggs either have moderate amounts of yolk (**mesolecithal**), as for lampreys, chondrostean fishes, and amphibians, or instead are laden with large amounts of yolk (**macrolecithal**), as for most fishes, reptiles, birds, and monotremes. Eutherian mammals, having a placenta to nourish the embryo, no longer need yolk, and their eggs have returned to the microlecithal condition

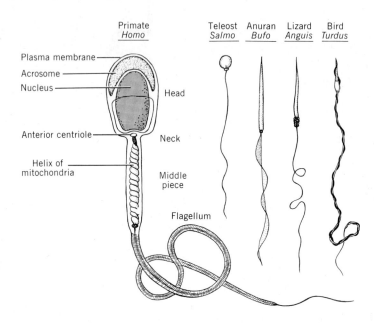

Figure 5-1   REPRESENTATIVE VERTEBRATE SPERM CELLS.

(Figure 5-2). Yolk is a complex material that includes proteins, phospholipids, and neutral fats.

The yolk is usually concentrated toward one side of the egg, the **vegetal pole,** leaving a region of clearer cytoplasm and the nucleus at the other side, the **animal pole.** Any yolk present is toward the vegetal pole, or the largest yolk platelets, at least, are in the vegetal hemisphere. Eggs with such asymmetrical distribution of yolk are **telolecithal** (= end + yolk). Metabolic activity is highest at the animal pole, and there may be a gradient in the distribution of pigment.

The egg is surrounded by a delicate **vitelline membrane.** Eggs of therian mammals are also enclosed in a thicker **zona pellucida** and a large **corona radiata,** or layer of adherent cells from the ovarian follicle. The eggs of other vertebrates may be enveloped, after ovulation, in jelly layers (amphibians), albumen (birds), and horny, membranous, or calcareous capsules or shells (many fishes, reptiles, birds).

Penetration of egg membranes and egg by a sperm cell is a complicated process involving both enzymatic and physical interactions between sperm acrosome and egg cortex, each of which undergoes striking changes. Entry of the sperm into the egg restores the diploid number of chromosomes, and activates the egg both to become refractory to the entry of additional sperm and to initiate development of the embryo.

Figure 5-2   Section of MAMMALIAN OVUM after ovulation and before fertilization.

The **zygote,** or fertilized egg, is transformed by cell division called **cleavage** into **CLEAVAGE** a multicellular embryo called a **blastula.** During cleavage the individual daughter cells are termed **blastomeres.** The process of cleavage, and the structure of the blastula, are both closely related to the amount of yolk present (Figure 5-3).

The microlecithal eggs of Amphioxus have **total** (or holoblastic) cleavage, which means that the cleavage furrows penetrate the entire yolk. Furthermore,

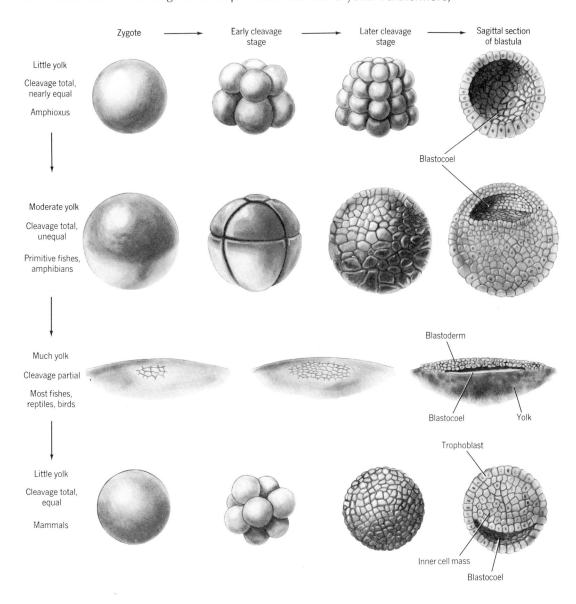

Figure 5-3 PRINCIPAL TYPES OF CHORDATE CLEAVAGE. Membranes, shells, and polar bodies omitted. Relative sizes only approximate.

cleavage is **equal,** because all blastomeres are of about the same size at any given time. The resultant blastula is a hollow ball of cells with a cavity called the **blastocoel.**

Cleavage of mesolecithal eggs is also total, but the greater amount of yolk in the vegetal hemisphere is an impediment that retards cell division. Consequently, development is slower there, blastomeres near the vegetal pole are larger than those near the animal pole, the blastocoel is displaced into the animal hemisphere, and cleavage is said to be **unequal.**

The yolk mass of macrolecithal eggs is simply too great to be penetrated by cleavage furrows. Cleavage is therefore **partial** (meroblastic) and is limited to the relatively small yolk-free region at the animal pole. A cellular **blastoderm** comes to be separated from the uncleaved yolk by a narrow cavity.

Cleavage is again total and equal for the microlecithal eggs of mammals, though the orientation of cleavage furrows is less regular than for Amphioxus. The blastula is distinctive in having a superficial layer of cells, the **trophoblast,** which surrounds an **inner cell mass** (Figure 5-3). The blastocoel is displaced toward the vegetal pole. Regardless of size and shape, the vertebrate blastula consists of a single tissue layer made up of several hundred cells with polarity that relates to the axes of the future body.

GASTRULATION AND MESODERM FORMATION

The blastula is converted to an embryo called a **gastrula** by various processes collectively called **gastrulation.** The gastrula has at first two, and then three, tissue or **germ layers.**

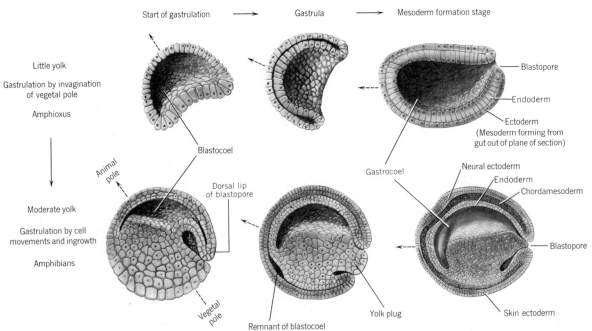

Figure 5-4 SOME TYPES OF CHORDATE GASTRULATION that are uncomplicated by a large yolk mass. Sagittal sections.

Again, Amphioxus exhibits the simplest and most primitive form of gastrulation. By the process of **invagination** the vegetal hemisphere of the blastula folds inward and extends to underlie the tissue layer of the animal hemisphere (Figure 5-4). In doing so, it obliterates the blastocoel and forms a double-walled cup. The lips of the cup then approach one another, as though a purse string were being drawn, leaving only a small opening, the **blastopore.** The embryo now has a new cavity, the **gastrocoel.** These tissue movements cause the embryo to rotate, in response to gravity, so that the animal-vegetal axis is horizontal, rather than vertical, and the dorsal surface of the embryo is uppermost. The outer germ layer is now called **ectoderm.** Because the adult gut tube will form from much of the inner tissue layer of the embryo, it is called the **archenteron,** or primitive gut. However, as shown in Figure 5-5, the notochord forms from the dorsal wall of the archenteron, and the **mesoderm,** or middle germ layer, forms from a series of dorsolateral outpocketings of the archenteron. The cavities of the pockets become **coelom.** This kind of mesoderm and coelom formation is called **enterocoely** (= gut + hollow). It is not found in vertebrates, yet seems to link chordates and echinoderms. The remaining part of the archenteron is **endoderm.**

Blastulas derived by total, but unequal cleavage can scarcely gastrulate by invagination because the yolk-laden blastomeres of the vegetal hemisphere are in the way. Instead, cells roll inward (a process called **involution**) at the site of the future blastopore (Figure 5-4). They extend into the blastocoel as a second tissue layer. Surface cells migrate to replenish the supply at the blastopore. Ingrowth is most rapid at what will be the dorsal lip of the blastopore, but ultimately cells stream inward on all sides. As in Amphioxus, the blastocoel is gradually obliterated as a gastrocoel is formed. The embryo then has two tissue layers, though the ventral part of the archenteron is swollen with yolk. The roof of the archenteron is called **chordamesoderm** because it forms the notochord in the midline, and a series of paired, right and left blocks of mesoderm, or **somites,** dorsolaterally (Figure 5-6). There is no outpocketing from the gut tube. Instead, coelom forms by cavitation or **schizocoely** (= split + hollow) in the somites.

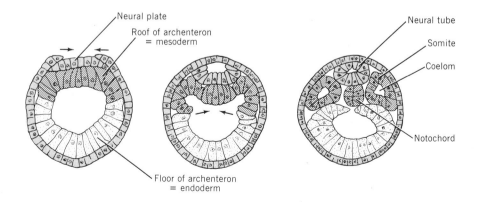

Figure 5-5 Successive stages in the ENTEROCOELIC MESODERM AND COELOM FORMATION OF AMPHIOXUS as seen in cross sections.

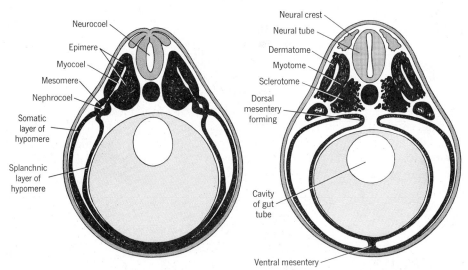

Figure 5-6 DIAGRAMS OF NEURULATION AND EARLY DIFFERENTIATION OF THE MESODERM as seen in cross sections of an embryo having a moderate amount of yolk.

The remaining part of the innermost tissue layer is endoderm. It reconstitutes itself dorsally after the formation of notochord and mesoderm.

If cleavage has been partial, gastrulation is further complicated by the mass of yolk. (The following account is generally accurate for reptiles and birds; fishes may differ.) The second tissue layer, or endoderm, forms under the blastoderm

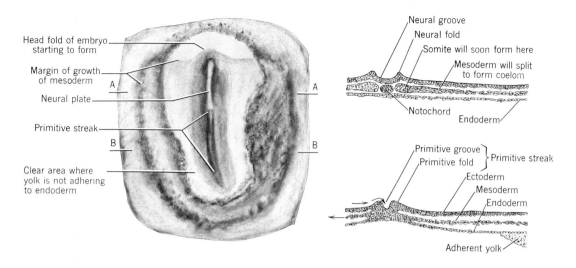

Figure 5-7 MESODERM FORMATION AND START OF NEURULATION IN THE CHICK. Left, surface view after about 17 h of incubation. Right above, cross section at level A-A; right below, cross section at level B-B.

by the process of **delamination:** Some cells detach themselves from the original layer of cells and then aggregate into a second sheet. Some mesoderm may also form by delamination, but most is produced by another process: A lengthwise thickening called the **primitive streak** forms on the posterior part of the germinal disc (Figure 5-7). Surface cells stream toward the streak where they involute and spread out between the original two tissue layers as mesoderm. The primitive streak is posterior to the **embryonic area** where the embryo is starting to develop and may represent a modified blastopore. Cells that involute and then move directly forward from the primitive streak will form the notochord, and along side it some mesoderm also swings forward into the embryonic area. Mesoderm that is lateral and posterior to the streak will contribute to the fetal membranes. Coelom formation is by schizocoely.

Thus far we have noted that mesoderm may form by separation from the gut tube, by delamination, and by primitive streak. Another method of mesoderm formation is general and important: Cells may detach themselves from existing tissue layers and migrate individually as **mesenchyme.** These branched cells ultimately aggregate in the head, limbs, and other places to form mesodermal tissues.

When the mammalian embryo reaches the blastula stage, maternal fluid indirectly enters the blastocoel causing it to enlarge greatly. (Figure 5-3 depicts an early stage before the blastocoel expands.) The embryo is now called a **blastocyst.** Endoderm forms by delamination from the inner cell mass and spreads to line the trophoblast. The flattened inner cell mass is now a blastoderm, and it forms a primitive streak to produce the notochord and most of the mesoderm for the fetal membranes and trunk of the embryo. Much mesoderm is also produced as mesenchyme. Coelom forms by schizocoely.

NEURULATION AND NEURAL CRESTS

The gastrula is converted to a **neurula** by processes called **neurulation.** These events, which overlap with the formation of the germ layers as described above, establish the central nervous system. We have seen that chordamesoderm extends forward in the midline from the dorsal lip of the blastopore or primitive streak. Chordamesoderm has the important function of causing, or **inducing,** the overlying ectoderm to thicken into a **neural plate** (Figure 5-6). Longitudinal **neural folds** form along the margins of the plate and arch inward to fuse in the midline. The resulting **neural tube** encloses the **neurocoel.** This establishes the dorsal hollow nerve cord that characterizes all chordates. (The process of neurulation may differ in fishes.)

As the neural folds come together, some cells are pinched off in the angles between the sinking neural tube and overlying ectoderm. These loose cells aggregate for a time in the spaces between adjacent somites where they are called **neural crests** (Figure 5-6). As we shall see, neural crests have diverse, important derivatives.

ESTABLISHMENT OF THE BODY PLAN

As neurulation progresses, embryos derived by total cleavage (Amphioxus, amphibians) lengthen so that a head and tail are established. Yolk-laden cells may distend the belly area, but are enclosed within the body contour. Embryos derived by partial cleavage may be thought of as consisting for a time of three germ layers spread "face down" on the uncleaved yolk the way a bearskin rug is spread on a floor. Soon a **head fold** lifts up, like a mounted head on the bearskin rug. A **tail fold** follows, and then lateral body folds. The embryo is then broadly joined to the yolk by a **body stalk,** but is otherwise free (see figure on p. 211).

As the body lengthens, the archenteron is drawn out into a gut tube. The blastopore (or an equivalent locus) becomes the anus, a mark of all Deuterostomia (see p. 28). Later a mouth opening ruptures anteriorly. The endoderm will form the lining of the gut and its derivative organs (lungs, liver, pancreas, some endocrine glands, yolk sac, allantois, urinary bladder), and will contribute to the formation of gametes.

The outer, or somatic ectoderm will form the superficial and germinative layers of the skin and their derivatives (glands, hair, feathers, claws), the lens of the eye, olfactory organ, inner ear, part of the pituitary gland and lining of the mouth, and will contribute to the formation of teeth and certain scales. The sensory ectoderm, or neural tube, will form the spinal cord and brain, motor nerves, retina of the eye, and part of the pituitary. Neural crest ectoderm forms ganglia, sensory nerves, part of the skull, part of the adrenal gland, and some pigment cells, and it contributes to the formation of some teeth and scales.

With minor exceptions, the mesoderm forms the muscular, skeletal, circulatory, and urogenital systems. Chordamesoderm contributes the notochord. Mesoderm derived mostly from the archenteron or primitive streak flanks the notochord. This divides into a segmented dorsal somite or **epimere,** a small partly segmented **mesomere,** and a sheetlike, unsegmented **hypomere** which wedges its way down between ectoderm and archenteron (Figure 5-6). Coelom may be present in each of these units, but expands only in the hypomere where it separates a lateral **somatic layer** from a medial **splanchnic layer.** The somatic layer will line the body cavity and bud off mesenchyme that contributes to the bone and muscle in the limbs. The splanchnic layer forms some head muscles, heart, mesenteries, and muscles of the gut (see p. 204). Where ectoderm and somatic mesoderm join, as in the body wall, they are called **somatopleure;** where endoderm and splanchnic mesoderm join, as in the wall of the gut, they are called **splanchnopleure.**

The mesomere forms only urogenital organs. The epimere further differentiates into a lateral **dermatome,** a **myotome,** and a medial **sclerotome** (Figure 5-6). The dermatome becomes mesenchyme that spreads out to form the dermis of the skin and such hard tissues as are derived therefrom. The myotome forms muscles of the spine, throat, and eye, and muscles and skeleton of the appendages. The sclerotome forms more or less of the vertebrae, depending on taxon.

All of these developmental steps are carried further in chapters to follow.

If such yolk as is present in the egg cleaves and is directly incorporated into cells of the body (amphibians), no fetal membranes are needed. The yolk supply being limited, hatching is early and the larva starts to feed.

FETAL MEMBRANES AND PLACENTATION

If the yolk does not cleave, and eggs are laid in water (most fishes), then an extraembryonic circulation is needed to absorb the yolk and carry it into the body. To accomplish this, a membrane that includes all three germ layers extends over the surface of the yolk from the body stalk. This **yolk sac** quickly becomes vascularized and functional. Since respiration and excretion are by direct contact with the environment, no other fetal membrane is needed.

Requirements of the embryo are much more stringent when development is on land within a shell (reptiles, birds), and hatching is delayed. The extraembryonic mesoderm derived from the primitive streak soon splits, thus establishing an extensive extraembryonic coelom (Figure 5-8). The resulting splanchnopleure is adjacent to the yolk and forms a yolk sac. The overlying somatopleure lifts into a **head fold of the amnion,** and later also into lateral and tail folds, all of which converge and fuse over the embryo. This creates an outer **chorion** and an inner **amnion.** The latter contains **amniotic fluid** which bathes the embryo and provides a sheltered space in which it can grow. Somewhat later, as the embryo becomes larger, a vesicle grows out of its hind gut and extends into the extraembryonic coelom. This splanchnopleuric vesicle forms the **allantois,** which grows out under the egg shell (from which it is separated only by the chorion and egg shell membranes), becomes vascularized, and serves the embryo for respiration. It is also a receptacle for excretory wastes, absorbs albumen, and takes some minerals from the shell.

Eutherian mammals nourish their young in the uterus by physiological exchange between fetal and maternal blood streams. An organ that performs such a function is a **placenta.** It makes yolk superfluous and, hence, eutherian eggs are secondarily microlecithal. (See p. 308 for other ways that the maternal body of other vertebrates may nourish the fetus.) The fetal membranes derived from somatopleure (chorion, amnion) do not become vascular, and therefore cannot support a placenta. Both yolk sac and allantois may become vascular, and either, or both, may contribute. The formation and nature of the fetal membranes and

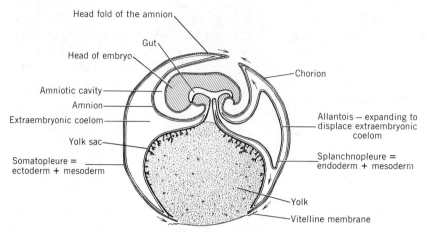

Figure 5-8   Stylized FETAL MEMBRANES OF THE BIRD early in development. Compare with figure on page 211.

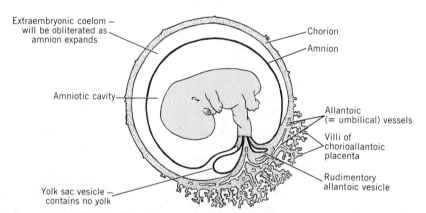

Figure 5-9   FETAL MEMBRANES OF THE PRIMATE TYPE illustrated by the human at about 6 weeks gestation.

placenta vary greatly among mammals. A short summary serves our purpose here, although the evolution of these structures is a fascinating story in itself (see Luckett in References at the back of this book).

The yolk sac is the most variable of the membranes. It is the principal fetal contribution to the placenta in marsupials and some rodents. It is large and persists to birth in carnivores, is first large and then lost in ungulates, and is first small and then rudimentary in primates (Figure 5-9). The allantois makes the fetal contribution to the placenta of most mammals. The allantoic circulation of reptile and bird is homologous with the umbilical circulation of such mammals (see figure on p. 260). The vesicle of the mammalian allantois ranges from large to vestigial. Since the chorion lies between allantois and uterus, it is incorporated into the placenta (which is, therefore, described as **chorioallantoic**), although most of its tissue erodes away in some taxa. The amnion, which varies less than the other membranes, becomes a thin, but tough sac. The amniotic fluid bathes the fetus, gives it freedom to move, and protects it from pressures on the maternal abdomen. Because of their common possession of an amnion, reptiles, birds, and mammals are collectively called **amniotes.** The **umbilical cord** of mammals is equivalent to an attenuated avian body stalk.

Integument and its Derivatives

The integument and its derivatives comprise an exceedingly varied and adaptable organ system. By weight, the system is often the largest of the body, and no other performs as many functions. It provides physical protection to delicate tissues below—particularly where most needed, as under the feet. It contributes importantly to water balance. Thus, amphibians can absorb water through the skin, even taking moisture from damp air or soil, whereas the skin of desert reptiles is resistant to water loss. Dissipation of heat to the environment may be increased by the dilation of superficial vessels and by the evaporation of sweat, whereas conservation of heat is increased by fat deposits and the erection of insulating hair or feathers. The integument guards against the entry of most injurious organisms and materials, and provides coloration essential to identification, aggressive and sexual behavior, and camouflage. It serves locomotion through friction pads, the interlocking of scales or claws with the substrate, the provision of airfoils, and in other ways. Respiratory exchange occurs through the damp skin of amphibians, even supplanting lungs in some salamanders. Secretions of skin glands may contribute to attraction or repulsion, nutrition of the young, excretion of salts and urea, and thermoregulation. The integument houses many sense organs, contributes to the contours of the body, screens out injurious energy waves, and may also store fat and glycogen, provide significant support and defense to the body, and synthesize vitamin D.

Because the integuments of the various vertebrates are so diverse, the system tells the morphologist much about the habits and environments of animals, and enables the systematist to identify most vertebrates. Some mammals can be identified by single hairs, and some birds from several feathers. However, because the integument is so responsive to habit and environment, it has so far told morphologists less than some other organ systems about phylogeny. Nevertheless, progress is being made.

DEVELOPMENT AND GENERAL STRUCTURE OF SKIN

The skin of all vertebrates has two principal layers, a superficial **epidermis** and a deeper **dermis** (Figure 6-1). The epidermis is derived from the ectoderm on the surface of the embryo. The dermis is derived from the dermatome supplemented by contributions from the lateral and ventral somatic mesoderm. Cells from these sources migrate as mesenchyme to distribute themselves evenly under the ectoderm. Some neural crest cells also invade the developing dermis.

The epidermis is stratified into two or more layers. The deepest layer rests on the dermis and consists of closely packed, discrete cells. It is called the **stratum germinativum** because its daughter cells are pushed outward and, as they mature, are transformed to become the more superficial cells of the skin. The layer or layers of the epidermis that are superficial to the stratum germinativum are exceedingly varied according to taxon. Most are (or will become, or have been) secretory in nature and, as classified by W. B. Quay, fall into two general categories, mucous cells and proteinaceous cells. The former produce various types of mucus, some kinds of poisonous secretions and, in some fishes, **photophores** (light-producing cells). The proteinaceous line of epidermal cells may produce slime, poisons, substances eliciting alarm reactions, enamel, and possibly some photophores. The principal product of this cell line, however, is the horny material **keratin,** which is the main constituent of feathers, hair, claws, reptilian scales, and also of the dead outermost layer, or **stratum corneum** of the dry skin of tetrapods ("keratos" and "cornu" both = horn). Two molecular types of keratin, designated α- and β-keratin, are recognized. The α-keratin tends to be more flexible, though the functional difference between the two kinds is not well-known.

The dermis is usually thicker than the epidermis. It has relatively fewer kinds of cells and is characterized by a meshwork of fibers. Most abundant are **collag-**

Stratum corneum
Melanophore
Cells in transition
Stratum germinativum
Basal membrane
Mesenchyme, present during development
Blood vessel
Collagenous fibers and smooth muscle fibers

Constant migration of cells
Inductions during development
Migration of cells to form bones or scales

Figure 6-1 SECTION OF GENERALIZED VERTEBRATE SKIN.

enous fibers, which are tough, straight, and made up of fibrils. The fibrils are oriented in a three-dimensional network. They are linked to a ground substance that resembles plasma protein and contains mucopolysaccharides. **Elastic fibers,** which are fewer in number, are wavy, nonfibrillar, and branched.

The dermis commonly has an outer, vascular, **stratum spongiosum** and a deeper, thicker, **stratum compactum.** These merge with one another and also bridge across to secure the skin to the connective tissue covering the muscles of the body wall. Smooth muscle fibers may be present in the dermis. Fat commonly invades the dermis or is deposited between the skin and the body. The glands of the skin are derived from the epidermis but usually penetrate into the dermis which then adds supportive tissue to them.

Pigment cells are called **chromatophores.** They are derived from neural crest cells and occur, according to taxon, in any level of the skin, but tend to concentrate near the epidermal–dermal boundary. Chromatophores of the epidermis are particularly characteristic of **homeotherms** (animals maintaining constant temperature) and are of one type called **melanophores.** They have numerous migratory organelles, termed melanosomes, which contain the pigment melanin. Melanin is black, brown, or red. Color imparted by these cells may be constant or may be responsible for **morphological color change,** which is seasonal, age-related, or otherwise relatively slow change.

Chromatophores of the dermis are found almost exclusively in **poikilotherms** (animals having variable body temperature). They may maintain constant color, cause morphological color change, or cause **physiological color change,** which is relatively rapid, as when a fish or lizard adapts its color to that of an altered substrate. There are three types of dermal chromatophores: Melanophores are similar to those of the epidermis. **Iridophores** have organelles called reflecting platelets that are oriented in stacks and contain crystalline deposits (chiefly of guanine) which scatter or reflect light. These cells are commonly large. **Xanthophores,** which are yellow, and **erythrophores,** which are red, have their pigments (pteridines and carotenoids) in organelles called pterinosomes.

These various kinds of chromatophores can be structurally and physiologically interrelated in the achievement of certain color effects. Their complex control may include influence by hormones of the pituitary, thyroid, gonads, and adrenals, and in some poikilotherms by the nervous system as well. Color is "used" by the various vertebrates for concealment, for making themselves conspicuous (e.g., as warning, social releaser, or sexual attractant), for control of heat absorption and conservation, for protection of the nervous system or gonads from light, and for control of the synthesis of vitamin D.

The epidermis and dermis are separated by a thin basal membrane (Figure 6-1). During development of the skin and of its derivatives, inductions occur across the membrane between the germinal epithelium and the mesenchyme. Even if an integumentary derivative incorporates only epidermal tissue (e.g., horny scales, feathers, hair), or only dermal tissue (e.g., certain bones), both layers are essential to the formation of (probably) all derivatives, and some of them (e.g., teeth, fish scales) incorporate tissue from each layer. In the absence of underlying

GENERAL DEVELOPMENT OF SKIN DERIVATIVES

dermis, the embryonic epidermis degenerates, and experiments show that the kind of epidermal derivative formed (whether scale, feather, hair, or other) is controlled by the nature of the associated dermis. The influence of the epidermis on the dermis may be less universal, yet epidermis apparently triggers the dermal contributions to teeth and fish scales.

INTEGUMENT OF FISHES: EMPHASIS ON DERMAL DERIVATIVES

SOFT STRUCTURES   The soft part of the integument of extinct agnaths and placoderms is, of course, unknown. The epidermis of CYCLOSTOMES is thin and has several kinds of glands, all of which are unicellular. Most numerous are **mucous glands** of two types. **Club glands** produce slime of fibrous protein. **Granular glands** probably discharge at the surface of the body, but their function is not yet known. Keratin is absent. A thin noncellular **cuticle** covers the epidermis. The dermis, which may be thinner than the epidermis, consists largely of a fibrous layer which contains collagenous fibers but no elastic fibers. There is no trace of scales.

The skin of jawed FISHES is usually thin and glandular. It fits tightly over the body (Figure 6-2). With some exceptions, keratin is entirely absent. The replacement of worn epidermis is constant. Mucous glands of one or another type are nearly always abundant. They are usually unicellular but may also be multicellular. The slimy mucus they secrete cleans the body and produces a cuticle which prevents the entry of foreign material, assists in osmoregulation, and reduces resistance as the fish swims. Granular and club glands are also common. Some fishes have **poison glands** associated with fin spines; others have multicellular light organs which may even be provided with tiny lenses and reflectors. The dermis, though still thin, is divided into a stratum compactum and a stratum spongiosum (except when it covers the fins, where it is reduced to a basal membrane).

DEVELOPMENT AND STRUCTURE OF HARD TISSUES   The scales and nonglandular integumentary appendages of fishes are largely of dermal origin, whereas those of tetrapods are largely of epidermal origin. The most complex derivatives of the integument of fishes are hard scales and denticles of various kinds. Before studying their nature and phylogeny it is desirable to know about the tissues of which they are constructed.

Figure 6-2   SECTION OF THE SKIN OF A TELEOST.

In historical perspective, the description and classification of the scales and hard integumentary appendages of fishes have been complicated by various factors: (1) Fish scales—particularly those of fossil fishes—include many kinds of hard tissues. (2) The various hard tissues grade into one another and combine in many ways. (3) Certain virtually identical tissues seem to develop from different germ layers. This throws doubt on their homology and has led to repeated use of such vague designations as "enamel-like dentine." (4) Some of the terms applied to scales describe gross shape (e.g., cycloid, rhomboid), others identify one of the tissues present (ganoid, cosmoid), and still others indicate a particular combination of tissues (palaeoniscoid, lepidosteoid). This has resulted in a confusing lack of parallel terms. However, difficulties are being overcome as biochemical analysis, physiological studies, and electron microscopy reveal details in the mechanisms of deposition and resorption of hard tissues.

Significance is attached to the embryonic origin and interactions of the precursors of hard tissues. Some neural crest cells join mesenchyme from the dermatome and contribute to the dermis of the skin and the tissues of the gums. Specifically, they form papillae which induce enamel organs from the overlying ectoderm. These organs, in turn, induce the papillae to form dentine if this substance is to develop at all. Bone may also be induced in this way. Finally, dentine induces the enamel organ to produce enamel (these tissues are defined below). If dentine is not deposited, then the enamel organ (or its equivalent—the term is not apt in this instance) may form horny scales or any of the other derivatives of the ectoderm described earlier in this chapter. There is, therefore, a basic similarity in the mechanism of development of teeth and all skin derivatives regardless of ultimate hardness and principal germ layer of origin. Proponents of these views believe that continuous armor could be formed under the influence of an extensive interaction between neural crest derivatives in the dermis and ectoderm which has the potential to form various hard tissues.

The Scandinavian paleontologists Ørvig and Stensiö have studied the minute structural detail of scales and fragments of fossil armor. They postulate that the ancestral vertebrate, and also the young of ostracoderms and placoderms, had tiny scales, each called a **lepidomorium** (= scale + division). Larger scales and armor were formed, they believe, by the aggregation of lepidomoria edge to edge (producing cyclomorial scales) or in onionlike layers (producing synchromorial scales). Some paleontologists believe that the sequence of fossils does not support this theory, but the two general hypotheses need not be entirely mutually exclusive.

There are three principal kinds of hard tissue: enamel, dentine, and bone. **Enamel** is the hardest tissue of the body. It is shiny, translucent in thin sections, and composed of elongate crystals of hydroxyapatite [$3(Ca_3PO_4)_2 \cdot Ca(OH)_2$]. In therian mammals it is prismatic. Internal cells and tubules are absent. Only about 3% of the tissue is organic. Enamel occurs only on teeth and superficial denticles, scales, or armor plates, and is usually external to any other hard tissues present. It is produced only by ectoderm—even at the back of the mouth where the ectoderm has migrated into position. Growth is by accretion on the inner surface of an enamel organ. Hence, enamel cannot be altered or replaced once it has been deposited. **Ganoine** is an enamel characterized by thick deposition in successive waves of growth which create a laminar structure.

Dentine is harder than bone and usually softer than enamel. The chemical composition of its inorganic salts is the same as that of enamel, but the content of organic fibers is typically about 25%. The generative cells usually, but not invariably, remain external to the hard tissue; their processes then penetrate the dentine via dentinal tubules. Dentine occurs only in teeth, denticles, scales, and external armor. It is present unless secondarily lost, and lies internal to enamel and usually external to bone where those tissues are also present. It is produced only by the outer surface of a mesodermal papilla which, in turn, occurs adjacent to the boundary of mesoderm with ectoderm. Dentine can be altered little, and only at its generative surface. Various types of dentine are recognized. **Osteodentine** is organized in **osteons** (columns having cylinder-within-cylinder construction) that are usually interspersed in a matrix of bone. (Osteons that occur in dentine are also termed denteons.) **Orthodentine** has no true osteons or bony matrix but instead is either laid down in a superficial compact layer (pallial dentine) or in layers that are concentric around a central pulp cavity (circumpulpar dentine). **Cosmine** (not a parallel term) is dentine having characteristic tufts of tiny canals that radiate upward and outward from a succession of small vascular centers distributed more or less in a plane that is parallel to the surface of a scale. Very hard dentine is termed **enameloid,** and may be difficult to distinguish from enamel.

Bone has about the same organic content as dentine, usually occurs internal to dentine if each is present, and develops in a deeper and less restricted part of the dermis. It usually has internal bone cells (osteocytes) located in small vacuities (lacunae) which intercommunicate by small canals (canaliculae) (Figure 6-3). Acellular (or anosteocytic) bone is common, however, in pteraspids and teleosts. The ancestral relationship between cellular and acellular bone is still in dispute. Bone (like osteodentine) may be characterized by osteons (here also called Haversian systems), but when it is adjacent to internal and external surfaces it is usually deposited in laminar sheets (like orthodentine). Bone may be compact or vascular and spongy. It may have few intrinsic collagenous fibers or many which are more or less layered and make the bone soft and flexible. Other collagenous fibers (Sharpey's fibers) may penetrate bone or dentine from adjoining connective tissue to bind scales together and anchor them to the substrate. The lamellae and osteons of bone can be resorbed and replaced at any internal or external surface, even if osteocytes are absent. Growth includes reorganization as well as accretion.

Hard tissues not found in the integument (cartilage, some kinds of bone) are described in Chapter 8.

PHYLOGENY OF BONY SCALES AND THEIR DERIVATIVES   Hard tissues appear to have been primitive for vertebrates. The earliest known fishes and jawless vertebrates are nearly always *more* heavily armored than the descendants in their respective lineages. Enamel, dentine, and bone were all present in fragments of

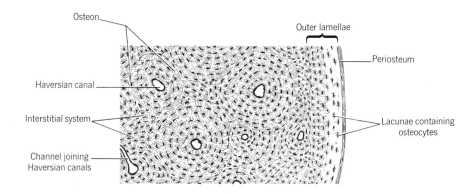

Osteon

Outer lamellae

Periosteum

Haversian canal

Interstitial system

Lacunae containing osteocytes

Channel joining Haversian canals

Figure 6-3   CROSS SECTION OF COMPACT BONE.

armor from the Ordovician period. Primitive armor may have served as an ion reservoir, for protection, for osmotic control, or perhaps to make the body heavy.

The early presence of hard tissues leaves one puzzled as to the origin of the heavy and complicated armor of ostracoderms from the (apparently) naked integument of protovertebrates. Surely there must have been an intermediate step. The lepidomorial theory, and the theory that extensive dermal papillae underlying extensive enamel organs could have formed large armor plates, are attempts to reconstruct that step.

Armor shields of ostracoderms and placoderms (cephalaspids, pteraspids, arthrodires, antiarchs) differ only in size from the coarse scales found elsewhere on their bodies. In section, armor shows three principal layers. The surface is composed of dentine (often reduced in placoderms) which may be capped with enamel completing surface projections called **denticles.** A middle layer is composed of bone that is riddled with anastomosing channels for small blood vessels and sensory pits. The basal layer is lamellar bone with fewer vascular channels.

Anaspids have no armor shields, and their scales have regressed in that only the basal layer is retained. It is generally assumed that the naked skin of cyclostomes resulted from further degeneration and that their ancestors had armor plates. However, the record does not exclude the possibility that their ancestors had small scales—or none.

Cosmoid scales differ in no fundamental respect from the ancient armor just described—the same basic layers are present. The term has gained wide usage, however, and is helpful for describing somewhat more advanced scales that are typically smaller, thinner, and characterized by having dentine of the cosmoid type. The surface of the scale is usually sculptured by the enamel of the denticles ("cosmoid" = ornamented). Thus each scale resembles a unit of the superficial layers of the armor shown in Figure 6-4. The scales may be **cycloid** (roundish in outline) or **rhomboid** (a parallelogram in outline). Cycloid scales are usually imbricated (overlapping). Rhomboid scales often overlap at their inner margins, but are fitted edge to edge at their outer surfaces. Cosmoid scales are found on the posterior parts of the bodies of some placoderms and on crossopterygians and early dipnoans. (The modern crossopterygian, *Latimeria,* has large cycloid scales with reduced surface layers; modern dipnoans have thin scales without cosmine.)

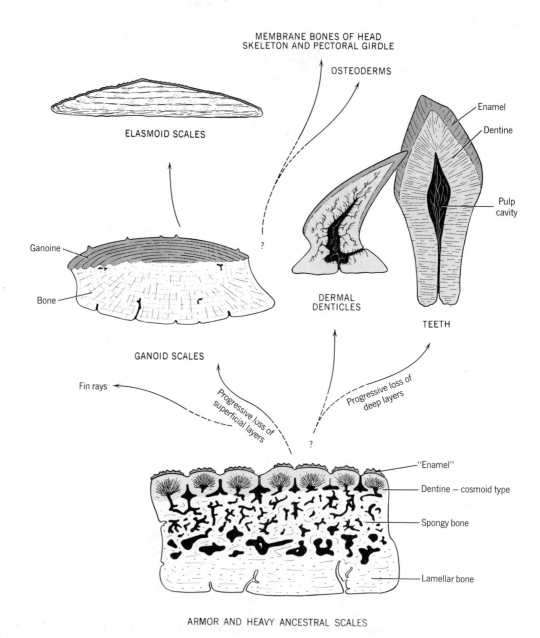

Figure 6-4 STRUCTURE AND RELATIONSHIPS OF DERMAL SCALES AND DERIVATIVES.

Ganoid scales are thick rhomboid structures that evolved from cosmoid scales. There are two principal types. The more primitive is called a **palaeon-iscoid scale.** The surface is thickened during successive periods of growth by laminations of the enamel called ganoine. Cosmoid dentine is retained under the ganoine. The base of the scale is lamellar bone pierced by vascular canals. This type of scale is found on primitive extinct actinopterygians and on the surviving chrondrostean, *Polypterus.* The second type of ganoid scale was derived from the first and is called a **lepidosteoid scale.** The ganoine is the same. The cosmine is deleted. The bony base is acellular, and the canals, though present, are no longer vascular. This type of scale is found on more recent chondrostean and holostean ray-finned fishes (but not on modern chondrosteans, which have degenerate scales) and also, surprisingly (since this is a rather advanced scale), on fishes of the extinct class Acanthodii.

Elasmoid scales were derived from ganoid scales of the lepidosteoid type. They are restricted to teleosts. The basal layer, which now forms the bulk of the scale, remains acellular but is laced by collagenous fibers coursing in various directions. The resulting bone (called isopedine) is somewhat flexible and soft. The ancestral ganoine is absent, and in its place is a thin surface glaze derived from the enamel organ. Elasmoid scales are thin, imbricated, and cycloid or **ctenoid** (having comblike projections on the exposed margins).

Isolated **dermal denticles** or placoid "scales" are confined to elasmobranchs and some aberrant placoderms. They evolved from cosmoid scales or possibly from the integument of a placoderm stock that sidestepped armor plates and heavy scales entirely. They lack the bony basal layer of scales and are always small and usually isolated. A central pulp cavity is surrounded by dentine, and this is capped by a tissue that is usually considered to be ectodermal enamel.

Various hard structures of tetrapods are derived from the bony scales of fishes. Caecilians have bony ossicles buried in the skin. **Osteoderms** are plates of bone which are derived from scales and located under the horny scutes of crocodilians, some lizards, and some extinct labyrinthodonts and reptiles (Figure 6-5). Some bones in the shells of turtles probably had the same origin. (Other bones in the shells of turtles are flattened ribs.) Splintlike bones that lie in the muscles of the ventral abdominal wall of crocodilians (but not lizards) are also seemingly derived from dermal scales. They are termed **gastralia.** (The bones in the shells of armadillos are of similar nature, but they evolved secondarily, long after their ancestors had lost all ossifications in the skin.)

Other hard structures have also evolved from scales. Teeth certainly evolved about as dermal denticles evolved. Various bones of the roof of the skull and pectoral girdle represent armor plates that lost their enamel and dentine and sank below the skin to join bones originating in the internal skeleton. The fin rays of bony fishes are also regarded as derivatives of scales. These structures will be described in later chapters.

Figure 6-5 OSTEODERM FROM THE NECK SKIN OF A LARGE CROCODILE. Actual size 7½ × 10½ cm.

INTEGUMENT OF TETRAPODS: EMPHASIS ON EPIDERMAL DERIVATIVES

SKIN OF AMPHIBIANS, LIVING AND EXTINCT   The epidermis of living amphibians is thin (typically five to eight cell layers), but in response to contact with the air it has a particular mucopolysaccharide which apparently helps control desiccation, and a stratum corneum with α-keratin (Figure 6-6). Only the outermost cell layer is dead, however, and this is lost every few days, sometimes in large patches. The sloughing is under hormonal control.

Amphibians have two kinds of multicellular, alveolar (flask-shaped) glands which originate from the epidermis and grow down into the dermis. Their products reach the surface by ducts. Abundant mucous glands secrete continuously and spontaneously to clean and lubricate the skin and to keep it moist so that cutaneous respiration will be possible. Granular glands are under nervous or hormonal control. They secrete an acrid milky fluid that is distasteful, and in some instances very toxic to predators. Granular glands are grouped together in the "warts" of toads. The amphibian dermis is two-layered and may be provided with lymph spaces and muscle fibers.

Terrestrial amniotes are able to withstand abrasion and desiccation largely because of keratinized derivatives of the epidermis. Having only a thin layer of dead keratinized cells, modern amphibians must instead seek moist habitats or use behavioral adaptations to avoid drying out. It is probable that the skin of terrestrial labyrinthodonts was thicker, drier, and more like that of some modern reptiles. Many had bony ossicles in the skin, and these are usually associated with a heavily keratinized epidermis.

SKIN OF REPTILES: HORNY SCALES   Reptiles use keratin (and lipids) of the epidermis to "airproof" their skin. The adaptation involves the fundamental patterning of the skin in that the distinctive epidermis forms a complete body covering of horny scales. Joints between scales are merely regions where the horny material is thin and folded (Figure 6-7). The epidermis of lepidosaurs, as described chiefly by Maderson, is of particular complexity and interest. In these animals an

Figure 6-6   SECTION OF THE SKIN OF AN AMPHIBIAN.

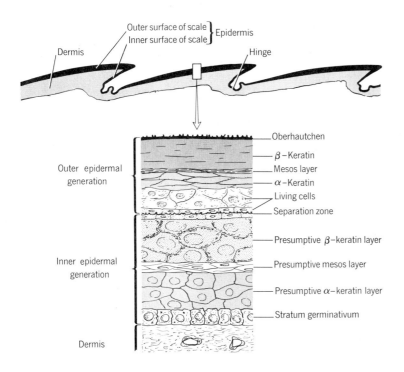

Outer surface of scale ⎫ Epidermis
Inner surface of scale ⎭

Dermis

Hinge

Oberhautchen

β–Keratin

Mesos layer

Outer epidermal
generation

α–Keratin

Living cells

Separation zone

Presumptive β–keratin layer

Inner epidermal
generation

Presumptive mesos layer

Presumptive α–keratin layer

Stratum germinativum

Dermis

Figure 6-7 SECTION OF THE SKIN AND EPIDERMIS OF A SQUAMATE REPTILE shortly before a molt.

entire "generation" of the epidermis is sloughed as a single unit. This occurs at least several times a year. Let us enter the cycle just after such a molt in what is termed the resting stage. The epidermis now consists of the stratum germinativum and an **outer epidermal generation** that characteristically has five layers. From the outside inwards there is first a thick, dead, acellular layer heavily keratinized by β-keratin. The surface of this layer is called the *oberhautchen* and has microscopic spicules. Under this β layer is a thin mesos layer of unknown significance and then a moderately thick layer of loose, dead, anucleate material having α-keratin. Below this are two layers of living cells: an outer layer, which will later be taken into the α layer, and an inner layer, which will later become clear and create the separation leading to sloughing.

At the end of the resting stage, the germinal epithelium rapidly proliferates the various layers of an **inner epidermal generation.** As these mature, they separate from the innermost layer of the outer epidermal generation, and sloughing follows.

The keratinous plate on the outer surface of a large flat scale is called a **scute.** The scutes of crocodilians and chelonians are not shed. Growth adds keratinized material over the entire inner surface of a scute, thus compensating for wear. Each wave of growth extends beyond the previous margin of a scute to form the familiar concentric rings of the turtle shell (Figure 6-8).

Figure 6-8   CARAPACE OR SHELL OF A DESERT TORTOISE, *Gopherus*, SHOWING SCUTES WITH GROWTH LINES. Dorsal view; anterior to left.

The reptilian dermis is thin. Mucous glands are absent, as they are also from the skins of the other truly terrestrial tetrapods. Scent glands of various types (generation glands, preanal glands, femoral pores, etc.) occur variously on the tail (some lizards), cloacal area (most squamates), thighs (lizards), and under the jaws (crocodilians). Their secretions influence social behavior. Some bones found in the skin of reptiles were mentioned above on page 95.

INTEGUMENT OF BIRDS: THIN SKIN AND FEATHERS   Following a different strategy, birds have over most of the body a thin, weakly keratinized skin which is loosely joined to the underlying tissues. It is appendages of the skin—the feathers—which are heavily keratinized. The lower leg and toes, however, are covered by horny scales similar to those of archosaurs. These are not shed. The **beak** is also heavily keratinized. The **shellbreaker** of birds and some reptiles is an elevation on the beak or rostrum which helps the hatchling to break out of its shell (see the figure on p. 19). (The egg tooth of snakes and lizards serves the same purpose, but is a real tooth.) The **spurs** of gamecocks are horny spines covering bony cores.

With rare exceptions, glandular derivatives of the avian skin are restricted to a large, branched, alveolar **uropygial gland** above the tail ("uropygium" = tail + rump) which secretes an oil used by the bird to preen its feathers. It is most developed in water birds.

Although intermediate stages are still largely speculative, there is biophysical, developmental, and anatomical evidence that feathers evolved from the epidermal scales of reptilian ancestors. They contain β-keratin, which is the same type of keratin that occurs in the outer surface of the scales of archosaurs. The keratin of avian skin, and also of the thinner, undersurfaces of archosaurian scales, is α-keratin. It is probable that feathers evolved to provide insulation and only later became adapted as airfoils.

There are several principal kinds of feathers and various intergrades. Most familiar and complex are the **contour feathers** which give the bird its external form and provide airfoils for flight (Figure 6-9). Few animal structures are so exquisite in design. The axis has a hollow, proximal (toward the body) **quill** and a solid, distal (away from the body) **shaft** or rachis. The vane is made up of **barbs,** which branch from opposite sides of the shaft, and smaller **barbules,** which branch from the barbs. The barbules on the distal side of each barb have on their edges **hooklets** which engage the proximal barbules of the adjacent barb. The resulting web is strong, light, and flexible. If disturbed, the elements of the vane may separate, but they do not break. The integrity of the feather is restored by preening with the modified edges of the bill, which reengages the hooklets.

Wing feathers (remiges) and tail feathers (rectrices) are enlarged and stiffened contour feathers. (Primary remiges are attached to the manus; secondary remiges are attached to the arm.) A variation of the contour feather lacks hooklets and therefore has no firm vane but is instead fluffy. Many birds, including the more primitive orders, have double contour feathers—a principal feather joined at its base by a shorter, softer feather called the **aftershaft.**

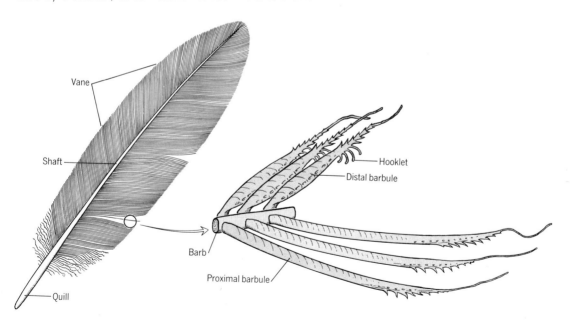

Figure 6-9 STRUCTURE OF A CONTOUR FEATHER.

Figure 6-10   A DOWN FEATHER.

Contour feathers are evenly distributed over the bodies of several kinds of birds (probably the primitive condition), but usually are restricted to feather tracts called **pterylae.** The feathers spread from the pterylae to cover the intervening areas. The conformation of pterylae is of use to systematists.

Down feathers have little or no shaft. Long barbs branch from the base of the feather and there are no hooklets. The resulting feather is small and soft (Figure 6-10). Down feathers, hidden by the contour feathers, are widely distributed and not restricted to pterylae. Their function is insulation: No material is superior to the down of the eider duck as filling for lightweight sleeping bags.

Bristles are derived from contour feathers by the partial or complete deletion of the vane. They are short stiff feathers which may screen foreign objects from the nostrils (hawks, blackbirds), increase the effective gape of the mouth (flycatchers), or form eyelashes (ostrich).

The colors of feathers come from two sources. Yellow, orange, red, brown, and black are the result of specific pigments introduced into the feather during its development. White results entirely from the microstructure of the feather. Blue, green, and iridescent hues result from a combination of black, yellow, or other pigment with microstructure that reflects only part of the light.

Feathers are molted and replaced once or (less commonly) twice a year. Most species drop the feathers one at a time so that function is not impaired, but ducks and some other birds lose most of the flight feathers at one time.

The development of a feather starts with a hummock of mesoderm, the **dermal papilla,** which is covered by ectoderm. This structure sinks into the skin, thus forming a narrow depression, or **feather follicle,** all around its base (Figure 6-11). The feather is formed only by ectoderm, but the ectoderm not only must be nourished by the vascular mesoderm, but also activated by it. Experiments show that in the absence of mesoderm, no feather can form, and that in the presence of a papilla, ectoderm that does not normally form a feather may do so.

A superficial keratinized **feather sheath** surrounds erupting feathers and subsequently sloughs away. At the base of the follicle the germinative layer forms a **collar.** The barbs of a down feather grow straight upward from the collar within the sheath. An early step in the formation of a contour feather is the development of a shaft as an outgrowth from one point on the collar. Barbs form first as branches of the shaft, and then as outgrowths of the collar itself that migrate onto the base of the shaft as it lengthens. When the sheath breaks away from the maturing feather the barbs unfold to the right and left of the shaft to change the cylindrical, embryonic feather into the flat, mature feather.

SKIN, SCALES, AND INTEGUMENTARY GLANDS OF MAMMALS   The skin of mammals is relatively thick—particularly the dermis, from which leather is made. However, the thickness varies greatly according to species and location on the body (and sometimes also according to season). The epidermis thickens where the hair is sparse and also in areas subject to pressure and abrasion, such as

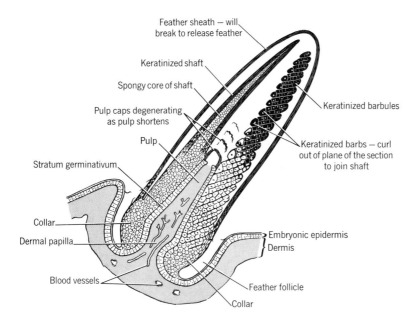

Feather sheath — will break to release feather

Keratinized shaft

Spongy core of shaft

Pulp caps degenerating as pulp shortens

Pulp

Stratum germinativum

Collar

Dermal papilla

Blood vessels

Keratinized barbules

Keratinized barbs — curl out of plane of the section to join shaft

Embryonic epidermis

Dermis

Feather follicle

Collar

Figure 6-11 SECTION OF A DEVELOPING CONTOUR FEATHER.

footpads, the kneepads of camels and warthogs, and pads on prehensile tails. Between the stratum germinativum and stratum corneum there may be one or more transitional layers, the most common of which is a thin **stratum granulosum** (Figure 6-12). Bundles of smooth muscle in the dermis are related to hair follicles.

The stratum corneum may form horny scales, as on the tails of opossums and beavers. **Claws** are strong keratinized structures which wrap around the tapering terminal bones of the digits. The tip and upper and lateral parts comprise the **unguis** and are harder than the underside (subunguis). Hooves are derived from claws. The unguis of the horse's hoof is built up of compacted horny tubules. The entire hoof spreads somewhat under the impact of a footfall. The shell of the armadillo has a heavily keratinized epidermis as well as bony dermal ossicles. The unique pangolin (mammalian order Pholidota) has markedly overlapping scales on its dorsal surface (see the figure on p. 467). These scales, which may be more than an inch long, are shed one at a time and replaced in larger sizes as the animal grows. The **baleen plates** of whales are lathlike outgrowths of the buccal epithelium that serve as strainers during feeding (see the figure on p. 572).

Sweat glands (also called sudoriferous glands) are unique to mammals. Many species have a million or more of these small glands distributed over the entire body. Others have fewer and restrict them to the muzzle or soles of the feet. Still others, including whales and sea cows, which have no use for them, have none. Sweat glands are tubular, simple (not lobulated), and coiled at their inner ends. There are two kinds, which differ somewhat in structure and nature

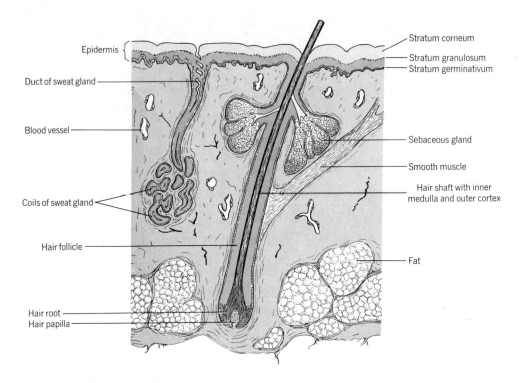

Epidermis {

Duct of sweat gland

Blood vessel

Coils of sweat gland

Hair follicle

Hair root
Hair papilla

Stratum corneum

Stratum granulosum
Stratum germinativum

Sebaceous gland

Smooth muscle

Hair shaft with inner
medulla and outer cortex

Fat

Figure 6-12 SECTION OF THE SKIN OF A MAMMAL.

of secretion. They develop in the embryo from cords of ectoderm which sink into the dermis. The evaporation of sweat from the surface of the skin helps to prevent overheating of the body and opposes slipping of foot pads over the substrate. Salt, urea, and some other wastes are excreted in the sweat. Glands in the eyelids (Moll's glands) which open near the eyelashes, and the wax glands of the external ear are considered to be enlarged and modified sweat glands.

Sebaceous glands are also limited to mammals. One or more of these branched alveolar glands drains into each hair follicle. They also occur without relation to hair on the nipples, lips, and genitalia. Their oily secretion dresses the hair and prevents excessive drying of thin skin. Lanolin, used as a base for cosmetics, is refined from the sebaceous secretion of sheep. Modified sebaceous glands (Meibomian glands) occur in the eyelids, where their secretion films over the eyeball and normally prevents overflow of the tears.

Many mammals have **scent glands.** There is wide variation in their nature and distribution. They may serve for defense, recognition, or sexual attraction. The glands may be located in the anal region (weasel family), on the face (bats, antelopes), on the back (kangaroo rat), on the feet (some artiodactyls), or indeed, on any other part of the body. Some scent glands are said to be derived from sebaceous glands and others from sweat glands.

Only mammals have **mammary glands,** which secrete the milk to suckle the young. The first indication of the development of the glands is the appearance in the embryo of a pair of epidermal ridges, the **milk lines,** which extend length-wise from the chest to the inguinal region. At intervals along the lines where the adult mammae will ultimately form, ectoderm sinks into the dermis and branches into solid cords. In females these cords enlarge at maturity, pushing under the skin and becoming compound (lobulated) and alveolar. Much of the human breast is fat. The gland becomes active at parturition under the influence of ovarian and pituitary hormones.

The number of mammae correlates with the number of young per litter and varies from one pair to about a dozen pairs. The mammae may be on the chest (primates, elephants, bats, sea cows), in the inguinal area (ungulates), or at intervals in between (rodents, carnivores). Mammary glands appear to be phylogenetic derivatives of primitive sweat glands.

Each milk gland sends numerous ducts to the surface. In monotremes these emerge in depressions whence the young lap up the milk. This probably approaches the primitive condition. Usually the site of emergence of the ducts is elevated into a **nipple** which the young can hold in the mouth and suck. In ungulates it is the skin circling the point of emergence that is elevated forming a hollow **teat** (Figure 6-13).

HAIR The phylogenetic origin of hair is obscure. There is no indication that hair evolved from reptilian scales. Where hair and scales occur together (as on tails of rats, shells of armadillos, and backs of pangolins), the hair grows between the scales, the pattern of the scales imposing a pattern on the distribution of the hair. A similar pattern of hairs often occurs where scales are lacking. Evidence is presented by Maderson to support the hypothesis that hairs arose from reptilian

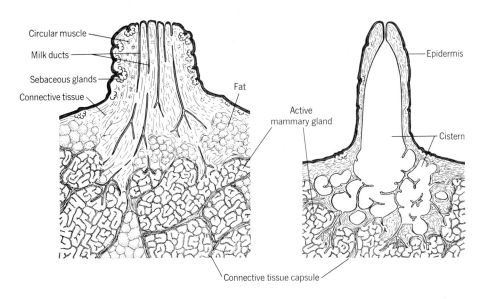

Figure 6-13 SECTION OF THE NIPPLE AND ASSOCIATED TISSUES OF A PRIMATE (left) AND THE TEAT OF AN ARTIODACTYL (right). Not drawn to the same scale.

sensory appendages of the mechanoreceptor type that were located between scales and contributed to thermoregulatory behavior. It is postulated that such structures multiplied sufficiently to become useful as an insulatory body covering. Some mammal-like reptiles may well have reached such a stage of evolution.

A typical hair has an expanded root and a shaft which, below the skin, is hidden in an epidermal sheath or **hair follicle** (see Figure 6-12). One or more sebaceous glands usually drain into the cleft between the hair shaft and adjacent tissues. The follicle slants at an angle to the skin surface, and a tiny smooth muscle runs downward from the outer part of the dermis to insert on it at such an angle that contraction elevates the follicle and its hair. This merely gives a human "goose flesh," but for most mammals it deepens the fur coat, thus increasing the effectiveness of the fur in displays and as an insulator.

In section, hairs are seen to be made up of two or three layers. Of greatest structural importance is the **cortex,** which is relatively dense and contains the pigment of the hair. Around the outside are microscopic scales that form the **cuticle.** The size, shape, and overlapping pattern of these scales vary from species to species. Mammalogists sometimes take advantage of this individuality when they wish to identify hairs from owl pellets or droppings of carnivores. Coarse hairs also have a central pith or **medulla** consisting of shrunken dead cells and air spaces.

Guard hairs are the relatively long straight hairs that give a pelt its apparent color and texture. These hairs are often grouped by twos and threes. Parting the contour hairs of most mammals (particularly of "fur-bearers") reveals shorter hairs which are very fine and numerous. These are wool hairs or **underfur.** They are usually somewhat flattened in cross section, and this makes them wavy. They often occur in groups of a dozen or more about the base of each guard hair. Underfur traps innumerable air pockets that provide insulation and may prevent water from penetrating to the skin.

Extra long and coarse hairs comprise eyelashes and manes and are found on the tails of ungulates. Whiskers, or **vibrissae,** are even coarser hairs that are specialized as tactile organs. The stiff shaft of a whisker serves as a lever that pivots at the surface of the skin to translate the slightest movement to the root. The bulb at the lower end of the root is surrounded by erectile tissue rich in nerve endings. The heaviest "hairs" of all are **quills.** Quills are hollow, but can be very stiff. The cuticle at the tip of a porcupine quill is modified to produce tiny, effective barbs.

Most mammals molt once or twice a year, the winter and summer coats often being different in density, quality, and color. The new pelage usually comes in first at one or several locations and spreads over the body in a pattern characteristic of the species.

Mesoderm plays a less prominent role in the formation of a hair than it does in the formation of a scale or feather. A solid cord of ectoderm sinks into the dermis. The walls of the cord become the double-layered **root sheath.** A small dermal papilla forms at the enlarged base of the cord. The ectodermal cells over this papilla proliferate to form the hair itself, which pushes outward through the sheath cells to emerge from the skin.

TRUE HORN
Sectioned to show core

PRONGHORN

ANTLER

After shedding   In velvet

Figure 6-14 SOME HORNS AND ANTLERS.

HORNS AND ANTLERS The horns and antlers of tetrapods are of various types. **Rhinoceros horn** is composed of keratinized fibers about ½ mm in diameter which are compacted into a solid structure tough enough (as some animal collectors have learned) to punch holes in army trucks. Growth is from the epidermis, where it covers many small dermal papillae at the base of the horn. The horn is evergrowing and is not shed.

Giraffe "horns" are merely knoblike projections from bones of the skull. They are permanently covered by skin.

The **antlers** of the deer family are also bony outgrowths of the skull, but they are shed and replaced each year. The hard, compact bone is covered by skin ("velvet") only during growth (Figure 6-14). When full size is attained, the circulation to the velvet is cut off, thus causing its death and ultimate sloughing. At the end of the breeding season, the bone at the base of an antler, just below a rough expansion called the burr, is weakened, and the antler is shed. Antlers are of varied shapes, often large, and usually branched in mature animals.

Pronghorns are limited to the American artiodactyls of that name. Again, there are bony projections from the skull which are covered by skin. However, instead of producing hair, the skin forms horn. The bony core is permanent and the horny cap is shed and regrown each year.

The **true horns** of cattle and antelopes (and of some dinosaurs and chameleons, among reptiles) have bony cores which are vascular and may contain extensions of the frontal sinuses. Over these cores are horny sheaths of epidermal origin. The horny substance is not filamentous like that of the rhinoceros horn and horse's hoof. The growth is by internal deposit so that the core keeps slipping outward, and growth rings may be seen around the base. The horn is "permanent," but there is some exfoliation of old horn.

Rhinoceros horn and the sheaths of true horns are alike in that each grows only by the accumulation at the basal end of hard material that cannot subsequently change its shape. In this they resemble claws, tusks, and mollusk shells. When such structures grow at equal rates on all sides, they grow straight. More often the rate of growth is unequal around the base. If the point of minimum growth lies opposite the point of maximum growth, then the structure always forms a logarithmic (equiangular) spiral. The rhinoceros horn is an example. If the point of minimum growth does not lie opposite the point of maximum growth, then a helical (corkscrew) spiral in space is superimposed on the flat logarithmic spiral. The ram's horn is an example.

PHYLOGENY?   As indicated earlier in this chapter, a reasonably satisfactory phylogeny can now be prepared for the various kinds of bony scales and related structures. The evolutionary trend has been for their reduction, both in bulk and complexity. Attempts have also been made to construct a phylogeny of other integumentary structures, but the task is made difficult by multiple origins, evolutionary plasticity, parallelism, and convergence. Considerations of paleontology, development, innervation, and function have been of little help. Quay believes that the various mucous glands of the aquatic vertebrates probably originated at various times, evolved along separate lines, and are not homologs of any glands of reptiles, birds, and mammals. Similarly, it is not known to what extent the granular glands of cyclostomes, fishes, and amphibians are related. The general correspondence of the keratinized skin derivatives of terrestrial vertebrates seems more evident, but truly homologous integumentary structures may not exist above the class level in modern vertebrates.

7

Teeth

Teeth have an importance for vertebrate morphology that is out of proportion to their contribution to the bulk of the body. This is true for several reasons: First, their durability has made them a significant part of the fossil record. Second, they are so adaptive that the diet of most animals can be approximated from their teeth. Third, great variation of structural detail among kinds of vertebrates combined with relative stability of structure within kinds makes teeth invaluable in systematics; experts can identify most species of mammals by a single cheek tooth. Finally, in spite of their adaptation to diet, teeth can often be used to trace the general course of evolution within, and among, genera, families, and orders. For these various reasons the teeth have been the subject of much study.

ORIGIN AND STRUCTURE

The origin of ancient integumentary armor was mentioned in Chapter 6, and illustrated on page 94. Such armor had surface denticles of enamel (or enameloid, according to some authorities) and dentine which merged into bone below. Teeth appear to have evolved from denticles released from armor near the margins of the mouth as ossification in the integument was gradually reduced.

The part of a mature tooth that is above the root and ultimately subject to wear is the **crown.** The **root** is hidden below the gum and is usually anchored to a jawbone. The **pulp cavity** within contains blood vessels and nerves (Figure 7-1). The bulk of a tooth usually consists of orthodentine. The bases of the teeth of some fishes contain a vascular osteodentine that may merge into the jawbone.

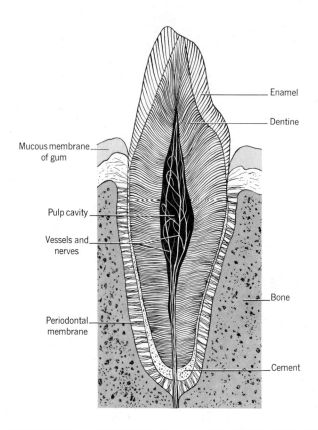

Enamel

Dentine

Mucous membrane of gum

Pulp cavity

Vessels and nerves

Bone

Periodontal membrane

Cement

Figure 7-1 LONGITUDINAL SECTION OF A MAMMALIAN INCISOR TOOTH.

Unworn crowns are covered by enamel which rarely exceeds 2 mm in thickness even in large animals. The nature of these hard tissues was described in Chapter 6. Roots of teeth that are held in sockets are covered by a thin layer of **cement.** (Some teeth that are specialized for grinding also have cement on the crowns. See the figure on p. 581.) Cement is a nonvascular bone that has no osteons and is usually acellular. It is rich in collagenous fibers and is softer than dentine.

DEVELOPMENT Knowledge of the process of tooth development facilitates an understanding of the mechanism of replacement and of the origin of complicated teeth from simple ones. The first step in the formation of teeth occurs in the embryo when a fold of ectoderm forms along the margins of the mouth and penetrates into the underlying mesoderm of the gums as a double-layered wall called the **dental lamina** (Figure 7-2). At intervals along the dental lamina, hummocks of tissue push into its inner edge causing it (as seen in section across the jaw) to look like an inverted goblet. This tissue may be regarded as mesoderm, but is derived from neural crest cells that have migrated from their site of origin (see p. 82), so is also termed **mesectoderm** or **ectomesenchyme.** The double-walled bowl of the goblet now constitutes the **enamel organ.** The cells of its inner layer (ameloblasts) will form enamel. Each mesodermal hummock, or **dermal papilla,** develops on its surface

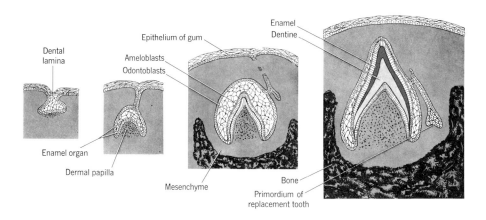

Figure 7-2 Diagrammatic sections showing DEVELOPMENT OF A THECODONT TOOTH.

the cells (odontoblasts) that will form dentine. The entire unit is called a **tooth bud.**

There is a reciprocal induction between the enamel organ and its papilla: each is necessary for the proper function of the other. The shape of the crown of the future tooth is determined by the conformation of the interface between enamel organ and papilla at the time the hard tissues are deposited. Until that time, the interface is gradually sculptured by differential pressures and growth rates of the various parts of the tooth bud. If there is to be more than one cusp, deposition of enamel and dentine starts at what will be the tip of the principal cusp. As maturation proceeds, the developing tooth slowly climbs up toward the surface of the gum. The enamel organ regresses ahead of the emerging crown, and no further enamel can then be deposited. The formation of dentine continues after the tooth becomes functional. Cement forms only in the presence of dentine. The papilla becomes the pulp.

The teeth of cartilaginous fishes are anchored to the skin by collagenous fibers (Sharpey's fibers) that run into the dentine from the dermis. The teeth of most vertebrates, however, are more or less fixed to the bones of the jaws. Often the outer margin of each jawbone forms a thin wall having on its inner (lingual) side a series of hollows to accommodate the teeth. Each tooth touches the bone only with the outer (buccal) surface of its root. It may be joined to the jaw by collagenous fibers or by cement. This mode of attachment, which may be the primitive one, is called **pleurodont** (= side + tooth) (Figure 7-3).

Some other teeth scarcely have any roots and abut against the rim of the jawbones to which they are joined by a continuum of hard tissue. This form of attachment, which has evolved independently several times, is called **acrodont** (= summit + tooth). Still other teeth have their roots held in sockets (alveoli) in the jawbones. This is the **thecodont** (= sheath + tooth) condition. There are intergrades among these kinds of attachment.

ATTACHMENT AND REPLACEMENT

Skin
Mandible

TEETH ANCHORED TO SKIN   PLEURODONT

ACRODONT   THECODONT

Figure 7-3 Sections of jaws showing TYPES OF TOOTH ATTACHMENT.

Replacement of teeth is necessary to provide for growth and to compensate for wear and accidental loss. Even before a first tooth is fully functional, a new tooth bud forms to initiate the development of its replacement. As the second tooth matures, the root of the first is resorbed, thus causing it to loosen and fall away. Replacements for pleurodont teeth form either just lingual or just anterior to the roots of the old teeth and move into position after the old teeth are shed. Replacements for thecodont teeth form directly under the roots of the old teeth.

Most vertebrates replace their teeth continuously, generation following generation, for as long as they live. Such animals are said to be **polyphyodont** (= many + to grow + tooth). Most mammals have only two generations of teeth and are said to be **diphyodont.** Some acrodont teeth, and some that fuse to form large tooth plates, are not replaced. However, many of the animals that seem to have only one generation of teeth do have one or more other generations that are shed before birth or resorbed in the embryo.

Polyphyodont teeth are not replaced at random, seemingly because that could result in temporary impairment of function. Two conditions usually are maintained. First, the teeth are in two sets, the even-numbered teeth constituting one set and the odd-numbered teeth another. Replacement in one set is out of phase with replacement in the other. Accordingly, since adjacent teeth are at different stages of the growth cycle, vacant positions are flanked by functional teeth. Second, neighboring teeth in a set (alternate teeth in the mouth) are usually at slightly different developmental stages. Every tooth tends to be a little more mature than the next anterior tooth of its set.

Edmund and, more recently, Osborn have proposed developmental models to explain the observations. Osborn suggests that the migration of inductive ectomesenchyme into the jaws from behind initiates the first wave of tooth development from posterior to anterior. If each developing tooth somehow inhibits the development of adjacent teeth, that would account for the odd and even sets. Large teeth develop more slowly than small ones, and this influences the schedule of individual tooth replacement for animals having teeth of different sizes.

FROM DENTICULATE ARMOR TO HETERODONTY CYCLOSTOMES have conical, horny teeth on the oral funnel and "tongue." Some extinct AGNATHA had small to medium-sized bony plates in the mouth that were roughened by surface denticles. Among PLACODERMS, the predaceous arthrodires had on the margins of their jaws bony plates with jagged edges which sometimes contained dentine (see figure on p. 48). All of these structures are tooth analogs, but are homologs only in the general sense of derivation from the same developmental tissues (see p. 91).

EVOLUTION OF TEETH

The teeth of OTHER FISHES are highly varied, and evolutionary trends are scarcely found. Typically their teeth are numerous, conical or bladelike (Figure 7-4), and **homodont** (= same + tooth), or all of about the same size and shape. They are carried on the margins of the jaws and, in ray-finned fishes, may also be on the roof of the mouth, fifth gill arch, and tongue. Primitively, bony fishes may have been pleurodont and polyphyodont, but most are now somewhat or entirely acrodont, and such teeth often are not shed. The teeth of sharks are anchored to the skin and are continuously replaced as new teeth migrate up over the margins of the jaws from the inside (Figure 7-3). Departures from these general conditions include reduction or loss of teeth, fusion to form permanent crushing plates (chimaeras, dipnoans, some rays), and development of multiple cusps (cladoselachians, pleuracanths) or whorls (acanthodians).

The teeth of most CROSSOPTERYGIANS (but not the surviving coelacanth) resemble those of ray-finned fishes in most respects, but have one distinctive and important characteristic: The enamel and dentine are folded so as to form complicated patterns as seen in cross section (Figure 7-4). This structure, termed **labyrinthodont,** strengthens the tooth and makes it resistant to wear. Labyrinthodont teeth survived for 100 million years, first among amphibians (hence their name, LABYRINTHODONTIA) and then STEM REPTILES (extinct anapsids).

MODERN AMPHIBIANS tend to have fewer teeth than their ancestors (none in toads) and they are small, simple, pleurodont, and no longer labyrinthodont. They are supported by pedicels of dental origin to which they are attached by

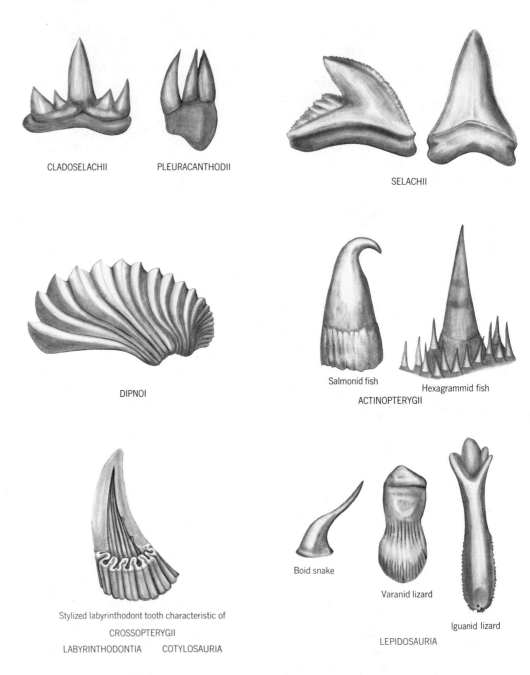

CLADOSELACHII   PLEURACANTHODII

SELACHII

DIPNOI

Salmonid fish   Hexagrammid fish

ACTINOPTERYGII

Stylized labyrinthodont tooth characteristic of

CROSSOPTERYGII

LABYRINTHODONTIA   COTYLOSAURIA

Boid snake

Varanid lizard

Iguanid lizard

LEPIDOSAURIA

Figure 7-4 CHARACTERISTIC TEETH OF VARIOUS TAXA.

zones of soft tissue (Figure 7-5). The phylogenetic and functional interpretation of this unique pattern remains to be clarified.

Many REPTILES, and also the extinct TOOTHED BIRDS, are homodont. All forms of tooth attachment are observed: Some snakes and many archosaurs are

Figure 7-5 DIVISION OF TEETH OF LISSAMPHIBIA INTO CROWNS AND PEDI-CELS shown by the caecilian, *Gymnopis*.

(and were) thecodont, various reptiles are acrodont, and many lizards are at least partly pleurodont. Most reptiles are polyphyodont, though acrodont teeth often are not replaced. Turtles are **edentate,** or toothless.

Various reptiles, in several orders, but particularly among MAMMAL-LIKE REPTILES, have teeth of different form and function at different parts of the tooth row. Such dentition is **heterodont** (= different + tooth). This significant advance was passed on to MAMMALS, and is related to chewing.

SOME CONSEQUENCES OF CHEWING   Most nonmammals use the teeth only for securing and holding food, which is then quickly swallowed. MAMMALS also secure and hold food, but in addition they usually shear, crush, or grind their food. These new functions are of the utmost importance because they speed digestion and greatly increase the diversity of foods—particularly plant foods—that can be eaten. In order to be sheared, crushed, or ground, it is desirable for food to be retained in the mouth and chewed. This necessitates (*1*) modification of the ancestral jaw articulation and palate (see discussion on p. 136), (*2*) cheeks to hold food in the mouth, (*3*) a tongue capable of positioning the food, and (*4*) several kinds of teeth. Mammals are heterodont unless they have secondarily reverted to homodonty in response to a diet of fish or insects (e.g., toothed whales, armadillos). Other distinctive features also relate to heterodonty: Mammalian teeth are thecodont because such teeth can best withstand shearing forces without being loosened. They are carried only at the margins of the jaws, possibly because elsewhere they would occlude with less force and precision. Specialization of the teeth leads to increased size, complexity, strength, and wearing ability. These, in turn, correlate with reduction in number of teeth functional at one time and reduction in number of sets of replacement teeth.

Most mammals are diphyodont, but this statement needs qualification. Typical mammals have a first set of teeth which consists of the temporary milk teeth plus the permanent molars. The fact that all these teeth belong to one generation is obscured by the circumstances that they erupt in sequence and may or may not be deciduous. The second generation of teeth includes all permanent teeth except the molars. (Some marsupials replace only one tooth in each jaw, and moles replace none.)

NUMBERS AND KINDS OF TEETH   Typical mammalian dentitions have three or four kinds of teeth. **Incisors** are adapted for securing food and sometimes for grooming (Figure 7-6). They may be conical spikes for holding insects or flesh, or simple blades for cutting plant stems. They are relatively small teeth and are single-rooted. Their function requires that they be forward in the mouth; the upper incisors are rooted in the more anterior of the two bones of the jaw (the premaxilla). Numbers of teeth are expressed in terms of one side of the mouth only and are commonly written as a fraction, the count of upper teeth forming the numerator and the count of lower teeth forming the denominator. Incisors of eutherian mammals are always 3/3 unless secondarily reduced. This count is exceeded by some marsupials.

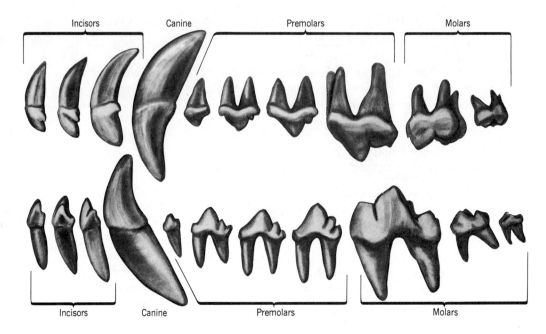

Figure 7-6 HETERODONT MAMMALIAN DENTITION illustrated by the teeth of the dog.

The **canines** are next posterior in the mouth. They are simple spikelike teeth with single roots. If not secondarily modified, they are long and strong and serve for holding and piercing in relation to both feeding and fighting. Canines always number 1/1 if not reduced. The alveolus for the upper tooth lies in, or just posterior to, the suture between the two bones of the upper jaw.

All replacement teeth behind the canines are called **premolars,** and all nondeciduous teeth of the first generation are called **molars.** Molars tend to be larger than premolars and to have more cusps and roots. The numbers of premolars and molars among ancestral therian mammals were probably 4/4 and 7-8/7-8, respectively. The primitive counts for eutherian mammals are considered to be 4/4 and 3/3.

The distinction between premolars and molars is sometimes unsatisfactory. As a practical matter it may not be possible to tell from an adult skull which teeth have been replaced. The entire tooth row forms a series such that teeth tend to resemble their immediate neighbors. Specialization may make the kinds of teeth indistinguishable: Artiodactyls have an incisiform lower canine and horses have molariform premolars. Furthermore, the distinction between premolars and molars breaks down if the replacement generation of teeth is incomplete or absent. For these reasons it is often convenient to speak of these teeth collectively as **cheek teeth.**

The numbers and kinds of teeth are expressed by a dental formula which consists of the numerical fractions already noted written in sequence starting with the incisors. Thus, the formula for the wolf is 3/3, 1/1, 4/4, 2/3; for the deer is 0/3, 0/1, 3/3, 3/3; and for man is 2/2, 1/1, 2/2, 3/3.

MORE ABOUT CHEEK TEETH   The cheek teeth of primitive mammals (Jurassic mammals, marsupials, insectivores) and of some more advanced orders, as well, have several cusps (see the posterior teeth, figure on p. 142). The upper molars usually have three principal cusps arranged in a triangle; the lower molars commonly have five principal cusps. Accessory cusps may also be present. The cusps are intricately arranged to interlock when the mouth is closed. Such teeth are admirably adapted for diets of flesh or invertebrates.

Speculation about the origin of multicuspid teeth from unicuspid teeth has been the subject of much debate. The theory of Bolk is now of only historical interest. His **concentration theory,** presented in a series of papers in the first decade of this century, postulated a fusion of teeth derived from two different replacement generations. The **concrescence theory** of Kükenthal and Röse was formulated in the 1890s. It holds that multicuspid teeth originated by the fusion of adjacent unicuspid teeth. This does actually happen in some fishes, but does not appear to have been a usual event. The favored theory, called the **differentiation theory,** was presented in the 1880s by Cope and Osborn and was revised and extended some years later by Gregory. This theory, for which there is abundant paleontological and embryological evidence, states that the single cusp of the ancestral tooth gradually became supplemented by two secondary cusps that formed from the side walls of the tooth crown. Additional smaller cusps may form from a ridge (cingulum) that encircles the base of the crown. The patterns of cusps are various and intricate. Furthermore, cusps may become joined in various ways by ridges (see figure on p. 581). All cusps and lophs are named, and the terms distinguish between corresponding cusps of upper and lower teeth. Unfortunately, some of the terms in wide use are not apt because they do not express true homologies between upper and lower teeth.

The description and functional analysis of the principal kinds of cheek teeth is a particularly interesting subject. It will be postponed until Chapter 28 because it relates to adaptation to diet rather than to the broad sweep of evolutionary change among major taxa, which is our concern in this part of the book.

8

Head Skeleton

The internal, jointed skeletal system of vertebrates is unique in the animal kingdom and is the most important of all organ systems in the study of vertebrate morphology. It is conservative enough in general pattern to show the broad outlines of vertebrate phylogeny: Homologous bones and evolutionary trends are readily demonstrated in skeletons characterizing the successive major taxa. Furthermore, the skeleton plays a central functional role. Wide variations have always been superimposed on the gradually evolving general pattern, and the skeleton has been sufficiently plastic to respond to the particular habits of the various animals. Therefore, it also provides reliable information about the specific adaptations of vertebrates: Postures and locomotor adaptations are accurately revealed, and other adaptations are sometimes indicated.

Because of its hardness and durability, the skeleton (including the teeth) becomes fossilized relatively often, and this contributes virtually (though not quite) all of our knowledge of past vertebrate life. Paleontology can be regarded as the comparative anatomy and morphology of extinct animals. No other science has contributed so much to our knowledge of vertebrate evolution, and it is, of course, limited almost entirely to the study of hard parts. Fortunately for the paleontologist, of all organ systems the skeletal system tells the trained observer the most about *other* organ systems: Most muscles take their origins and insertions on bones, and often leave tuberosities or scars to show the positions and extent of these contacts; the important cranial nerves reveal their sizes and courses by the foramina they traverse in the skull; the relative development of the different parts of the brain may be revealed by the braincase; nasal chambers, orbits, and otic cavities give some information about the sense organs they house; the nature and distribution of sensory canals on the head may be shown in detail; even some blood vessels leave traces on the skeleton.

Moreover, of all organ systems this one is the easiest to preserve, store, and demonstrate. It is, therefore, a good one to teach and learn about.

MORE ABOUT HARD TISSUES

Hard tissues have been described as we have come to them: horn, enamel, dentine, and bone in Chapter 6, and cement in Chapter 7. As we turn to the deeper skeleton, another hard tissue, cartilage, must be added and more should be said about the development and nature of bone.

Cartilage is a tissue that combines hardness with some flexibility. It is present in several invertebrates but is particularly characteristic of vertebrates. Its cells (chondrocytes) are dispersed in spaces (lacunae) in a matrix of glycoprotein (chondromucin) and fibers. Chondrocytes form at the surface membrane of the tissue, or perichondrium, and gradually round up, grow, and become wider spaced as the matrix they produce pushes them farther apart (Figure 8-1). Cartilage is usually derived from mesoderm but, curiously, in part of the head and in virtually all of the gill region it may also be derived from neural crests, which are initially ectodermal.

Hyaline cartilage has relatively few contained fibers and is translucent in thin sections. Its surfaces are exceedingly smooth, as where it caps the bones at movable joints. **Fibrous cartilage** contains a heavy meshwork of collagenous fibers which make it cushionlike and tough, as where it separates the vertebrae of the lower back. **Elastic cartilage** is rich in elastic fibers and consequently has flexibility and resilience, as in the external ear. **Calcified cartilage** contains deposits of calcium salts which make it hard and firm. It is common in the skeletons of elasmobranchs. Unlike bone, calcified cartilage is not remodeled once it has formed. There are intergrades among these types of cartilage.

In terms of ontogeny, there are two kinds of bone. As explained in Chapter 6, bone that is associated with integumentary structures such as armor, scales, denticles, and osteoderms ossifies directly from the dermis. Bones of the ancestral integument that came to lie below the skin, such as the roofing bones of the skull, also ossify directly from deeper mesoderm. This kind of bone is called **dermal bone** or **membrane bone.**

The other kind of bone does not ossify directly from membranous tissue but instead gradually replaces cartilage that has formed earlier. Thus, in the embryonic chondrocranium the cartilaginous skeleton is eroded away just ahead of the deposition of bone. Such bone is termed **replacement bone** or **endochondral bone.** Its development is described further on page 169. Once the two kinds of bone have formed, they are identical. The distinction is useful, however, for establishing the origins and homologies of various elements of the skeleton.

A property of bone and cartilage that conditions their other physical properties is **heterogeneity.** Disregarding minor impurities and imperfections, cast

Figure 8-1 HYALINE CARTILAGE.

iron, ceramics, and glass are homogeneous, or uniformly the same everyplace and in all directions. Wood, by contrast, has grain that makes resistance to bending and splitting different in different planes. Similarly, bones have lamellae and osteons with specific orientations and may be compact or spongy (see p. 92). The internal cells and fibers within cartilage are unevenly arranged. Consequently, one cannot speak of *the* strength of, for instance, bone, but can report only the approximate strength of a given type of bone when loaded in a given way relative to the orientation of its components.

Cartilage and, particularly, bone are further heterogeneous in that each is a composite material consisting of two dissimilar components. In each instance, one component is an intricate meshwork of oriented collagenous fibers and sometimes of elastic fibers. The other component is a glycoprotein (for cartilage), or hydroxyapatite (for bone). The physical properties of the composite are unlike those of either component taken alone and are not the sum or average of the two taken together. Thus, collagenous fibers are soft, flexible, and very resistant to elongation, whereas hydroxyapatite is extremely hard, brittle, and resistant to compression. Fine grain contributes strength because threadlike strands of a material are many times stronger per unit of cross section than bulk material. Thus the composite bone is more rigid than collagenous fibers, more flexible and resistant to fracture than the mineral, and stronger for its weight and more versatile in the kinds of loads it can withstand than either component alone. Engineers make composite materials from fibers of glass in resin, boron in aluminum, tungsten in copper, and others.

The presence of lacunae in bone (see figure on p. 93) also increases its strength. When a microfracture extends into a lacuna it tends to stop there instead of enlarging. Within skeletal elements, bone, but not cartilage, is usually spongy. The spicules of the meshwork are oriented so as to maximize strength in proportion to weight in ways explained on pages 398 to 403. Finally, the fibers in the organic matrix of bone are woven or stratified in complex patterns that increase resistance to breakage.

We can speculate with reasonable confidence that as the remote ancestor of chordates became bilaterally symmetrical and motile, it was essential that, in order to be effective, its feeding mechanism and certain of its sense organs become concentrated at its anterior end. Since the mouth was also associated with respiration, the gills developed anteriorly. Inevitably, the nervous system responded to this concentration of sensory and motor structures by enlarging to form a brain. A head skeleton then evolved to house and protect the brain and sense organs, and to contribute to the efficiency of the feeding and respiratory mechanisms.

ORIGIN, FUNCTION, AND SEGMENTAL NATURE OF THE HEAD

Segmentation may have evolved as the head evolved. In any event, it is believed that the head, as well as the body, of the ancestral vertebrate was segmental. Specialization and loss, or partial loss, of head segments has obscured the evolutionary story, however, and it is difficult to learn how many somites, or embryonic segments of the mesoderm, are represented in the head, and which adult structures were derived from each. This problem was a principal field of research and speculation among morphologists from the turn of the eighteenth century through the first decades of the nineteenth century, and still receives attention. A recent contributor, Jollie, believes that the vertebrate head skeleton incorporated one somite anterior to the mouth, two more between mouth and ear, and two and a half behind the ear. (Half sclerotomes may recombine in the embryo—hence the half segment.) Fishes may add one, two, or several additional segments to the back of the head.

COMPONENTS OF THE HEAD SKELETON   Since the head skeleton is complicated, both as to origin and structure, it is helpful to find parts that can be described one at a time. The logical components to select are the chondrocranium, visceral skeleton, and dermal elements, because the structures that these contribute to the complete skull are at first rather distinct, both in their phylogeny and their ontogeny.

The **chondrocranium** supports the brain and organs of special sense. The **visceral skeleton** (or splanchnocranium) supports the gill arches and their derivatives. The **dermal elements** (or dermatocranium) complete the relatively superficial framework of the skull. The chondrocranium and visceral skeleton may remain cartilaginous or become bony, but they are always present. The dermal skeleton is always bony and is usually present, but has been secondarily lost by several major vertebrate groups. Authorities are divided on the relative antiquity of these components. Regardless of which may have evolved first, the earliest known jawless vertebrates already had all three of them.

The term "skull" is sometimes used for all of the skeleton of the head. However, it is also used—and will be used here—for the single unit that forms the braincase and upper jaw and houses the nose and ear. The word is useful, but inexact.

CHONDROCRANIUM AND DERIVATIVES   Most organs of the body can continue to function when subjected to moderate pressures by contacts with the environment or by locomotor or digestive activities. This is not true, however, of the central nervous system or organs of special sense. The spinal cord of primitive vertebrates derived some protection from the stiff notochord. The larger brain was shielded above by bony scales or plates that formed from the dermis of the skin and was supported below and on the sides by a trough of hyaline cartilage. Capsules and rods of cartilage that protected the sense organs and stiffened the rostrum merged with this brain trough to form the chondrocranium (= cartilage + cranium). This structure has persisted in all vertebrates ever since its origin some 550 million years ago. Animals that do not have bony heads (cyclostomes and cartilaginous fishes) have relatively complete and heavy chondrocrania. Most vertebrates that do have bony heads retain a completely cartilaginous chondrocranium only in larval or fetal life. In adult life these animals replace at least part of the more delicate cartilage by bones that ossify within the cartilage.

Homologies of the major features of the chondrocranium of the different vertebrates have been established by noting their relations to such conservative landmarks as the notochord, hypophysis, and cranial nerves and blood vessels. Homologies of its numerous and complicated little rods and vacuities are more difficult to establish; seemingly, some will never be deciphered. Embryonic recapitulation has not been very helpful. The homologies of the bones that ossify in the chondrocranium are more satisfactory.

The chondrocranium is a single unit. It is composed, however, of numerous elements which are distinct in the embryo. We will arrange these in five groups that are relatively large and constant (Figure 8-2).

The **notochord** lies within, just above, or just below the base of the developing chondrocranium. It may remain free or be obliterated, but usually its sheaths become cartilaginous and join the chondrocranium. The contribution it makes is small but important because its presence in the head may predate that of all other skeletal structures there, and because the constant position of its anterior end, just posterior to the hypophysis, serves as a reference point.

Anterior to the notochord is a pair of bars called the **trabeculae** (= small beams). The hypophysis is located between their posterior ends. Sometimes, in lower vertebrates having broad heads, the two bars remain free and widely separated (the platytrabic condition); sometimes they are joined by a plate of cartilage; and usually they fuse anteriorly to form a narrower Y-shaped structure

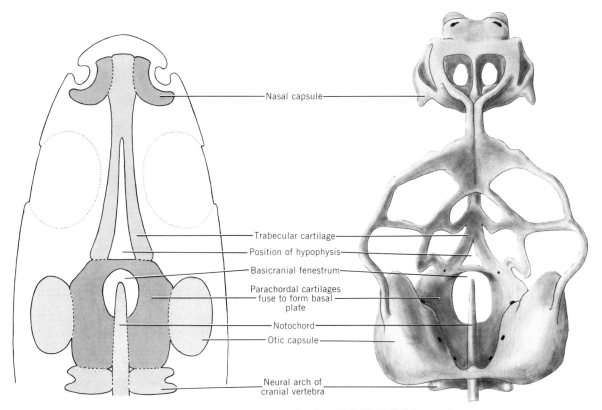

Nasal capsule

Trabecular cartilage
Position of hypophysis
Basicranial fenestrum
Parachordal cartilages fuse to form basal plate
Notochord
Otic capsule

Neural arch of cranial vertebra

Figure 8-2 COMPONENTS OF THE VERTEBRATE CHONDROCRANIUM seen in dorsal view. Left, generalized early embryo; right, slightly simplified from the 25 mm stage of development of the lizard *Lacerta*.

having its leg pointing forward (the tropitrabic condition). The trabecular part of the chondrocranium is related to the forebrain, nasal capsules, orbits, and rostrum. In older embryos it may be complicated by curved outgrowths that form an elaborate framework. Neural crests contribute to the formation of the trabeculae.

Behind the trabeculae, and flanking the notochord, another pair of cartilages form which are called (because of their position) **parachordal cartilages.** These soon merge above, below, or around the notochord to form the **basal plate.** The anterior part of the plate commonly encloses a large vacuity, and its lateral margins are pierced by foramina for the exit of cranial nerves. The side walls of the chondrocranium are usually incomplete. They consist of varied and some-times complicated, pillars and rods that fuse to the basal plate at its lateral mar-gins. It is relatively clear that the basal plate is of segmental origin from the sclerotomes of embryonic head somites. It is considered to be in series with the bases of vertebrae. At the back of the basal plate are one or two projections called **occipital condyles** which articulate with the first free vertebra of the spine. Their interesting origin from vertebral elements will be explained in Chapter 9.

A fourth contribution to the chondrocranium consists of one or more pairs of arches that rise up from the posterior angles of the basal plate to flank or encircle the spinal cord just where it enters the skull. These may be considered to be the **neural arches of cranial vertebrae.** In embryos of some fishes they sometimes even bear short ribs. The separate arches merge in late embryos and are then known collectively as the **occipital arch** (occiput = back part of the skull).

A final contribution to the chondrocranium is the cartilaginous **sense cap-sules** that house the nasal chambers and inner ear. The nasal capsules join the anterior ends of the trabeculae. The otic, or auditory, capsules join the margins of the basal plate just in front of the occipital arch. The eyes are also enclosed in capsules of cartilage. These remain free, however, because if they joined the chondrocranium, the eyes could not be moved independently of the head.

Bones typically develop in the chondrocrania of six of the nine classes of vertebrates (all except Agnatha, Placodermi, and Chondrichthyes). The names and positions of these bones are best learned from specimens and from Figure 8-3. No one list of such bones serves for all vertebrates having ossified skulls; most of the bones are rather constant, but some (including the mesethmoid and orbitosphenoids) are of variable occurrence. In the higher vertebrates, the otic bones and certain of the basal bones tend to fuse together.

VISCERAL SKELETON AND DERIVATIVES   The **pharynx** is a somewhat ex-panded part of the digestive tube lying between the oral cavity and the esophagus or stomach. In the earliest stages of chordate evolution the pharynx became perforated laterally by paired gill slits which served in feeding and respiration. A part of the pharynx still contributes to the feeding mechanism of all the ver-tebrates that have jaws, and the respiratory function of the pharynx is retained by all jawless vertebrates, fishes, and larval amphibians. When tetrapods ceased to use the pharynx for respiration, much of its musculature and skeleton became

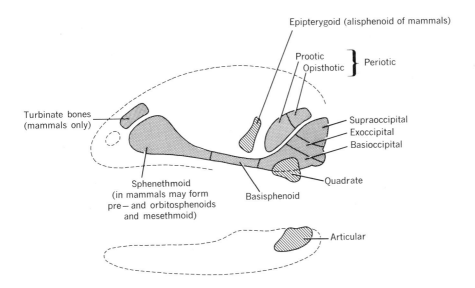

Epipterygoid (alisphenoid of mammals)

Prootic

Opisthotic } Periotic

Turbinate bones
(mammals only)

Supraoccipital

Exoccipital

Basioccipital

Sphenethmoid
(in mammals may form
pre— and orbitosphenoids
and mesethmoid)

Basisphenoid

Quadrate

Articular

Figure 8-3 REPLACEMENT BONES OF THE SKULL AND MANDIBLE AT THE EARLY TETRAPOD STAGE OF EVOLUTION (except as noted). Ossifications of the chondrocranium are shaded; those of the mandibular arch are hatched. All bones are paired except those under the brain, and the supraoccipital.

available for other uses and were adapted for the control and support of the tongue, vocal apparatus, and related structures. Thus, the pharynx has had a long and varied history.

The first protochordates may have had a dozen pairs of gills. The number was increased to as many as 100 by some protochordates (amphioxus), so that food particles could be entrapped on sticky gill bars as water was strained through the pharynx. The earliest vertebrates may have also included some filter-feeders, but the most ancient well-known vertebrates subsisted on larger food and did not require such a specialized pharynx. They reduced the count to 5–15 pairs of gill slits. By the time jaws evolved, the number had been further reduced and stabilized at about 6 (more in some sharks and fewer in some tetrapods).

Between each typical gill pouch, and also anterior to the first and posterior to the last, are bars of tissue that consist of skeletal elements (largely derived from neural crests), muscles, and respiratory filaments together with the nerves and vessels serving these structures. Each such bar is called a **visceral arch**—"visceral" because it forms largely from a specialized part of the gut tube. The basic number of visceral arches for jaw-bearing vertebrates is one more than the primitive number of pouches, or 7.

The relationship between the segmentation of the visceral arches and the segmentation of the head (discussed above) is of interest. The mesodermal part of the digestive tube forms in the embryo from the hypomere. Unlike the somites above it, the hypomere is not segmented. It follows that the muscles and skeleton of the visceral arches do not exhibit the primary segmentation of the body. The presence of serial gills, however, imposes on the derivatives of the pharynx a secondary segmentation which comes to be functionally integrated with the primary segmentation of the cranial nerves, yet need not be just the same in all vertebrates.

It is probable that the ancestral vertebrate had a preoral somite (primary segmentation), but the presence of a preoral visceral arch (secondary segmentation) is doubtful, or at best controversial. The first constant arch, therefore, is just behind the mouth. It is numbered 1, and the others follow in sequence. Each gill slit, or pouch, is given the same number as the arch that lies just anterior to it (Figure 8-4).

No other part of the vertebrate skeleton, except possibly the notochord itself, is as ancient in origin as the visceral skeleton. Each gill bar of jawless vertebrates is supported by a single cartilaginous rod. It may angle and bend, and usually joins its fellows above and below the gill slits, but it is not jointed. The skeleton of each visceral arch of gnathostomes is jointed. The basic number of paired segments is seemingly four per arch. Of these, the middle two, called the **epibranchial** (above) and the **ceratobranchial** (below), are relatively important for our story. There are also unpaired, midventral segments between the lower ends of the gill bars, which are considered, on somewhat slim evidence, to have been derived by remote ancestors from the paired gill bars.

The first visceral arch enlarges to become the jaws, if jaws are present, and is then called the **mandibular arch.** Its epibranchial forms the upper jaw and takes the name **palatoquadrate.** The ceratobranchial presumably forms the lower jaw and is called the **mandibular cartilage.** The jaws become more or less firmly anchored to the chondrocranium to provide needed bracing (see below), and the second visceral arch, or **hyoid arch,** may be recruited to help in the support. The epibranchial is its key element and is called the **hyomandibula.** Succeeding arches, which serve primarily in respiration, are called **branchial arches** (branch

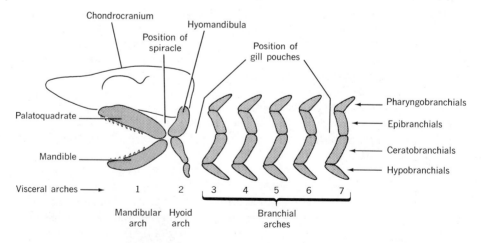

Figure 8-4 PRIMITIVE VISCERAL SKELETON represented by a stylized elasmobranch. Compare with Figure 8-8.

= gill). Thus, the third visceral arch is also the first branchial arch. This terminology is clarified by Figure 8-4.

Ossifications that form in the visceral cartilages are, like those of the chondrocranium and nearly all of the postcranial skeleton, replacement bones (Figure 8-3). As noted in the following section, most of the bones that become functionally related to the jaws are not replacement bones and, therefore, are not derived from the visceral skeleton. The true first arch bones are usually small, yet no other bones of the body have a more engrossing and unexpected history. More of this later in the chapter.

CONTRIBUTIONS FROM THE INTEGUMENT   The rigid scales or heavy armor of most ostracoderms, placoderms, dipnoans, and crossopterygians was the most complete and solid over the head where it supported the teeth and shielded the roof of the brain, the delicate gills, and other soft tissues. At first these integumentary bones were variable as to size and pattern: Anaspids had small scales; cephalaspids had huge solid shields; arthrodires had large plates composed of individual elements joined by sutures. By the time the bony fish stage of evolution was reached, the elements tended to stabilize as medium-sized pieces. The crossopterygians developed a general pattern of bones that was to persist throughout tetrapod history. The various classes have modified the basic pattern, however, by fusions, deletions (particularly where the gills were lost, at the back of the head, and between head and pectoral girdle), and sinking the bones below the skin and even underneath certain muscles.

In order to establish the homologies of these numerous bones it has been necessary for morphologists to use as many clues as possible. The paleontological sequence of various changes has indicated trends. Attention has been given to the relation of these bones to ossifications in the chondrocranium and to nerves, vessels, sense organs, and the opening often present on the top of the head that accommodates the pineal organ. The sensory canals of aquatic forms have been helpful because they follow a definite pattern and usually leave grooves, pits, or tunnels in the underlying bones. The number of ossification centers, and the sequence of their appearance in the young animal, have also been useful because a particular bone usually ossifies from a given number of centers that appear in constant sequence in relation to the centers of other bones and to the regions of the brain. Indeed, the several parts of the brain induce specific ossifications adjacent to them. Since the characteristics of some centers are known to vary somewhat, however, this clue must be interpreted with caution. In these various ways it has been possible to establish with confidence the homologies of nearly all the bones derived from the integument. The names and positions of the more constant bones are shown in Figure 8-5.

RELATIONS OF THE CRANIAL COMPONENTS

The recognition of these components of the head skeleton (chondrocranium, splanchnocranium, dermatocranium) is more than a mere convenience. As we have seen, they may retain a degree of structural independence in the adult. This is particularly true of anamniotes. The skeleton of the gills must be discrete as long as the gills are functional, and the dermal bones often remain superficial where they meet edge to edge to form a continuous covering. Furthermore, in

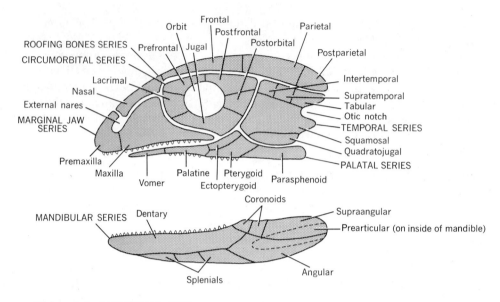

Figure 8-5 PRINCIPAL DERMAL BONES OF THE SKULL AND MANDIBLE AT THE EARLY TETRAPOD STAGE OF EVOLUTION. All bones are paired except the parasphenoid. (The sclerotic bones of the optic capsule are not shown.)

the evolution of the various groups of animals, the bones of the skull have sometimes all become more and more strongly ossified and, in contrast, have sometimes all regressed. At other times, however, derivatives of the several components have evolved independently. Thus, it appears that the chondrocranium and visceral skeleton were emphasized by cartilaginous fishes and cyclostomes at the same time that the dermal skeleton was regressing. Conversely, the replacement bones of the skull, and these only, tended to regress in late cephalaspids, lung fishes, some primitive ray-finned fishes, and late labyrinthodonts. Evidently, replacement and dermal bones may respond differently to evolutionary forces even though they are visually identical.

In spite of original and potential independence, the replacement and dermal bones of higher vertebrates become intimately associated. Enlargement of the brain forces the chondrocranial ossifications outward, whereas enlargement of jaw musculature forces many dermal bones inward. These components meet and jointly form a single, firm unit.

When the first visceral arch was converted to jaws it became mechanically necessary to brace the arch more firmly than had been the case while it functioned only for respiration and filter-feeding. This was first accomplished by attaching the palatoquadrate cartilage to the chondrocranium. Later, the hyomandibula of the second arch became a strut to further brace the jaw. Subsequently, the hyomandibula served alone, or the dermal bones covering the palatoquadrate joined other dermal bones to provide a firm anchor. The number, position, and firmness of the various points of attachment of the jaws to the remainder of the skull have been various. In fact, there are more than a dozen

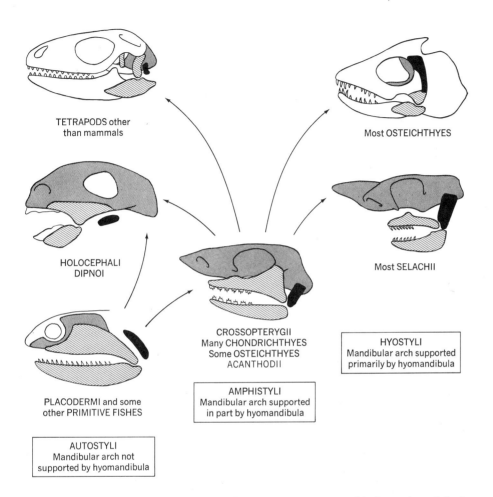

Figure 8-6   PRINCIPAL TYPES OF JAW SUSPENSION. Mandibular arch and derivatives are hatched, hyomandibula and derivatives are black, chondrocranium is shaded, teeth and extent of dermal bones are indicated in outline.

named patterns of jaw suspension. To further complicate the story, similar types of suspension have evolved independently in several instances. Properly interpreted, the relatively simple terminology proposed long ago by Thomas Huxley is still adequate. The terms, together with the animal groups to which they apply and the relationships among them, are shown in Figure 8-6. Some aspects of the story are amplified in the following pages.

JAWLESS VERTEBRATES: INNOVATIONS AND VARIETY   The chondrocranium of Agnatha usually remains cartilaginous throughout life and is, therefore, poorly known for most extinct forms. In cephalaspids, however, it was often covered on all surfaces by a thin veneer of bone.

The visceral skeleton of agnaths is a continuous unit of cartilage: The individual arches tend to join above and below the gill slits, and the unit joins the

EVOLUTION OF THE HEAD SKELETON

chondrocranium (Figure 8-7). All arches are branchial in function. Their number is variable, but they are numerous relative to the arches of other vertebrates. (The specialized mouth parts of cyclostomes are supported by cartilages that are not derived from the visceral skeleton and are not represented in other vertebrates.)

The dermal head skeleton of agnaths ranges from armor shields (cephalaspids, pteraspids) through small scales (anaspids) to absence (cyclostomes). The dermal elements cannot be homologized with the cranial bones of other vertebrates.

PLACODERMS: ENTER JAWS   The chondrocranium of placoderms was similar to that of agnathous vertebrates in being either entirely cartilaginous or partly ossified.

Important advances had been made in the visceral skeleton. Most significant, the first arch had made the transition from gill support to jaws. Some of the predaceous arthrodires had formidable jaws indeed. Replacement bones often formed in the large palatoquadrate. This structure was autostylic (= self-supported), as it was attached to the chondrocranium by ligaments and not the hyomandibula. The second arch remained a typical branchial arch. As in higher vertebrates, the branchial arches of at least some placoderms were each supported by several skeletal elements.

The dermal skeleton usually consisted of heavy head and thoracic shields that were joined by hinges (arthrodires and antiarchs). Individual dermal bones of placoderms cannot be homologized with those in the armor or skulls of other vertebrates.

CARTILAGINOUS FISHES: SPECIALIZATION AND REGRESSION   Cartilaginous fishes have lacked bones throughout their long history. To provide the protection that dermal bones gave to the brains of ostrocoderms and placoderms, the chon-

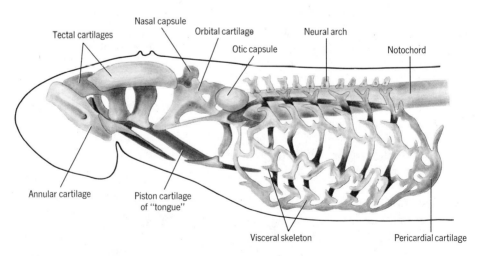

Figure 8-7   HEAD SKELETON OF A CYCLOSTOME shown by the lamprey, *Petromyzon*, as seen in left lateral view.

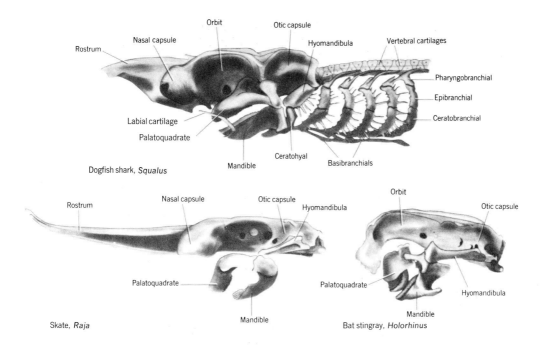

Figure 8-8 HEAD SKELETONS OF SELECTED CARTILAGINOUS FISHES seen in left lateral view. Above, chondrocranium, visceral skeleton, and anterior vertebrae; below, chondrocranium and jaws with their support.

drocranium became unusually solid, with complete side walls and a roof (Figure 8-8). Although the chondrocranium never ossifies, it is sometimes so hardened with granules of calcium salts that it cannot be cut with a knife. Biology classes often study the chondrocranium of the shark, but it is not typical of that structure for vertebrates in general.

Jaw suspension is usually amphistylic (pleuracanths, cladodoselachians, some sharks). It may also be hyostylic, however (other sharks), and in Holocephali, as that term implies, jaw suspension is strengthened for a shellfish diet by fusing firmly with the braincase—a form of autostyly. There are usually six, but as many as eight, postmandibular arches. Typical branchial arches have four paired skeletal elements, and there are also unpaired median ventral cartilages. The segmentation and configuration of the visceral skeleton of cartilaginous fishes are considered to be generally primitive for all fishes.

BONY FISHES: DIVERSITY AND COMPLEXITY   In no other class is the skull so varied and complex as in the bony fishes. This is not unexpected, since the class is so large, yet other parts of the body do not exhibit as much diversity. Much of the variety results from adaptation to many diets, body shapes, and habits, which concern us less here than evolutionary trends. It is important that the three subclasses have somewhat different skulls, for only one of them can include the ancestors of tetrapods, and the skull helps us to recognize which subclass this is.

The chondrocrania of the earliest known bony fishes were relatively well-

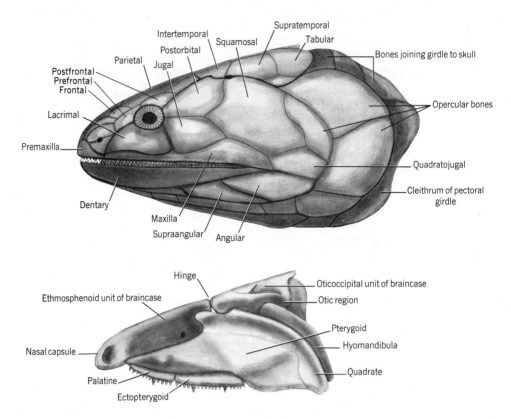

Figure 8-9   HEAD SKELETON OF *Eusthenopteron,* A CROSSOPTERYGIAN OF THE RHIPIDISTIAN GROUP. Superficial dermal bones are shown in the upper drawing; the hinged chondrocranium and some other deep parts of the skeleton are shown in the lower drawing.

ossified, often, seemingly, in one unit without sutures. From this start the dipnoans had already regressed by Mesozoic times so that only the exoccipital bones ossify in an otherwise cartilaginous braincase. Crossopterygians also came to have a rather cartilaginous chondrocranium that always has a striking and unique feature: It is in two pieces (Figure 8-9). An anterior (or ethmosphenoid) unit is derived from the trabeculae and supports the orbits and rostrum. Joined to this by a movable hinge is a posterior (oticoccipital) unit that is derived from the parachordal cartilages and supports the brain. This crossopterygian trademark relates to the gape and power of their jaws. Typical ray-finned fishes have braincases that are more or less completely ossified with separate bones.

The visceral skeleton is more nearly the same among the three subclasses. Dermal bones support the teeth, hence the mandibular arch usually regresses. Several replacement bones may form in the palatoquadrate of bony fishes. Of these, the quadrate has the most evolutionary significance. It forms the upper part of the hinge of the jaw of bony fishes, amphibians, reptiles, and birds. The only replacement bone in the lower jaw is the small articular which forms the lower part of the hinge.

The second arch is not branchial in function. Its large hyomandibula usually supports the jaw (except in dipnoans) which, therefore, is hyostylic (ray-fins) or amphistylic (crossopterygians). There are usually five branchial arches. The gill arches may be bony (most ray-fins) or cartilaginous (dipnoans) and do not differ importantly from those of placoderms and cartilaginous fishes.

With few exceptions, the bony fishes have complete dermal skeletons of small to medium-sized bones (Figure 8-10). A series of bones joins the pectoral girdle to the skull. A movable bony operculum covering the gills is distinctive. Each of the major taxa of bony fishes has its own pattern of dermal bones. The pattern of the crossopterygians, and only this pattern, can be homologized with the pattern of amphibians. This is of primary importance and is one of several lines of evidence supporting crossopterygian ancestry for all tetrapods. The dermal bones of the other major groups of fishes are assigned names that are also applied to the bones of tetrapods, but for them the relationships are, in fact, quite uncertain.

AMPHIBIANS: CONSERVATIVISM OR REGRESSION   When the progenitors of amphibians crawled out of the water they lost from their skulls such piscine characters as gill bars and operculum. Evolution of the tetrapod skull was then by no means finished—important changes were still to come to the palate, skull roof, and jaw mechanism—yet a trend toward stability of form had been initiated by crossopterygians that was to continue through all their descendants. There is less variation of skull structure in all the tetrapods together than in the bony fishes alone. Labyrinthodonts bequeathed the crossopterygian skull to the first reptiles without making striking changes. Surviving amphibians, on the other hand, have somewhat specialized skeletons. The skulls of extinct labyrinthodonts are more within the main stream of evolution than are those of most modern amphibians (Figure 8-11).

Ancestral amphibians had a nearly full complement of replacement bones derived from the chondrocranium; only the supraoccipital was missing. Lissamphibia lack even such usually prominent bones as the basioccipital and basisphenoid. The exoccipitals are retained by these animals largely to provide the paired occipital condyles. The more primitive labyrinthodonts had a single occipital condyle formed by the basioccipital.

The most significant changes that occurred in the skull during the transition from fish to amphibian concerned the visceral skeleton. The quadrate of the upper jaw now articulated with the squamosal without an "assist" from the hyomandibula. Thus, jaw suspension had become autostylic (again), as it was to remain throughout subsequent vertebrate evolution. Most extinct amphibians probably rested their large heads on the ground and conveyed vibrations in the substrate to the inner ear by bone conduction through the jaw and other structures. This mechanism was improved by incorporating a reduced, stubby hyomandibula into the line of sound transmission. The bone was then an ear ossicle and is assigned a new name, **stapes,** or columella.

Ventral elements of the hyoid arch support the tongue as they do also in fishes, but the tongue of amphibians is larger and more muscular. The more posterior arches function as branchial arches only in larvae and in those few

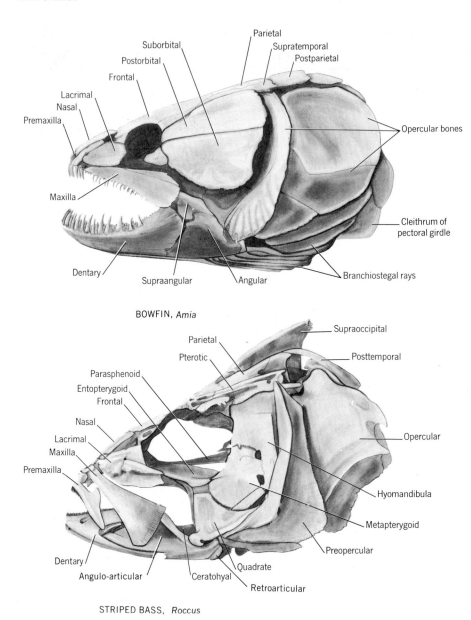

Parietal
Suborbital
Supratemporal
Postorbital
Postparietal
Frontal
Lacrimal
Nasal
Premaxilla
Opercular bones

Maxilla

Cleithrum of
pectoral girdle

Dentary
Supraangular
Angular
Branchiostegal rays

BOWFIN, *Amia*

Supraoccipital
Parietal
Pterotic
Posttemporal
Parasphenoid
Entopterygoid
Frontal
Nasal
Opercular
Lacrimal
Maxilla
Premaxilla
Hyomandibula
Metapterygoid
Preopercular
Dentary
Angulo-articular
Quadrate
Ceratohyal
Retroarticular

STRIPED BASS, *Roccus*

Figure 8-10   HEAD SKELETONS OF RAY-FINNED FISHES of the infraclass Holostei (above) and Teleostei (below).

species that retain gills as adults. Otherwise, they are reduced in number to three and converted to help support the tongue and newly evolved larynx. The tracheal rings may have the same origin. The complex of bones that moves the tongue and suspends the larynx from the base of the skull is called the **hyoid apparatus.**

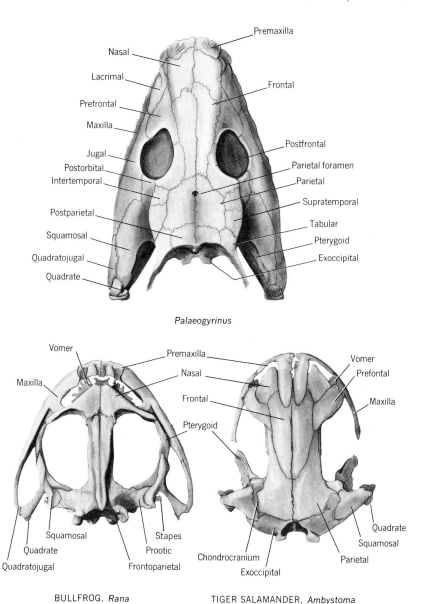

Figure 8-11 AMPHIBIAN SKULLS OF THE SUBCLASS LABYRINTHODONTIA (above), AND ORDERS ANURA (below, left), AND URODELA (below, right).

Of the dermal bones there is less to say. The operculum is lost, and the pectoral girdle is no longer joined to the skull. There has been some tendency to simplify the pattern of bones, but in general, amphibians have not introduced innovations.

REPTILES AND BIRDS: VARIATIONS ON THE BASIC PLAN   Reptiles are important to our story for modifications introduced by one or more of their lineages in relation to the structure of ear ossicles, palate, and jaw mechanism. The skulls of birds are sufficiently similar to those of their reptilian ancestors, so that we can here consider birds to be merely specialized reptiles.

In these classes, the replacement bones of the chondrocranium vary widely in configuration, but a full complement is typical. In birds, all of them fuse with one another and with other bones of the braincase before, or soon after, hatching, leaving no trace of sutures in the adult (Figure 8-19). Reptiles and birds have one occipital condyle.

The visceral skeleton of birds and most reptiles remains essentially the same as for amphibians: Quadrate, articular, and some cartilage form from the first arch; stapes from the second arch; hyoid apparatus chiefly from the second arch, but also from the third and sometimes fourth arches; larynx and tracheal rings probably from the sixth and seventh arches (Figure 8-12).

Mammal-like reptiles modified the first arch derivatives in a way that proved to be significant. As the dentary bone of the lower jaw became ever larger, the articular became smaller and was displaced until it finally lost its position as the lower element in the hinge of the jaw. Likewise, the quadrate became smaller and at last lost its position as the upper element of the hinge (Figure 8-13). A forward shift of the hinge helped to release these bones and create a new jaw articulation between the dentary and squamosal. The articular and quadrate now became ear ossicles as the hyomandibula had done 100 million years before. The articular is given the new name **malleus** and the quadrate the new name **incus.** The stapes now articulates laterally with the incus and medially with the inner ear. This classic evolutionary sequence was first pieced together from elusive clues derived largely from embryology and comparative anatomy. More recently, abundant paleontological evidence has confirmed even the details of the story. These transitions are of particular interest because they relate to the

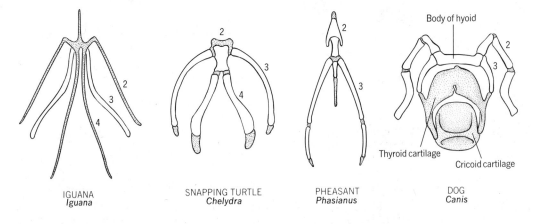

Figure 8-12   VISCERAL SKELETONS OF SELECTED TETRAPODS shown in ventral view. Numbers indicate visceral arch of origin. Stippled elements are cartilaginous.

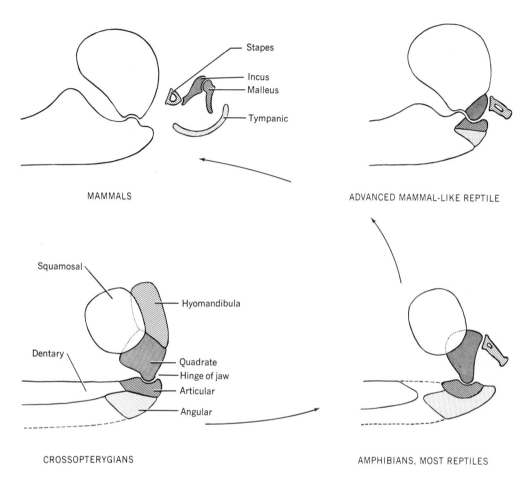

MAMMALS

ADVANCED MAMMAL-LIKE REPTILE

— Stapes
— Incus
— Malleus
— Tympanic

Squamosal
— Hyomandibula
Dentary
— Quadrate
— Hinge of jaw
— Articular
— Angular

CROSSOPTERYGIANS

AMPHIBIANS, MOST REPTILES

Figure 8-13 PHYLOGENY OF THE ARTICULATION OF THE JAW AND ASSOCIATED STRUCTURES.

recognition of the first mammals. Conveniently, but arbitrarily, mammals were long distinguished from their immediate reptilian ancestors on the basis of having an ossicle called the malleus instead of a jawbone called the articular. This distinction is no longer satisfactory for some fossils that are on the borderline.

The dermal bones of reptiles vary somewhat in number and may be heavy or delicate. Variations in configuration and relationship are more striking, however, and differences in the relation of bones on the roof and side walls of the skull to the muscles of the jaws are the basis for the naming of the subclasses of reptiles. Amphibians and fishes accommodate their jaw muscles beside the braincase and under the superficial roof bones of the skull which may be continuous or notched. This arrangement persisted in the stem reptiles—long since extinct. Subsequent reptilian lineages developed more powerful jaws. Presumably in response to the resultant increased and altered stresses, the cranial vault strengthened in some places and weakened in others. Ultimately, one or two pairs of openings formed in the temporal region of the skull, and jaw muscles then moved out through them and took origin in part from their margins. It is evident that this process occurred independently several times, because the fenestration (fenestra = window) evolved at different places on the skull in different groups. Names for the various skull types consist of the combining form "apsid," meaning arch, plus a prefix meaning "not," "two," "wide," etc., as appropriate.

Turtles are related to the ancestral, **anapsid** stem reptiles, and like them have no true temporal openings (Figures 8-14, 16, 17). (Sea turtles demonstrate this condition well, but other turtles have formed an emargination into the skull roof bones from behind.) Synapsida and Euryapsida each have one pair of temporal openings, but in each it is bordered by different bones. The mammalian condition was derived from the **synapsid** arrangement of mammal-like reptiles merely by enlarging the opening. **Diapsid** skulls have the same opening as synapsid skulls and also another that is dorsal to it. This pattern is found in extinct archosaurs, surviving archosaurs (crocodilians, which have small openings, but must usually be used to illustrate the condition to students), and also in the primitive lepidosaur, *Sphenodon* (Figures 8-15, 16, 17). Other surviving lepidosaurs and birds have modified the primitive diapsid pattern by the deletion of one or more of the arches of bone from below or between the original openings (Figures 8-14, 17, 19).

Another change in cranial architecture that was introduced by certain reptiles and later became universal among mammals is likewise related to the feeding mechanism. Amphibians and most reptiles bolt their food. It is unimportant to them, therefore, that respired air comes through the nares into the forward part of the mouth. Other reptiles, and particularly mammal-like reptiles, tore, crushed, or even chewed their food before swallowing. To avoid interrupting their breathing, it became necessary to deliver inspired air into the pharynx behind the chewing mechanism. This was doubly important for those that were becoming homeotherms and were therefore increasing their respiratory rates. The paired vomers, which formerly had bordered the internal nares behind, gradually merged, migrated posteriorly, and moved dorsally above the new, more posterior internal nares (Figure 8-18). The parasphenoid was lost, thus "getting out of the way" of the relocated nares. The pterygoids shortened to the rear. Meanwhile, shelflike processes of the maxillas, and later also of the palatines, grew to the midline in front of the shifting nares to form a new, or **secondary palate.** (Actually, secondary palates evolved independently several times. Of animals other than mammal-like reptiles, some turtles have an incomplete secondary palate and crocodilians have a complete one.)

The palate of crossopterygians and early labyrinthodonts was complete and solid. The palates of later labyrinthodonts, surviving amphibians, most reptiles, and birds have lateral vacuities of larger or smaller proportions that lighten the skull without much loss of strength and, for some of these animals, enable protruding eyes to be withdrawn into the head on occasion to escape harm or assist in swallowing.

Birds and some reptiles are able to move the palate forward and backward on the braincase by pivoting the quadrates to and fro and by providing a hinge

SEA TURTLE, *Chelonia*

TEGU LIZARD, *Tupinambis*

Figure 8-14 REPTILE SKULLS AND MANDIBLES OF THE SUBCLASSES ANAPSIDA (above) AND LEPIDOSAURIA (below).

in the roof of the skull in front of the orbits. This action elevates the upper jaw, which increases the gape of the mouth. Skulls that have this mechanism are said to be kinetic and are described further on pages 561 and 562. Ornithologists divide birds into four or more groups on the basis of the relationships of the palatal bones involved in this mechanism.

Figure 8-15 REPTILE SKULL OF THE SUBCLASS ARCHOSAURIA shown by the alligator.

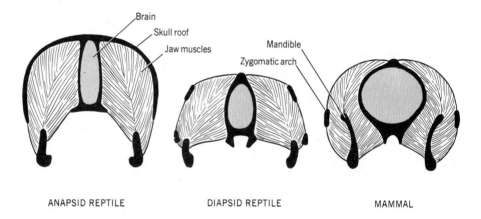

ANAPSID REPTILE   DIAPSID REPTILE   MAMMAL

Figure 8-16 SOME RELATIONSHIPS BETWEEN JAW MUSCLES AND THE HEAD SKELETON shown by cross sections of the head at the epipterygoid-alisphenoid level of the skull. Diagrammatic, but based on the sea turtle, *Chelonia*, the tuatara, *Sphenodon*, and the dog, *Canis*.

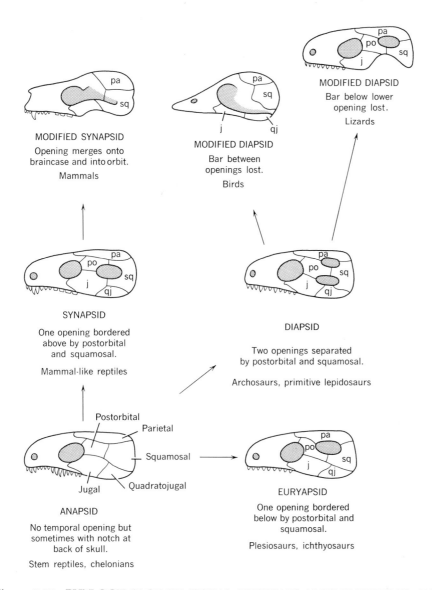

Figure 8-17 PHYLOGENY OF TEMPORAL OPENINGS AMONG REPTILES AND THEIR DESCENDANTS.

MAMMALS: SOME FURTHER MODIFICATIONS The mammalian skull is at the same time exceedingly variable in regard to adaptive features such as strength and proportions, and conservative in regard to basic plan. Temporal architecture, jaw suspension, ear ossicles, and secondary palate remain about as inherited from the most advanced synapsid reptiles (Figures 8-20, 21, 22). The brain is, of course, larger than for other vertebrates, a factor that contributes to complete functional integration of the replacement and dermal bones of the braincase.

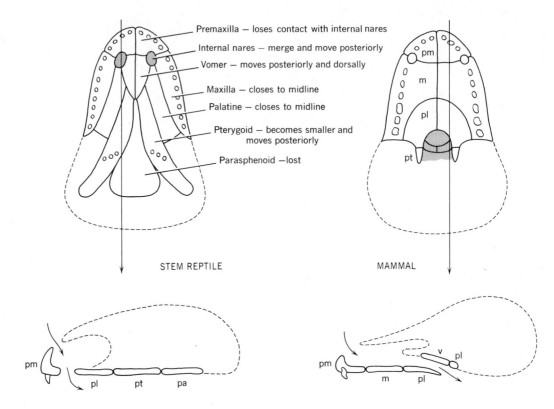

Premaxilla — loses contact with internal nares

Internal nares — merge and move posteriorly

Vomer — moves posteriorly and dorsally

Maxilla — closes to midline

Palatine — closes to midline

Pterygoid — becomes smaller and moves posteriorly

Parasphenoid — lost

STEM REPTILE

MAMMAL

pm, m, pl, pt

pm, pl, pt, pa

pm, m, pl, v

Figure 8-18 EVOLUTION OF THE SECONDARY PALATE. Palatal views above; parasagittal sections below. Dashed lines indicate parts of skull not involved in this evolutionary sequence. Lower arrows show path of inspired air.

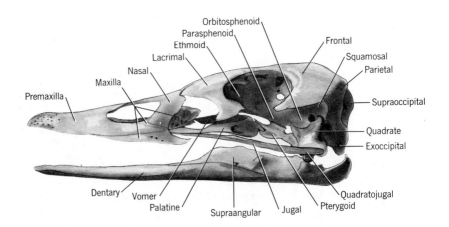

Orbitosphenoid
Parasphenoid
Ethmoid
Lacrimal
Nasal
Maxilla
Premaxilla
Frontal
Squamosal
Parietal
Supraoccipital
Quadrate
Exoccipital
Dentary
Vomer
Palatine
Supraangular
Jugal
Pterygoid
Quadratojugal

Figure 8-19 SKULL AND MANDIBLE OF A BIRD shown by the swan, *Cygnus.* Sutures of the braincase are obliterated in the adult.

Opossum, *Didelphis*

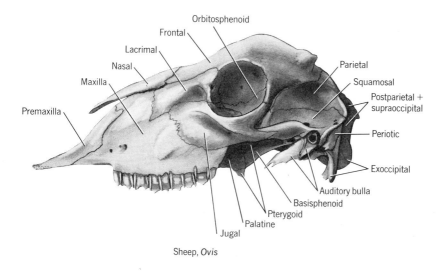

Sheep, *Ovis*

Figure 8-20   MAMMALIAN SKULLS OF THE PRIMITIVE ORDER MARSUPIALIA (above) AND THE ADVANCED ORDER ARTIODACTYLA (below).

There are again two occipital condyles as for most amphibians. Cranial sutures are nearly always more in evidence than for birds and usually less in evidence than for reptiles. Certain combinations of bones are particularly prone to fuse early in life. Common fusions include the postparietal with the supraoccipital, basioccipital with the exoccipitals to form a single occipital, the otic bones (prootics and opisthotics) with the squamosals to form temporals, and the four sphenoid bones (basi-, pre-, ali-, orbito-) with one another to form a single sphenoid.

Mammalian nasal structure is distinctive. The anterior bony nares have merged to form a common opening. A relatively large nasal chamber is in part filled (as in some birds) by delicate scrolls of bone, the conchae, or turbinates, which are outgrowths from the inside walls of the maxillas, nasals, and ethmoids.

Premaxilla

Maxilla

Palatine

Vomer

Jugal

Presphenoid

Alisphenoid

Basisphenoid

Carotid foramen

Squamosal

Stylomastoid foramen

Mastoid process

Basioccipital

Paroccipital process

Palatine fissure

Palatine foramen

Internal nares

Zygomatic arch

Anterior lacerate foramen

Foramen rotundum

Alar canal (not present in cat)

Foramen ovale

Postglenoid foramen

Tympanic bulla

Posterior lacerate foramen

Hypoglossal foramen

Figure 8-21   SOME BONES AND FEATURES OF THE SKULL OF A WOLF.

These are covered with epithelium and serve to warm and clean inspired air before it reaches the lungs. Their presence is believed to be associated with endothermy. It is, therefore, of interest that some mammal-like reptiles possessed incipient turbinates *before* the articular became a malleus!

A trend toward simplification of the skull by the loss of bones runs all through tetrapod evolution. Bones that are characteristic of amphibians and reptiles and are not represented in mammals include the prefrontals, postfrontals, postorbitals, quadratojugals, parasphenoid, and all dermal bones of the lower jaw except the angular, prearticular, and dentary (which is retained as the single jawbone). The ectopterygoid usually fuses with the pterygoid. The prearticular contributes a process to the malleus; the angular moves over from the angle of the jaw to contribute to a uniquely mammalian structure, the tympanic bulla, which helps to enclose the middle ear. Details of the construction of the bulla are used in the classification of carnivores.

Reptilian bones that occur in mammals under new names may be summarized as follows:

Reptilian bone	Mammalian homolog
articular + prearticular	malleus
quadrate	incus
angular	tympanic
epipterygoid	alisphenoid
sphenethmoid	presphenoid + orbitosphenoid
vomer (paired)	vomer (unpaired)

It is doubtful that mammals have evolved any really new bones, but the entotympanic, which contributes to the tympanic bulla, seems to be a candidate.

SUMMARY OF PRINCIPLES AND TRENDS Many evolutionary mechanisms and principles discussed earlier in the book are illustrated by the phylogeny of the head skeleton. *Recapitulation* is seen in the ontogeny of the principal components of the skull and in the number and nature of ossification centers. *Serial homology* is exemplified by the head somites. *Inductions* have been demon-

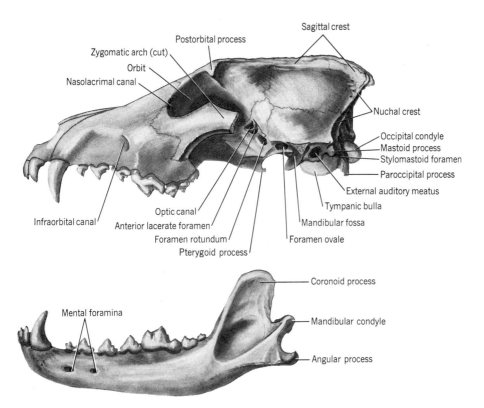

Figure 8-22 SOME FEATURES OF THE SKULL AND MANDIBLE OF A WOLF.

strated between regions of the developing brain and specific cranial ossifications. *Neoteny* is the probable explanation of gradual loss of ossification in the heads of some fish taxa. *Convergence* is manifest, for instance, in the crushing jaws of dipnoans and the unrelated holocephalians. There are classical examples of *homology* (e.g., the hyomandibula and stapes) and *analogy* (e.g., the quadrate-articular jaw joint and the squamosal-dentary jaw joint). *Preadaptation* for auditory function is seen in the bones released at the posterior end of the synapsid jaw.

Many *evolutionary trends* can be identified. Early vertebrates tended to reduce the number of visceral arches and to stabilize their number. A tendency to reduce overall ossification is observed in various taxa, and from fishes to mammals many specific bones dropped out, particularly around the orbit, at the back of the skull, on the temporal area, and on the lower jaw. This has contributed to a relative shortening of the postorbital part of the skull.

The head skeleton is of great importance to phylogeny and systematics at all levels but is particularly noteworthy for the study of fish taxa (patterns of dermal bones, jaw suspension), the subclasses of reptiles (temporal fenestration), and the reptile-mammal transition (origin of ear ossicles, secondary palate, nasal structure).

FEATURES OF THE SKULL

This chapter has emphasized evolutionary trends of the isolated skull. But the skull is not isolated. To be fully understood it must also be related to other organ systems. The vertebrate skull is so complex and the range of its variation is so great that its named features are legion. Both student and professor must limit themselves to the major parts of skulls representing the major taxa, leaving the details to specialists. I have selected for illustration (Figures 8-21, 22) some fea-

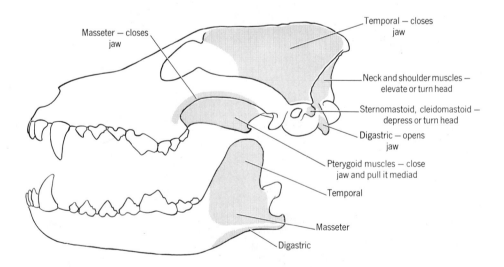

Figure 8-23 RELATIONS OF THE SKULL AND MANDIBLE OF A WOLF TO VARIOUS MUSCLES.

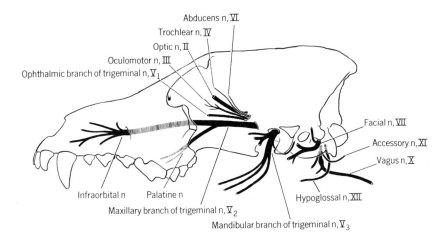

Figure 8-24   RELATIONS OF THE SKULL OF A WOLF TO VARIOUS NERVES.

tures of the mammalian skull and mandible that are relatively prominent and constant and that correlate with structures to be presented elsewhere in this book. Figures 8-23 and 8-24 show processes and crests in relation to muscles, and foramina in relation to cranial nerves.

Finally, although the elements present in a skull and the basic relationships of these parts are determined by a long evolutionary history, the relative sizes and configurations of these parts are related to specific adaptations of (chiefly) the feeding mechanism, brain, and sense organs. The importance of these factors to the architecture of the skull will be noted in other chapters.

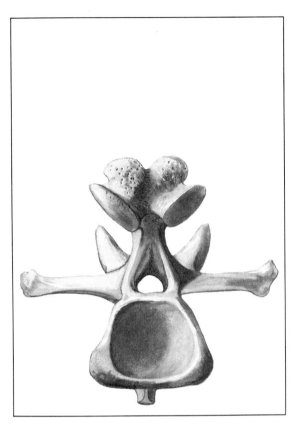

9

Body Skeleton

The functions of the body skeleton are to protect the viscera, contribute (in amniotes) to ventilation of the lungs, serve as a store for various minerals, give rigidity to an otherwise soft body, and, importantly, to provide the series of firm, hinged segments that are essential, in conjunction with the muscles, for locomotion. An external skeleton, like that of an insect or crab, also serves in locomotion, but for large animals, particularly if terrestrial, such a skeleton would need to be prohibitively cumbersome and heavy in order to avoid fracture from contacts with the environment. The reception of external stimuli, thermoregulation by homeotherms, and integumentary respiration by certain aquatic forms may also be better served with an internal skeleton.

STRUCTURE AND DEVELOPMENT OF VERTEBRAE

GENERAL STRUCTURE The vertebral column is older than any other part of the postcranial skeleton except the notochord. Nevertheless, it is not so ancient as the major features of the soft organ systems, and is virtually absent in the oldest known vertebrates. Its structure and functions diversified slowly. We shall start with an account of the features of a typical vertebra to gain the terminology necessary to discuss the evolution of vertebrae.

The principal part of a vertebra is the spool-like body or **centrum,** which lies just below the spinal cord, where it surrounds, restricts, or replaces the notochord (Figure 9-1). The centrum may consist of one or two elements (rarely more—in fishes). If a tetrapod centrum has two elements (many extinct amphibians, few amniotes), then the more anterior is called an **intercentrum** and the

Figure 9-1   SOME FEATURES OF VERTEBRAE. Left side views.

more posterior (which may be paired) is called a **pleurocentrum** (Figure 9-6). If a tetrapod has only one central element, it may be the intercentrum (some extinct amphibians) or the pleurocentrum (amniotes). As explained below, it is not yet clear if these terms are appropriate for fishes.

Extending dorsally from the centrum are pillars that flank the spinal cord, one on each side, and join above the cord to complete the **neural arch.** A **neural spine** may extend from the summit of the neural arch. Of frequent though less universal occurrence is a similar **hemal arch** which extends ventrally from the centrum to surround blood vessels. It may be continued by a **hemal spine.** Neural and hemal spines are related to muscles that move the axial skeleton. Of like function are a variety of processes that project, according to species, from walls of the vertebra. Centra may accommodate the **capitulum,** or head of a rib, either with a process (parapophysis) or with a concavity. The part of a rib called the **tuberculum** (Figure 9-1) is supported by a **diapophysis.** Any lateral process is a **transverse process.** The most prominent of these is usually in the position of the diapophysis but on vertebrae lacking ribs. Adjacent vertebrae always articulate by their centra (if the centra are complete), and in tetrapods they also articulate by processes carried by the neural arches and called **zygapophyses** (= joining + processes). The **prezygapophyses** on the anterior aspect of one vertebra articulate with the **postzygapophyses** on the posterior aspect of its neighbor. Prezygapophyses face upward or inward, whereas postzygapophyses face downward or outward. This is useful to remember for determining which end of a single vertebra is which. (A ventrolateral process may be called a basapophysis, and a median ventral process a hypapophysis. Some mammals have a metapophysis on the neural arch close to the prezygapophysis, or an anapophysis near the postzygapophysis.)

The shapes of the articulatory surfaces at the ends of the centra are of evolutionary, functional, and systematic importance. If each surface is concave, the centrum is said to be **amphicoelous** (= on both sides + hollow). Such centra actually touch only at the periphery of the intervertebral joint (Figure 9-2). Limited

Figure 9-2 VERTEBRAE SHOWING VARIOUS SHAPES OF CENTRUM, and other features, as seen in posterior view.

motion is permitted in any direction. Within the joint is a space filled by connective tissue, cartilage, or remnants of the notochord. Adjacent spaces may be joined by a perforation through the interior of the centrum. Other kinds of centra are concave anteriorly and convex posteriorly, the bulge of one vertebra fitting into the hollow of the next. Such a centrum is **procoelous** (= front + hollow). Conversely, an **opisthocoelous** (= behind + hollow) centrum is concave posteriorly and convex anteriorly. Joints between procoelous or opisthocoelous vertebrae permit motion in any direction (except as modified by the zygapophyses) and they resist dislocation. Some centra have flat ends and are said to be **platyan** (or acoelous). These centra withstand compression and limit motion (unless the intervertebral joints are provided with thick fibrous disks). Still other centra have saddle-shaped ends. They are called **heterocoelous** (= different + hollow). They allow vertical and lateral flection, but prevent rotation around the axis of the spine. There are intergrades among these types, combinations, and occasionally doubling of the concave and convex surfaces. The functional aspects of vertebral structure are interpreted further in Part III.

DEVELOPMENT AND HOMOLOGY   Vertebrae are complex structures, and the variety of their construction is extreme. The recognition of homologous parts is, therefore, very important so that evolutionary sequences can be identified. Contributions have come from comparative anatomy, paleontology, and embryology, and much has been learned. Nevertheless, uncertainty remains in regard to some important points.

The differentiation of the embryonic somite into dermatome, myotome, and sclerotome was described in Chapter 5 (see figure on p. 80). Mesenchyme from the segmental sclerotomes usually distributes itself around the notochord to form a continuous cylinder, the **perichordal tube.** This tube has nodal thickenings, or rings, at levels that will come to be intervertebral. Flanking the perichordal tube is additional sclerotomal mesenchyme which in amniotes, but doubtfully in anamniotes, forms caudal half segments that are conspicuously more cell-dense than the cranial half segments. It has long been taught that the caudal half segment of one sclerotome joins the cranial half segment of the adjacent sclerotome to become the anterior half of a vertebral centrum that is, therefore, intersegmental in relation to the primary segmentation of the nearby myotomes. Thus, the axial muscles, in their simplest form, come to span from one centrum to the next, as is necessary to flex the spine. The relation of neural and hemal arches to the half sclerotomes is less clear, but they seem to be associated primarily with the caudal halves. Spinal ganglia line up with the cranial halves, and intersegmental blood vessels between halves. Only at this stage does chondrification occur in all the primordia. The cartilage is continuous for a time, but under the inductive influence of spinal ganglia, notochord, and myotomes it breaks to establish the individual vertebrae. The cartilaginous vertebrae are later usually replaced by bone, which may be supplemented by dermal bone developed in or around the notochordal sheaths, or elsewhere.

That half sclerotomes recombine in so regular a way is now questioned by some morphologists (see the article by Verbout listed in the References). If most, or all of each vertebra develops from the dense caudal half sclerotome, but then expands toward the septa between myotomes, then the needed alternation of muscles and vertebrae is established without the recombination of sclerotomal segments. These various views remain controversial.

Also debated is the relation between the embryonic segmentation of the sclerotome and the adult segmentation of some vertebrae (particularly of labyrinthodont amphibia) into intercentrum and pleurocentrum. Some investigators believe that the entire centrum is homologous among the various vertebrate taxa regardless of the number of elements observed, and that the equivalence of the elements among classes is now only tentative.

EVOLUTION OF THE SPINE

BEGINNINGS: NOTOCHORD WITH SUPPLEMENTAL CARTILAGES   The notochord is persistent in adults of jawless vertebrates, placoderms, pleuracanths, cladoselachians, chimaeras, acanthodians, dipnoans, crossopterygians, and most members of the lower orders of ray-finned fishes. In contrast to the embryonic notochord (more often seen in the laboratory), the adult chord is large and resilient. It has an outer elastic sheath that merges with a tough, inner, fibrous sheath.

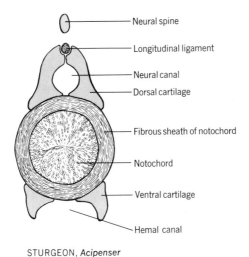

Neural spine
Longitudinal ligament
Neural canal
Dorsal cartilage
Fibrous sheath of notochord
Notochord
Ventral cartilage
Hemal canal

STURGEON, *Acipenser*

Figure 9-3 CROSS SECTION OF THE NOTOCHORD AND SPINAL CARTILAGES OF A CHONDROSTEAN FISH.

The stiffness of the notochord enables muscles of the body wall to flex the body (as required for the vertebrate manner of swimming) rather than shorten it. The notochord of amphioxus is itself a contractile organ. Its contraction increases its stiffness, particularly during fast swimming. It is probable that the notochords of ancestral vertebrates had the same capacity.

All these vertebrates (except hagfishes) have neural arches; many have hemal arches, at least in the tail; and some have arch bases or other elements which flank, but do not restrict, the notochord (see Figure 9-3, and the figure on p. 128). It is probable that the various cartilages evolved where tendons inserted on the notochordal sheaths. Limited ossification of these elements is present in arthrodires and bony fishes. This kind of spine is primitive for vertebrates in general.

ADVANCED FISHES: THE SPINE TAKES OVER   The next evolutionary steps were for the notochord to be replaced by centra and for elaboration of arches and processes. Selachians among cartilaginous fishes, teleosts among ray-finned fishes, and also some other bony fishes evolved vertebrae with firm centra that articulate with one another. The deeply amphicoelous centra restrict or interrupt the notochord, thus destroying its integrity and function. Such a spine is stronger than the notochord and provides greater anchorage for muscles.

Variation of structure among the fishes is considerable. SELACHIANS are unique in having not only the centra but also the arches in continuous contact (Figure 9-4). Such an arrangement would virtually immobilize a bony spine, but the cartilaginous nature of the selachian spine provides the slight flexibility that is needed. Each vertebral centrum may have several separate pieces, and there are two principal dorsal arches (a neural plate and an intercalary plate, or interneural). The centrum chondrifies in the notochordal sheaths (a **chordal centrum**). Calcification of the centra is common, but instead of being continuous, the hard tissue is arranged in concentric cylinders around the central axis and in

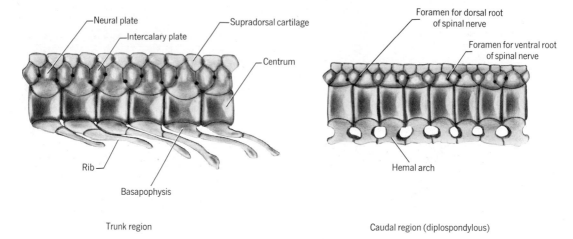

Trunk region

Caudal region (diplospondylous)

LEOPARD SHARK, *Triakis*

Figure 9-4   SECTIONS FROM THE TRUNK (left) AND CAUDAL REGIONS (right) OF THE SPINE OF AN ELASMOBRANCH in left lateral view. As the nerve foramina show, the caudal section is diplospondylous.

pillars in the positions of the primitive arch bases. Details of the pattern are various and have systematic value (Figure 9-5).

The vertebrae of BONY FISHES usually have one central element, a neural arch with spine, and, in the tail, a hemal arch with spine (Figure 9-2). The centrum usually ossifies directly from mesenchyme surrounding the notochord (a **perichordal centrum**), but may also form from cells that invade the chordal sheaths or from cartilaginous precursors.

There is little regional differentiation in the vertebal column of either cartilaginous or bony fishes. The first vertebra may be slightly modified for articulation with the skull. Behind the coelomic cavity a caudal region is distinguished by the absence of rib facets and the presence of enlarged hemal arches with spines. Some bony fishes (including the bowfin commonly seen in the laboratory) and many cartilaginous fishes have a curious doubling of vertebrae in the caudal region, giving two (sometimes even more) complete vertebrae per muscle segment and per pair of spinal nerves (Figure 9-4). The origin and significance of this condition, called **diplospondyly,** have been much discussed but remain obscure. Perhaps it increases flexibility. Rarely the centrum is single and the neural arches are doubled. Arches may be fused to the centra or free.

AMPHIBIANS: VARIED SOLUTIONS TO NEW PROBLEMS   Few organs were so affected by the change from aquatic to terrestrial life as was the spine. Formerly

Figure 9-5   CROSS SECTIONS OF THE CENTRA OF TWO ELASMOBRANCHES, the white shark, *Carchorodon* (above) and angel shark, *Squatina* (below), showing calcified cartilage in black.

it resisted only the stresses imposed by strong axial muscles; now the axial musculature gradually became reduced but there were new stresses, imposed by gravity, which acted largely in a different plane. Formerly the paired appendages had not been related to the spine; now the slowly strengthening limbs transmitted their support to the axis of the body. Formerly it was sufficient for the spine to be nearly uniformly flexible throughout its length; now it had to resist bending in some places and provide new mobility elsewhere. The requirements were for vertebrae with firm centra, intervertebral joints that could facilitate or restrict motion according to location, processes that could increase the leverages of muscles, and a more intimate relation with the girdles. All this took many millions of years to accomplish. Indeed, it remained for the amniotes to complete the changes. Nevertheless, the credit for most of the transformations, and nearly all of the evolutionary experimentation that tested the innovations, must be given to the extinct amphibians. Various taxa of labyrinthodonts are named and largely defined by the nature of their vertebrae.

The story of the evolution of amniote vertebrae starts with the particular crossopterygians (Rhipidistia) that were ancestral to tetrapods. Their vertebrae were nearly identical with those of the earliest amphibians (Ichthyostegalia) (Figure 9-6). The notochord was persistent. There was a neural arch, a large intercentrum encircling the notochord, and small paired pleurocentra. The spine was still fishlike in function, but it contained the three ossifications that were to feature in all subsequent vertebrate evolution.

Two principal lineages diverged from this common origin. One led first to a group of labyrinthodonts (Rhachitomi; = vertebra + cut) that replaced the notochord by a large intercentrum and a pair of small, unfused pleurocentra, and thence to a group (Stereospondyli; = solid + vertebra) that retained only the intercentrum. This lineage had no descendants. The other lineage either increased the size of the pleurocentrum to equal status with the intercentrum (the condition in the Embolomeri; = wedge + parts), or enlarged the pleurocentrum to the near or complete exclusion of the intercentrum (amniotes).

The vertebrae of modern amphibians remain an enigma. The pattern of development gives no useful information about the homologies of vertebral elements. Their centra (which always consist of a single unit) are probably broadly homologous to each other and to those of other tetrapods.

Amphibians, like other tetrapods, have zygapophyses to strengthen the spine and control its flexibility. Centra are amphicoelous, procoelous, or opisthocoelous. Regional differentiation of the spine is minimal for tetrapods but exceeds that of fishes. The first (and sometimes the second) vertebra is modified to give increased mobility to the head. A **cervical vertebra** is somewhat set off by the reduction or absence of ribs. **Trunk vertebrae** bear ribs (Anura excepted). A single **sacral vertebra** is enlarged to articulate (via its fused rib) with the pelvic girdle. Typical tail, or **caudal vertebrae** lack zygapophyses, and have hemal arches. Anura, however, have no free caudal vertebrae but have instead a rodlike **urostyle** which is probably derived by the fusion of several caudal vertebrae (Figure 9-7).

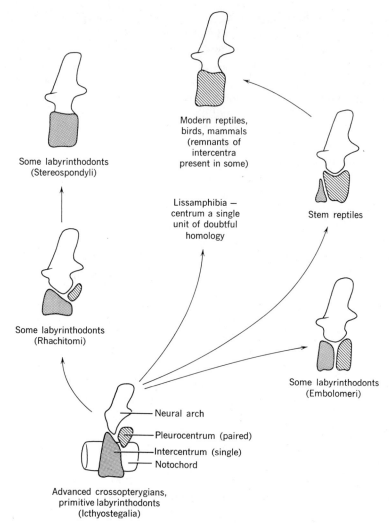

Figure 9-6 PHYLOGENY OF TETRAPOD VERTEBRAE as seen in left lateral view.

TOAD, *Bufo*

AMNIOTES: STRENGTH AND SPECIALIZATION The centrum of amniotes is a pleurocentrum. Stem reptiles inherited vertebrae with large pleurocentra and small intercentra. Subsequent amniotes retain only one functional intercentrum (in the first vertebra of the neck). Some reptiles and several mammals have vestigial intercentra elsewhere along the spine. Amniotes have strong zygapophyses (rarely they are doubled for added strength) except in the caudal region. The first

Figure 9-7 VERTEBRAL COLUMN OF AN ANURAN showing in dorsal view the single cervical vertebra, short trunk, absence of free ribs, single sacral vertebra, and fusion of caudals to form a urostyle.

two cervical vertebrae are specialized to support the skull (of which more under the next subheading).

Because most REPTILES have more distinct necks than amphibians, they have more distinct cervical regions; and because their limbs are stronger, they have two (sometimes more) sacral vertebrae instead of one. Centra are usually procoelous but also take other shapes. Hemal arches are retained only in the caudal region where they are separate bones shaped like Y's. The ancestral intercentra with which they once articulated having disappeared (or merged with them?), they articulate with the spine at the intervertebral joints. Such arches are called **chevron bones** because of their shape.

Because BIRDS have a locomotor specialization in common—flight—they have more specialized and more uniform spines than other tetrapods. The spine of Archaeornithes is transitional between those of their reptilian ancestor and Neornithes. No other class has so many cervical vertebrae—commonly, 15–20 (10 in Archaeornithes)—and these vertebrae are distinctive for being heterocoelous. Fused to the 2 sacral vertebrae of the reptilian ancestor are 10–20 trunk and caudal vertebrae (5–6 in Archaeornithes) and the pelvises to form a solid unit called a **synsacrum** (Figure 9-8). The remaining trunk vertebrae number only 4–6, and there may even be fusions among some of these. The result is a short, rigid back. The free caudal vertebrae of Neornithes have been reduced to (usually) 6 or 7; Archaeornithes had a long bony axis to the tail. At the end of the tail of

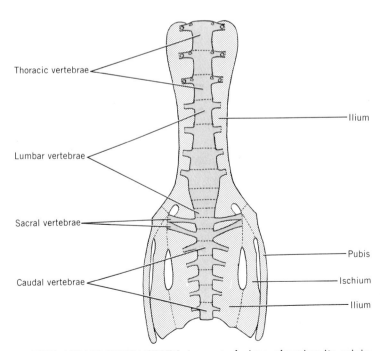

Figure 9-8   THE AVIAN SYNSACRUM, in ventral view, showing its origin.

modern birds is a unique blade of bone called the **pygostyle** which supports the tail feathers. It represents the fusion of 4–7 vertebrae. Even more embryonic somites form and disappear again in recapitulation of the long ancestral tail. Some of these features are seen in the lower figure on page 555.

MAMMALS are unique in forming bony, platelike caps, or **epiphyses,** at the ends of their centra posterior to the first intervertebral joint. These usually fuse to the centra when the animal is mature. They are derived from the segmental thickenings of the mesenchyme (perichordal tube) that surrounds the embryonic notochord. Their function relates to the growth process. With rare exceptions (some edentates and sirenians), mammals have 7 cervical vertebrae whether the neck is short (dolphin) or long (giraffe). There are about 20 trunk vertebrae. Unlike the condition in other classes, these are sharply divided, by presence or absence of ribs, into anterior **thoracic vertebrae** and posterior **lumbar vertebrae,** (see figure on p. 498). To assign isolated vertebrae to region, look for rib facets. Lumbar vertebrae tend to have larger centra, shorter and stouter neural spines, and longer transverse processes than thoracic vertebrae. (We shall see on p. 462 and 497 that from a functional viewpoint there are other ways to divide the trunk vertebrae.) Mammals have 3 or more sacral vertebrae fused to form a **sacrum.** Chevron bones are usually restricted to the base of the tail.

CRANIO-VERTEBRAL JOINT   The occipital process of the skull of fishes usually resembles the end of a typical centrum. The head seldom has more mobility on the spine than one vertebra has on another. The cranio-vertebral joint is like an intervertebral joint.

Amphibians are the first vertebrates to have a neck, however short, and linked with this advance is a more flexible cranio–vertebral joint. The second vertebra is unchanged or only somewhat enlarged.

Reptiles and birds have a ball-and-socket joint between the small first vertebra, now called the **atlas,** and the single occipital condyle. The second vertebra, called the **axis,** is enlarged, and the joint between the atlas and axis is specialized.

The atlas of mammals has large zygapophyses, winglike transverse processes, scarcely any centrum, and no neural spine (Figure 9-9). Its articulation with the two condyles of the skull permits hingelike up and down motion. The axis has a bladelike neural spine and a large centrum. The joint between atlas and axis permits side-to-side motion and rotation around the axis of the spine.

In adults of some extinct amphibians, in reptiles, and in the embryos of many tetrapods, small structures with no apparent function are found between the skull and the neural arch of the atlas. These form the **proatlas.**

The story of how these changes came about is of particular interest. It has been pieced together from many embryological, paleontological, and comparative anatomical studies. In Chapter 8 it was noted that several ancestral vertebrae were incorporated into the vertebrate chondrocranium. It follows that the cranio-vertebral joint falls within the primordial spine, not beyond its anterior limit. At some time (or times?) in tetrapod evolution one vertebra at the cranio–vertebral boundary became reduced in size and partly fused to adjacent units of the series.

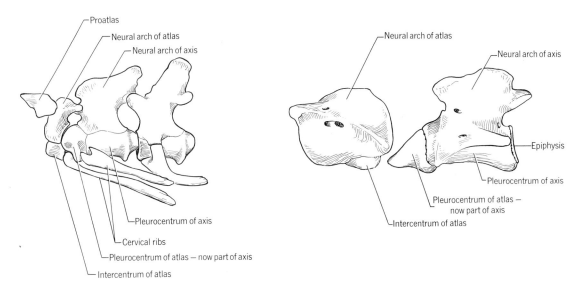

CROCODILE, *Crocodylus*

DEER FAWN, *Odocoileus*

Figure 9-9 ORIGIN OF THE ELEMENTS OF THE ANTERIOR CERVICAL VERTE-BRAE of a reptile and a mammal shown in left lateral view.

This was the proatlas. Its neural arch remains free in some labyrinthodonts and some reptiles. Otherwise, the parts of the proatlas may fuse to other vertebrae or to the skull (though there is no evidence of such fusion in modern amphibians).

The atlas of amniotes always retains its "own" neural arch and intercentrum. The axis not only retains its own neural arch and pleurocentrum, but also incorporates into its odontoid process the pleurocentrum of the atlas and even that of the proatlas if that element has not joined the skull (mammals, some reptiles). The intercentrum of the axis is lost (mammals) or also incorporated into the composite centrum of the axis.

Ribs are intersegmental splints of cartilage or replacement bone which articulate with the vertebrae. Originally they extended the entire length of the spine to increase the direct contact of axial muscles with the skeleton. As they lengthened they came to protect underlying viscera while retaining requisite flexibility and minimum weight. Only in amniotes are they specialized to contribute to the breathing mechanism.   **RIBS**

There appear to be two sets of ribs among the different vertebrates. Each set evolved in the septa of connective tissue that occur between successive muscle segments of the primitive body wall. The set called **dorsal ribs** (or intermuscular ribs) evolved where these septa intersect with the partition between dorsal (epaxial) and ventral (hypaxial) muscle masses. **Ventral ribs** (or pleural ribs) form between the hypaxial muscles and the lining of the coelom. These occur only in the trunk region (Figure 9-10).

Figure 9-10   CROSS SECTION OF A BONY FISH SHOWING POSITION OF DOR-
SAL AND VENTRAL RIBS in relation to muscle masses and body cavity.

The first ribs known are already complete and independent structures; paleontology gives no clue of their origin. It was formerly thought that ribs originated
as outgrowths from the sclerotomes. The sclerotomes probably do contribute to
the heads of ribs, but it is now thought, on the basis of their ontogeny, that they
are primarily new structures not derived from the spine. The oldest known tetrapod ribs had two heads, and this condition is retained by most ribs that are
strong. Weaker ribs (often more posterior on the trunk) may have a single head
as the result of either fusion or loss. The more ventral head, named the **capitulum,** articulates with the intercentrum, if that piece is present, and otherwise
with the pleurocentrum near the intervertebral joint. The dorsal head, or **tuberculum,** articulates with the diapophysis of the neural arch (Figure 9-1).

Jawless vertebrates and placoderms have no ribs. The functions of ribs were
performed in part by the armor of some of them, and none has centra with which
ribs could articulate. With few exceptions (e.g., Holocephalia) other fishes have
ribs. There is disagreement as to whether the short cartilaginous ribs of pleuracanths and selachians are of dorsal or ventral origin (Figure 9-4). The long bony
ribs that enclose the body cavity of ray-finned fishes are ventral ribs. Various
primitive but unrelated fishes (e.g., salmon, *Polypterus*) have both ventral and
smaller dorsal ribs. The ribs of tetrapods are in the position of ventral ribs; however, consensus now is that they are the ancestral dorsal set.

The short cervical ribs of tetrapods may articulate firmly with the vertebrae
(labyrinthodonts, many reptiles), fuse with the vertebrae (birds, mammals), or be
lost (turtles). Free caudal ribs are retained by most labyrinthodonts and some
reptiles but otherwise are either lost or indistinguishably fused to the transverse
processes. The same is true of the lumbar ribs of mammals. Some Anura have
free ribs, but many retain only fused sacral ribs. Urodela have short ribs with
unique articulations. The ribs of amniotes are in two pieces—a principal ossified
segment and a shorter **sternal rib** which ossifies in birds but is otherwise usually
cartilaginous. The more anterior thoracic ribs of most amniotes articulate with
the sternum (breastbone) via their sternal segments (see figure on p. 444). Ribs
of birds and of some reptiles have **uncinate processes** to provide anchorage for
shoulder muscles (see figure on p. 555).

STERNUM   The **sternum** is a midventral skeletal element that usually articulates with the
more anterior thoracic ribs (Figure 9-11). Its functions are to strengthen the body
wall, help protect the thoracic viscera, accommodate muscles of the pectoral
limb and, in some amniotes, aid in ventilating the lungs. Most of these functions
are not relevant to fishes, and only tetrapods (and not all tetrapods) have a sternum. There is wide structural variation among species and relatively great individual variation among animals of a kind. The sternum is rarely present in fossils
of amphibians and reptiles because it was cartilaginous. It contributes little to our
knowledge of vertebrate evolution.

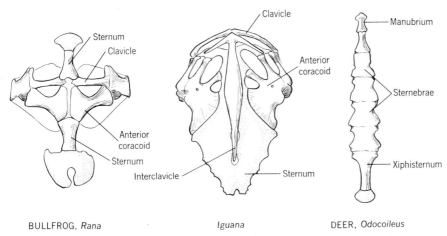

Figure 9-11 VARIED STERNUMS AND RELATED BONES OF THE PECTORAL GIRDLE seen in ventral view.

The sternum forms from either paired or midventral primordia (or both) which are now generally regarded (following considerable debate) as new structures not derived from the pectoral girdle or ribs. (The rib ends of mammalian embryos inhibit ossification of the sternum.)

It is probable that the most primitive tetrapods had no sternum. It is a simple cartilaginous plate in most living Urodela and is missing in others. In Anura it ranges from poorly developed to well-ossified. The sternum is cartilaginous, and is often large in lizards and crocodilians. It is absent in snakes and turtles. Modern birds have a huge ossified sternum with a prominent keel to give origin to flight muscles (see again figure on p. 555). Mammals are unique in having the sternum divided into a linear series of about half a dozen bony segments.

ORIGIN OF APPENDAGES

Speculation over the origin and evolution of appendages dates back nearly to the publication of the *Origin of Species*. One fanciful theory, later disproved by evidence from both embryology and paleontology, derived paired appendages from gill arches. In the early 1880s several anatomists postulated that the primitive vertebrate had continuous paired lateral fins from the gills to the anus, and that from there a single continuous fin ran in the midline around the tail and forward along the back to the head. It was considered that the fins of familiar fishes represent such portions of the ancestral continuous fins as remained after intervening deletions were established. The long lateral body folds of amphioxus, the broadly based paired fins of cladoselachians, and the embryology of musculature in the fins of sharks were cited in evidence. Some of this support must now be abandoned: The lateral folds of amphioxus are not considered homologous with true fins; cladoselachians are no longer regarded as close to the ancestral vertebrate stock. However, some anaspids may have had fin folds. Modern knowledge of fossil fishes, and recent experiments on the induction of accessory limbs in amphibian larvae, support the conclusion that appendages do always tend to form where the continuous fin folds were postulated to be. In truth, we still do not know just how appendages evolved. We hope for the discovery of fossils that will provide answers, but the complete story may be forever lost in antiquity.

MEDIAN FINS   The **dorsal fin**(s), located along the middorsal line; the **anal fin**(s), between anus and tail; and the **caudal fin** comprise the median fins. They occur in nearly all jawless vertebrates and fishes. (Larval amphibians, adults of many Urodela, and some amniotes that are highly specialized for aquatic life have secondarily evolved analogous structures.)

DORSAL AND ANAL FINS   Dorsal and anal fins function to prevent the body from yawing (turning around the vertical axis) and rolling (turning around the longitudinal axis).

The primitive condition probably was for each fin to be supported within the contour of the body by a series of rodlike **radials,** or **pterygiophores** (= fin + bearer of), arranged one per body segment (Figure 9-12). Commonly, the number is reduced (or increased) and segmental arrangement is lost. Each ptery-giophore is usually divided into two or more pieces. The proximal piece is often conspicuously larger than those more distal and is then called a **basal.** Ptery-giophores sometimes articulate with neural and hemal spines, but they are probably not derived from vertebrae.

The exposed membrane of the fin may originally have been supported only by dermal scales in the covering skin—those on the leading edge being larger than the rest. The fins of cephalaspids and some placoderms apparently were of this nature. The fins of more advanced fishes are supported internally by a series of slender **fin rays.** Fin rays of cartilaginous fishes are slender, unsegmented, and horny and are called **ceratotrichs** (= horn + hair). Those of bony fishes are slightly broader, segmented, and bony and are called **lepidotrichs** (= scale + hair). Higher teleosts have in the dorsal fin only six or fewer lepidotrichs which have enlarged and become rigid. The leading edge of one or more median fins of many fishes (pteraspids, acanthodians, most cartilaginous fishes) is stiff-

Fin rays = ceratotrichs

Pterygiophores

LEOPARD SHARK, *Triakis*

Fin rays = lepidotrichs

Pterygiophores

CARP, *Cyprinus*

Figure 9-12 DORSAL FIN SKELETONS OF A CARTILAGINOUS FISH (left) AND A BONY FISH (right).

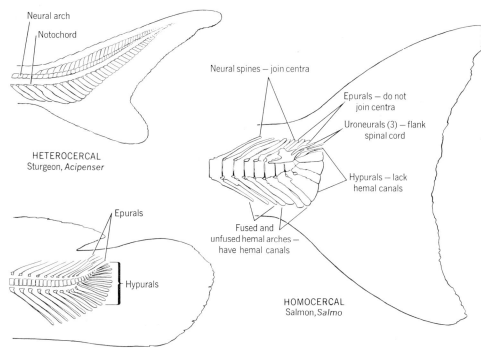

Figure 9-13 SHAPE AND STRUCTURE OF THE TAILS OF SOME BONY FISHES.

ened by a stout ray which serves as a cutwater and sometimes also (or instead) for defense or display. This ray may be an enlarged lepidotrich, but in cartilaginous fishes it is derived from one or more dermal denticles. Dorsal and/or anal fins may be long and continuous (some cyclostomes, pleuracanths, some teleosts), single (most ray-finned fishes), double (many selchians and sarcopterygians), multiple (*Polypterus*), or absent.

CAUDAL FIN   If the spine is straight to the tip of the tail, then dorsal and ventral lobes of the tail are of about equal size and the fin is said to be **diphycercal** (= double + tail). If the spine tilts upward and enters the dorsal lobe, then the dorsal lobe is longer than the ventral lobe, most of the fin membrane is ventral to the axis of the tail, and the tail is termed **heterocercal** (Figure 9-13). If the spine enters a larger ventral lobe, the tail is **hypocercal** (see anaspids in the figure on p. 44). If all the fin membrane is posterior to the spine, then dorsal and ventral lobes are about equal and the tail is said to be **homocercal.** Several intergrades and modifications are also recognized.

The ancestral tail may have been diphycercal, but the most primitive tails of record are heterocercal (cephalaspids, placoderms, most Chondrichthyes, the more primitive Osteichthyes of each subclass). Few vertebrates have hypocercal tails (anaspids). From the heterocercal tail there evolved the (secondarily?) diphycercal tail (cyclostomes, pleuracanths, later sarcopterygians, *Polypterus*) and homocercal tail (nearly all teleosts). These tail shapes can be identified in the illustrations for Chapter 3.

The caudal fin of ray-finned fishes, unlike the other median fins, is supported within its fleshy base by several modified neural arches and spines called **epurals** (= upon + tail), and more numerous modified hemal arches and spines called **hypurals** (= below + tail). These and related structures are further defined by Figure 9-13. The membrane of the fin is stiffened by fin rays corresponding in structure to those of the dorsal and anal fins of the same fish. Lepidotrichs of the caudal fin are usually branched.

STRUCTURE AND EVOLUTION OF GIRDLES

GIRDLES OF FISHES   The pectoral girdle is older, larger, and more complicated than the pelvic girdle. It includes one or more elements of cartilage or replacement bone and several dermal bones derived from ancestral scales or armor plates.

PLACODERMS illustrate the initial stages in the evolution of the pectoral girdle. A cartilaginous fin base was related to overlying plates of the dermal skeleton. The large pectoral girdle of CARTILAGINOUS FISHES is distinctive in two respects: Dermal elements are absent (presumably because of regression), and right and left halves of the girdle are fused in the midline. The result is a U-shaped girdle of one piece called the **scapulocoracoid.**

In BONY FISHES the replacement element, also termed scapulocoracoid, may ossify in one or several units. The dermal bones are identified in Figure 9-14. They join the girdle to the skull. This anchors the girdle in a manner that is not available to cartilaginous fishes (because they lack the requisite bones) and,

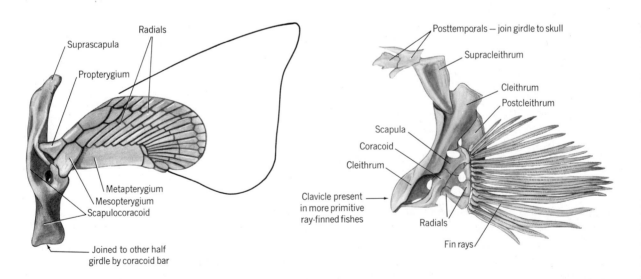

LEOPARD SHARK, *Triakis*

ROCKFISH, *Sebastodes*

Figure 9-14 LEFT PECTORAL GIRDLE AND FIN SKELETON OF AN ELASMO-BRANCH (left) AND A TELEOST (right).

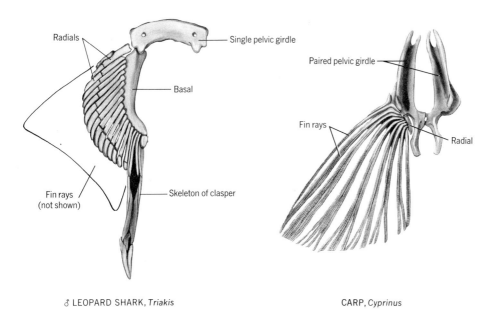

Radials

Single pelvic girdle

Basal

Paired pelvic girdle

Fin rays

Fin rays
(not shown)

Skeleton of clasper

Radial

♂ LEOPARD SHARK, *Triakis*   CARP, *Cyprinus*

Figure 9-15   PELVIC GIRDLE AND LEFT FIN SKELETON OF A CARTILAGINOUS FISH (left) AND A BONY FISH (right) seen in dorsal view.

hence, few bony fishes follow cartilaginous fishes in having the halves of the girdle joined in the midventral line. The **cleithrum** is the basic dermal element. The **clavicle** is lost in higher teleosts. The bones between the cleithrum and skull vary in number.

Placoderms had small pelvic fins at best, and a pelvic girdle has not been described. The pelvic fins of other fishes are weakly supported by a single skeletal element on each side of the body. They are bony except in cartilaginous fishes and dipnoans. The two pieces are usually separate but may overlap or articulate with one another, and in cartilaginous fishes are joined across the midventral line by a bridge of cartilage (Figure 9-15).

GIRDLES OF TETRAPODS   The **pectoral girdle** of PRIMITIVE AMPHIBIANS differed from that of fishes in that the replacement bones were larger and the dermal bones reduced (Figure 9-16). All bones dorsal to the cleithrum were lost (except in one transitional group of primitive labyrinthodonts); thus the contact with the skull was broken and the head was freed to turn on the evolving neck. Some dipnoans and crossopterygians had a small **interclavicle** which united the two half girdles in the midventral line. This bone enlarged in labyrinthodonts, probably to compensate for loss of anchorage of the girdle to the head. There were two replacement bones: a dorsal **scapula** and a ventral bone which for the moment I shall call the coracoid. Among MODERN AMPHIBIANS, Urodela have no membrane bones at all in this girdle, and Anura have no interclavicle and usually lack the cleithrum (Figure 9-11).

Clavicle (both
bones plus interclavicle
= the furcula)

Scapula has
a spine

Coracoid process —
derived from
posterior
coracoid

MAMMALS

BIRDS

Interclavicle
variable

(Cleithrum
present in
some forms)

Anterior
coracoid

Posterior
coracoid

LIZARDS

STEM REPTILES, SYNAPSIDS,
MONOTREMES

Suprascapular
cartilage

Cleithrum
(variable)

Cleithrum

Scapula

Glenoid
cavity

Clavicle

Anterior
coracoid

Interclavicle

LABYRINTHODONTS

ANURANS

Figure 9-16 PHYLOGENY OF THE TETRAPOD PECTORAL GIRDLE. Left lateral
views.

STEM REPTILES, SYNAPSIDS, and MONOTREMES are alike in having a full
complement of bones. Interclavicle and clavicle are present, and the cleithrum
is present (for the last time) in the more primitive reptiles. The scapula is large
and there are two coracoids. It is agreed that the single coracoid of amphibians
did not split; instead, a new bone was added behind the original coracoid. We
refer, therefore, to an **anterior coracoid** (or **precoracoid**) and a **posterior coracoid** (or simply **coracoid**). OTHER REPTILES, including turtles, lepidosaurs, and

archosaurs, usually lose the posterior coracoid and at least some of the membrane bones.

BIRDS have a bladelike scapula that is oriented parallel to the spine. The anterior coracoid is large and articulates firmly with the sternum. The posterior coracoid has been lost. The two clavicles fuse ventrally to form the wishbone or **furcula.** The interclavicle may be incorporated into the furcula of some birds, but probably is often absent.

When only one coracoid is present (labyrinthodonts, turtles, crocodilians, lizards, birds), it is usual to refer to it as *the* coracoid without qualification. We have seen, however, that with respect to tetrapod phylogeny it is the anterior coracoid.

The clavicle is the only membrane bone retained by MAMMALS, and even this bone may be missing. It is the *anterior* coracoid that is completely lost this time. The posterior coracoid ossifies independently in the fetus and then fuses to the scapula to form the **coracoid process** of that bone. The scapula is unique in having a **spine.** The spine represents the anterior border of the ancestral bone, so in fact it is the anterior border of the mammalian scapula that is new. The ventral end of the spine is continued as the **acromion process** to articulate with the clavicle.

The **pelvic girdle** of tetrapods is much enlarged over that of fishes and is relatively uniform in basic structure (Figure 9-17). Each half of the girdle is a single cartilaginous unit in the embryo, but three bones are constant in the adult. These are a dorsal **ilium** which articulates with one or more sacral vertebrae, an anterior **pubis,** and a posterior **ischium.** The bones of one side commonly fuse in the adult to form the **innominate bone.** One or both of the ventral bones of the two sides usually articulate or fuse across the midventral line; the contact is called the **pelvic symphysis.**

Primitive AMPHIBIANS had a solid girdle shaped like a triangle with the ilium forming the apex. The pubis can be distinguished from the ischium by having a foramen (obturator foramen) which accommodates a nerve. The atypical girdle of frogs has a long, anteriorly inclined ilium. The pubis of modern amphibians is cartilaginous.

The girdle of REPTILES takes various shapes but is like that of labyrinthodonts in basic plan. The contact with the spine is firmer. A large pubo-ischiadic fenestrum is usually present between the two ventral bones.

The pelvic girdle of BIRDS is distinctive. It is large and firmly attached to the synsacrum. The long ilium extends both anterior and posterior to the socket for the femur, or **acetabulum.** The pubis is turned backward below the ischium. There is no symphysis.

MAMMALS have a long and expanded ilium which extends only forward from the acetabulum. The large obturator fenestrum represents both the obturator foramen and the pubo-ischiadic fenestrum of the ancestor. A symphysis is nearly always present. Monotremes and marsupials have **epipubic bones** of doubtful origin that articulate with the pubic bones and extend forward in the ventral body wall.

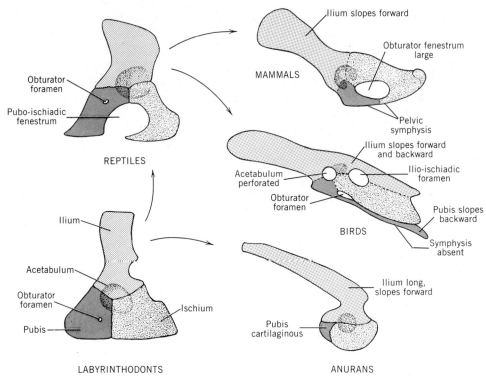

Figure 9-17 PHYLOGENY OF THE TETRAPOD PELVIC GIRDLE. Left lateral views.

PAIRED FINS The function of paired fins is usually to prevent the body from pitching (turning around the transverse axis) and rolling, and (particularly in higher fishes) to brake forward motion. The function and position of fins is discussed further in Chapter 26.

Among AGNATHOUS VERTEBRATES, cyclostomes have no trace of paired appendages, and the same was true of pteraspids. Anaspids had lateral folds or pectoral spikes of questionable relationship with fins (see figure on p. 44). The spikes were of dermal origin, superficial, and not motile. Cephalaspids had pectoral lobes behind the lateral wings of the cephalic armor. They were muscular, but seem to have had no internal skeleton.

PLACODERMS and ACANTHODIANS experimented with newly acquired paired appendages that varied from stiff fins (arthrodires) to hinged arms (antiarchs) and multiple spines (acanthodians) (see illustrations in Chapter 3). A complex dermal skeleton was sometimes supplemented by a cartilaginous internal skeleton.

At this point in the evolution of fins, the relative importance of the dermal and internal skeleton was reversed, the latter dominating the former. The ancestral internal skeleton probably consisted of a series of parallel radials. These were probably segmented, with relatively heavy basals. Paired fins of CHONDRI-CHTHYES (except pleuracanths) depart little from this plan, although the detailed pattern varies widely (Figures 9-14, 9-15). (If there are three basals, they are termed pro-, meso-, and metapterygia.) The distal part of the fin is supported by ceratotrichs.

The ACTINOPTERYGII are named according to fin structure; **actinoptery-gium** = ray + fin. There is a proximal row of bony radials and a distal series of lepidotrichs which, like those of the caudal fin, are branched and segmented.

SARCOPTERYGII are also named according to fin structure; **sarcopterygium** (= flesh + fin) refers to the fleshy nature of the bases of these fins. The ancestral fin skeleton is modified—probably by reorienting the row of basals—to establish an axis. There are two principal kinds of sarcopterygium. In dipnoans and pleuracanths (the latter are not classified as Sarcopterygia but have similar fins), the radials are biserial; there is a series of radials on each side of a median axis. This fin was once thought to be the most primitive and accordingly was named an **archipterygium** (Figure 9-18). The term remains but is no longer apt. In Crossopterygii the radials are uniserial; there is a series on one side of the axis. This fin, called a **crossopterygium** (= fringe + fin), was unquestionably ancestral to the tetrapod limb.

ORIGIN OF LIMBS The transformation from fin to limb is another subject that has been long debated. Early theories were that fishes evolved in times of drought and used their limbs either to crawl between pools of water or to burrow in mud. It is now reasonably certain that tetrapods evolved in humid areas where water was abundant. They learned to "walk" on the bottom with strong lobe-shaped fins—as some modern fishes do—and occasionally crawled in shallow water and onto damp shores to escape aquatic enemies or find terrestrial food.

There is no question that limbs evolved from a crossopterygium, and there is general agreement about the homologies between the proximal segments of the fin axis and the proximal limb bones. Opinion differs, however, on which digit represents the continuation of the fin axis (the fourth or fifth digit is favored), the number of segments of the ancestral axis that have been retained (probably four or five), which fin radials became what foot bones, and (for the pectoral appendage) which way the fin twisted to assume the tetrapod stance. The transition shown in Figure 9-18 is an educated guess.

The paired fins of fishes always have about the same functions and vary in basic structure according to class or subclass. The pectoral fins are the stronger and more firmly related to the axial skeleton. The limbs of tetrapods, by contrast, have a variety of functions but retain a unity of basic structure. The pelvic limbs are the stronger and more firmly related to the axial skeleton. The basic skeletal structure of limbs is indicated in Figure 9-18. (The terminology shown is used worldwide by comparative anatomists, but other terms are used by human anatomists and mammalogists to identify the particular variations of the foot bones of their materials.) The bones of the wrist comprise the **carpus;** those of the ankle comprise the **tarsus.** Carpal and tarsal bones are known collectively as **podials.**

Humerus (Femur)

Ulna (Fibula)
Radius (Tibia)
Radiale (Tibiale)
Intermedium
Ulnare (Fibulare)
Series of centralia
Series of distal carpals (Tarsals)
Series of metacarpals
    (Metatarsals)
Phalanges

1

2   5

3   4

STYLIZED LIMB OF PRIMITIVE TETRAPOD

TRANSITIONAL STAGE — HYPOTHETICAL

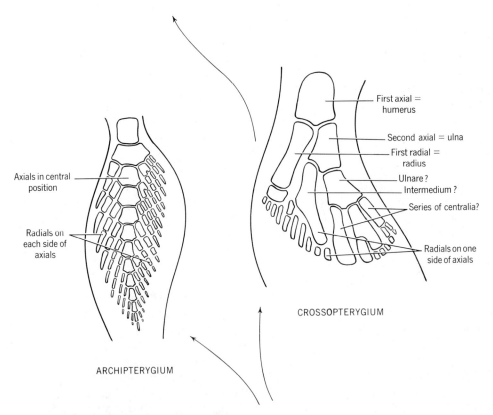

Axials in central
position

Radials on
each side of
axials

ARCHIPTERYGIUM

First axial =
    humerus

Second axial = ulna
First radial =
    radius

Ulnare ?
Intermedium ?

Series of centralia?

Radials on one
side of axials

CROSSOPTERYGIUM

Figure 9-18 PHYLOGENY OF THE PECTORAL APPENDAGE OF SARCOPTERYGII AND TETRAPODS. Left appendage in dorsal view. Corresponding terminology for the hind leg of a primitive tetrapod shown in parentheses.

Of the podials, the centralia are least constant. The forefoot is called the **manus** and the hind foot the **pes.** Metacarpal and metatarsal bones are known collectively as **metapodials.** The skeletal patterns of the various tetrapod feet are derived from the primitive pattern by deletions and fusions which can usually be verified by embryonic development. The derivations of some patterns representative of major taxa are shown in Figures 9-21 and 9-22.

STRUCTURE, DEVELOPMENT, AND GROWTH OF LONG BONES   The larger bones of tetrapod appendages are called **long bones.** A typical long bone has a cylindrical shaft called the **diaphysis** (= through + growth) which contains the marrow cavity, and at each end an enlargement called the **epiphysis** which articulates with adjacent bones. Epiphyses may be cartilaginous or bony. If bony, they are spongy within. Bony epiphyses usually fuse with their respective diaphyses at maturity.

The first embryonic primordium of a long bone is a condensation of mesenchyme. This then forms a one-piece, cartilaginous model of the future bone. Ossification of the diaphysis begins as a thin veneer of intramembranous bone deposited around the center of the model by the limiting membrane, or **periosteum** (Figure 9-19). Spicules of bone penetrate the cartilage and replace the matrix, which is gradually destroyed. Soon the incipient marrow cavity is sur-

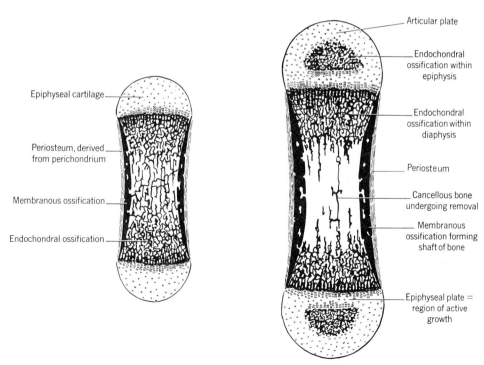

Figure 9-19 HISTOGENESIS OF A MAMMALIAN LONG BONE at two developmental stages as seen in longitudinal section.

rounded by solid lamellae of bone arranged like the layers of orthodentine. The smallest mammals may retain this structure, but usually longitudinal canals are eroded within the developing bone tissue and the cavities reossify with the cylinder-within-cylinder construction of osteons. Only at the surface of the bone is lamellar structure retained into full maturity (see figure on p. 93). Epiphyses, by contrast, ossify (if at all) from one or more internal centers which expand outward to replace the cartilage. Their contained spicules are largely lamellar. The mature bone is largely endochondral, but partly dermal in origin.

Bones grow by a complicated and wonderful process which is best known for reptiles and mammals. The diaphysis increases in length only at the cartilaginous plates that separate it from its epiphyses. If (as in mammals) these plates are obliterated at maturity, an increase in length stops. Growth in the transverse diameter of the shaft is accomplished by the deposition of bone at its outer surface and erosion of its inner surface to enlarge the marrow cavity. Some tendonous insertions must migrate to maintain their proportionate positions as a bone grows. The periosteum stretches like a sheet of rubber as the bone enlarges within; it must slip over the surface of the bone since the latter grows without stretching. Some further points are shown in Figure 9-20.

The configuration of mature bones is greatly influenced by mechanical interaction with the developing muscles. In the absence of normal muscles and muscle activity the skeleton is abnormal.

EVOLUTION OF LIMBS   AMPHIBIANS (other than the legless caecilians) usually have short limbs splayed to the sides of the body. The trunk is usually lifted from

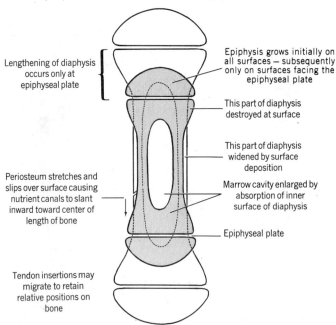

Lengthening of diaphysis occurs only at epiphyseal plate

Epiphysis grows initially on all surfaces — subsequently only on surfaces facing the epiphyseal plate

This part of diaphysis destroyed at surface

This part of diaphysis widened by surface deposition

Periosteum stretches and slips over surface causing nutrient canals to slant inward toward center of length of bone

Marrow cavity enlarged by absorption of inner surface of diaphysis

Epiphyseal plate

Tendon insertions may migrate to retain relative positions on bone

Figure 9-20 MORPHOGENESIS OF A MAMMALIAN LONG BONE between two developmental stages as seen in longitudinal section.

Figure 9-21 LEFT CARPUS AND METACARPUS in dorsal view. Distal carpals shown by light shading, proximal carpals by dark shading, metacarpals, centralia, and pisiform by no shading. There is doubt regarding the homologies for birds.

the ground when the animal walks, but only with difficulty. In urodeles undulations of the spine may be used to twist the girdles, thus helping to advance the limbs. Epiphyses are of hyaline cartilage and fit like corks into the ends of the bony shafts (most amphibians) or are calcified and fit over the ends of the shafts like match heads (Anura). The podials of modern amphibians are often cartilaginous. The principal joint of the foot is between the podials and metapodials. There are only four digits on the hand and four or five on the foot. One to three phalanges are present in each toe. The marrow cavities of the long bones of

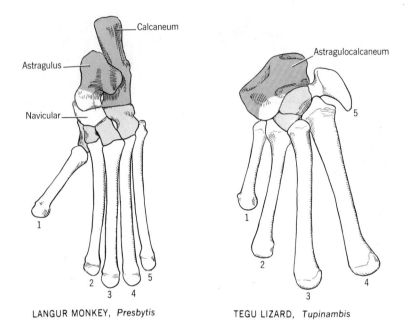

LANGUR MONKEY, *Presbytis*

TEGU LIZARD, *Tupinambis*

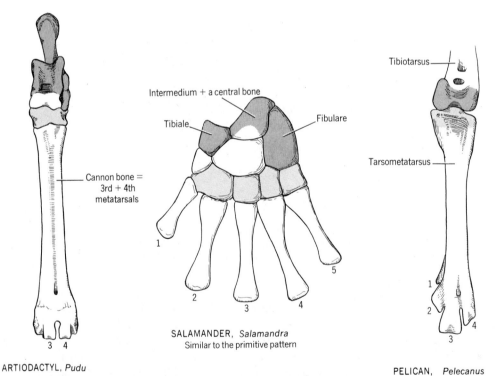

SALAMANDER, *Salamandra*
Similar to the primitive pattern

ARTIODACTYL, *Pudu*

PELICAN, *Pelecanus*

Figure 9-22   LEFT TARSUS AND METATARSUS in dorsal view. Distal tarsals shown by light shading, proximal tarsals by dark shading, metatarsals and centralia by no shading.

amphibians and higher vertebrates produce blood cells—a function not performed by the skeleton of fishes.

Many REPTILES still have the limbs positioned far to the sides of the body, but some dinosaurs and mammal-like reptiles placed their feet well under the body. The limbs are usually stronger than for amphibians, and the hind limbs are often disproportionately large. Epiphyses are usually cartilaginous but may ossify in lizards. An "extra" bone, the **pisiform,** may be added to the outside of the carpus, and the tibiale is no longer a free bone in the tarsus. The joint of the foot is often between podials. The **phalangeal formula** (number of phalanges per digit starting with digit 1) is 2·3·4·5·3 for the manus and 2·3·4·5·4 for the pes, if segments and digits have not been lost through specialization or degeneration.

The limb structure of BIRDS is uniform and specialized, as shown in Figures 9-21 and 9-22. Epiphyses are cartilaginous in immatures and virtually absent in adults. The principal digit of the wing is customarily identified as number 2, although there is embryological evidence that it is number 3, and the question is not yet resolved. The phalangeal formula of the foot is 2·3·4·5·0.

MAMMALS have bony epiphyses on each end of the long bones, on the distal ends of the metapodials, and on the proximal ends of all but the terminal phalanges. The pisiform is retained. In the tarsus the fibulare forms the heel bone, or **calcaneum** (Figure 9-22). The tibiale joins the intermedium, and the resultant large bone, called the **astragulus,** lies partly over the calcaneum. The ankle joint is between the astragulus and tibia. The ancestral articulation between fibula and

Figure 9-23 LIMB SKELETON OF A MATAMATA TURTLE, *Chelys.* Foreleg to left; hind leg to right. Dorsal views of left legs.

Scapula

Dorsal border

Supraspinous fossa

Posterior angle

Infraspinous fossa

Posterior border

Spine

Acromion

Coracoid process

Glenoid cavity

Greater tubercle

Head

Deltoid tuberosity

Humerus

Olecranon fossa

Lateral epicondyle

Condyle

Olecranon

Trochlear notch

Radial notch

Attachment of ligament
joining radius and ulna

Radius

Ulna

Grooves for tendons
of extensors and abductors

Styloid process of radius

Styloid process of ulna —
pivots on carpus during
pronation and supination

Supraspinatus —
extends shoulder

Rhomboideus

Infraspinatus — rotates
and abducts arm

Teres major —
flexes shoulder

Trapezius and spinodeltoid

Acromiodeltoid

Biceps — flexes elbow

Triceps (part) — extends elbow

Infraspinatus

Supraspinatus

Triceps (part)

Brachialis — flexes elbow

Deltoids — flex shoulder

Anconeus — extends elbow

Radial carpal extensor

Extensors of carpus
and digits

Anconeus

Triceps

Supinator

Figure 9-24   FEATURES AND FUNCTIONS OF THE FORELEG SKELETON OF THE
DOG. Lateral view of left leg. Not all muscle attachments are shown.

calcaneum is reduced or lost. Fusions among the tarsals are common. The basic
phalangeal formula is 2·3·3·3·3. Features of the appendicular skeleton are identified in Figures 9-24 and 9-25, and are related to articulations and muscles. The
olecranon process of the ulna is particularly characteristic of mammals.

Sartorius — flexes hip and extends knee
Tensor fasciae latae — flexes hip and extends knee
Middle (above) and deep gluteals — extend hip
Rectus femoris — extends knee
Rectus abdominis — compresses abdomen
Biceps femoris — extends hip, knee, and ankle
Semitendinosus — extends hip and ankle
Semimembranosus — extends hip
Quadratus femoris — extends and rotates hip
External obturator — rotates leg outward
Adductors
Gluteals
Vastus lateralis — extends knee
Adductors
Gastrocnemius — extends ankle, flexes knee
Long extensor of digits
Popliteus — flexes knee
Patellar ligament = insertion of extensors of knee
Extensors of ankle and flexors of digits
Tibialis anterior — rotates and flexes foot
Short fibularis — extends ankle

Crest
Innominate bone
Tuberosity
Acetabulum
Attachment of ligament joining femur and pelvis
Obturator fenestrum
Pelvic symphysis
Head
Greater trochanter
Femur
Facet for sesamoid in gastrocnemius
Patellar groove
Lateral condyle
Groove for tendon of long extensor of digits
Crest
Facet for head of fibula
Fibula
Tibia
Lateral malleolus
Medial malleolus

Figure 9-25   FEATURES AND FUNCTIONS OF THE HIND LEG SKELETON OF THE DOG. Lateral view of left leg. Not all muscle attachments are shown.

Nodules of bone tend to form where tendons play over joints. These are called **sesamoid bones** ("sesamoid" = having the shape of a seed). The largest sesamoid is the **patella,** or kneecap. The central tendons of certain kinds of muscles tend to ossify—particularly in birds (they are encountered in a turkey drumstick).

**SOME
MISCELLANEOUS
BONES**

A bone called the **baculum** (or os penis) is present in the penis of carnivores, bats, insectivores, rodents, colugos and some primates (Figure 9-26). Its size and shape vary widely among species and it is therefore useful in systematics. A corresponding but much smaller bone may be present in the female clitoris. Additional small bones are found here and there among tetrapods: in the eyelids of crocodilians, in the crest of a bird, in the snout of pigs, at the base of the external ear of some rodents, at the base of the aortic arch of some artiodactyls, and so on. These are of more functional than evolutionary significance.

Figure 9-26   BACULUM of the maned wolf, *Chrysocyon,* in right lateral and ventral views.

10

Muscles and Electric Organs

The muscular system (studied in conjunction with the skeleton) is of primary importance in the analysis of locomotor mechanisms, athletic performance, and procedures in physical therapy. It is of secondary importance for establishing phylogenetic relationships. Homologies are nearly always evident among the muscles of related orders and lower taxa. Homologies often can be established among related classes, but there are exceptions, and the muscular system is not often useful for verifying phylogenies determined on the basis of other organ systems.

Techniques used in the study of muscles include dissection, observation of relationships with bones, nerves, and other muscles, high speed and X-ray cine-photography, the use of force plates, strain gauges, and accelerometers, mechanical analysis (see Chapter 22), electomyography, and a range of physiological and biochemical methods.

FUNCTION AND GROSS STRUCTURE OF MUSCLES

Nearly all functions of the body are in part muscular. Without muscles, vertebrates could not move, their tissues soon would be starved or poisoned, and the products of their glands could not be distributed. Humans could not read, speak, or write to communicate their thoughts.

Muscles accomplish all this by doing only one thing—by creating tension along the axis of their fibers, which tends to shorten their substance. Active muscles usually do shorten, thus moving a bone or constricting a space. They may also prevent motion by opposing gravity or the pull of other muscles. With little or no motion, they may function to cause a part to become more rigid. Furthermore, they may offer controlled resistance to extrinsic forces that tend to stretch them out.

The muscular system also has secondary functions: It contributes importantly to the maintenance of the body temperature of homeotherms and, because of its bulk, distributes the weight of the body, influences the contours of the body, and offers protection to some of the viscera.

There is much variation in the shapes of muscles. Since each muscle fiber contracts only along its length, it is usually most effective for the fibers of one muscle to lie approximately parallel, thus establishing a longitudinal axis for the muscle as a whole. A *cylindrical* or *straplike form* might be taken, therefore, as a fundamental shape. Several muscles of the throat are usually of this nature. If one end of a muscle is tapered to insert on the skeleton, the shape of the muscle becomes that of a *teardrop*. Several muscles of the hip, thigh, and upper arm are commonly of this sort. Most muscles of the limbs are tapered at each end and hence are *spindle-shaped*. One end may be divided into two or more parts (or "heads") and the fleshy part of the muscle is seldom symmetrical because it must fit against other muscles. Some muscles are spread out as *sheets* (most abdominal muscles), and others are flat at one end and gathered at the other, thus becoming *fan-shaped* (various shoulder muscles). Muscles that surround orifices have curved fibers and are *washer-shaped;* muscles that surround spaces (stomach, uterus) have fibers oriented in various directions and are more or less *hollow spheres*.

Some spindle-shaped muscles (particularly of birds) do not have their fibers oriented in the long axis of the muscle but instead have fibers that slope inward to insert on a central tendon. In longitudinal section such muscles may look like feathers and hence are called **pinnate.** As explained on pages 413 and 414, pinnation increases force of contraction but reduces shortening. Some muscles have several converging central tendons and are multipinnate (see the subscapularis muscle in the figure on p. 475).

Where muscles attach directly to the skeleton, the connective tissue that surrounds and pervades them is continuous with connective tissue surrounding the bones. Where muscles do not impinge directly on the skeleton, they are joined to bones by **tendons,** which are tough cords of closely packed, parallel, collagenous fibers. (**Ligaments** join bone to bone. Their collagenous fibers are somewhat less regular and they include elastic fibers.) Some muscles (e.g., several abdominal muscles of mammals) do not attach to bones but instead distribute their forces over broad areas by means of strong flat sheets of connective tissue. These sheets are called **aponeuroses.** The loose connective tissue that binds muscle to muscle, and skin to muscle, is called **fascia.**

When a muscle contracts, it pulls equally on each end. Commonly, one attachment is relatively free to move and is then called the **insertion.** The relatively fixed attachment is the **origin.** These terms must be used with caution, however; tables of muscle origins and insertions can be misleading. Which end of a muscle is the more movable depends on posture, the activity of other muscles, and contacts with the environment. When one lifts an object from a table, the proximal end of the biceps is its origin; when one chins oneself on a bar, the distal end is the origin. During the propulsive phase of a limb's motion, its muscles move the body on the limb, not the limb on the body; contrary to usual terminology, it is then the distal ends of the muscles that are relatively fixed. Often neither end of a muscle is fixed, and sometimes neither moves.

It is necessary to have a vocabulary for describing the actions of muscles. **Flexors** are commonly defined as muscles that reduce the angle between adjacent bones; **extensors** increase the angle. These definitions usually serve but can be misleading. Considerations of position, phylogenetic origin, and innervation (as explained below) make it clear that flexors of a hind limb swing joints to the rear, whereas extensors swing them forward. At the hip, such muscles may increase or decrease the angle between adjacent bones, depending on posture. Also, since the elbow and knee bend in opposite directions, the situation is reversed for the forearm, flexion bends the elbow forward, not to the rear. Furthermore, some muscles of the back that straighten (hence extend) the spine from a hunched position also, on further contraction, arch (hence flex) the spine. If confusion threatens, it is well to avoid one-word designations of muscle function. **Adductors** move parts inward toward the sagittal plane of the body or axis of a limb; **abductors** move parts outward away from the body or axis. Opening and closing the fingers and clapping the hands use these sets of muscles alternately. **Levators** raise and **depressors** lower such parts as the jaw or shoulders. Paired fins may alternatively be said to be depressed or adducted. **Protractors** push a part (such as the tongue or an entire limb) away from its base and **retractors** draw it back. **Sphincters** constrict openings (mouth, duct orifices) and **constrictors** compress spaces (pharynx, abdomen); they are opposed by **dilators. Rotators** turn parts about their long axes (spine, limbs). The rotators that turn the soles of the hands or feet upward are **supinators;** those that turn them downward are **pronators.**

For every action there is an opposing or restoring action. Opposing muscles are called **antagonists.** Rarely does one muscle contract alone. Muscles that supplement one another are called **synergists.**

Muscles are given names that describe their actions (levator maxillae, flexor digitorum, adductor mandibulae), shapes (biceps, rhomboideus, trapezius), positions (temporalis, pectoralis, gluteus), or attachments, the origin being named before the insertion (geniohyoid, sternomastoid, cleidobrachialis). Most muscles were first named for the human; the terms are not always apt when applied to homologs in other animals. Nevertheless, it is easier to remember the terms when their derivations are understood.

Three types of muscle are distinguished by differences in histology and physiology (Figure 10-1). **Smooth muscle** is found in the skin, blood vessels, urogenital system, respiratory channels, and alimentary tract and the ducts of its derivatives. Its spindle-shaped, uninucleated cells lack striations. They have filaments that are oriented obliquely to their long axes and insert on their walls. **Cardiac muscle** is restricted to the heart. Its striated fibers are branched and divided into nucleated

FINE STRUCTURE, CONTRACTION, AND PHYSIOLOGY

SMOOTH	SKELETAL	CARDIAC
Not striated	Striated	Striated
Spindle-shaped	Cylindrical	Cylindrical
Not branched	Not branched	Branched
Nucleus central	Nuclei peripheral	Nuclei central
No disks	No disks	Intercalated disks
Relatively involuntary	Voluntary	Relatively involuntary
Slow	Fast	Fast
Contraction not inherent	Contraction not inherent	Contraction inherent

Figure 10-1   THE THREE TYPES OF MUSCLE TISSUE.

units by spaced intercalated disks. Like smooth muscle, it is relatively involuntary, although some yogis, and some persons using biofeedback apparatus, have gained limited control of such muscles.

The remaining muscles of the body are **skeletal muscles,** and it is they that concern us in this chapter. Each skeletal muscle is surrounded by a tough envelope of connective tissue, the **epimysium** (= upon + muscle), which is continuous with tendons and with such fascia as may be present (Figure 10-2). The epimysium also merges with septa, collectively called the **perimysium,** which penetrate the muscle and divide it into bundles of fibers. Such bundles may be distinct, making the muscle stringy, as for the rhomboideus of many mammals, or may be virtually absent. The perimysium, in turn, is continuous with a net of connective tissue, the **endomysium,** which surrounds the sarcolemma or limiting membrane of the individual muscle fibers. This continuous system of connective tissue gives muscles their shape and strength, and transmits their forces to origin and insertion.

If a fragment of muscle is macerated and pulled apart with fine needles, the hairlike **muscle fibers** can be seen. Although scarcely visible to the naked eye, they may be as long as the whole muscle. They have multiple, peripheral nuclei and are cross-banded (hence the term ''striated'' muscle). The muscle fibers are penetrated by transverse tubules which open to the outside and by tiny anastomosing canals called the **sarcoplasmic reticulum.** These function in both transport and transmission.

When muscle fibers are stained and viewed with a light microscope it is seen that each is made up of dozens of finer strands, called **myofibrils,** each of which is about 0.001 mm in diameter. When enlarged some 400,000 diameters by the electron microscope, it is found that each myofibril is made of many filaments of two kinds.

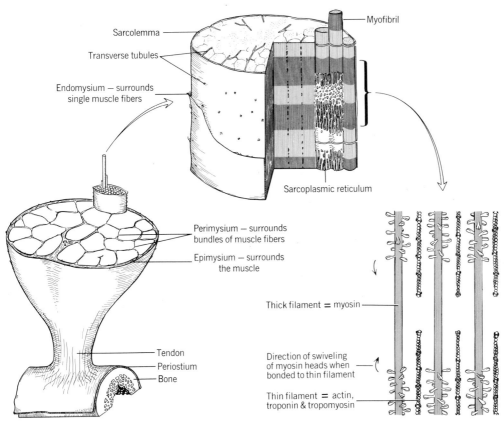

Figure 10-2   STRUCTURE OF SKELETAL MUSCLE. Nuclei and mitochondria omitted.

Thick filaments are about 0.01 μm in diameter. They are composed of a protein called **myosin.** The molecules of myosin have long slender tails arranged in cylindrical bundles, thus establishing the axes of the filaments. The short, globular, and bifid heads of the myosin molecules project outward at various angles from the filament axes. The various molecules of each filament are oriented in opposite directions. Hence, a filament has heads projecting at each end, but is smooth in the middle (Figure 10-2).

Thin filaments have half the diameter of thick filaments and consist of the proteins **actin, tropomyosin,** and **troponin.** Actin molecules are small and globular, and in each filament are arranged like beads in double strands that twist about one another. Tropomyosin molecules are long and thin and lie on the surface of the actin strands. Troponin molecules are small and are spaced along the tropomyosin molecules.

Contraction occurs when myosin heads bond to actin and then swivel so as to draw the thick and thin filaments past one another. Tropomyosin and troponin control the active site on the actin molecule. Normally tropomyosin blocks this site. During activation of a fibril the action of troponin induces a change in the orientation of tropomyosin such that the actin binding site is exposed and myosin is able to react with it, thus forming a cross-bridge. The energy source is adenosine triphosphate. Bonding takes place only in the presence of calcium ion released by the sarcoplasmic reticulum following nervous stimulation. The bonds release when these ions are pumped back into the reticulum. The overlap of thick and thin filaments is responsible for the striations of skeletal muscle; the banding changes during contraction as the overlap of the filaments increases.

Each muscle fiber contracts completely or not at all. Usually, several muscle fibers are innervated by a single nerve fiber, and together they form a **motor unit.** Graded response by an entire muscle results from the recruitment of varying numbers of motor units. The tone, or **tonus** of resting muscles is caused by the contraction of a minimal number of motor units.

A single adequate stimulus causes a skeletal muscle to **twitch,** or contract briefly followed by relaxation. If a second stimulus follows before relaxation is complete, there is a **summation** of tension (Figure 10-3). Repeated stimuli in rapid succession cause steady, maximum tension, or **tetanus.**

Contraction is said to be **isometric** if contraction causes tension but shortening of the muscle is prevented. It is **isotonic** if tension does not increase because the muscle shortens with constant resistance. In the living animal most muscles shorten little on contraction. Tension is maximum when the muscle is stretched somewhat beyond resting length before contraction starts (Figure 10-4). The velocity of contraction falls off sharply as load increases.

Striated muscle fibers are classed as red, white, or intermediate according to the amount of myoglobin they contain. **Red fibers** are responsible for the dark meat of poultry. Compared to white fibers, red fibers have more myoglobin, store

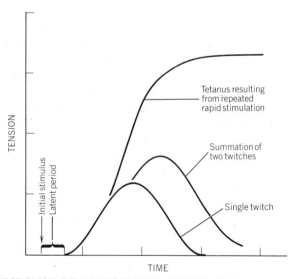

Figure 10-3   RESPONSE OF A MUSCLE TO STIMULATION.

Figure 10-4 LENGTH-TENSION CURVE (left) AND LOAD-VELOCITY CURVE (right) FOR SKELETAL MUSCLE.

more oxygen and lipids but less glycogen, have a richer blood supply, are smaller in diameter, tend to contract slower, and are slower to fatigue. Red fibers tend to be abundant in postural muscles (birds) and muscles used for constant but moderate activity (slow-swimming muscles of some fishes). **White fibers** have the opposite characteristics and are abundant in muscles used in bursts of activity (flight muscles of pheasants). Most muscles have a combination of red, white, and intermediate fibers.

Force (or tension) of muscular contraction is a property of great importance in the analysis of bone-muscle systems. It is commonly stated that the maximum force a muscle can exert is equal to the force of contraction of one of its fibers times the number of fibers, and therefore is proportional to the cross-sectional area of the muscle. This is only an approximation. It assumes that all muscle fibers are parallel to each other and to the axis of the tendon of insertion, that all fibers are of the same size, and that all have equal and constant force of contraction.

The first condition, all fibers parallel, is approximated by some straplike muscles that do not taper at origin or insertion (coracomandibularis of shark and thyrohyoid of cat). The fibers of most muscles gather somewhat at the ends of the muscles where they angle in toward the emerging tendons. The strength of a muscle that is partly or entirely pinnate (see p. 178) is roughly proportional not to a level cross section of the muscle as a whole but to a section that cuts all fibers at right angles, regardless of their orientation. Such a cut is easy to conceive, but difficult to accomplish.

The second condition, all fibers of the same size, is rarely met. Most muscles contain red, white, and intermediate fibers, and the proportions vary among different muscles of the same animal and between the same muscles of different animals. White fibers have about twice the diameter of red fibers and hence are less closely packed. Each type of fiber tends to be larger in large animals.

FORCE, WORK, AND TORQUE

The third condition, equal and constant force of contraction by all fibers, never pertains. We have just seen that tension is proportional to length at the time of stimulation. Furthermore, not all fibers of a muscle contract simultaneously, even in a maximum effort. The motor units take turns. The proportion of all units that contract in a maximum effort is subject to alteration by conditioning and physiological factors.

As a consequence of these variables it is never possible to determine accurately from nonliving material the actual or relative strengths of muscles. Moreover, cross-sectional areas of muscles are difficult to measure and are influenced by degree of stretching and method of preservation. Morphologists are turning more and more to experimentation with living material. However, it is also difficult to measure forces in live bone-muscle systems under conditions that closely resemble normal behavior. Often the investigator is obliged to settle for approximations of actual circumstances. This need not cause dismay. The living body is so complex, adaptable, and variable that when the researcher can learn how the mechanism works and the order of magnitude of the parameters involved much has been learned.

Force is the product of mass times acceleration. It is expressed in dynes (the push needed to cause 1 g to accelerate at 1 cm/sec^2) or newtons (the push needed to cause 1 kg to accelerate at 1 m/sec^2). However, since morphologists (and many engineers) usually work with a uniform gravitational field, it is adequate to consider force to be a push or pull that causes motion (or must be resisted to prevent motion) and to express force in kilograms or pounds.

Returning, then, to strength, a striated muscle can deliver to its tendon a force of about 3 kg/cm^2 (42 lb/in.^{2}), but ranging from 1 to 8 kg/cm^2 of its cross-sectional area taken at right angles to the fibers, but let us remember that this is an approximation.

With this relationship in mind, contrast a mouse and a hypothetical animal shaped like a mouse but having the size of an elephant. In each instance, the load on the muscles is proportional to the mass of the body, which is proportional to the cube of any linear dimension. Thus, as body length increases x times (from ordinary mouse to giant mouse), body weight increases x^3 times. But the forces the muscles can exert are roughly proportional to areas and will thus increase only x^2 times. If body size were significantly increased without altering body proportions, the load on the locomotor system would increase faster than its capacity to provide support. This important principle is basic to some interpretations of the structure of vertebrate giants made on pages 426 to 429.

The **contraction distance,** or amount of shortening possible for a striated muscle having approximately parallel fibers may be as much as 30% of its resting length, or nearly 60% of its stretched length. The longer a muscle, the greater the contraction distance. Some muscles are maximally stretched and contracted (e.g., muscles of distensible tongues), but since muscles are physiologically most effective at intermediate lengths, leverages and behavior are usually such as to restrict the shortening of muscles well within their theoretical maxima.

Work is force times the distance through which it acts. Since, in living systems, loads usually change as structures move, it is difficult to measure work accurately. To the extent that force is roughly proportional to the cross-sectional area of a muscle, work is roughly proportional to its volume (or mass). If a muscle contracts isometrically, it does no work although it does expend energy. The concept of force times time is then more useful than that of work for estimating nutritional requirements and onset of fatigue.

The **lever arm** of a muscle is the perpendicular distance from its line of action to the pivot of the motion caused by its contraction. Thus, if a muscle inserts on a bony lever (e.g., the heel bone) at right angles, then the distance from the insertion to the pivot (the ankle joint in some circumstances) is the lever arm of the muscle. The turning force, or **torque** of the action is then the force delivered to the insertion times the length of the lever arm. The leverages provided by the bone-muscle system are of the utmost importance in determining function. These, and related concepts are developed further in Chapter 22.

It is both convenient and instructive to arrange the many muscles into groups for study. Several methods of grouping suggest themselves, and the student can use each with benefit. First, all muscles of one region of the body can be studied together. Thus, the muscles of the spine, forelimb, head and neck, etc., can be studied in turn. What muscles would be seen in a cross section of the thigh? What muscles would be cut in passing from the breast to the lung? What muscles have origin or insertion on the scapula? This approach is efficient at the operation table and dissection table, particularly with large animals.

A second method groups together muscles of like function. What muscles extend (protract) the forelimb? Or turn the head in a specified way? Or maintain standing posture? One discovers that a single muscle (e.g., pectoralis) can have several actions, and that some actions (e.g., swinging the thigh to the rear) can be accomplished alternatively by two or more muscles (though not with identical efficiency). This approach is practical for the functional morphologist, behaviorist, and physical therapist.

A third method is of significance for comparative studies. The various muscles can be arranged in major categories on the basis of embryonic origin. Such categories have somewhat independent phylogenies, relate to the positional and functional groups noted above, and can be distinguished by nervous innervation. Let us identify them by tracing their origins and innervations.

The mesoderm of the early embryo is differentiated into a dorsolateral, segmented epimere, a small mesomere, and a ventrolateral, unsegmented hypomere (see figure on p. 80). As has been explained in previous chapters, the epimere further differentiates into dermatome, myotome, and sclerotome. The sclerotome forms no muscles. The dermatome forms much of the skin, including any intrinsic smooth muscles which may be present there. The myotome and the hypomere are the sources of virtually all other muscles of the body.

As development proceeds, the myotomes behind the head and pharynx form much of the musculature of the body wall, or **axial muscles.** In most fishes, the axial muscles of each side of the body are clearly separated by a membranous partition, the **lateral septum,** into dorsal **epaxial muscles** and ventral **hypaxial muscles** (Figure 10-5). On the trunk, but not on the tail, it appears (on the basis

CATEGORIES OF MUSCLES

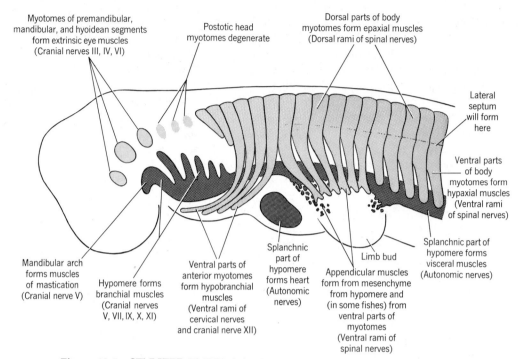

Myotomes of premandibular, mandibular, and hyoidean segments form extrinsic eye muscles (Cranial nerves III, IV, VI)

Postotic head myotomes degenerate

Dorsal parts of body myotomes form epaxial muscles (Dorsal rami of spinal nerves)

Lateral septum will form here

Ventral parts of body myotomes form hypaxial muscles (Ventral rami of spinal nerves)

Mandibular arch forms muscles of mastication (Cranial nerve V)

Hypomere forms branchial muscles (Cranial nerves V, VII, IX, X, XI)

Ventral parts of anterior myotomes form hypobranchial muscles (Ventral rami of cervical nerves and cranial nerve XII)

Splanchnic part of hypomere forms heart (Autonomic nerves)

Limb bud

Appendicular muscles form from mesenchyme from hypomere and (in some fishes) from ventral parts of myotomes (Ventral rami of spinal nerves)

Splanchnic part of hypomere forms visceral muscles (Autonomic nerves)

Figure 10-5   STYLIZED EMBRYO SHOWING THE DIFFERENTIATION OF MUSCLE GROUPS FROM THE MYOTOMES AND HYPOMERE as identified on page 80. Ultimate innervation of the muscle groups is given in parentheses.

of experimental studies done chiefly on the chick) that in higher vertebrates the hypomere also contributes importantly to the formation of hypaxial muscles. Epaxial muscles dorsoflex the spine and are innervated by dorsal rami of spinal nerves. Hypaxial muscles ventroflex the spine and support the body wall. They are innervated by ventral rami of spinal nerves. When epaxial and hypaxial muscles of one side of the body contract together, the spine is flexed to that side.

The epaxial parts of three pairs of head myotomes form the **extrinsic eye muscles.** These myotomes are identified with the premandibular, mandibular, and hyoidean segments of the ancestral head. The extrinsic eye muscles are innervated by the third, fourth, and sixth cranial nerves. The remaining head myotomes lie behind the ear and are transitory or absent in embryos of surviving vertebrates above the cyclostomes.

Below the pharynx, from the pectoral girdle to the jaw, are muscles which were derived phylogenetically by the forward migration of hypaxial muscles originally located on the trunk. Because of their position they are called **hypobranchial muscles.** Their innervation by the twelfth cranial nerve and by ventral rami of cervical nerves indicates their posterior origin.

There is evidence from the embryology of sharks that the muscles of the fins, or **appendicular muscles,** form as extensions from the hypaxial muscles of the body wall. This may be the primitive condition. In tetrapods, however, the appendicular muscles form in place from mesenchyme derived, at least in part, from the hypomere. Innervation is by ventral rami of spinal nerves.

The hypomere has important muscular derivatives other than the hypaxial and appendicular muscles already noted. At the level of the trunk it splits to enclose the coelomic cavity. The outer (somatic) layer forms no further muscle. The inner (splanchnic) layer forms the **heart** and the **muscles of the viscera.** At the level of the pharynx, there is no persistent coelom and the hypomere does not split into layers. It is instead interrupted by the pharyngeal slits and becomes associated with the visceral skeleton. The **branchial muscles,** which develop from this portion of the hypomere, are striated and voluntary, but are, nevertheless, in series developmentally with the smooth muscle of the gut. Branchial muscles are innervated by cranial nerves, the fifth nerve serves the important jaw muscles of the mandibular arch, the seventh nerve serves the hyoidean arch, and the ninth nerve the first branchial arch. The remaining arches are served principally by the tenth nerve, but the eleventh nerve and ventral rami of cervical nerves may also innervate branchial muscles.

In summary, the major categories of muscles are the axial musculature (which has epaxial and hypaxial divisions), extrinsic muscles of the eye, hypobranchial muscles (which are derived from hypaxial muscles), appendicular musculature (which has dorsal and ventral divisions), muscles of the gut, and branchial muscles (which are serially related to the visceral skeleton).

BASES FOR ESTABLISHING HOMOLOGIES In order to trace the evolution of individual muscles one must have criteria for recognizing homologous muscles in different taxa. Within orders it is usually possible, and within families it is nearly always possible, to recognize equivalent muscles on the basis of position and relationships with other muscles and with the skeleton. Thus, the supraspinatus muscle of mammals always occupies the supraspinous fossa of the scapula and inserts on the greater tubercle of the humerus. In some instances, however, the criterion of positional relationships fails: Reptiles do not have a supraspinous fossa; the attachments of a muscle may change with evolution to such an extent that the action of the muscle is materially altered; adjacent muscles of similar action may fuse; a muscle may disappear; and ancestral muscles (unlike bones of the skeleton) tend to split in the course of evolution to become several muscles.

Paleontological evidence of homology is sometimes provided by a series of fossil bones having muscle scars that evince the migration, fusion, or loss of a particular muscle. The progressive change of certain muscles in the feet of extinct horses has been learned in this way. However, the paleontologist must be careful to avoid making unjustified assumptions and is usually dependent on the muscles of surviving animals to guide analysis.

It was shown above that embryology is useful for establishing major categories of muscles. Since the pattern into which the initial muscle masses split tends to be less specialized than that of the adult, embryology is also useful for homologizing specific muscles. This approach has been applied to birds and some other vertebrates and deserves further study.

The criterion of muscle homology that has received the most attention is that

EVOLUTION OF MUSCLES

of nerve supply. In the last decade of the nineteenth century, a German anatomist postulated an invariable relationship between peripheral nerves and the muscles they innervated. Various authors have now studied nerve-muscle relationships in detail (notable among them have been Howell, Romer, and Haines), and it is agreed that nerve supply is an important criterion of homology but that instances are known of muscles that have evolved nerve relationships differing from those of their evolutionary precursors.

The comparative myologist is well-advised to use as many criteria of homology as are available to him.

MUSCLES OF PRIMARY SWIMMERS   The muscular system of CYCLOSTOMES (particularly of lampreys) is more simple and more primitive than that of other vertebrates. A lateral septum is lacking, so the prominent axial musculature is not divided into epaxial and hypaxial divisions. The segmentation of the body is clearly evident: Each myotome contributes one muscle segment, or **myomere.** An axial skeleton being virtually absent, the short fibers of the myomeres insert on partitions of connective tissue, the **myosepta,** which lie between successive myomeres. Myomeres and septa are thrown into gentle folds that are scarcely more complicated than those of amphioxus. The ventral portions of those myomeres lying close behind the pharynx turn somewhat forward, foreshadowing the hypobranchial musculature.

Appendicular muscles are, of course, absent, and, since jaws are lacking and the visceral skeleton is constructed in one unit, related branchial muscles are not prominent. There is an elaborate musculature associated with the specialized mouth and tongue. These are dissimilar in lampreys and hagfishes and cannot be homologized with muscles of higher vertebrates.

Extrinsic eye muscles develop from head myotomes as in other vertebrates. However, their number and position are such as to make it difficult to homologize them with the analogous muscles in other classes.

The musculature of JAWED FISHES is more advanced, yet remains less complex than that of tetrapods. Strong axial muscles, which flex the spine and tail from side to side in swimming, are divided into epaxial and hypaxial portions by a horizontal septum. Dorsal ribs, if present, lie in this septum. The myomeres, although straight in the embryo, become more angled than in cyclostomes. This extends the action of each over several body segments (see Figure 10-6 and the figure on p. 519). In some fishes, strands of connective tissue extend between myosepta, thus forming tendons that may be stiffened by bones. Strong longitudinal tendons extend to the tails of many fast-swimming fishes.

Straplike hypobranchial muscles extend from the pectoral girdle to the visceral arches. One of them, the **coracomandibularis,** assists in opening the jaw (Figure 10-7). The hypobranchial muscles have become distinct from the hypaxial muscles from which they evolved, but they retain the longitudinal orientation imposed by their forward migration.

The girdles of fishes lie firmly anchored within the axial musculature. Appendicular muscles have evolved with the fins and are divided into a dorsal mass

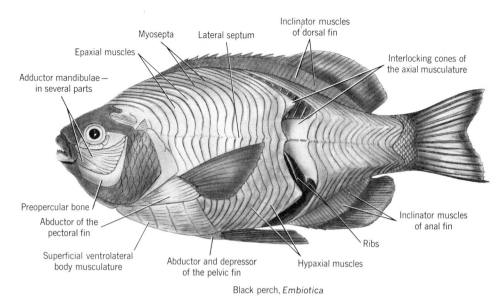

Black perch, *Embiotica*

Figure 10-6 MUSCULATURE OF A TELEOST with two myomeres removed to show the shape of the myosepta.

of extensors (or abductors, or levators—all these terms are used) that move the fins upward or forward, and a ventral mass of flexors (adductors, depressors) that move them downward or backward. These muscles, which may be somewhat subdivided, have origins on the girdles and adjacent fascia and insert on the radials of the fins.

The pharyngeal morphology of some sharks suggests that ancestral fishes, having homogeneous visceral arches, had simple and serial branchial muscles. Superficially, sheetlike **dorsal** and **ventral constrictors** were apparently nearly continuous over the gill area. These compressed the pharynx in the active phase of ventilating the gills. A series of **levators** ran from fascia above the pharynx to the several arches and served to lift the gill bars. (Hypobranchial muscles performed the complementary role of depressing the pharyngeal skeleton.) Deeper series of **adductors** closed the internal angles of each Σ-shaped visceral arch (see figure on p. 124). The adductor closing the middle angle between the epibranchial and ceratobranchial elements probably was the most important and the ventral adductor the least important. The regularity of this ancestral pattern is much altered among the various surviving fishes according to the type of jaw suspension, feeding mechanism, and presence or absence of spiracle and operculum. Muscles of the first two arches are the most specialized. The levator is retained to lift the palatoquadrate cartilage (except in some autostylic fishes). The middle adductor of the mandibular arch is much enlarged to become the **adductor mandibulae,** which closes the jaws. The ventral constrictors of the mandibular and hyoid arches form the sheetlike **intermandibularis** muscle, which lies between the mandibles and raises the floor of the mouth. Muscles of the branchial arches are relatively unspecialized in Chondrichthyes. The levators,

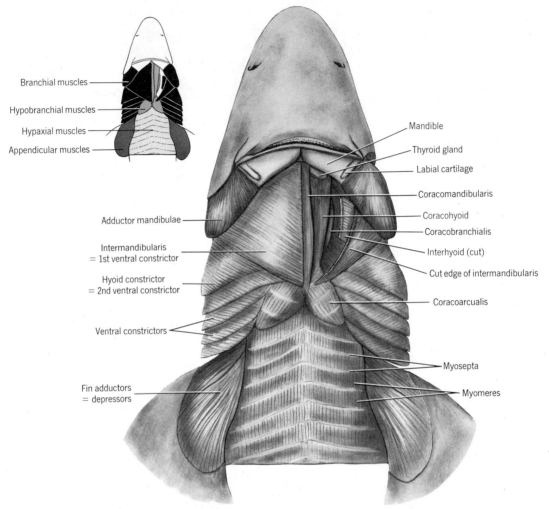

Branchial muscles

Hypobranchial muscles

Hypaxial muscles

Appendicular muscles

Adductor mandibulae

Intermandibularis
= 1st ventral constrictor

Hyoid constrictor
= 2nd ventral constrictor

Ventral constrictors

Fin adductors
= depressors

Mandible

Thyroid gland

Labial cartilage

Coracomandibularis

Coracohyoid

Coracobranchialis

Interhyoid (cut)

Cut edge of intermandibularis

Coracoarcualis

Myosepta

Myomeres

Figure 10-7 ANTERIOR VENTRAL MUSCULATURE OF AN ELASMOBRANCH shown by the shark, *Squalus.*

however, tend to mass over the more posterior gills as the **cucullaris,** or **trapezius** muscle (Figure 10-8). Branchial muscles of the gills of Osteichthyes are usually reduced to remnants of the ventral constrictors.

Fishes have six extrinsic eye muscles (Figure 10-9). Four **rectus muscles** (anterior, posterior, superior, inferior) have their origins close together deep in the posterior part of the orbit. These rotate the eye around the longitudinal and vertical axes of the head. The superior and inferior **oblique muscles** have their origins deep in the forward part of the orbit. They rotate the eye around its optical axis (the transverse axis of the head). Four of the muscles (anterior, superior, inferior recti, and inferior oblique) are derived from the premandibular myotome and are innervated by the third cranial nerve (Figure 10-5). The superior oblique

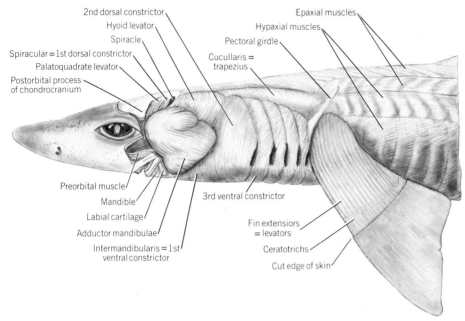

Figure 10-8 ANTERIOR LATERAL MUSCULATURE OF AN ELASMOBRANCH shown by the shark, *Squalus*.

is derived from the mandibular myotome and is innervated by the fourth cranial nerve. The posterior rectus is derived from both mandibular and hyoidean myotomes and is innervated by the sixth cranial nerve.

The shark is particularly favorable for the study of the muscles of fishes.

AXIAL AND HYPOBRANCHIAL MUSCLES OF TETRAPODS   Several general trends are evident in the evolution of the axial musculature of tetrapods. In fishes, these muscles, being the propulsive muscles, are the most massive of the body. As limbs take over the propulsive role, their muscles enlarge and the axial musculature diminishes. The axial skeleton of tetrapods, in contrast, becomes firmer in order to play a new supportive role and, concomitant with this trend, the remaining axial musculature becomes more intimately related to the skeleton and adds to its functions dorsoflection and ventroflection of the spine, which are rarely marked in fishes. Myosepta regress and disappear, and many muscles develop long fibers that span from two to many vertebrae. Furthermore, certain muscles tend to form sheetlike layers, and others become associated with the pectoral girdle.

The axial musculature of AMPHIBIANS is transitional between that of fishes and reptiles (Figures 10-10 and 10-11). Epaxial muscles (called the dorsalis trunci)

Figure 10-9 EXTRINSIC EYE MUSCLES OF VERTEBRATES exemplified by a shark. Left orbit.

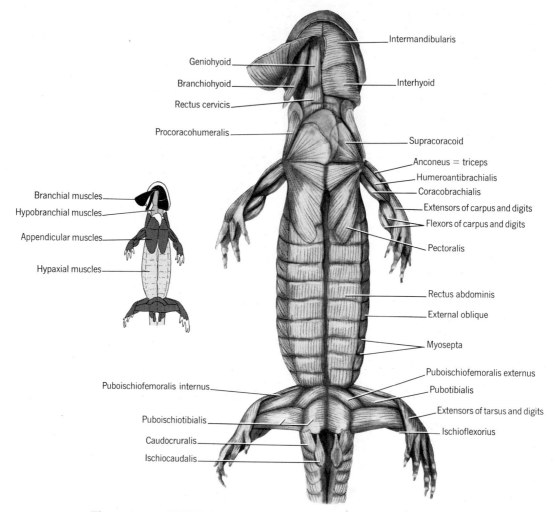

Figure 10-10   VENTRAL MUSCULATURE OF A URODELE shown by the tiger salamander, *Ambystoma.*

are conservative; myosepta are still present and are nearly vertical instead of angled, as in fishes. Hypaxial muscles are equally primitive on the tail; on the trunk they are more advanced and are commonly classified in three groups. The **subvertebral group,** located below the transverse processes of the vertebrae, is small in mass and flexes the spine. The **rectus abdominis** muscle (or group) runs lengthwise along the ventral body wall between the two girdles. It ventroflexes the body and supports the viscera. The **lateral group** is located on the flanks and is the most differentiated part of the hypaxial musculature. As in other tetrapods, it breaks into (usually) three sheetlike layers, each having its fibers oriented in a different direction. Together they support and compress the body wall. The layers,

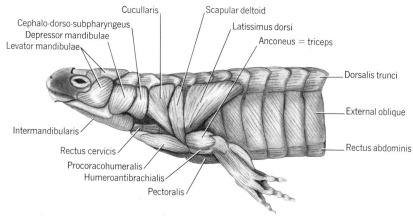

Figure 10-11 ANTERIOR LATERAL MUSCULATURE OF A URODELE shown by the tiger salamander, *Ambystoma.*

in order from outside in, are the **external oblique** muscle, the **internal oblique,** and the **transversus.**

The pelvic girdle of amphibians is stronger than that of fishes, but since it gains an articulation with the spine it does not require muscular support. The pectoral girdle, in contrast, no longer articulates with the head and does not establish an articulation with the spine. Several muscles evolve to hold this girdle to the trunk.

The hypobranchial muscles of amphibians are, in general, like those of cartilaginous fishes. The muscles of the fleshy tongue are, however, added derivatives that are of the utmost importance to the biology of the class.

Amphibian muscles are seen to best advantage in Urodela, such as *Ambystoma* or *Taricha.* They are not seen to advantage either in Anura or the commonly studied *Necturus,* where they are specialized because of its neotenous gills and small limbs.

The axial musculature of REPTILES and of MAMMALS are alike in basic design. Epaxial muscles of the tail, though without myosepta and much reduced in bulk, resemble those of amphibians; those of the cervical region tend to form layers on the now more flexible neck. Epaxial muscles of the trunk become exceedingly complex and varied in detail.

The hypaxial musculature of the cervical and caudal regions is similar to that of amphibians, though often with longer tendons. On the trunk, the subvertebral group of hypaxial muscles is much reduced, except in the lumbar area. Behind the thorax the lateral group remains essentially as for amphibians. More anteriorly, however, the ribs, now enlarged, penetrate and alter this group of muscles. The transversus is excluded from the thorax and the external and internal obliques become, respectively, the external and internal **intercostal muscles,** which contribute to the new function of ventilation of the lungs. (In various reptiles—particularly turtles—ventilation is assisted by other muscles that are unique.)

Muscles of the shoulder derived from the lateral group of hypaxial muscles are more prominent in amniotes than in amphibians. They include the **serratus,** which runs from the ribs to the scapula to suspend the thorax, sling fashion, from the girdle; the **levator scapulae;** and the **rhomboideus.**

The muscular **diaphragm,** found only in mammals, is apparently also of hypaxial origin. Its nerve, the phrenic nerve, branches from ventral rami of cervical nerves because the embryonic diaphragm originates anterior to the adult position.

Hypobranchial muscles remain much as in amphibians, but are somewhat altered by the regression of the gills, and complicated by the enlargement of the larynx.

Large lizards (but not turtles) are favorable for the demonstration of the muscles of reptiles.

The trunk of BIRDS is short and relatively rigid as an adaptation to flight. The tail is short and the tongue is usually bony. Consequently, the axial and hypobranchial musculature is much reduced.

APPENDICULAR MUSCLES OF TETRAPODS   Muscles of the PECTORAL LIMB of tetrapods have three general sources. First, one or several muscles are contributed by the branchial musculature. These are innervated by cranial or cervical nerves and are identified under the next subheading. Second, as noted under the previous subheading, several muscles are contributed by the axial musculature. These are innervated by ventral rami of spinal nerves that do not join the network of nerves at the base of the limb called the **brachial plexus.** Finally, most appendicular muscles of tetrapods are derived directly from appendicular muscles of fishes. These are also innervated by ventral rami of spinal nerves, but these nerves each join the plexus before entering the appendage.

When the appendicular nerves of fishes emerge from their plexuses, they tend to be arranged in a more dorsal group that runs to the dorsal mass of fin elevators and a ventral group that runs to the ventral mass of fin depressors. The appendicular muscles of adult tetrapods are numerous and complex, yet in the embryo they differentiate from dorsal and ventral masses in recapitulation of the ancestral piscine condition, and in the adult the many individual muscles usually can be identified as derivatives of the dorsal or ventral mass by their relationship to nerves emerging from the dorsal or ventral part of the respective plexus. In some instances, however, the ancestral functions of extension (for dorsal mass derivatives) and flexion (for ventral mass derivatives) become reversed.

Homologies of the muscles of the pectoral limb are well-established. The larger and more constant of the **dorsal group muscles** are as follows: (*1*) The **latissimus dorsi,** present in all tetrapods and large in mammals. (*2*) A muscle or muscle complex that becomes the strong **subscapularis** of mammals. The **teres major** is believed, on embryological evidence, to be derived from the subscapularis by splitting. (*3*) The **dorsalis scapulae** of amphibians is an arm abductor. It persists in higher classes and usually forms two or more muscles called **deltoids.** (*4*) In all tetrapods the **triceps** is the powerful extensor of the lower arm. It extends from the shoulder to the ulna where, in mammals, it inserts on the

Figure 10-12 VENTRAL MUSCULATURE OF A SQUAMATE REPTILE shown by the iguanid, *Iguana.*

olecranon process. (5) The numerous **extensors of the manus and digits** take origin largely from the lateral epicondyle of the humerus. Their actions are to elevate the wrist and straighten the digits.

The principal **ventral group muscles** of the pectoral limb are the following: (1) The **pectoralis,** present in all tetrapods and sometimes somewhat divided. This powerful adductor is the largest flight muscle of flying vertebrates (Figure 10-14). (2) The **supracoracoid** muscle of the lower classes is importantly altered in birds and mammals. In birds it shifts to the sternum, under the pectoralis, and inserts over a bony pulley onto the upper surface of the head of the humerus, thus serving to elevate the wing (Figures 10-14 and 10-15). In mammals the

Figure 10-13 ANTERIOR LATERAL MUSCULATURE OF A REPTILE shown by the crocodilian, *Caiman.*

insertion on the humerus is retained, but as the coracoid bone regresses, the origin of the muscle shifts to the scapula. The embryonic muscle grows out on each side of the spine of the scapula to become both the **supraspinatus** and the **infraspinatus.** (3) The **coracobrachialis,** large in amphibians, becoming smaller in mammals, is an arm flexor or adductor. (4) The flexors of the lower arm are the **brachialis** and (in amniotes only) the **biceps.** (5) The numerous **flexors of the manus and digits** have origin largely on the medial epicondyle of the humerus. In general, they depress the wrist and curl the digits. They may also stabilize the hand in subtle ways.

With minor exceptions, the muscles of the PELVIC LIMB of tetrapods are all derived from appendicular muscles of piscine ancestors. Dorsal and ventral group muscles are again recognized, and each group is commonly classified into prehip and posthip portions. In some tetrapods each of the four resulting units has its own nerve, but often the nerves combine or invade each other's "jurisdictions." Homologies between muscles of reptiles (particularly Lepidosauria) and those of mammals are relatively satisfactory; those for Lissamphibia and birds are more provisional than the commonly adopted terminologies indicate.

Dorsal group muscles of mammals include the following: (1) The **gluteal muscles** on the hip. (2) The extensors of the shank run from the girdle and femur to insert by a tendon on the crest of the tibia, with the patella, where present, guiding the tendon over the knee joint. The individual muscles can be homologized among the amniotes; collectively they are called the **quadratus femoris.**

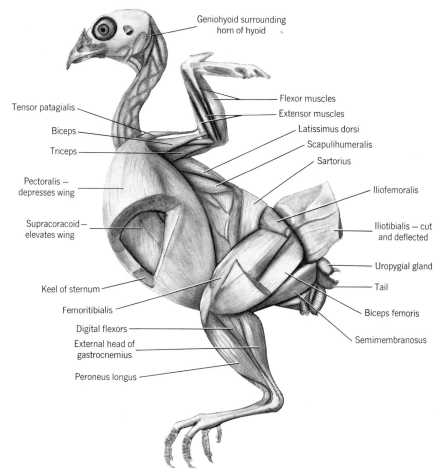

Figure 10-14 LATERAL MUSCULATURE OF A BIRD shown by the Japanese quail, *Coturnix*.

(4) The **extensors of the pes and digits** lie on the lateral and anterior surface of the shank.

  Ventral group muscles include the following: (*1*) The **obturator muscles** which lie immediately inside and outside of the pelvic fenestrum of that name. They are adductors and rotators. (*2*) The **femoral adductors,** which run between the ventral part of the girdle and the femur. (*3*) Flexors of the thigh and shank including the **gracilis, semimembranosus, semitendinosus,** and **biceps femoris.** (*4*) The **caudofemoralis** is an important flexor of the thigh in reptiles, but with reduction of the tail, whence this muscle has its origin, only several small derivatives remain in mammals. (*5*) **Flexors of the pes and digits** lie on the medial and posterior surface of the shank. Most of them send long tendons around the ankle joint to insert under the foot, but in mammals the strong **gastrocnemius** inserts on the newly evolved heel bone (calcaneum).

BRANCHIAL MUSCLES OF TETRAPODS   Branchial muscles of the first (mandibular) arch are essentially as for fishes. The ancestral adductor (or levator) mandibulae become divided into one or more **pterygoid muscles,** which run from the inside of the mandible to the pterygoid region of the skull, a **temporalis,** which runs from the coronoid process of the jaw to the temporal area, and a complex **masseter,** which runs from the outside of the mandible to the zygomatic arch and adjacent parts of the skull. The variations of these muscles are closely related to feeding habits. The intermandibularis, called **mylohoid** in mammals, is retained from the ventral constrictor series of fishes.

The principal muscle of the second arch of all tetrapods except mammals is the **depressor mandibulae,** which supplements or replaces hypobranchial muscles as the opener of the jaw. In mammals this muscle is lost and the mouth is opened by a new muscle, the **digastric,** which is derived from the ventral constrictors of both the first and second arches. Accordingly, it is innervated by both the fifth and seventh cranial nerves. Another second arch muscle of interest is the **stapedial muscle.** This tiny muscle controls the motion of the stapes—a second arch bone.

The trapezius (or cucullaris) of fishes is a muscle of moderate size derived from the more posterior gill levators. This muscle persists in tetrapods but enlarges and splits to become several muscles that assume the new functions of moving the scapula on the thorax and turning the head. In mammals these muscles

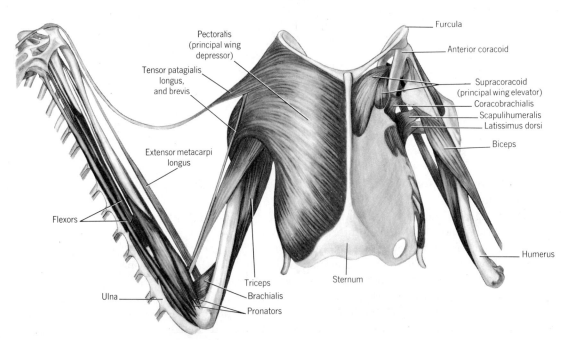

Figure 10-15   SOME FLIGHT MUSCLES OF A SOARING BIRD, the golden eagle, *Aquila,* seen in ventral view with the pectoralis removed on the left side. (Drawn from an air-dried dissection and hence somewhat shrunken.)

Figure 10-16 VENTRAL MUSCULATURE OF A MAMMAL, as seen in the cat. Sternomastoid, pectoralis complex, and tensor fasciae antibrachii removed on the right. See also pages 174 and 175.

Sternomastoid

Clavotrapezius

Anterior thoracic trapezius

Acromiodeltoid

Clavobrachialis

Brachialis

Spinodeltoid

Triceps

Insertion of cutaneous maximus (cut)

Posterior thoracic trapezius

Latissimus dorsi

External oblique

Lumbodorsal fascia

Tensor fasciae latae

Middle gluteus

Fascia lata

Superficial gluteus

Biceps femoris

Gastrocnemius

Temporalis

Cartilage of ear (cut)

Rhomboideus capitis

Splenius

Levator scapulae

Pectoralis

Supraspinatus

Rhomboideus

Infraspinatus

Ventral serratus

Spinalis dorsi

Longissimus

Dorsal serratus

Intercostal

Iliocostalis

Longissimus

Multifidus

Sartorius

Caudofemoralis

Vastus lateralis

Adductor femoris

Semimembranosus

Semitendinosus

Internal obturator

Soleus

include at least two muscles that retain the name **trapezius,** and also the **sternomastoid** and **cleidomastoid.** (In man, at least, the branchial origin of these muscles is no longer evident in ontogeny.) The other tetrapod muscles of branchial origin are small. They are the muscles of the larynx and constrictors of the throat.

EXTRINSIC MUSCLES OF SKIN AND EYE IN TETRAPODS   Muscles that run from underlying tissues to insert on and move the skin are not found in fishes or amphibians and are rare in reptiles. Snakes are a notable exception; their locomotor apparatus may include separate muscles to move each ventral scute. Birds have a muscle to tense the skin on the leading edge of the wing. Extrinsic skin muscles are most characteristic of mammals. A derivative of the second arch constrictor is the **constrictor colli,** a thin superficial muscle over the ventral and lateral parts of the neck (Figure 10-12). In mammals this muscle becomes a complex of facial muscles collectively known as the **platysma.** Facial muscles reach their highest development in man. The ancestral innervation, by the seventh cranial nerve, is retained.

A second muscle of the skin, the **cutaneous maximus,** is derived from the latissimus dorsi and pectoralis. Although vestigial in man, this is often an extensive muscle over the trunk where it may serve to curl the body (echidna) or become subdivided for flicking insects from the skin (horse).

The six extrinsic eye muscles of fishes are retained in tetrapods with remarkably little variation (Figure 10-9). However, the eyeball usually can no longer be rotated around its optical axis and one or more additional muscles evolve by splitting from one of the preexisting muscles. From the posterior rectus develops a **retractor bulbi,** of from one to four parts, which pulls the eyeball deeper into its socket. This action is protective and may also aid in swallowing. The muscle is marked in amphibians and some reptiles, but is lacking in many mammals. From the superior rectus there develops a muscle to lift the upper lid, and other muscles form to move the nictitating membrane.

ELECTRIC ORGANS

Electric organs are found in some 500 species of fishes belonging in seven families of Chondrichthyes and Osteichthyes. It is possible that richly innervated areas on the head shields of cephalaspids were also electric organs, though this interpretation is rejected by many paleontologists. The organs may be on the tail (electric skate, some teleosts), on the fins (electric ray), under the skin (a teleost), behind the eye (star-gazer), or over much of the trunk (electric eel). They are usually derived from muscle cells (hence their inclusion in this chapter), but their origin from glandular and nervous tissue is not ruled out in every instance. Diversity of occurrence, location, structure, and also physiology indicate that electric organs are ancient specializations that evolved independently several times and have undergone convergence.

Figure 10-17 DORSAL MUSCULATURE OF A MAMMAL as seen in the cat. Trapezius muscles, clavobrachialis, latissimus dorsi, lumbodorsal fascia, tensor fasciae latae, and biceps femoris removed on the right. See also pages 174 and 175.

Figure 10-18   ELECTRIC ORGAN OF THE ELECTRIC EEL, *Electrophorus*. Left, cross section of the body; right, columns of electroplaxes.

Many fishes are only weakly electric. The electric ray, however, can develop 50 A (the ampere is a measure of the amount of current delivered) and the electric eel can produce more than 500 V (the volt is a measure of the driving force of the current). Shocks of 2000 W have been recorded (the watt, a measure of power, is the product of current flow times force).

Communication, orientation, and the detection of prey are the most common functions of electric organs—particularly for fishes living in murky water. The organs of some species serve also for offense or defense; even large fishes can be electrocuted by the more powerful discharges. Electric fishes emit constant discharges and are highly sensitive to the disturbances that objects produce in the electric fields near their bodies. The sense organs that monitor the electric field are derived from the lateral line system and are located at the bases of deep pits in the skin (see p. 360 and 361).

The functional unit of an electric organ is the **electroplax,** a large, coin-shaped, multinucleated cell (Figure 10-18). Usually one flat surface is minutely folded; mitochondria concentrate under this membrane. The other flat surface is richly innervated. Hundreds or thousands of electroplaxes are stacked to form a column, and many columns are commonly present in one organ. In the resting state, an electric potential develops between the inside (negative) and outside of each electroplax. When the organ is fired by its nerve, the potentials are momentarily reversed (at least in some species) so the current exceeds the resting potentials. The organ may either be "wired" largely in series (+ pole of one cell to − pole of adjacent cell, which gives maximum voltage as is desirable for freshwater species) or largely in parallel (+ pole to + pole, which gives maximum amperage).

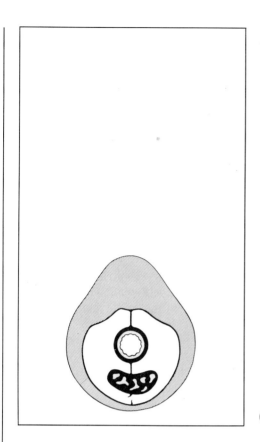

11

Coelom and Mesenteries

Coelomic cavities are the spaces that surround the heart, lungs, digestive system, and certain urogenital organs. The coelom, unlike cavities of the nervous and respiratory systems, occurs within tissues of mesodermal origin. The function of coelom is to allow the internal organs to move freely and to change their relative sizes and positions as is required when the heart beats, the lungs fill and empty, the digestive tract passes food, and the pregnant uterus enlarges. The lining of the coelom, which covers the body walls and envelops the viscera, is a **serous membrane** composed of flat cells that secrete a fluid to lubricate the organs so they can slip easily past one another.

Mesenteries extend across the coelom from body wall to viscera. They are sheets of serous membrane strengthened by thin layers or bands of collagenous and elastic fibers. Mesenteries that join one organ to another are called **omenta** or "ligaments" (the latter is an unfortunate term, since these are not true ligaments). Mesenteries support the internal organs, without restricting function, and transmit nerves and vessels. In mammals they are commonly sites of fat storage.

The coelom has been of relatively little importance to the vertebrate morphologist because its structure is too simple and constant to contribute importantly to functional or evolutionary analysis. Mesenteries correlate with systematics at the class level, and also with marked postural differences (e.g., of dog, man, sloth). The detailed configurations of mesenteries are often too complicated to decipher without embryological analysis.

DEVELOPMENT, EVOLUTION, AND RECAPITULATION

In amphioxus, echinoderms, and other invertebrate deuterostomes, the coelom forms from a series of pouches that pinch off from the dorsolateral wall of the gut tube, a process called enterocoely. This may have been the ancestral method of coelom formation in vertebrates. However, all surviving vertebrates form coelom by the cavitation or splitting of initially solid mesoderm, the process of schizocoely (review p. 79 to 81).

Early embryos of vertebrates may have small and transitory coelomic cavities in the myotomes (**myocoels**), mesomeres (**nephrocoels**), and questionably in the sclerotomes (see figure on p. 80). Nephrocoels become the adult renal capsules (see p. 283); myocoels have no known function or derivatives. The coelom of the hypomere is the **splanchnocoel,** but since it is large and persistent and gives rise to all the coelomic cavities of the adult, it is usually called simply the coelom.

This coelom splits the hypomere into an inner **splanchnic layer** and an outer **somatic layer.** As the coelom expands, right and left splanchnic layers move toward one another. They either come together in the midsagittal plane of the body or encounter the endoderm of the gut tube and its diverticula. Where they come together dorsal to the gut, they form the **dorsal mesentery;** ventral to the gut they form the **ventral mesentery,** and between the gut and its derivatives they form omenta. Parts of the ventral mesentery quickly degenerate, thus causing right and left coelomic cavities (from right and left hypomeres) to become confluent (see Figure 11-1 and figure on p. 80).

The parts of the splanchnic layer of the hypomere that encounter the gut tube and its derivatives form the smooth musculature, connective tissue, and serous membranes of the various organs. The somatic layer of the hypomere remains in contact with the body wall. It forms the serous membrane that lines the outer part of the coelom and, as noted in Chapter 10, may form mesenchyme that contributes to hypaxial and appendicular musculature.

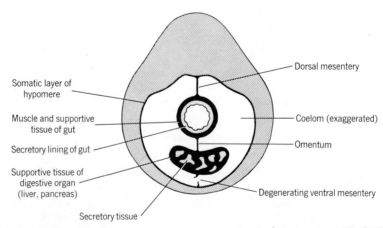

Figure 11-1 DERIVATIVES OF THE HYPOMERE IN RELATION TO THE GUT AND COELOM. This developmental stage follows the last shown on page 80.

Since coelomic cavities may occur in the ventral portions of the visceral arches, and also in the developing tail musculature of embryos, it may be inferred that the general coelom was once more extensive. However, the functional coelom of surviving vertebrates extends only from the level of the posterior part of the pharynx to the cloaca. Initially, in both phylogeny and ontogeny, the gut tube is straight and is supported by straight and continuous dorsal and ventral mesenteries. This simple structure becomes much complicated in adults of most vertebrates by partitioning of the coelom, and by deletions, folding, and fusions of the mesenteries. We will follow only the principal trends.

The heart of fishes lies anterior to the pectoral girdle and ventral to the posterior gill chambers. In HAGFISHES a **transverse septum** extends upward from the ventral body wall posterior to the heart, partly separating an anterior **pericardial cavity** from a larger **peritoneal cavity** (Figure 11-2). In selachians the developing transverse septum temporarily separates the two cavities and then secondarily develops small orifices. Lampreys and other fishes retain a complete septum (see figure on p. 212).

These basic relationships have not been modified by URODELES. The small pericardial cavity remains far forward where it is separated by a transverse septum from the principal coelom, which may now be called a **pleuroperitoneal cavity**

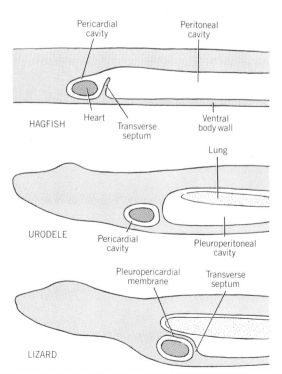

Figure 11-2 PARTITIONING OF THE COELOM OF POIKILOTHERMS. Mesenteries not shown. See also the figure on page 207.

because slender lungs are present. These are anchored to the lateral body wall by **lateral mesenteries.**

The heart of OTHER TETRAPODS lies at the level of the pectoral girdle or posterior to the girdle. The lungs are dorsal to the heart. (Parts of the liver intervene between heart and lungs in birds and some reptiles but not in mammals.) The heart is separated from the lungs (and liver if present) by more or less horizontal partitions that have their origin in the embryo as folds in the serous membrane of the right and left lateral body walls. These grow out to join in the midline of the body. They are called lateral mesocardia (birds) or **pleuropericardial membranes.** Posteriorly they join the transverse septum to form the adult pericardial membrane, or **pericardium.**

CROCODILIANS, SOME LIZARDS, AND BIRDS partition the pleuroperitoneal coelom into additional cavities by complex outgrowths and fusions of the mesenteries. The thoracic air sacs of birds separate a ventral oblique septum from a dorsal pulmonary diaphragm that is supplied with striated muscle. Their lungs grow up against the dorsolateral body walls, thus obliterating the pleural cavities in the adult.

In the partitioning of their coelom, embryonic MAMMALS resemble first early fishes (incomplete partition, posterior to heart, consisting of the transverse septum) and then reptiles (pericardium derived from transverse septum and pleuropericardial membranes). Mammals then separate paired **pleural cavities** from the peritoneal cavity by a **diaphragm.** The ventral portion of this organ comes from the transverse septum. The dorsal portion is derived from the dorsal mesentery and from still another pair of outgrowths from the lateral body wall, the **pleuroperitoneal membranes** (Figure 11-3). The striated muscle of the diaphragm comes from cervical myotomes (and hence is innervated by cervical nerves) because it develops at that level in the embryo and then migrates posteriorly as the neck lengthens and the lungs expand.

The pericardial cavity of anurans and reptiles is bordered dorsally by the pericardial membrane and ventrally by the body wall. In birds, the growing liver forces its way between body wall and membrane, thus wrapping the pericardial membrane nearly around the heart. In mammals, growth of the lungs does the same thing, this time obliterating the ancestral reptilian condition by removing the pericardial cavity entirely from the body wall (Figure 11-3). Right and left pleural cavities are then separated by dorsal and ventral mesenteries and by the pericardial membrane. The combined partition is called the **mediastinum** and is unique to mammals.

The mesenteries of ANAMNIOTES remain relatively straight and complete (dipnoans), are nearly absent except for portions related to stomach and liver (lampreys, selachians), or range between these extremes. Fusions and moderate folding are usual. Mesenteries of fishes are commonly pigmented—supposedly to protect light-sensitive gonads.

The embryonic liver of TETRAPODS starts to grow within the transverse septum. As it enlarges, it bulges out of the septum posteriorly and finally separates more or less completely from the developing diaphragm, trailing the **coronary**

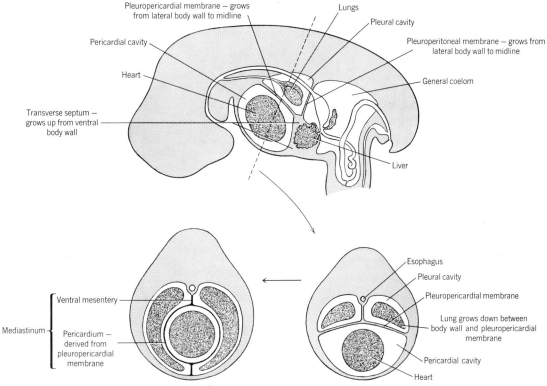

Figure 11-3 PARTITIONING OF THE COELOM IN A MAMMAL seen in sagittal section (above) and cross sections (below).

ligament behind. As the liver grows out of the septum, it grows into the ventral mesentery. The part of the ventral mesentery extending from liver to ventral body wall is the **falciform** (= sickle-shaped) **ligament;** the part between liver and gut tube is the **lesser omentum.** A small portion of another part of the ventral mesentery may anchor the bladder to the body wall. Between liver and bladder the ventral mesentery of tetrapods is missing. The dorsal mesentery is more complete and much complicated by folding and fusions. Between the stomach and body wall the dorsal mesentery of mammals becomes extended into a saclike **omental bursa** (= membrane + purse).

12

Digestive System

FUNCTION AND GENERAL STRUCTURE

The digestive system functions to (*1*) receive ingested food, (*2*) store it temporarily, (*3*) reduce it physically, (*4*) further reduce it chemically, (*5*) absorb the products of digestion, and (*6*) hold temporarily and then eliminate undigested wastes.

The general structure of the gut relates to contours of the body (particularly in fishes) and to the general nature of the functions noted. Food is taken into the mouth with the aid of teeth and jaws, tongue (some agnaths and tetrapods), forefeet (some mammals), lips (some mammals), beak (birds), or pharyngeal suction (many fishes, amphibians, and turtles). The nature of these mechanisms relates closely to feeding habits in ways described in Chapter 28.

Most vertebrates are intermittent feeders; when food is available it must be taken into the body faster than it can be digested. If the food is bulky, or if feeding and drinking are infrequent and rapid, quantities of food and water must be stored temporarily. The principal storage organ is the stomach. Some birds have a storage sac off the esophagus. Many rodents, and some other mammals, have internal or external cheek pouches (opening, respectively, inside or outside the lips). If digestion is slow (as for diets of coarse vegetation), much digesting food must be retained at one time and the storage capacity of the entire tract is greatly increased.

Physical reduction of food—especially of roughage—is necessary to release nutrients from undigestable components and to increase the surface contact between food and digestive juices. Physical reduction is accomplished by the (*1*) chewing, rasping, or grinding of oral teeth, pharyngeal teeth (some fishes), or stomach (gizzard of many birds); (2) moistening, softening, and dissolving of food by fluids of the mouth, stomach, and intestine; (3) churning and mixing by peristalsis (anterior to posterior waves of contraction), reverse peristalsis, and segmentation (dividing motions) of the stomach and small intestine; and (4) emulsification of fats by secretions of the liver.

Chemical reduction of food is accomplished principally in the stomach and small intestine by enzymes produced in those organs or in the pancreas. Since the chemical nature of foodstuffs eaten by the various animals is similar, it is not surprising that the enzymes provided and the glands secreting them are also similar in the different vertebrates. Animals that employ bacterial fermentation as an aid to digestion (ungulates, some marsupials) must provide long storage in the stomach or caecum.

Absorption of the end products of digestion requires great surface contact between the digested food and the intestinal epithelium. This is accomplished by (1) a long intestine, (2) folds or pleats in the lining of the gut, (3) microscopic villi on the lining of the tract, and (4) smaller microvilli (revealed by the electron microscope) in portions of the tract.

The digestive system has been less significant for comparative morphology than have most other organ systems. Its structure is at times useful in systematics, and it reveals general dietary habits, but some of the organs are too constant in structure to offer a phylogenetic story, whereas others vary without evident correlation with phylogeny or function. Perhaps future discoveries will alter this evaluation.

DEVELOPMENT OF THE GUT

The developmental process of gastrulation provides the early embryo with an inner germ layer, the endoderm, which is nearly spherical if there is little yolk, and sheetlike if there is much yolk (see figures on p. 78 and 81). In either event, as the embryo lengthens, the endoderm is drawn out into a tube by processes shown in Figure 12-1. This tube becomes the lining of the gut. At each end it breaks through to the ectoderm, thus establishing oral and anal openings. The midgut of embryos provided with a large yolk mass (fishes, reptiles, birds) is continuous with the **yolk sac,** which envelopes and gradually absorbs the yolk (see figure on p. 84).

Initially the gut tube is straight, or nearly so, but soon it folds and coils and establishes outgrowths, or diverticula, which become the lining and secretory cells of associated organs (Figure 12-1 and the figure on p. 207). The derivatives of the complicated pouches that form in the oral cavity and pharynx are respiratory or endocrine in function and will be described in Chapters 13 and 20. Close behind the pharynx of tetrapods a ventral diverticulum foreshadows the respiratory system. Several diverticula develop posterior to the expanding stomach. These become the liver, gall bladder, pancreas, and their ducts. Near the posterior end of the gut of amniotes, a ventral diverticulum grows rapidly to become the fetal membrane called the **allantois.**

The muscular and connective tissue associated with the gut and related organs (and therefore most of their bulk) is of mesodermal origin. The differentiation of this mesoderm is related to the ontogeny of the coelom, as described in the last chapter.

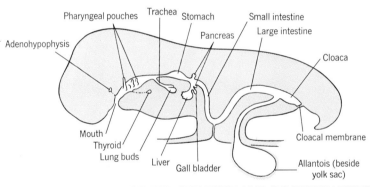

Figure 12-1 DEVELOPMENT OF THE GUT TUBE AND ITS DERIVATIVES IN AN AMNIOTE. Other organs and coelom are not represented. A later stage is shown on page 207.

MOUTH AND ORAL CAVITY   The complicated structure of mouth, oral cavity, and pharynx cannot conveniently be discussed under the heading of any one organ system. The teeth have been described (Chapter 7) as has the evolution of the secondary palate in relation to chewing and breathing (Chapter 8). Respiratory and glandular derivatives of the pharynx will be presented in Chapters 13 and 20, and various specializations that relate to feeding will be noted in Chapter 28.

The ANCESTRAL VERTEBRATE was likely a filter-feeder. Like amphioxus and the larva of the lamprey, it probably had a small mouth, virtually no oral cavity, and a large pharynx specialized to remove microscopic food particles from water by trapping them in mucus covering numerous gill bars (see figures on p. 31 and 46). The mucus was probably moved to the intestine by cilia.

STRUCTURE AND EVOLUTION

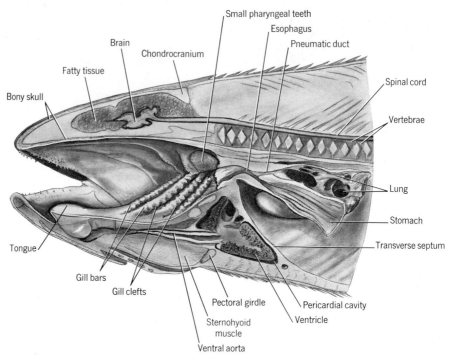

Figure 12-2 SAGITTAL SECTION OF THE HEAD AND ANTERIOR BODY OF A HOLOSTEAN FISH, the bowfin, *Amia*.

The AGNATHA have given up filter-feeding, but because they have no jaws or true teeth they ingest small or soft food and have correspondingly small oral cavities. Mouth parts may be adapted for nibbling (most ostracoderms) or specialized for clinging to a host and abrading its flesh (cyclostomes). Anaspids, some cephalaspids, and cyclostomes have a rasping organ on the floor of the oral cavity that is called a tongue but is not homologous with the tongue of higher vertebrates.

The mouths and oral cavities of CARTILAGINOUS and BONY FISHES are extremely varied. Although fleshy lips are absent, the mouth parts may be highly protrusible or otherwise specialized (of which more in Chapter 28). The oral cavity and pharynx are usually distensible. Gill bars are often provided with food strainers, grinding mills, or teeth (see Figure 12-2 and the figure on p. 572). The basal elements of the visceral skeleton support a firm tongue that is little movable, yet may be provided with teeth and is the partial homolog of the tetrapod tongue. Since the food of fishes is always wet, no further lubrication need be provided and oral glands are restricted to scattered mucous cells.

In general, the evolutionary trends among TETRAPODS have been first to increase oral lubrication and then to add limited physical and chemical digestion. They have oral cavities of moderate or large size, depending on feeding habits.

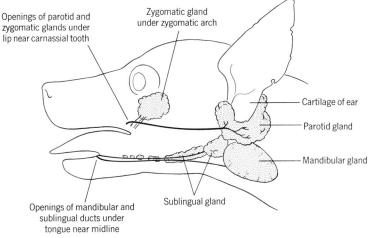

Openings of parotid and zygomatic glands under lip near carnassial tooth

Zygomatic gland under zygomatic arch

Cartilage of ear

Parotid gland

Mandibular gland

Openings of mandibular and sublingual ducts under tongue near midline

Sublingual gland

Figure 12-3 ORAL, OR SALIVARY, GLANDS OF THE DOG.

A tongue is present and is supported by derivatives of the second and third (sometimes also fourth) visceral arches. The tongue is usually fleshy and highly movable, but is relatively firm and fixed in many birds and some reptiles. It may function outside the mouth in securing food (see figure on p. 575), or inside the mouth in manipulating food during chewing, in swallowing, and sound control. Mammals have lips that are moved by derivatives of the platysma muscle. Lips may assist in taking food into the mouth.

Tetrapods have multicellular oral glands with ducts (Figure 12-3). They are named according to position as labial, lingual, palatine, nasal, maxillary, mandibular, pharyngeal, etc. However, the number, distribution, and detailed structure of these glands is diverse, and correspondence of name or position does not necessarily indicate homology. All tetrapods that are not secondarily aquatic require the secretions of oral glands to lubricate dry food. For some tetrapods the oral secretions seem to have no other function, but they facilitate tasting in frogs, and in some mammals (and to a lesser degree in certain other forms) a starch-digesting enzyme is present. Furthermore, various animals have evolved special functions for oral glands. Certain glands of some snakes, lizards, and shrews become poison glands, blood-feeding bats secrete an anticoagulant, and the nasal glands of some marine birds and reptiles migrate to the orbits where they function in salt excretion.

FINE STRUCTURE OF THE GUT IN GENERAL   The fine structure of the alimentary canal is basically similar throughout its length. The gut is constructed in layers (Figure 12-4). The innermost principal layer is the **mucosa.** It consists of a surface **epithelium,** a deeper **lamina propria,** and a **muscularis mucosae.** Cells of the epithelium may be squamous, but usually are columnar with basal nuclei. Interspersed among the less specialized cells are mucus-secreting goblet cells and (according to region of the tract) unicellular or multicellular glands that

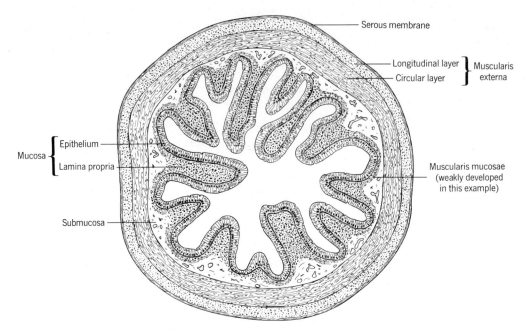

Figure 12-4   GENERAL STRUCTURE OF THE GUT TUBE as seen in cross section.

secrete digestive juices. The epithelium has many folds when the tract is empty. These flatten out as the gut is distended. Microscopic projections called villi may be present on the folds, and numerous microvilli sometimes further increase the surface of the cells. The lamina propria is a network of loose tissue that underlies the epithelium and fills the cores of the villi. Next is the muscularis mucosae (not always present), a thin layer of smooth muscle which controls motions of the lining of the gut that are independent of the gut tube as a whole.

The second principal layer is the **submucosa,** which is a conspicuous stratum of loose connective tissue containing nerves, capillaries, lymphatic ducts and nodules, and ganglia of the parasympathetic nervous system. The larger crypts and glands of the epithelium may subtend into the submucosa.

Outside the submucosa is the third prominent layer, the **muscularis externa.** It consists of smooth muscle. An inner portion is called the circular layer because its fibers are arranged in a tight spiral around the gut tube. Their contraction lengthens and constricts the gut. An outer portion is called the longitudinal layer. Its fibers are nearly longitudinal and serve to shorten the tract. The coordinated action of the two layers of the muscularis accomplish peristalsis and segmentation.

The pharynx, rectum, and part of the esophagus are bound directly to adjacent structures; other parts of the tract lie in the coelom and are enveloped in serous membranes.

ESOPHAGUS AND STOMACH   The histology of the esophagus is distinctive in two ways. The much folded and highly distensible epithelium consists of stratified squamous cells, which are cornified in animals that swallow coarse food. The muscularis externa of the anterior part of the esophagus (particularly in mammals) may have striated (hence voluntary) instead of smooth muscle fibers.

The lining of the stomach is divided into several regions that are distinctive in fine structure and function. They may be sharply demarcated or may merge. The most anterior, called the **esophageal region,** is of variable occurrence. Its fine structure is identical to that of the esophagus, but origin from the esophagus is doubtful. Like the esophagus, its only secretion is mucus. A **cardiac region** is present in mammals only. It likewise secretes only mucus, but the cells of its epithelium are columnar. The digestive region of the stomach is the **fundus.** Its lining is thickened by a dense layer of straight tubular gastric glands. Their open mouths form microscopic pits. The columnar cells of the glands are of several kinds. They secrete an enzyme that initiates protein digestion; hydrochloric acid, which provides the acidity necessary for the action of this enzyme; and sometimes also a fat-splitting enzyme. The necks of the glands secrete mucus. The stomachs of mammals also secrete renin, which coagulates milk. The most posterior region of the stomach is the **pyloric region.** Its coiled tubular glands secrete mucus.

The muscularis externa of the stomach is relatively thick. At the anterior end of the stomach, circular fibers form the **cardiac sphincter;** at the posterior end, the **pyloric sphincter** controls passage of chyme (digesting food) into the intestine.

Neither esophagus nor stomach can be identified in filter-feeders and AGNATHA. The esophagus of FISHES is usually short and sometimes merges into the stomach (Figure 12-2). Commonly it has distinctive pleats or papillae directed toward the stomach. It is ciliated in most selachians. The stomach of fishes is usually either straight or bent into a J or U shape (Figure 12-5). It is particularly large in selachians.

The esophagus of AMPHIBIANS is short, ciliated, and well-supplied with mucous glands. REPTILES tend to have a longer esophagus because of the increased length of the neck and more developed lungs. The esophagus tends to be ciliated if soft food is eaten, but is cornified in some turtles (see figure on p. 577). The stomach remains simple and straight or gently curved in most amphibians and reptiles (Figure 12-6), but is rounded and very muscular in crocodilians.

The esophagus of BIRDS is long. Its lining is usually cornified. At least some members of many families of birds have a permanent dilation of the lower part of the esophagus to serve as a storage organ called a **crop** (Figure 12-6). The dilation is usually ventral to the esophagus and sharply set off, but it may be dorsal instead and less distinct. The stomach of birds is in two parts. The anterior part, derived from the fundus but called the **proventriculus,** is very glandular and produces digestive enzymes. The posterior part corresponds to the pyloric region and is called the **ventriculus,** or **gizzard.** It may be exceedingly muscular for grinding coarse food (sometimes with the aid of pebbles eaten by the bird). Proventriculus and ventriculus are least distinct in carnivorous birds and most sharply demarcated in granivorous species.

The esophagus of MAMMALS is long, devoid of cilia, and cornified in such

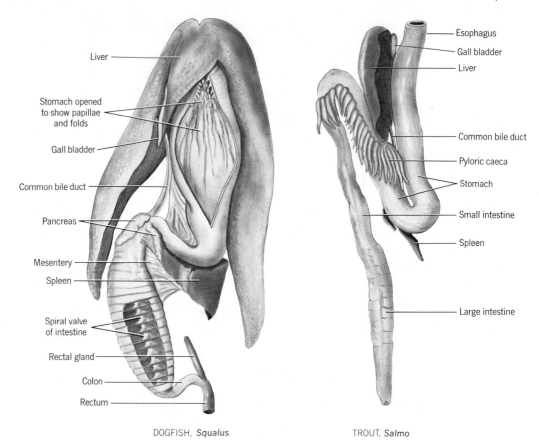

Liver

Stomach opened
to show papillae
and folds

Gall bladder

Common bile duct

Pancreas

Mesentery

Spleen

Spiral valve
of intestine

Rectal gland

Colon

Rectum

Esophagus

Gall bladder

Liver

Common bile duct

Pyloric caeca

Stomach

Small intestine

Spleen

Large intestine

DOGFISH, *Squalus*   TROUT, *Salmo*

Figure 12-5   DIGESTIVE TRACTS OF AN ELASMOBRANCH (left) AND A TELEOST (right) that feed, respectively, largely on flesh and insects. Ventral views.

roughage eaters as artiodactyls, perissodactyls, and rodents. The stomach may be simple and saclike (man, many rodents, some insectivores, carnivores) or complexly compartmentalized (see below). An esophageal region is often present and is sometimes extensive. The entire stomach of monotremes is cornified, and most of the stomach of ruminants (cud chewers) is of the esophageal type. A cardiac region, found only in mammals, is characteristic but not universal.

INTESTINE AND CAECA   Digestion is completed and foodstuffs absorbed in the anterior part of the intestine. Here the columnar epithelium forms folds and microscopic villi which, in turn, have microvilli on their exposed surfaces (Figure 12-8). Both unicellular and tubular multicellular glands are abundant in the intestinal lining. Their secretions contain a variety of enzymes, but seemingly more in higher vertebrates than in fishes. Hormones (secretin, pancreozymin, cholecystokinin, enterogastrone) that influence the activities of stomach, liver, and pancreas may also be secreted. Secretions of the liver and pancreas empty into a portion of the gut close behind the stomach, called the **duodenum.** The combined intestinal juice is alkaline.

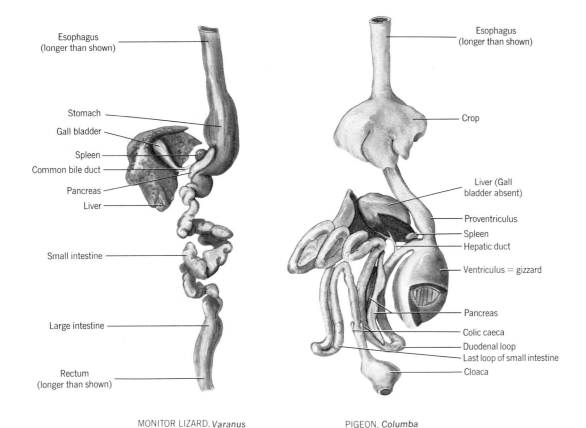

Esophagus (longer than shown)

Stomach

Gall bladder

Spleen

Common bile duct

Pancreas

Liver

Small intestine

Large intestine

Rectum (longer than shown)

Esophagus (longer than shown)

Crop

Liver (Gall bladder absent)

Proventriculus

Spleen

Hepatic duct

Ventriculus = gizzard

Pancreas

Colic caeca

Duodenal loop

Last loop of small intestine

Cloaca

MONITOR LIZARD, *Varanus*   PIGEON, *Columba*

Figure 12-6 DIGESTIVE TRACTS OF A REPTILE (left) and a GRANIVOROUS BIRD (right). Ventral views.

Water is absorbed and feces formed in the shorter posterior part of the intestine. Villi and microvilli are usually absent from this region. Mucous cells are abundant, and lymph nodules are often present in the submucosa. The part of the intestine that passes out of the coelom and through the pelvic girdle is called the **rectum.** The posteriormost end of the gut of many vertebrates is a **cloaca,** or common chamber for wastes of the digestive and urinary systems and for products of the gonads (see figure on p. 306).

Anterior and posterior regions of the intestine are relatively distinct in tetrapods where they are called, respectively, the **small** and **large intestines** because of the larger diameter of the latter. Furthermore, tetrapods have one or two pouch-like diverticula at the juncture between small and large intestines. These are called **colic caeca.** Primitively their function may have been merely to increase the surface of the gut; now they function variously for storage, fermentation, or vitamin concentration.

The intestine of CYCLOSTOMES runs straight from pharynx to cloaca, and regional differentiation is slight. These features, seen also in amphioxus, are

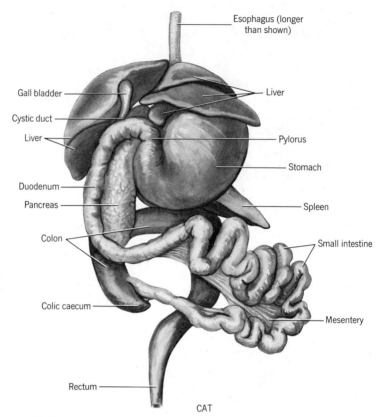

Esophagus (longer than shown)

Gall bladder

Liver

Cystic duct

Liver

Pylorus

Duodenum

Stomach

Pancreas

Spleen

Colon

Small intestine

Colic caecum

Mesentery

Rectum

CAT

Figure 12-7 DIGESTIVE TRACT OF A CARNIVOROUS MAMMAL. Ventral view.

doubtless primitive for vertebrates. In lampreys, a single marked fold of the intestinal mucosa runs lengthwise of the gut in a gentle spiral. This is the analog, if not also the homolog, of the spiral valve described below.

The nature of the digestive tract of SELACHIANS is relatively constant. The gut, now too long to run straight through the coelom, is N-shaped. One angle lies in the long stomach and one in the intestine. The three limbs run lengthwise in the body. Most of the posterior limb comprises the **spiral intestine,** or **spiral valve** (Figure 12-5). This part of the gut is large in diameter and tapers at each end. The mucosa is thrown into a single prominent fold as in the lamprey, but here the attachment of the fold to the wall of the intestine spirals more tightly as it runs down the tract. The structure of the organ resembles a spiral staircase in a circular tower. The number of turns ranges among the species from 5½ to 50. The functional advantage is great increase in the surface area of the epithelium. An equivalent increase could be achieved by lengthening and coiling the entire

Figure 12-8 INTESTINAL EPITHELIAL CELLS showing microvilli.

intestine, but the spiral structure of the mucosa is better adapted to the long slender body cavity of the fish. A short rectum joins the cloaca. A dorsal appendage from the rectum is called the **rectal gland.** Its function is salt excretion (see p. 290). CHIMAERAS have no cloaca, and there is no rectal gland although similar glandular tissue is found in the wall of the rectum.

Turning back to PLACODERMS, it is of importance that one fossil of an antiarch shows the imprint of a spiral intestine. This complicated organ is clearly primitive among vertebrates.

The intestine of BONY FISHES is more variable than that of cartilaginous fishes. It is rarely straight, commonly thrown into one or two S curves, and occasionally coiled. Its length may be less than that of the body but is usually somewhat longer than in cartilaginous fishes and reaches twelve body lengths in some species. A spiral intestine is present in all modern bony fishes except teleosts (and several of these have remnants of the valve). Ray-finned fishes are distinctive for another structure that increases the surface of the intestine; adjacent to the stomach the intestine develops diverticula called **pyloric caeca** (Figure 12-5). Most fishes have scores or hundreds of caeca, but some have few and several have none. The tubular caeca may open into the intestine individually or may cluster to form a compound organ. Histologically they resemble the adjacent intestine. Among bony fishes, only dipnoans and the surviving crossopterygian have a cloaca. There is no rectal gland.

Tadpoles have long coiled intestines, but adult AMPHIBIANS have relatively short and simple digestive tracts ranging in length from ½ to 3¼ times the length of the body. As in other tetrapods, a coiled small intestine is set off from a shorter large intestine. At the boundary between the two there may be a single small colic caecum. This structure is apparently vestigial in many surviving amphibians, yet its presence in the class must be considered primitive. A cloaca is present.

The intestine is straight in most snakes and amphisbaenians but is otherwise moderately coiled in most REPTILES. Its length usually ranges from ½ to 2 times the body length but tends to be longer in turtles. Small and large intestines are distinct. A dorsal colic caecum of small or moderate size is present in many species but has secondarily been lost by others.

The duodenum of BIRDS always forms a long narrow loop that lies ventral in the body cavity and is tightly joined to the pancreas. The remainder of the small intestine is relatively long and forms various complicated patterns of folds and coils that are constant within families. The large intestine is short, nearly straight, and villous. It is in a dorsal position. Two colic caeca are usual—rarely one or more than two are present. These are fingerlike in form, often of considerable length, and join the gut laterally or ventrally. A cloaca is present and has a dorsal diverticulum called the **cloacal bursa** that functions in the formation of antibodies.

The gross morphology of the intestine of MAMMALS is highly variable and correlates with systematics only in a general way. The intestine may be as short as 2–6 body lengths (many insectivores and carnivores) or as long as 20–25 body lengths (some artiodactyls and marine mammals). A duodenal loop is usually present but is less tightly bound than in birds. Furthermore, compared to birds, the pattern of folding of the small intestine is less regular and the large intestine tends to be longer and more bulky—particularly in herbivores. Several unrelated mammals have paired colic caeca. This may be the primitive condition, but a single ventral caecum is the rule. It may be small (even secondarily absent) but can be two or more times the length of the body.

Although some taxa have distinctive features (spiral valve, pyloric caeca, gizzard, rectal gland), evolutionary trends are few in the gut tube. Functional adaptations, however, are readily found.

FUNCTIONAL ADAPTATIONS OF THE GUT   The morphology of the digestive tract correlates sufficiently well with function so that the approximate eating habits and diet of a vertebrate usually can be determined from its digestive system, though the structure of the system in some fishes and marine mammals remains puzzling.

If the vertebrate feeds on nutritious food (which digests quickly), if the food is ingested as small particles, and if feeding is slow but frequent or nearly continual, then there need be no temporary storage of food, and a short gut is adequate. Many filterers and strainers, parasites that rasp the flesh of their hosts, and some feeders on algae, nectar, mollusks, and insects are in this category. They have relatively small stomachs or no stomach at all (cyclostomes, holocephalians, dipnoans), and the gut tends to be short and relatively straight (though it is unexpectedly long in baleen whales). Nectar-feeding birds have the entrance and exit to the muscular stomach close together so that nectar can pass directly into the intestine. The blood eaten by vampire bats passes directly from esophagus to intestine, backing up into the stomach only when the foregut is full.

Carnivores, scavengers, and fish eaters also eat nutritious foods, so their guts also tend to be short (except as noted below), but these animals often eat large meals very quickly. Large prey may be swallowed whole, food may be available only after a successful hunt or wait, or it may be necessary to gulp a meal to prevent a fellow carnivore or scavenger from robbing a meal. Consequently, capacious temporary storage is essential.

The esophagus of most such animals is remarkably distensible. The fishes tend to have long straight stomachs. Carnivorous reptiles and mammals have a simple but distensible stomach. Most of the birds have crops that are pleated when empty and enormous when full. Their gizzards are relatively thin-walled and nonmuscular. The small intestine of most carnivores and scavengers is relatively uniform and short, but in piscivorous vertebrates other than fishes it may be curiously long (penguins, pinnipeds, some toothed whales), thick-walled, and small in internal diameter. The hind gut of flesh eaters is short, and caeca are small or absent. See the viscera of the carnivores shown in Figures 12-5 and 12-7.

Scales, bones, feathers, and hair are avoided by some of these animals but are swallowed by many. These may be digested (small bones), regurgitated (by raptorial birds), or passed through the gut.

Somewhat similar to the requirements of carnivores are those of feeders on swarming insects. Anteaters have a roomy stomach, which in some species is heavily cornified and muscular (one pangolin has keratinized pyloric teeth). Their salivary glands are very large (see figure on p. 575). Their guts are moderately short.

Leaves and stems are relatively low in food value. Accordingly, they must be eaten in large quantity, and storage is needed. Furthermore, the cellulose wall of plant cells cannot be broken down chemically by enzymes of vertebrates, so thorough mechanical grinding in the mouth or stomach is required. Finally, if the cellulose itself is to be utilized as food, bacterial fermentation is essential and this, in turn, means that the food must remain many hours in a spacious fermentation chamber.

Plant-eating birds have a very large crop that is positioned relatively far posterior and ventral in the body so that its mass, when full, will not interfere with the bird's stability. Furthermore, there is a muscular gizzard to substitute for teeth in grinding the food. The crop of the hoatzin is also muscular, has a horny lining, and shreds the leaves that the bird eats.

Temporary storage is provided by cheek pouches in some rodents, primates, and bats. Having neither pouches nor crop, most herbivorous reptiles and mammals have a large stomach. The stomach is huge in artiodactyls, sirenians, kangaroos, langurs, and sloths, and is complex in that it is divided into several compartments having somewhat different functions (Figure 12-9). It may also have been highly modified in some extinct herbivorous archosaurs.

The gut of herbivores is relatively long, particularly if abundant roughage is eaten, or if the animal is large. The intestine is much coiled, often in set patterns, and the long hind gut is of large diameter. The caecum (or caeca) is always large but is particularly capacious if the stomach is relatively small as in perissodactyls, swine, elephants, and various rodents (Figure 12-10). The folds, villi, and microvilli of the digestive tube increase its surface at least 600-fold. The epithelial lining of the gut is subjected to considerable wear and is constantly replaced; the turnover time is 2–3 days.

Importantly, most herbivores have commensal populations of bacteria that feed on the cellulose fraction of the food. The byproducts are fatty acids, amino acids, and vitamins that are then digested by the herbivores. Most such fermentation takes place in the stomach if that organ is complex, and otherwise in the intestine and caecum.

LIVER AND GALL BLADDER   The vertebrate liver is unique to the subphylum and varies little among the classes. It is the largest organ of the body. The functions of the liver are many and diverse: It is a storage depot for carbohydrate and (particularly in cyclostomes and fishes) for fats. It converts protein to carbohydrate or fat with the release of nitrogenous waste, which is transported to the gills or kidneys for elimination. It elaborates much of the yolk that the maternal body transfers to the growing eggs. The embryonic liver (also the adult organ of fishes) produces blood cells, and the adult liver removes "old" red cells from the blood stream. Various toxicants can be removed from the blood by the liver, and substances needed for clotting are released. Several vitamins are manufactured or stored. Finally, the function that relates the liver to digestion—bile is secreted

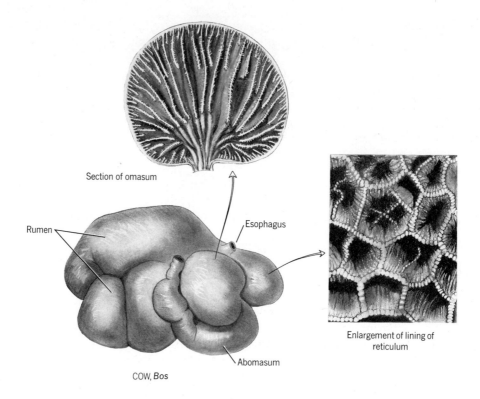

Section of omasum

Rumen

Esophagus

Enlargement of lining of reticulum

Abomasum

COW, *Bos*

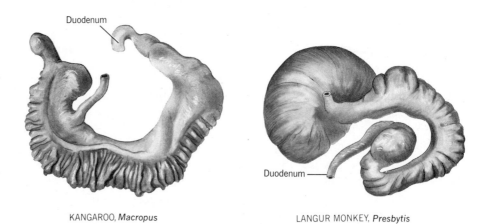

Duodenum

Duodenum

KANGAROO, *Macropus*

LANGUR MONKEY, *Presbytis*

Figure 12-9 COMPLEX STOMACHS OF HERBIVOROUS MAMMALS IN THREE ORDERS.

into a duct system and delivered to the duodenum where it emulsifies fats, thus making them digestible by pancreatic enzymes.

Esophagus

Duodenum

Glandular part
of stomach

Colon

Descending coils
of long, large
intestine

Ascending coils
of large intestine

Small intestine

Colic caecum
very large

Rectum

MUSKRAT, *Ondatra*

Figure 12-10   GUT OF AN HERBIVOROUS RODENT.

Liver and gall bladder develop from one or (usually) two ventral **hepatic diverticula** from the gut tube just posterior to the stomach (Figure 12-1). With few exceptions (birds), the more posterior diverticulum forms the gall bladder, and the anterior diverticulum (which may be somewhat paired in fishes) branches and expands to become the liver. However, liver tissue may also originate in adjacent tissues and migrate *toward* the gut, in which case the secretory tissue will be of mesodermal as well as endodermal origin.

Blood coming to the liver from the viscera in the hepatic portal system seeps through specialized capillaries called **sinusoids** before collecting in the hepatic veins to exit from the organ (Figure 12-11). The sinusoids are suspended in a labyrinth of tunnels that weave within a meshwork of interconnecting cellular walls. Sphincters control blood flow in the sinusoids. The walls between sinusoids are one cell thick in mammals and some birds; they are two cells thick in other vertebrates. Adjacent cells are held together by the tiny bile channels that follow their borders and also by projections that fit like snap fasteners into holes in

Figure 12-11   STEREOGRAM OF A FRAGMENT OF MAMMALIAN LIVER.

adjacent cells. In higher vertebrates the vessels and ducts of the liver are arranged into polyhedral units called **lobules** (Figure 12-12).

The hepatic diverticulum of amphioxus is a single hollow organ that lies against the right side of the gut. Its fine structure and functions are unlike those of the vertebrate liver with which it is homologous, if at all, only in a general way.

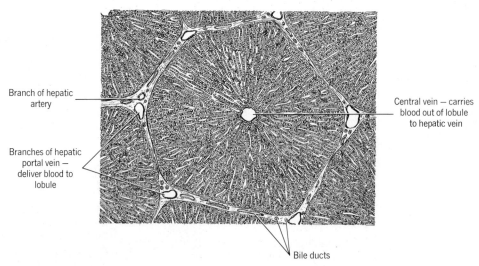

Figure 12-12   ONE LIVER LOBULE AND PARTS OF ADJACENT LOBULES as seen in cross section.

The liver tends to be lobed, and two principal lobes are frequent. However, the lobes may be side by side or one in front of the other, and the organ can have no lobes or several arranged in a variety of patterns that are without known functional or systematic significance.

Bile is usually prevented from entering the gut unless chyme is leaving the stomach. At other times it backs up into the **gall bladder,** which stores it temporarily and concentrates it as much as ten times. The **cystic duct** from the gall bladder joins the **hepatic duct** from the liver to continue to the duodenum as the **common bile duct.** The amount of bile secreted in relation to body weight varies widely among vertebrates. The presence and size of the bladder correlates with rate of bile secretion, intermittence of feeding, and fat content of the diet. The organ is always present in carnivores. It is lacking in the adult lamprey, several teleosts, and in certain herbivores distributed in five families of birds and six orders of mammals.

PANCREAS   The pancreas is a pale-colored organ that lies adjacent to the duodenum. It is found only in vertebrates, and all vertebrates have a pancreas. It is always a compound organ having both **exocrine** (secreting into a duct system) and **endocrine** (secreting into the blood) functions. About half a dozen enzymes are present in the pancreatic juice. These are able to digest nearly all foodstuffs. The endocrine secretions, insulin and glucagon, are essential for control of the intermediate metabolism of carbohydrates.

The pancreas has its origin in aggregations of cells in the wall of the embryonic foregut at the level of the hepatic diverticula. These form a dorsal pancreatic diverticulum and a pair of ventral diverticula. Growth brings the branching diverticula together and they contribute in various combinations to the adult organ. In mammals, the dorsal diverticulum contributes part of the exocrine tissue and seemingly all of the endocrine tissue. It is apparent from this complicated ontogeny why the pancreas may have one, two, or three ducts with considerable individual as well as systematic variation.

The exocrine tissue of the pancreas is tubular, branched, and alveolar (Figure 12-13). Individual cells are pyramidal. Those at the ends of the alveoli contain

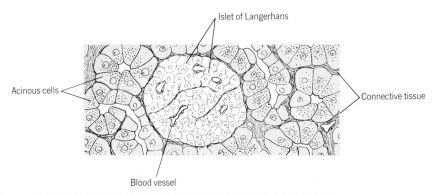

Islet of Langerhans

Acinous cells

Connective tissue

Blood vessel

Figure 12-13   SECTION OF THE PANCREAS OF A MAMMAL.

granules that are converted to enzymes when released. The endocrine tissue consists of several kinds of cells that form small scattered aggregations called **islets of Langerhans.**

Amphioxus has no pancreas, but isolated cells of the exocrine type are found in the wall of the gut around the opening of the hepatic diverticulum. In cyclostomes the pancreatic cells form aggregates under the duodenal epithelium. Dipnoans do not have a discrete pancreas but the organ is present in the intestinal wall. The organ is free and compact in selachians. It is diffuse in teleosts, being scattered as flecks of tissue in the mesenteries or along the blood vessels that penetrate liver and spleen. The pancreas of birds is somewhat three-lobed and usually has three ducts. In other tetrapods the organ is more compact and usually has two ducts or one, which may join the common bile duct.

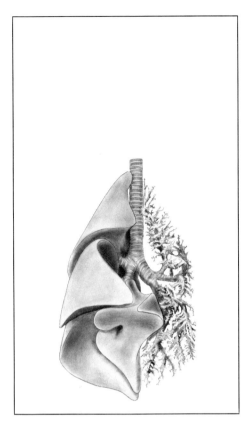

13

Respiratory System and Gas Bladder

All the cells of all vertebrates use oxygen and release carbon dioxide. The only way that these gases are moved in and out of tissues is by **diffusion.** Since all vertebrates are much too large for each cell to interact directly with the environment, certain organs, comprising a respiratory system, are specialized to accomplish the essential gaseous exchange with the environment in behalf of all of the body. Such exchange is termed **external respiration** and may take place in certain fetal membranes, at the surface of the skin, in gills, in lungs, or occasionally in some other place. Oxygen and carbon dioxide are transported between respiratory organs and other tissues by the circulatory system. They are then exchanged with the tissues in the respective capillary beds, a process termed **internal respiration.**

Diffusion of gases in the respiratory system (i.e., the efficiency of external respiration) is increased by (1) providing a large surface of contact between the environmental medium (water or air) and the blood, (2) reducing the thickness of the barrier between medium and blood, (3) maintaining the contact of each blood cell with the medium for an adequate time, and (4) establishing for each gas a large diffusion gradient between medium and blood. It is the function of the respiratory organ to provide the first three of these requirements. The fourth requirement is met by behavioral and structural adaptations for moving both medium and blood to, from, and within the respiratory organ. The pumping of water in gills, and of air in lungs is called **ventilation.** This chapter presents the structure, function, and evolution of respiratory and ventilation systems.

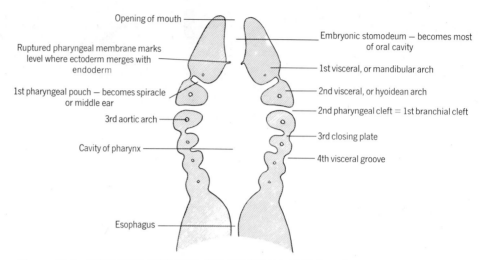

Figure 13-1   FRONTAL SECTION OF THE HEAD AND PHARYNX OF THE LARVA OR EMBRYO OF A JAW-BEARING VERTEBRATE at a developmental stage corresponding to that of the lower drawing on page 211. Compare also with the figure on page 124.

AQUATIC GAS EXCHANGERS

CUTANEOUS RESPIRATION   To judge by amphioxus and tunicates, the chordates that were ancestral to vertebrates had large and complicated gills. These seem to have evolved, however, more to accomplish filter-feeding rather than respiration. The animals were small, probably lived in well aerated and constantly moving water, and may have respired largely through their integuments.

That cutaneous respiration is feasible for vertebrates with damp bodies and low metabolism is demonstrated by many modern amphibians. They have a rich blood supply to the skin. In damp air or flowing water their needs for gas exchange are met, at least in part, by cutaneous respiration. One large group of salamanders has neither gills nor lungs as adults.

Nevertheless, efficient respiratory gills evolved from the pharynx very early in vertebrate history.

DEVELOPMENT AND STRUCTURE OF THE PHARYNX   Certain larval vertebrates have gills that develop from surface ectoderm and extend beyond the contour of the head. These are called external gills and are discussed in a subsequent section. Most gills lie within the contour of the head and are called internal gills. Internal gills develop in relation to the pharynx.

The pharynx is the portion of the foregut that lies between the developing oral cavity and the esophagus. It is lined by endoderm. The lateral walls of the embryonic pharynx develop six or more pairs of evaginations termed **pharyngeal pouches** (Figure 13-1). Opposite to the pouches, on the outside of the body, are shallow indentations called **visceral grooves.** Pouches and grooves are separated by thin partitions termed **closing plates.** Adjacent pouches are separated by **visceral arches,** each pair of which contains its respective pair of arteries or **aortic arches.**

Subsequent development depends on the presence or absence of jaws, gills, and related structures in the adult. In agnatha the first pharyngeal pouch becomes a typical **gill chamber.** In other fishes it is lost or is reduced in size and modified in function to become the cavity of the **spiracle.** Succeeding pouches of agnatha and fishes form gill chambers. The first pouch of tetrapods becomes the cavity of the middle ear; more posterior pouches form gill chambers in larval amphibians, but otherwise lose their pouchlike character after contributing glandular and lymphatic tissues from their epithelial linings (see Chapter 20). At the placoderm level of evolution, however, a posterior pair of pouches probably became the primordia of air bladders and lungs.

The closing plates of gill-bearing vertebrates rupture in the embryo to establish communication between the gill chambers and the outside of the body. Tetrapods retain the first closing plate as the eardrum; succeeding closing plates and visceral grooves of amniotes and of most adult amphibians lose their identity and leave no derivatives.

The first visceral arch (the mandibular arch) becomes the jaws of gnathostomes. The second (or hyoid) arch usually supports the jaws in fishes and contributes to the middle ear in tetrapods. As was explained in Chapter 7, the more posterior visceral arches support the **gill bars** of fishes and have various derivatives in tetrapods. The respiratory epithelium of gills develops from the margins of visceral arches near the positions of the embryonic closing plates; it is considered to be endodermal in cyclostomes but probably is more often ectodermal in gnathostomes. As explained in Chapter 9, musculature related to derivatives of visceral arches develops from the branchial portion of the hypomere; that of the floor of the pharynx is hypobranchial in origin.

GENERAL STRUCTURE AND FUNCTION OF GILLS   The basic structure of internal gills is remarkably similar for all fishes. Each gill bar consists of a part of the visceral skeleton, blood vessels derived from the corresponding aortic arch, the associated cranial or cervical nerve, intrinsic branchial muscles, and the related epithelium (Figure 13-2). The bars may or may not be extended by **gill septa** of supportive tissue. These are sometimes short (Figure 13-3) but in elasmobranchs they extend to the body surface, thus separating adjacent gill slits. On their pharyngeal (inner) margins most bars carry one or more rows of **gill rakers,** the function of which is to prevent food particles from entering the gill chambers.

A typical gill bar bears two rows of **gill filaments,** which are on opposite sides of the septum if a septum is present. Each row may be likened to a comb, with the long delicate filaments corresponding to the teeth of the comb. Each filament is stiffened by fine skeletal gill rays, and is subject to muscular control.

A bar with filaments (and septum if present) forms a partition between adjacent gill chambers. It follows that the anterior row of filaments faces into one chamber, whereas the other row faces into a succeeding chamber.

A single gill bar with its anterior and posterior rows of respiratory filaments is called a **holobranch** (= entire + gill). If a bar bears filaments on one surface only, it is a **hemibranch** (= half + gill). The filaments on the posterior surface

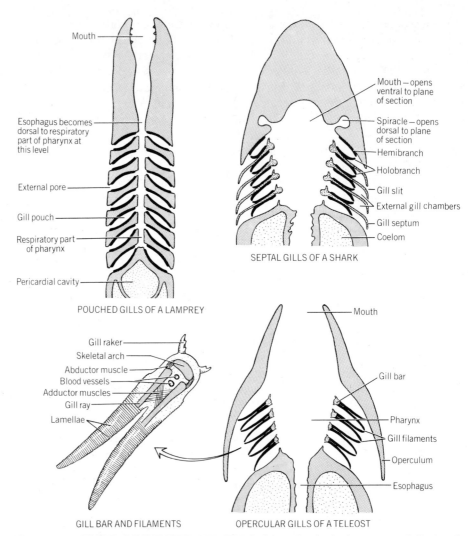

Mouth

Esophagus becomes
dorsal to respiratory
part of pharynx at
this level

External pore

Gill pouch

Respiratory part
of pharynx

Pericardial cavity

POUCHED GILLS OF A LAMPREY

Mouth—opens
ventral to plane
of section

Spiracle—opens
dorsal to plane
of section

Hemibranch

Holobranch

Gill slit

External gill chambers

Gill septum

Coelom

SEPTAL GILLS OF A SHARK

Gill raker
Skeletal arch
Abductor muscle
Blood vessels
Adductor muscles
Gill ray
Lamellae

GILL BAR AND FILAMENTS

Mouth

Gill bar

Pharynx

Gill filaments

Operculum

Esophagus

OPERCULAR GILLS OF A TELEOST

Figure 13-2   TYPES OF VERTEBRATE GILLS shown by frontal sections of the head and pharynx, and STRUCTURE OF A GILL BAR AND FILAMENTS.

of the mandibular arch (facing into the first or spiracular gill chamber) are often modified to serve a nonrespiratory function and then comprise a **pseudobranch** (= false + gill).

The gill filaments bear, on both their upper and lower surfaces, tiny parallel **gill lamellae** numbering in each series about 20/mm of filament but ranging at least from 11 to 31/mm of filament. Lamellae of adjacent filaments touch or even interdigitate. Thus water passing through the gills must seep through the minute crevices of the mesh thus formed. The total respiratory surface per unit weight of fish varies about tenfold according to the activity of the species. For a sea bass,

Figure 13-3 STRUCTURE OF A GILL BAR, GILL FILAMENTS, AND LAMELLAE showing the structural basis for large surface area, countercurrent exchange, and thin water-to-blood barrier.

which has an intermediate value, an individual weighing 20 kg has a respiratory surface of about 9.2 m² (60 ft² for a fish of 44 lb).

An **afferent branchial artery** (afferent = toward + to bear) enters each gill bar from below. As it extends upward it gives off **filamental vessels,** one of which loops to the apex of each filament and returns to drain into an **efferent branchial artery** (efferent = away + to bear), which continues upward and out of the gill.

Capillary beds in the lamellae arch between the loops of the filamental vessels. Each capillary is scarcely larger in diameter than a single red blood cell, and is constituted by the enfolding arms of pillar cells (Figure 13-3). The nuclei of the flat epithelial cells lie opposite the nuclei of the pillar cells so they will not retard diffusion. The resulting water-to-blood barrier is only about one μm thick, and each blood corpuscle is exposed to the water on both sides of a lamella.

Lamellae are oriented parallel to the stream of water, which always passes between them from the inside of the gill apparatus to the outside. Blood always flows in the lamellae from the outside of the apparatus inward. Consequently, when blood first enters the area of respiratory exchange, it contacts water that is already depleted in oxygen. This blood, however, has minimum oxygen content, so a gradient exists and oxygen immediately diffuses into the blood. As the blood continues through the lamellae its oxygen content increases, but so does the oxygen content of the ever "fresher" water with which it comes in contact. Thus, a diffusion gradient is maintained and gaseous exchange continues until the blood leaves the lamellae. The diffusion of carbon dioxide from blood to water is simultaneously favored by the same mechanism. The transfer of heat or diffusible materials between currents of gas or liquid passing one another in opposite directions is called **countercurrent exchange.** It was perfected several hundred million years ago in the gills and gas bladders of fishes and, as will be explained elsewhere in this book, is the basis for other important adaptations of vertebrates.

In summary, diffusion is favored in many ways. The combined surface area of the lamellae is enormous, and the combined surface area of lamellar capillaries is correspondingly large. The barrier between blood and water is exceedingly thin. The fine mesh of the gills slows the water stream, and both water and blood are slowed by the expansion of their combined respective channels at the site of exchange. This provides time for diffusion to approach equilibrium. Countercurrent exchange assures that the diffusion gradients for oxygen and carbon dioxide will scarcely diminish as diffusion proceeds. The hemoglobin of fish blood combines more readily with oxygen than does that of tetrapods. Finally, as explained in a following section, ventilation moves water through the gills in a nearly continuous flow.

The teleost gill is the most efficient respiratory organ in the vertebrate world. Up to 80% of the oxygen in water is removed. Efficiency is essential because water contains only about $1/30$th as much oxygen as air (depending on temperature), and diffusion is many thousand times slower.

The universal respiratory function of gills is exchange of oxygen and carbon dioxide between blood and environment. Gills also function importantly in excretion and osmoregulation. Bony fishes excrete nearly all their nitrogenous waste from the gills. It is transported in the blood as urea but is mostly converted in the gills to ammonia, which is lost by diffusion. Gills of freshwater bony fishes passively admit water and actively absorb salts; gills of marine bony fishes pass little water and actively excrete salts. Cartilaginous fishes are unique in that their gills are nearly impervious to urea, which is retained in high concentration in the blood and tissues. These differences among the fishes will be correlated with kidney function and osmoregulation in Chapter 15.

EVOLUTION OF INTERNAL GILLS   Gills are commonly classified in three categories according to arrangement and relation to supportive and protective tissues. Some gills, however, are intermediate in nature.

Pouched gills are characteristic of agnatha. The gill filaments are arranged over the surface of discrete, spherical or lenticular, pouchlike gill chambers (Figure 13-2). Each pouch may have its own external pore (never a large slit) to the surface of the body (cephalaspids, anaspids, lampreys), or the pouches on each side of the body may communicate with the outside by a common duct and pore (pteraspids, some hagfishes). Similarly, each pouch characteristically has its own internal pore to communicate with a typical pharynx (hagfishes, probably cephalaspids) or with a division of the pharynx that is separate from the path of food (lampreys).

The number of gill chambers ranges from 5 to 15, being variable even within orders. The evolutionary relationship of pouched gills to other types is not clear.

Septal gills differ from pouched gills in that the gill chambers tend to be larger, to communicate more widely with the pharynx internally, and to communicate with the outside of the body through vertical **gill slits** instead of by pores. The serial slits cause the supportive tissue that joins the gill bars to the surface of the body to be in the form of platelike **gill septa,** that share in the support of the gill filaments. This is the basis for the term **elasmobranch** (= plate + gill) that is applied to fishes with septal gills. These are Selachii, Cladoselachii, and Pleuracanthodii.

The first gill chamber and cleft of elasmobranchs is reduced to a spiracle. On the anterior face of the spiracular chamber is a vascular pseudobranch that receives oxygenated blood and apparently monitors it in some way before delivering it to the eye.

Elasmobranchs nearly always have a hemibranch on the anterior face of the first branchial chamber and then four holobranchs. There are four posthyoidean gill-bearing arches and five gill clefts. Three genera of sharks have six clefts and one has seven. Perhaps the greater number is the ancestral condition, but this is not known. In contrast to jawless vertebrates, stability, not variability, in number of clefts should be stressed for elasmobranchs.

Holocephalians have gills that are intermediate between septal gills and those of the next category. Septa are present but are much reduced so that effective serial gill slits are absent. Instead, a unique fleshy operculum covers the gills. These fishes are also distinctive for having an anterior hemibranch, three instead of four holobranchs, and then a posterior hemibranch. The fifth cleft is closed.

Opercular gills are characteristic of Osteichthyes. Septa are usually shorter than their filaments, and may be virtually absent so that only the gill bars remain to anchor the gill filaments. A bony operculum is present to protect the otherwise exposed filaments and to contribute to the pump and valve system of the gill apparatus.

A spiracle is present in the living coelacanth, and is usually present in chondrostei. These fishes have a pseudobranch on the anterior face of the spiracular chamber. Other bony fishes lose the spiracle but may retain the pseudobranch, which then fills a depression on the wall of the pharynx or moves back to the anterior face of the first branchial chamber. The structure of pseudobranchs is various and their function is not adequately known: They may have a reduced number of free filaments, may be covered by epithelium, or may assume a glandular character and sink well below the surface. They are thought to contribute to the special metabolic requirements of the eye and other organs in ways that include enzymatic control of the dissociation of carbonic acid in the blood.

A hemibranch is present on the anterior face of the first branchial chamber (posterior wall of hyoid arch) of Dipnoi and some Chondrostei, as it is also in elasmobranchs. Otherwise this hemibranch is missing in bony fishes (although,

AFRICAN LUNGFISH, *Lepidosiren*   SALAMANDER, *Pseudobranchus*

Figure 13-4   EXTERNAL GILLS in larvae of a fish and amphibian seen, respectively, in lateral and dorsal views.

as just noted, the pseudobranch of the spiracular chamber may move to a corresponding position when its "own" chamber is lost).

The functional gills of bony fishes usually consist of four holobranchs (starting with the third visceral, or first branchial arch). The complement is reduced to three, two, or even one holobranch (plus a posterior hemibranch in some instances) by some Dipnoi and some Teleostei for which the gas bladder has come to function in respiration.

Few trends can be noted in the evolution of gills—particularly when it is remembered that septal gills, as exemplified by elasmobranchs, are not in the ancestry of opercular gills. There has been a general tendency to reduce and to stabilize the number of gill bars and chambers. Within the Osteichthyes, but not the Chondrichthyes, there was an early trend to loss of the spiracle and loss of septa (but among amphibians, some larval caecilians and several salamanders still retain a spiracle!) Other variations relate to habit or are seemingly nonadaptive (for instance, the determination of which holobranchs will be retained as the series is reduced).

EXTERNAL GILLS   External gills develop from skin ectoderm of the branchial area but are not directly related to the visceral skeleton or branchial chambers. They are filamentous or featherlike, and their epithelium may be ciliated. Blood supply is indirectly from the second aortic arch if there is a single pair of gills and from several aortic arches if there are more. Muscles wave the gills to provide ventilation in quiet water.

External gills occur in the larvae of two of the three Recent genera of dipnoans, the chondrostean *Polypterus,* one teleost, and amphibians. They occur also in adults of **perennibranchiate** (= through the year + gill-bearing) urodeles, which, however, are neotenous in this respect (see Figure 13-4 and the figure on p. 20). The principal function of external gills is respiratory, but they develop early in ontogeny— before the pharyngeal slits are formed—and in some instances absorb yolk or other nutrients before the fish hatches. It is clear that external gills are a larval adaptation that evolved early in the phylogeny of bony fishes; they probably occurred in crossopterygians. There is no evidence that external gills ever occurred in adult fishes.

The terms internal and external gills are not always apt. Some larval rays have extensions from their "internal" gills that stream far out of the gill slits. Young tadpoles of anurans have "external" gills that are hidden under a fleshy operculum. Some older tadpoles have gills of uncertain homology.

VENTILATION OF INTERNAL GILLS

Since water contains much less oxygen than air, and diffusion is very much slower, effective ventilation of the gills is essential, yet the greater density and viscosity of water place a greater burden on the ventilation apparatus.

Ostracoderms probably took water into the pharynx by way of the mouth, passed it through the gill pouches, and expelled it from the head through the branchial pores. Cyclostomes also expel water through the branchial pores, but are unique in that the specialized use of the mouth for feeding and prehension precludes its constant use in respiration. Instead, water may enter and leave the gill pouches from the external branchial pores. Flow is thus tidal, though water probably passes the lamellae in only one direction. The visceral skeleton is a continuous, unjointed lattice of cartilage. Water is actively expelled from the pouches by muscular contraction; it reenters the pouches as the result of the inherent resilience of the visceral skeleton.

Sharks draw water into the pharynx through mouth and spiracles. This is accomplished largely by elastic recoil of the visceral skeleton following the previous contraction of the pharynx, though hypobranchial muscles may contribute. At the same time, branchial muscles enlarge the external gill chambers (identified on Figure 13-2) between the filaments and the gill slits. This suction pump closes the valvelike slits and pulls water through the gills even though the slits are closed. Next, oral and pharyngeal chambers are compressed, converting them from a suction pump to a pressure pump and moving water through the gills to the external gill chambers. When the pressure there surpasses the outside pressure, the gill slits open, water is forced out, and the cycle is completed. The entire sequence can be reversed, thus forcing jets of water out of spiracles or mouth to clean the gills or reject food. When rays rest on the bottom, their ventral mouths are against the substrate. They may then suck water into the pharynx only through their large, dorsally positioned spiracles.

Ventilation of opercular gills resembles that of septal gills. Again, the passage of water through the filaments is nearly continuous, even though the operculum opens and closes. A flap of tissue on the inside of the operculum near its margin functions as a valve to seal the external gill chamber when it is acting as a suction pump. Some active teleosts (tuna, mackerel) swim continuously and water constantly enters the open mouth, whence it is forced through the gills by the swimming process.

AERIAL GAS EXCHANGERS

ORIGIN AND DEVELOPMENT Warm or stagnant water retains much less oxygen than cold and moving water. Accordingly, the many fishes that live in swampy areas (probably including the ancestors of tetrapods) find it difficult or impossible to meet their needs for oxygen with gills alone. Carbon dioxide is more soluble in water, and diffuses more rapidly, so it is not the limiting factor. In oxygen-poor water, therefore, many fishes turn, at least in part, to aerial respiration. More than 20 genera of bony fishes are habitual air breathers and some suffocate if prevented from breathing at the surface. The fish can make fewer trips to the surface if gill respiration is retained to rid the body of carbon dioxide, but some of these fishes have reduced gill area and, when not needed, perfusion of the gills by blood is probably reduced. In times of drought gills are actually detrimental because they dry out easily.

Numerous fishes have "experimented" with gulping air into the pharynx, or pouches off the gill chambers, or gut where it comes in contact with specialized, highly vascularized epithelium. By far the most successful aerial gas exchanger, however, has been the lung.

Internal organs derived from the gut tube that are filled with air and function primarily in respiration are here called lungs, whether they are the familiar paired structures of tetrapods or the unpaired structures of certain fishes. Internal organs that are filled with gas but are not respiratory are called gas bladders. These occur only in bony fishes.

The respiratory system of amniotes develops from a single, ventral evagination of the gut tube close behind the pharynx. The initial primordium quickly bifurcates to form two primary lung buds, which, in turn, divide further to produce whatever respiratory tree is to be present. Mesoderm contributes the supportive tissues. The lungs of Dipnoi and the chondrostean *Polypterus* also develop ventral to the gut and may be single, bilobed, or paired. Lungs of air-breathing Actinopterygii (except *Polypterus*) develop dorsal to the gut and are single. The same is true of gas bladders. How can these differences be explained? The following observations are useful: (*1*) A fossil of an antiarch shows imprints of paired, ducted bladders or lungs. (*2*) The tissue that evaginates to form the respiratory system of amniotes takes its origin from paired areas of endoderm located just posterior to the pharyngeal pouches. (*3*) In anurans and caecilians among amphibians, the lungs do in fact develop from paired lateral evaginations of the gut tube.

From these clues it appears that: (*1*) The evolution of lungs extends back at least to the Devonian period. (*2*) Lungs were initially paired lateral organs that developed in series with the pharynageal pouches. (*3*) These ancient organs shifted ventrally to form the respiratory structures of tetrapods, dipnoans, and some primitive ray-finned fishes. (*4*) In most ray-finned fishes the ancient paired organs either shifted dorsally to merge over the gut and become a single bladder, or one of the initial lungs was lost leaving the other to shift around the gut to right or left. Embryology and the morphology of related blood vessels indicate that different groups of fishes made the shift in different ways.

LUNGS AND GAS BLADDERS OF FISHES   Because of its mode of origin, the embryonic lung or gas bladder is always joined to the gut by a **pneumatic duct.** If the organ is to function as a lung, the duct is retained (see figure on p. 212). Such a lung or bladder (and also a fish having this condition) is said to be

physostomous (= bladder + mouth). The surface area of fish lungs is much increased by pockets and partitions in their walls. A gas bladder (or fish) having no pneumatic duct is **physoclistous** (= bladder + closed). Physostomes are usually fishes that are relatively primitive or unspecialized and live in freshwater. Conversely, physoclists tend to be relatively specialized and to be marine.

Gas bladders and lungs of fishes may be located high in the body cavity above the animal's center of mass. This arrangement enables the fish to remain upright without expending muscular effort. Of fishes having the gas bladder below the center of mass, most must use their fins to counter a tendency to roll, but one catfish has solved the problem another way—it habitually swims upside down.

Gas bladders comprise about 4–11% of the body by volume. They may be long or short, straight or curved, simple or partitioned into two or three more or less distinct parts. Gas is secreted into the bladder from the blood. Rarely (and primitively?) an extensive part of the epithelium of the bladder is vascularized for this purpose. Typically, the secretory area is limited to one or several anterior **gas glands** where the epithelium is columnar and tightly folded on itself (Figure 13-5). Underlying each gas gland is a **rete mirabile** (= net + marvelous) consisting of as many as tens of thousands of uniquely long afferent capillaries, all oriented in the same direction, among which pass a like number of efferent capillaries. Gas gland and rete mirabile together are called a **red body.** In the rete the surface contact between blood approaching and leaving the bladder may be more than a square meter. This countercurrent multiplier exchanges gases from the outgoing capillaries to the incoming capillaries, thus maintaining maximum gas content in the gas gland. However, the amazing physiology of the gland cannot be explained by diffusion gradients alone. The oxygen content within the bladder of certain deep-sea fishes reaches nearly 1000 times the oxygen content of the surrounding seawater, and the partial pressure of oxygen in the bladder ranges as high as 200 atm.

Gas is resorbed from the epithelium of the posterior part of the bladder or from a portion thereof, which may be set off and then takes the name **oval body.**

Figure 13-5 GENERAL STRUCTURE AND BLOOD SUPPLY OF A PHYSOCLISTOUS GAS BLADDER.

Arterial blood reaches the red body from the celiacomesenteric artery (which delivers to the anterior viscera) and reaches the oval from the dorsal aorta. It leaves these structures, respectively, in the hepatic portal and posterior cardinal veins. This is in sharp contrast to the blood circulation of lungs. The red body is innervated by the vagus nerve and the oval body by the sympathetic nervous system. The lining of the gas bladder may be thin or fibrous. In some fishes the organ is more or less covered by striated muscle.

The most important function of virtually all physoclistous bladders, and also of some physostomous bladders, is that of **hydrostasis.** When this is true, the common term "swim bladder" is an appropriate substitute for gas bladder. The term "air bladder" is less apt; the mixture of oxygen, nitrogen, and carbon dioxide in the bladder approximates that of air when the fish is near the surface but changes as the fish descends, reaching nearly 90% oxygen in some instances.

By secreting or resorbing gas the fish can alter the volume of its bladder. This changes the mass per unit volume of the fish and allows it to adjust its density to that of the environment. Adjustment is under reflex control. It is not rapid enough to be used as a means of moving from less dense surface water to more dense deep water (or the reverse), but once the fish has changed its depth by swimming, it can maintain its level with the gas bladder. Some fishes make daily vertical excursions of as much as 1000 m. Fishes that inhabit shallow rapid streams, and marine fishes that remain at constant depth, tend to reduce or lose the gas bladder. Some deep-sea fishes have a bladder, and some do not. Bottom fishes such as flounders and halibuts have no bladder.

A second function of the gas bladder is **sound production.** The layman's conception of fishes as being silent is not generally correct; many fishes produce buzzing, rasping, squeaking, clicking, or other sounds. The bladder may serve as resonator for sounds produced by rubbing the pharyngeal teeth together or scraping certain bones together. Vibrations may be produced by action of the intrinsic or extrinsic muscles of the bladder itself, and some physostomes produce sound by the controlled passage of air between the bladder and gut.

Another function of the bladder is **sound** and **pressure reception.** The soft tissues of the fish cannot receive sound waves because their density is nearly the same as that of water. Compact bone may serve, but the compressible gas bladder is far superior and has been adapted for that purpose by several orders and many families of fishes. Sound vibrations received by the bladder must be transmitted to the inner ear in order to be sensed. This is accomplished either by paired long extensions of the bladder into the back of the skull (cods, herrings) or by paired chains of four (sometimes three) ossicles that impinge on the bladder posteriorly and on the perilymph of the internal ear anteriorly (minnows, catfishes, carps). The ossicles are derived from processes of anterior vertebrae and are called the **Weberian apparatus.**

EVOLUTION OF LUNGS FROM AMPHIBIANS TO MAMMALS   Typical lungs of tetrapods differ from fish lungs in being paired, in having a higher surface-to-volume ratio, in joining the ventral side of the gut tube by a duct termed the **trachea,** in receiving blood of low oxygen content in vessels that are related in

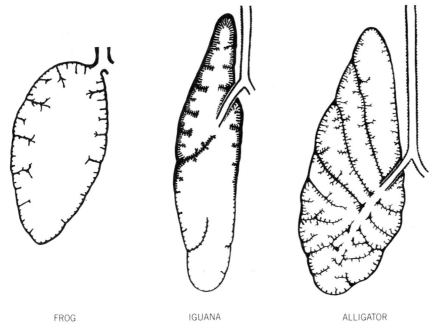

FROG   IGUANA   ALLIGATOR

Figure 13-6 LUNG STRUCTURE OF SOME LOWER TETRAPODS as seen in frontal section. Somewhat stylized and two dimensional. A finer order of branching is not visible at this scale.

development to the 6th aortic arches, and in returning oxygenated blood directly to the heart without prior mixing with blood from other organs. The principal evolutionary trend has been adaptation to increasing body size or metabolic rate by increasing the compartmentalization of the lung.

ANURA have large but short lungs (Figure 13-6). The interior of the lung is an open sac, but the walls have partitions of the first, second, and third order providing a total respiratory surface of about 1 cm^2/g body weight (15–20 cm^2 for a medium-sized frog). The very short trachea divides into two short **bronchi,** one leading to the apex of each lung. The epithelium of these ducts is ciliated and thus is able to clean the respiratory system. Cartilage may support the walls of trachea and bronchi against collapse. The opening from trachea to pharynx is called the **glottis.** It is a longitudinal slit that is usually flanked by a dorsal pair of **arytenoid cartilages** (arytenoid = cup-shaped), which support vocal cords, and a ventral pair (often fused) of **cricoid cartilages** (cricoid = ring-shaped). These cartilages are regarded as derivatives of posterior visceral arches of ancestors. Together they constitute the **larynx,** a structure characteristic of tetrapods. The larynx is joined by ligaments to the hyoid apparatus.

APODA, having long slender bodies, usually retain only the right lung. (Their larvae have both lungs.) Otherwise, their respiratory system resembles that of Anura.

The more terrestrial LABYRINTHODONTS probably had the most advanced lungs of the class, likely resembling those of large reptiles. Lungs of URODELA by contrast, have regressed; most species have lost their lungs entirely. Where present, they are long slender sacs with smooth walls. Their respiratory function is diminished (skin or gills serving instead) and they act as hydrostatic organs.

Lungs of REPTILES are large and varied. The degree of partitioning may be similar to that of frogs, but in order to maintain an adequate ratio of lung surface area to body weight (see p. 22 and 184) the larger reptiles (turtles, monitors, crocodilians) have lungs with many compartments that are heterogeneously partitioned. Trachea and bronchi are longer than for amphibians (much longer in snakes) and are supported by cartilaginous rings that may be closed or opened dorsally. Each bronchus enters its lung near the middle or near the anterior end, but not at the apex. Without further division, the bronchus continues within the lung, or opens into more or less through airways. The larynx again consists of cricoid and arytenoid cartilages (variously shaped and fused) that (except in snakes) are joined to the hyoid apparatus. Vocal cords are present only in some lizards. Many reptiles hiss by passing air slowly through a partly closed glottis, at the edge of which there may be an erectile, sound-producing flap. (Certain snakes hiss by rubbing the scales of adjacent body regions together and amplifying this sound via the inflated body.)

Lungs of MAMMALS have the same basic plan as those of reptiles but are much more finely and homogeneously divided. Therefore they are more efficient, as is required by the higher metabolic rate of these homeotherms. Lobulation of the lungs is variable and without systematic or evident adaptive import. Lobes may be absent (horse, whales, sea cows, some bats) but usually there are at least two lobes on the left and at least three on the right (Figure 13-7). The lobes may or may not be divided into lobules, which can be conspicuous or faint.

Of more functional importance is the nature of the respiratory tree. The trachea (like that of reptiles) is supported by rings of hyaline or fibrous cartilage, which are here incomplete dorsally. Elastic connective tissue joins ring to ring and completes the tube where cartilage is absent. The resultant structure is ideal for holding the airway open yet allowing the tube to twist as the neck is turned, change length with swallowing, and change diameter if there is marked alteration of internal pressure, as in coughing. The trachea is lined by ciliated epithelium. Smooth muscle and mucous glands are present in the walls.

The trachea divides into right and left primary bronchi, each of which enters its lung somewhat anterior and dorsal to the center. Within the lungs the primary bronchi divide into smaller bronchi, which divide again through numerous generations of branching. All internal bronchi are supported by cartilage which, however, may be in plates rather than rings. The airways are continued by branching **bronchioles** that are membranous, not cartilaginous. These, in turn, open into nonciliated **respiratory bronchioles,** where gaseous exchange begins, and finally, into **alveolar duct systems,** which are clusters of about 20 hemispherical **alveoli** all opening into a common terminal chamber (Figure 13-8).

Lung volume is nearly proportional to body size, but alveolar size (and hence respiratory surface) varies with metabolic rate. Total surface ranges from about 8 cm^2/g body weight (some primates) through 50 cm^2/g (mouse) to 100 cm^2/g

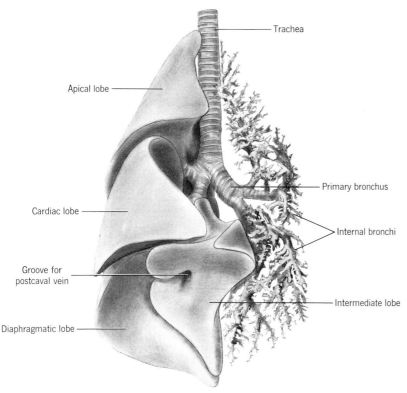

Trachea

Apical lobe

Primary bronchus

Internal bronchi

Cardiac lobe

Groove for
postcaval vein

Intermediate lobe

Diaphragmatic lobe

Figure 13-7 MAMMALIAN LUNG illustrated by the dog. Ventral view. Dissected on the left to show the bronchial tree.

(bat). In other terms, the respiratory surface of the human lung is about 70 m^2, or 40 times the surface area of the body, and a 20 kg dog has a respiratory surface that is 9 times greater than that of a moderately active fish of the same weight.

Pulmonary vessels may follow the airways, or the arteries only may follow according to species. Alveoli are richly supplied with capillaries where blood is separated from air only by the endothelium of the capillary, a basement membrane, and an exceedingly thin alveolar epithelium.

The large larynx is attached to the hyoid apparatus. Paired arytenoid cartilages help support and control the vocal cords. The cricoid cartilage is single. Two additional cartilages are present that are lacking in other vertebrates: a large ventral **thyroid cartilage** (thyroid = shield-shaped) and a cartilage in the **epiglottis.** The epiglottis is a stiff valvelike flap that guides air between posterior nares and glottis during respiration and, to keep food out of the respiratory system, closes the glottis during swallowing.

Figure 13-8 RESPIRATORY BRONCHIOLE AND RELATED ALVEOLAR DUCT SYSTEM OF A MAMMAL.

AVIAN LUNGS AND AIR SACS   Birds match their uniquely high activity and metabolic rate with a respiratory system that is unique in structure, function, and high efficiency. Avian lungs are small and compact, and instead of moving freely in pleural cavities they adhere to the dorsal body wall. They are nearly rigid and have a virtually fixed volume. Large air sacs join the lungs and serve to ventilate them (see below). In most of the lung air is not tidal, but instead moves in one direction. The air-blood circulation is crosscurrent in nature. Compared with mammals, there is more blood in the lung per unit volume, more of that blood is in capillaries, and the ratio of respiratory exchange surface to volume is about 10 times greater. Some birds can fly steadily at elevations at which mammals at rest become comatose!

The trachea, like the neck, is long. Indeed, in certain large birds it may form one or more loops near or within the sternum. Its resting diameter may vary at different levels.

Each **primary bronchus** enters its respective lung ventrally somewhat anterior to the center of the organ. Just inside the lung four secondary bronchi, or **ventrobronchi,** join the primary bronchus (Figure 13-9). These follow the ventromedial contour of the organ, branching as they go. Seven to ten **dorsobronchi** next join the primary bronchus. They branch over the dorsolateral surface of the

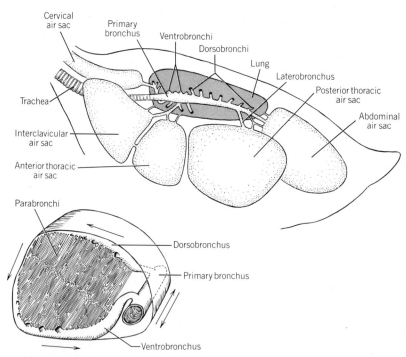

Figure 13-9   AVIAN RESPIRATORY SYSTEM. Stylized left lateral view of the entire system, and cross-sectional view of the left lung. Pathways of the neopulma not shown.

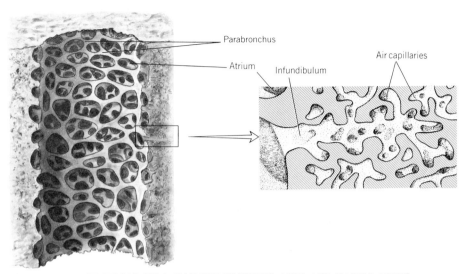

Figure 13-10 SECTION OF A PARABRONCHUS AND AIR CAPILLARIES.

lung. Ventrobronchi and dorsobronchi are connected by thousands of **parabronchi,** which are about 1 mm in diameter. The walls of parabronchi are honeycombed with pockets (atria) which, in turn, have alcoves (infundibula) from which branch interlacing and crosslinking **air capillaries** (Figure 13-10). These are a mere 3 to 10 μm in diameter, which is much less than the diameter of mammalian alveoli.

Lungs of birds function in relation to air sacs that are devoid of respiratory epithelium but contribute to ventilation of the system. The primary bronchi terminate in large paired **abdominal air sacs.** Extensions from these sacs penetrate the synsacrum, femur, and thigh muscles. Paired **posterior thoracic air sacs** join the primary bronchi by way of **laterobronchi;** paired **anterior thoracic air sacs** usually join the third ventrobronchus. These sacs also have secondary connections to the duct system by small recurrent bronchi. Smaller paired **cervical air sacs** pneumatize cervical vertebrae and muscles; they join the first ventrobronchus. Finally, an unpaired **interclavicular air sac** sends diverticula into the humeri and among muscles of the axillas and shoulders. The entire system is somewhat variable and exceedingly complex.

Birds have a larynx of characteristic form. Its cartilages are the same as those of reptiles. Vocal cords are absent. Sound is produced instead by a unique **syrinx** located at or near the bifurcation of the trachea (Figure 13-11). A portion of the outer or (more often) mesial wall of each bronchus (or less frequently the wall of the trachea, or of trachea and bronchi in combination) is a thin membrane supported at its margins by modified cartilaginous rings. This membrane impinges on air sacs and thus is free to vibrate in the stream of air passing through the system. The organ is sometimes asymmetrical and can be conspicuous in birds with resonant voices (Figure 13-12). Complicated intrinsic and extrinsic muscles

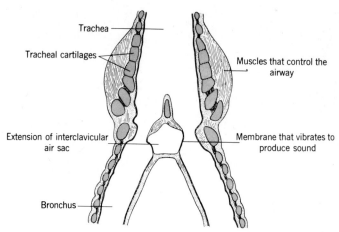

Figure 13-11 STRUCTURE OF THE SYRINX OF A EUROPEAN BLACKBIRD as seen in longitudinal section.

control tension of the membranes and also the characteristics of the resonating system. The muscles are variable among species and often are used in systematics.

VENTILATION OF LUNGS AIR-BREATHING FISHES ventilate their lungs much as they ventilate their gills. Even when present, internal nares are not used (they may function in olfaction), Air is gulped into the oral cavity using the pharynx as a suction pump. With mouth closed, the pharynx then acts as a pressure pump to force the air down the esophagus and into the pneumatic duct. Exhalation is usually passive. The mechanism is a two-cycle *pulse pump.*

Among tetrapods, ANURA retain this inefficient system, seemingly because it relates to their important vocalizations, but with some differences. The internal nares are functional for the first time in vertebrate history. With glottis closed, and nares open, air is sucked into the large buccopharyngeal space by lowering the throat. Then with glottis and nares both open, air escapes from the body under pressure from the resilient lungs. In the process the air passes through the top of the buccopharyngeal space but does not mix with the "fresh" air already stored in the ventral part of the space. The nares then close and raising of the throat forces the fresh air into the lungs via the open glottis. Thus, the tidal flow of respired air causes mixing of fresh air with some residual used air in the lung. The greater diffusivity of carbon dioxide allows much of this gas to escape through the moist skin, independent of the breathing apparatus.

Ventilation by *aspiration* is more efficient. Air is sucked into the lungs by negative pressure created in the thoracic or abdominal cavities. Exhalation is either passive or results from constriction of these cavities. Aspiration breathing

Figure 13-12 SKELETAL SUPPORT OF THE SYRINX OF A MERGANSER, *Mergus,* seen in ventral view.

frees the oral cavity and pharynx to adapt principally to feeding, is a one-cycle system and, because it is rapid enough to rid the body of carbon dioxide as well as supply it with oxygen, it lends itself to a simpler circulatory pattern and allows the skin to be dry.

Aspiration breathers tend to have strong ribs and strong intercostal and abdominal muscles. However, the nature of the pumping mechanism varies widely (and remains to be studied in many taxa). LABYRINTHODONTS were doubtless the first such breathers. Inhalation and most of exhalation is forced in SNAKES and results from the action of muscles inserting on the ribs and ventral skin. Any part of the long body can serve if other parts are distended by food. CROCO-DILIANS inhale when their diaphragmatic muscle (neither homolog nor analog of the mammalian diaphragm) pulls the liver toward the pelvis. The pistonlike liver, in turn, sucks on the lungs. In water there is enough pressure on the body so that exhalation can be passive; on land other muscles pull the liver forward. TURTLES inhale when muscles crossing the limb apertures of the shell enlarge the internal cavity. Exhalation may be passive or active.

Lungs of MAMMALS are inflated by the negative pressure that results from contraction of the external intercostal muscles and the dome-shaped diaphragm. (Additional muscles function in forced breathing.) Lungs deflate largely by their inherent elasticity, but internal intercostal and abdominal muscles act in forced breathing. Surface tension of the liquid film within the small alveoli would cause the lungs to collapse were a lipoprotein not secreted to reduce the tension.

When BIRDS are at rest, the sternum is actively rocked to change the volume of the air sacs—particularly of the posterior sacs. The uncinate processes of the ribs (see figure on p. 555) are lever arms for the strong external intercostal muscles used during inhalation. In all of the lungs of some birds, and in most of the lungs of the others (the part termed *paleopulmo*), air flows to the parabronchi in the dorsobronchi, ventilates the parabronchi in one direction only, and exits by the ventrobronchi. Air is, however, tidal in the air sacs, primary bronchi, and in part of the lung (*neopulmo*) of some fowl and songbirds. Perhaps only the neopulmo functions when the bird is at rest, the paleopulmo being recruited during activity. However, the function of the system during flight is scarcely known. It must be assumed that ventilation is increased enormously. The air sacs also serve as evaporative coolers during periods of activity.

14

Circulatory System

Animals must have a system of internal transport unless their size and structure place all their cells in close proximity to an aqueous environment. Vertebrates, being large and solid, have the most highly evolved circulatory system in the animal kingdom. It functions to transport respiratory gases, nutrients, metabolic wastes, hormones, and antibodies. It serves (in conjunction with the kidneys and some other organs) in maintaining the internal environment. It removes toxic and pathogenic materials from the body, and may function (with muscles and the integument) in temperature regulation. Furthermore, it has the capacity to repair leaks, compensate for damage, and respond with amazing versatility to the varying requirements of the moment.

The vertebrate circulatory system has two components, a **blood-vascular system** and a **lymphatic system.** The former consists of the heart, blood vessels, and blood. **Arteries** distribute blood from heart to tissues, and **veins** return blood from tissues to heart. Small arteries (arterioles) are joined to small veins (venules) by **capillaries** which form a network within the tissues. Capillaries are usually about 1 mm long, scarcely larger in diameter than a single red blood cell, and only a fraction of a millimeter distant from one another. Physiological exchange between blood and tissues takes place through their thin walls. In several places in the body (digestive organs, kidneys, hypophysis) blood that has passed a capillary bed elsewhere enters a second capillary bed before reaching the heart. The veins between two capillary networks constitute a **portal system.**

The blood-vascular system of vertebrates (unlike that of invertebrates, except annelids) is a continuum of ducts and is therefore said to be a **closed system.** However, fluid constituents of the blood leak out of the capillaries driven by diffusion, osmosis, and the hydrostatic pressure produced by the heart. Fluids would accumulate in the tissues and cause swelling were they not drained away by the second component of the circulatory system, the lymphatic system. Tissue fluids enter netlike or blindly ending **lymphatic capillaries** where they constitute **lymph.** Lymph passes slowly into larger and larger **lymphatic vessels** until it is discharged into the venous system at several points. Lymphatic capillaries in the gut are called **lacteals.** They absorb the digested long-chain fats; their lymph is whitish after a fatty meal (lacti = milky).

The heart provides little pressure to drive venous blood and none to drive lymph. Their passage is assisted by pressures on their vessels resulting from respiratory and other motions of the body and probably by their own intrinsic musculature. Most veins and lymphatic vessels of tetrapods have valves to prevent backflow. Many lower vertebrates also have lymph sinuses, some of which beat weakly as **lymph hearts.**

Ontogeny and phylogeny are more strikingly related in the circulatory system than in any other. Embryonic hearts, arteries, and veins of the higher vertebrates closely resemble the corresponding organs of remote ancestors. The pattern of circulation changes in development much as it must have changed in evolution, and it is fascinating to observe the sometimes marked changes in progress because the system functions continuously throughout.

The circulatory system also has more individual variation than any other. This is particularly true of the veins and is the combined result of a complicated ontogeny (which provides many opportunities for deviations) and the near functional equivalence of various departures from the norm.

Furthermore, the system is exceedingly adaptable. A piece of a vein grafted into an artery transforms structurally to become an artery. Nearly any vessel of the body can be tied off without serious inconvenience to the animal if the obstruction is made slowly enough for the system to compensate by enlarging alternate routes. Blood can, and does, flow in either direction in several vessels. Blood can be diverted toward or away from any given part of the body, the volume of circulating blood can be changed, and the rate of circulation can be varied about five-fold.

In spite of marked individual variation, parts of the system are useful in systematics. This is particularly true of the heart and major circuits at levels of the class and subclass, and of patterns of arteries in the limbs and basicranial area at levels of the order and family.

THE HEART   **DEVELOPMENT**   The circulatory system is the first of all organ systems to become functional during development. This is hardly surprising since differentiation and growth of all parts of the body are soon dependent on internal transport. The chick heart starts to pulsate at about 30 h of incubation when the body is not yet large enough to enclose the organ; the human heart becomes functional at 4 weeks when the embryo is scarcely 5 mm long.

Figure 14-1 FORMATION OF THE HEART IN AN EMBRYO HAVING MUCH YOLK.

The part of the splanchnic layer of the hypomere that is just posterior to the pharynx and ventral to the gut (or will soon come into that position—there are complicating changes of position in mammals) becomes markedly thicker on both sides of the body (Figure 14-1). These mesodermal folds approach one another in the midline and fuse to form a longitudinal tube. The tube is fixed to surrounding tissues at each end but otherwise becomes free as it passes through an expanded portion of the coelom. The free section establishes four chambers, that begin to contract in sequence and thus become the heart.

When the embryonic heart is first established it has two layers, an internal **endocardium** and an external **epimyocardium.** The endocardium of the mature heart has a thick layer of elastic connective tissue under an endothelial lining. The epimyocardium matures—while functioning—to have an external **epicardium** and a **myocardium.** The former becomes the enveloping serous membrane of the organ and is underlain by connective tissue. The myocardium is the muscle of the heart. Recall that cardiac muscle fibers are striated and branched and have central nuclei and intercalated disks (see figure on p. 180). It is involuntary and, as explained below, has the inherent capacity to contract rhythmically.

PRIMITIVE HEART: A SINGLE-CIRCUIT PUMP   Amphioxus has no heart. It has instead a pulsating vessel in the position where the heart evolved in vertebrates. This vessel approximates the embryonic primordium of the vertebrate heart and is apparently homologous.

The structure of the adult ancestral vertebrate heart can be inferred from the structure of the embryonic heart of descendants. It is a nearly straight tube having four chambers that contract in sequence. Importantly, it pumps a *single stream* of *unoxygenated* blood forward in the body (Figure 14-2). A thin-walled **sinous venosus** receives blood from the great veins and empties it through a simple **sinuatrial valve** into the **atrium.** The atrium, also thin-walled but muscular, expels blood through one or more rows of **atrioventricular valves** into a large thick-walled **ventricle.** The ventricle pumps into the **conus** (also called conus arteriosus and bulbus cordis), which looks like an enlarged artery and is lined with several rows of cup-shaped **semilunar valves.**

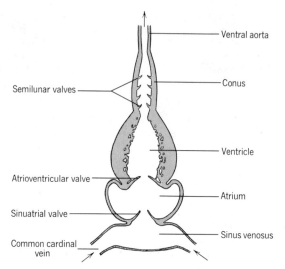

Figure 14-2 HYPOTHETICAL ANCESTRAL VERTEBRATE HEART as seen in frontal section. This structure is closely approximated in the ontogeny of all vertebrate hearts.

Hearts of CYCLOSTOMES and FISHES (except dipnoans) vary widely in detailed structure but depart little from the general ancestral plan. In fishes the heart is relatively far forward in front of the pectoral girdle and under the posterior gills (see figure on p. 212). The sinus venosus ranges from large (most sharks) to small (cyclostomes). The atrium is relatively large and usually shifts to a position dorsal to the ventricle (Figure 14-3). The inner layer of the ventricle is exceedingly spongy. In teleosts this chamber is conical, with apex pointing posteriorly. The conus varies from long and active as a pumping organ (cartilaginous fishes, holosteans, *Polypterus*, bowfin) to virtually absent (cyclostomes, teleosts). When large it contains two to eight rows of semilunar valves (some of which, however, may not span the channel of the organ). The conus prevents backflow of blood as the ventricle fills. Teleosts have a **bulbus arteriosus** within the pericardial cavity in the position of the conus of other fishes. This organ has smooth, not cardiac muscle, but is highly elastic and passively evens the flow of blood into the afferent branchial arteries.

The pericardial cavity of elasmobranchs, being bordered in part by the skeleton, is semirigid. Consequently, as the ventricle contracts, blood enters the sinus venosus and atrium partly by suction. Hearts of fishes are relatively small because fishes have relatively small blood volume. As expected, active fishes have larger hearts than their sluggish relatives. Cyclostomes and many fishes have accessory hearts or pumping mechanisms elsewhere in the body. Some of these are mentioned later in this chapter. Change in the cardiac output of fishes usually results from a large change in stroke volume and only a small change in heart rate.

LUNGFISHES TO REPTILES: INTERMEDIATE AND FACULTATIVE HEARTS We have just seen that the heart of most fishes pumps one stream of unoxygenated blood. The heart of birds and mammals pumps a stream of unoxygenated blood

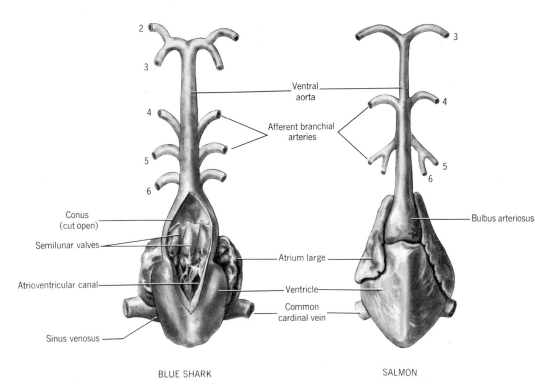

Figure 14-3 HEARTS AND RELATED VESSELS OF REPRESENTATIVE FISHES. Ventral views.

and also a completely separate stream of oxygenated blood. By contrast, the heart of dipnoans, ancestral crossopterygians, amphibians, and reptiles is intermediate in that it usually receives both "kinds" of blood, yet does not provide complete structural separation of the two streams, thus allowing mixing under some conditions.

Since the intermediate heart has probably survived for nearly 400 million years, and varies widely in both structural and functional details, it is not accurate to think of it as imperfect or merely a transitional evolutionary step toward a better mechanism. The amount and place of mixing of oxygenated and unoxygenated blood is varied by the animal according to circumstance. It is desirable to send unoxygenated blood to the lungs when they are functioning. However, it is preferable to send much of the same blood elsewhere if respiration is temporarily occurring primarily in gills (dipnoans in fresh water), or skin (submerged anurans), or simply because the lungs are not active (diving turtles and crocodilians). The heart of these animals is facultative, it adapts, at least in some degree, to a range of conditions not encountered by lungless fishes, birds, or mammals.

The intermediate heart has been much studied, but so far only in selected animals. The difficulty of analyzing the function of these hearts, particularly under stressful conditions, is very great. The following descriptions summarize current

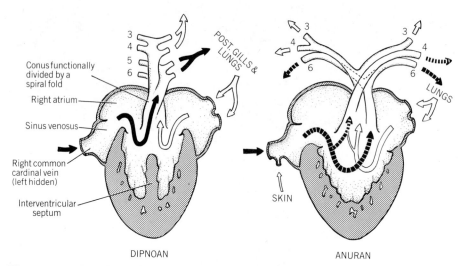

Conus functionally
divided by a
spiral fold

Right atrium

Sinus venosus

Right common
cardinal vein
(left hidden)

Interventricular
septum

DIPNOAN   ANURAN

Figure 14-4 DIAGRAMS OF INTERMEDIATE HEARTS. Ventral views of frontal sections. Somewhat distorted by two-dimensional representation—sinus venosus and pulmonary veins enter atria more dorsally; divisions of conus spiral in space. There is variation within the taxa represented. Aortic arch derivatives are numbered.

interpretations. Since both conformation and blood flow of the functioning heart are exceedingly complex and constantly changing, the figures are somewhat diagrammatic to provide clarity. This is particularly true of the divisions or derivatives of the conus.

The atrium of DIPNOI is partly divided by an **interatrial septum** into right and left chambers. The sinus venosus delivers unoxygenated blood to the right chamber, and the pulmonary veins supply oxygenated blood to the left (Figure 14-4). The ventricle (in genera that use the lung frequently) is partly divided by a large muscular **interventricular septum,** and studies show that mixing of the two streams of blood remains surprisingly low. The conus, which is large and no longer contractile, is also partly divided by a **spiral fold,** or flap of tissue that is anchored to the wall of the conus along a spiral path. Most unoxygenated blood is normally shunted to the 5th and 6th aortic arches, which deliver to the posterior gills and lungs. Oxygenated blood goes to the 3rd and 4th arches, which send it to the body.

The atrium of ANURA is completely divided into right and left chambers. Again, blood in the left chamber has usually been oxygenated in the lungs, and blood in the right is relatively unoxygenated. However, blood returning from the skin joins systemic veins, so there is mixing of blood on the right side (Figure 14-4). The ventricle is undivided (probably a primitive rather than degenerate condition), yet blood flows through it almost without eddies and mixing of the two streams in the ventricle is minimal. The conus has a spiral fold that normally shunts right side, unoxygenated blood to the lungs and skin, and left side, oxygenated blood to the right systemic arch, from which branch the arteries to the

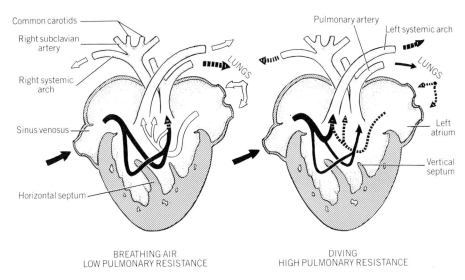

Figure 14-5 DIAGRAMS OF THE FACULTATIVE HEART OF THE TURTLE. Ventral views of frontal sections. Distorted by two-dimensional representation—sinus venosus and pulmonary veins enter atria more dorsally; divisions of conus spiral in space.

head and forelimbs. Both types of blood may enter the left systemic arch, the mix depending on the resistance in the pulmonary circuit.

It is not surprising that URODELA, being less dependent on pulmonary respiration, have a less effective double-circuit pump. Regression has gone so far in lungless salamanders that the interatrial septum may be lost.

The sinus venosus of reptiles varies from large (turtles) to small or vestigial, and again joins only the right side of a completely divided atrium. The ventricle of REPTILES OTHER THAN CROCODILIANS is incompletely divided into dorsal and ventral chambers by a **horizontal septum** and there is also a smaller vertical septum (Figure 14-5). Both atria enter the dorsal chamber. The conus (of *all* reptiles) is unique in that the embryonic organ divides completely into *three* channels, a pulmonary trunk and independent right and left systemic trunks. When these reptiles respire in air, pulmonary vessels are dilated, pulmonary resistance is low, and blood circulates as shown in the left-hand drawing of Figure 14-5. If it becomes disadvantageous to drive blood through the lungs, as when an aquatic reptile remains submerged long enough to deplete oxygen stored there, then pulmonary resistance increases, and the exit of blood from the heart is shifted to the right (right-hand drawing). (It is probable that an intermediate pattern can occur having a flow from the left atrium as in the right-hand drawing, and a flow from the right atrium as in the left drawing.)

CROCODILIANS have a complete interventricular septum, but there is a foramen between the bases of the two systemic trunks. The right systemic arch crosses over to join the left ventricle; the left systemic arch and pulmonary arch both join the right ventricle. (A "trunk" is a division of the conus; an "arch" is

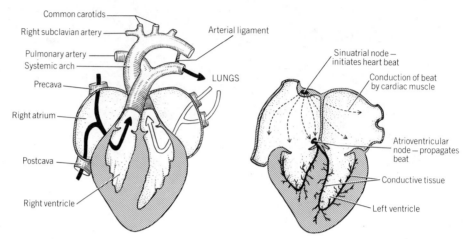

Figure 14-6   HEART OF A MAMMAL. Ventral views of frontal sections. Somewhat diagrammatic.

the continuation extending to the dorsal aorta.) Oxygenated blood from the left ventricle flows directly into the right trunk and also enters the left trunk through the foramen. Unoxygenated blood from the right ventricle flows directly into the pulmonary trunk, yet, in spite of the direct connection to the left systemic, does not normally enter because resistance is greater there. It is probable that when the animal has been submerged long enough to deplete the oxygen in its lungs, pulmonary resistance is increased and some blood *does* enter the left systemic trunk, thus bypassing the lungs.

HEARTS OF HOMEOTHERMS: A DOUBLE-CIRCUIT PUMP   Adult BIRDS AND MAMMALS have completely double circulations: a low-pressure pulmonary circuit using the right side of the heart and a high-pressure systemic circuit using the left side (Figure 14-6). The sinus venosus is vestigial in birds and absent in adult mammals, the embryonic sinus having merged with the right atrium. The atrium is divided and relatively smaller than in many fishes: The ventricle is divided and disproportionately strong on the left side. The embryonic conus divides into a pulmonary trunk joining the right ventricle and a systemic trunk joining the left. Unlike the condition in lower vertebrates, the systemic arch is single. It loops to the right in birds and to the left in mammals.

The developing mammalian heart recapitulates evolutionary stages, having a sinus venosus that joins the right atrium, incomplete interatrial and interventricular septa, and an undivided conus (Figure 14-7).

CONTROL OF THE HEART BEAT   The beat of the adult heart of amniotes is influenced by the autonomic nervous system, some hormones, and temperature. However, cardiac muscle has the inherent capacity to contract rhythmically. The embryonic ventricle is the first chamber to beat. When the atrium starts, it passes its own faster rhythm to the ventricle, and finally when the sinus venosus pulses, its still faster rhythm is followed by the entire heart. In lower vertebrates and

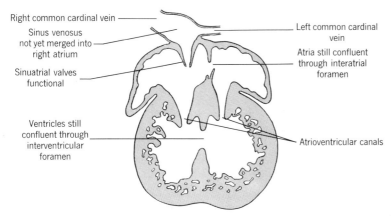

Right common cardinal vein

Sinus venosus not yet merged into right atrium

Sinuatrial valves functional

Ventricles still confluent through interventricular foramen

Left common cardinal vein

Atria still confluent through interatrial foramen

Atrioventricular canals

Figure 14-7 RECAPITULATION IN THE HEART OF A MAMMALIAN FETUS as seen in a ventral view of a nearly frontal section. Not seen in this plane of sectioning are the small pulmonary veins and the conus, which is starting to divide into pulmonary and systemic trunks.

embryos, a **sinuatrial node** develops in the sinus venosus and initiates the beat, which is then transmitted over the heart by the muscle tissue itself, not by nervous tissue. The node is derived from muscle cells but is unlike either muscle or nervous tissue. Its pacemaker mechanism is not known.

When the embryonic sinus venosus of homeotherms merges with the right atrium, it carries the sinuatrial node with it and a second, or **atrioventricular node,** forms (Figure 14-6). This node distributes the beat over the ventricle and, unlike the first, has a special conductive tissue to carry out its function.

SUMMARY OF CARDIAC EVOLUTION The ancestral vertebrate heart may have been straight, but known hearts are folded to place the atrium dorsal or anterior to the ventricle. The single-circuit heart is nearly symmetrical; others have marked asymmetry. The sinus venosus of intermediate hearts joins the right atrium, or the right side of an incompletely divided atrium. In birds and mammals it merges into the wall of the right atrium, contributing its node and valves, and possibly forming part of the coronary sinus, which drains heart muscle.

The atrium is partly divided in dipnoans and urodeles, and completely divided in other tetrapods. The incomplete interventricular septum of some dipnoans apparently is not homologous to the septum of tetrapods. The origin of the complete septum of crocodilians, birds, and mammals is debated. Holmes believes that these are broadly equivalent to each other and to the horizontal septum of turtles and lizards, although other tissue may augment the horizontal septum as it is converted.

The conus is partly divided by a spiral fold in dipnoans and anurans. It is completely divided into three trunks (internally if not also externally) in reptiles, and into two trunks in birds and mammals. This has long seemed to some morphologists to be a difficulty in deriving birds and mammals form known reptilian lineages. however, the embryonic conus of all these animals is at first undivided, then divided into two and, finally, in reptiles, into three channels. The adult two-channel condition could have evolved from the three-channel condition by neoteny or the deletion of a last developmental step.

BLOOD AND BLOOD-FORMING TISSUES

Vertebrate blood consists of blood cells of various types suspended in a fluid called **plasma.** Plasma is an aqueous solution of nutrients, metabolic wastes, salts, hormones, and proteins. Blood proteins are probably formed by the liver. They include albumens, the most abundant; fibrinogen, which contributes to the complicated clotting reaction; and several globulins, which respond to the entry of certain foreign materials into the body. Lymph is essentially a plasma with reduced protein content. Serum is the fluid remaining after blood has clotted.

Blood cells are of three principal kinds: red cells or **erythrocytes,** white cells or **leucocytes,** and **thrombocytes.** Erythrocytes occur only in blood vessels. They tend to be smaller than leucocytes, yet vary considerably in size, being relatively large in amphibians and small in mammals (Figure 14-8). They are usually flat, oval or round, and nucleated, but in nearly all mature mammals are round and enucleate. Erythrocytes are rich in the red protein hemoglobin, which combines readily with oxygen and is responsible for the efficiency of these cells in oxygen transport. They also transport carbon dioxide.

There are only 1% (mammals) to 10% (fishes) as many leucocytes as erythrocytes in the blood stream, but only leucocytes occur in the lymphatic system. These cells actively move through capillary walls and quickly aggregate at sites of local infection. Some leucocytes destroy foreign bodies by engulfing them—a process called **phagocytosis** (= to eat + cell + process). These cells are also involved in the immune response. It appears that leucocytes that enter the tissues may change function and become typical connective tissue cells; blood is usually classified as a special type of connective tissue.

Many varieties of leucocytes have been identified among the different vertebrates, and intergrades make classification difficult. Two main kinds are granulocytes and lymphoid cells. **Granulocytes** (also called polymorphonuclear

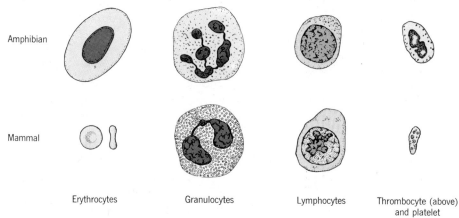

Amphibian

Mammal

Erythrocytes   Granulocytes   Lymphocytes   Thrombocyte (above) and platelet

Figure 14-8   REPRESENTATIVE VERTEBRATE BLOOD CELLS.

leucocytes) are large cells. The nucleus is subdivided into two or more lobes and the cytoplasm is highly granular. **Lymphoid cells** have a central, unlobed nucleus and lack cytoplasmic granules. Large lymphoid cells are called monocytes; small cells (which are much more abundant in all vertebrates) are lymphocytes.

Thrombocytes are small, nucleated, spindle-shaped cells of the blood stream. They occur in all vertebrates except mammals, which have instead enucleate cell fragments called platelets. When thrombocytes or platelets escape through a cut in a blood vessel they adhere to other tissues and disintregrate, thereby releasing a material that initiates the clotting process (thrombocyte = clot + cell).

Unlike other cells, blood cells are short-lived; they survive for days, weeks, or several months and then for some reason are destroyed. Consequently, they are produced and removed continuously from early embryonic life to death. For the most part, cell formation (**hemopoiesis**) and destruction occur in the same tissues. Many of these tissues also contribute to antibody production and clean the blood by filtration and the removal of many kinds of pathogens.

Solid, isolated masses of mesoderm, called **blood islands,** form on the yolk sac of the early embryo of most vertebrates. These produce the first blood of most vertebrates. Blood also forms in conjunction with vessels of the body generally, and vessels continue to produce red cells in adult fishes. The digestive tract, thymus, kidney, liver, and parts of the nasopharynx produce blood in the embryos only of some vertebrates and in embryos and adults of others. **Lymph nodes** occur sparingly in certain water birds and are otherwise limited to mammals where they are abundant along the lymphatic vessels in certain areas (Figure 14-9). They are whitish lumps of variable size and shape that produce quantities of lymphocytes.

Islets and cords of splenic tissue occur under the submucosa of the gut of hagfishes and within the lengthwise intestinal fold of lampreys. In other vertebrates the **spleen** is a discrete reddish organ located in the dorsal mesentery. It may be elongate or compact and relatively small (birds) or large (mammals). In

Figure 14-9   DIAGRAM OF A SECTIONED LYMPH NODE.

mammals it produces lymphocytes, destroys erythrocytes, and also stores quantities of erythrocytes for release when the body has need of them. In all other classes the spleen produces all kinds of blood cells. Ungulates and some other mammals also have **hemal nodes** that are small nodules of splenic tissue distributed along blood vessels of the gut, kidney, and liver.

There are two kinds of bone marrow: Yellow marrow occurs in the larger cavities of the long bones and is fatty; red marrow occurs in the ribs, vertebral centra, and epiphyses of long bones. The types are not always distinct. Red marrow is hemopoietic in tetrapods. In mammals it produces all types of blood cells except lymphoid cells; in the other classes it produces all types of blood cells.

In general, hemopoietic tissue consists of a matrix of connective tissue intimately related to a rich blood supply. The supportive framework often includes **reticular tissue,** the stellate cells of which are phagocytic and can differentiate into various kinds of blood cells. Commonly, the blood is carried in **sinusoids** that differ from capillaries in being larger and more irregular. They anastomose freely and carry venous blood. (An **anastomosis** is a netlike intercommunication of vessels.) Similar lymphatic sinusoids carry lymph in the lymph nodes.

BLOOD VESSELS   **DEVELOPMENT AND STRUCTURE**   The first visual indication of the formation of the circulatory system is the appearance of blood islands on the yolk sac. The peripheral cells of adjacent islands gradually become contiguous and form a network of tiny vessels. The deeper cells separate from one another and become blood cells.

Vessels soon form also from mesenchyme within tissues of the body. They are initially all of about the same size and together comprise a continuous network in relation to organs of rapid growth such as the central nervous system and eyes. Gradually certain channels enlarge or merge to become the first arteries and veins. All parts of the system are initially paired and symmetrical. However, the primordia of the heart and certain arteries fuse in the midline very soon after their appearance, and asymmetries are established early in certain veins.

Some major lymphatic vessels develop in relation to veins and subsequently become independent. More peripheral parts of the system probably have an independent origin.

Arteries, veins, and capillaries are at first indistinguishable histologically. Each is a tube formed from thin, flat endothelial cells that are loosely wrapped on the outside in a meshwork of connective tissue. This is the definitive structure of capillaries. Arteries and veins each retain these tissues as a **tunica interna** but add more peripheral tissues as they mature (Figure 14-11). Arteries develop a thick **tunica media** that usually consists of circularly arranged smooth muscle fibers, but in the largest arteries consists instead of yellow elastic fibers. These provide the strength and resilience needed to keep blood moving under high pressure. Arteries are completed by a thinner **tunica externa** of longitudinally

Figure 14-10   ARTERIOLE, VENULE, AND NETWORK OF CAPILLARIES.

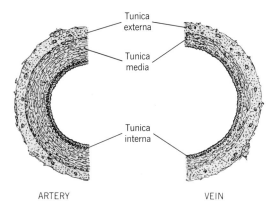

Figure 14-11 STRUCTURE OF BLOOD VESSELS AS SEEN IN CROSS SECTION.

oriented connective tissue. The scanning electron microscope reveals that at least some arteries of at least some mammals are lined by a meshwork of irregular projections that vastly increase their surface.

Veins are usually larger in diameter and thinner-walled than corresponding arteries. This relates to their function of containing most of the blood of the body and transporting it under relatively low pressure. Veins are also more variable in structure, however, and in some instances are much like arteries. The tunica media is typically thin and may be indistinct. The tunica externa, by contrast, is as thick or thicker than that of arteries. At death, the more muscular arteries squeeze most of the blood into the veins. When cut, the arteries stand open but empty, whereas the veins usually are either gorged with blood or collapsed.

Lymphatic capillaries resemble blood capillaries but are larger, more irregular in shape, and more permeable. Their endothelial cells are relatively large. Lymphatic vessels are constructed somewhat like veins, although the three layers tend to be less distinct.

THE INITIAL PATTERN OF ARTERIES   The pattern of vessels that comprises the initial functioning system of the embryo is basically the same for all vertebrates. Virtually the same pattern is also found in adult amphioxus. It can, therefore, confidently be considered primitive for vertebrate phylogeny as well as ontogeny (Figure 14-12). It should be learned as our point of departure.

The heart pumps blood forward under the pharynx in the **ventral aorta** (also called truncus arteriosus)—a vessel that is single where it leaves the heart but sometimes paired under the anterior branchial arches. In embryos, the ventral aorta distributes its blood to paired **aortic arches** that run upward through the visceral arches. In adults of lower vertebrates, the aortic arches have differentiated into proximal (toward the heart) **afferent branchial arteries,** gill capillaries, and distal **efferent branchial arteries.**

The principal distributing vessel of the body is the **dorsal aorta.** This vessel is paired throughout its length when it first differentiates. It may remain paired dorsal to the pharynx, but at more posterior levels immediately fuses in the midline. Blood entering the paired dorsal aortae from the anterior aortic arches

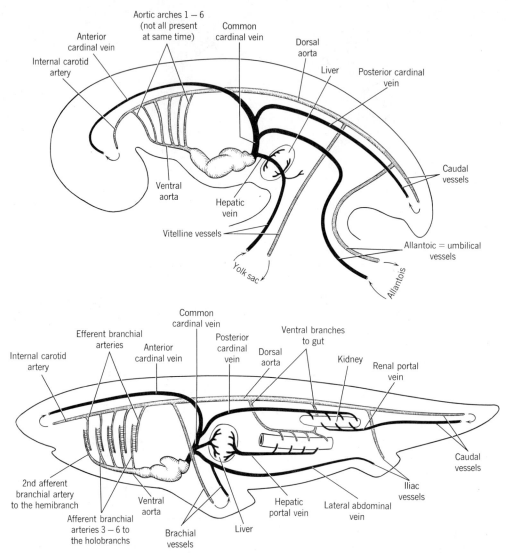

Figure 14-12   BASIC PATTERN OF THE VERTEBRATE CIRCULATORY SYSTEM as seen in an amniote embryo (above) and an adult fish at the shark level of evolution (below). Not shown: subcardinal and supracardinal veins, dorsal and lateral branches of dorsal aorta. All vessels are paired except dorsal and ventral aortae, caudal vein, and vessels of the gut.

(or anterior branchial efferents) runs forward and continues into the head in extensions of the aortae called **internal carotid arteries.** Blood entering the aortae from the posterior arches (or efferents) runs posteriorly into the unpaired dorsal aorta whence it is distributed by three sets of branches: Paired and segmental **dorsal branches** deliver blood chiefly to the body musculature, the largest branches serving the paired appendages. Paired **lateral branches** deliver to urogenital organs. Primitively these are numerous and segmental. **Ventral branches** deliver to the yolk sac and viscera. These branches are initially numerous but only vaguely segmental. They are unpaired with two exceptions: The **vitelline arteries,** which serve the yolk sac, and **allantoic arteries,** which serve the allantois, represent pairs of branches that are established before the embryonic primordia of the dorsal aorta have fused.

EVOLUTION OF ANTERIOR ARTERIES   Embryos of vertebrates nearly always have six pairs of aortic arches, but the first (located in the mandibular visceral arch) is always lost or reduced to remnants in the adult.

CYCLOSTOMES have a long ventral aorta and eight or more aortic arches. The dorsal aorta is single over the gills (lampreys) or paired (hagfishes). ACTINOPTERYGII except Chondrostei are advanced in that the second aortic arch, like the first, has been modified or lost (Figure 14-13). They are relatively primitive in that their afferent and efferent branchial arteries are simple and free of loops and branches. Arches 3-6 characteristically serve the four holobranchs. The efferent branchials tend to lie at the levels of the respective visceral arches. Over the pharynx are paired dorsal aortae that extend forward under the brain as internal carotid arteries. Jaws, orbits, mouth, and pseudobranchs are supplied by arteries of varied patterns, doubtful homologies, and many names.

DIPNOI and CHONDROSTEI vary among themselves but tend to differ from typical ray-finned fishes in that (1) the airbreathers among them have true pulmonary arteries that branch from the efferent segments of the sixth arches, (2) the second aortic arch is usually retained to serve the hemibranch on the hyoidean visceral arch, (3) the fifth and sixth afferents of the air breathers tend to branch from the ventral aorta just anterior to the partly divided conus that diverts into them the relatively unoxygenated stream of blood (see p. 252).

The phylogenetically important extinct rhipidistian fishes probably had simple efferent drainages as in ray-finned fishes. They certainly also had pulmonary arteries joining the sixth arch.

CHONDRICHTHYES also retain the second aortic arch to serve the hyoidean hemibranch. Thus, there are five (rarely more) functional afferent branchial arteries. There are typically four efferent branchials, each collecting from an efferent loop that drains the two hemibranchs facing into one gill chamber. Numerous shunts join adjacent loops. The efferents lie opposite the gill chambers, not the visceral arches as in ray-finned fishes.

Adult TETRAPODS lack both first and second aortic arches. A characteristic carotid system delivers blood to the head. It consists of (1) common carotid arteries derived from segments of the ancestral (and embryonic) paired ventral aortae, (2) external carotid arteries that supply the throat and ventral part of the head and possibly represent the paired ventral aortae anterior to the ventral roots of the third arch, and (3) internal carotid arteries that supply the brain and much of the head.

URODELA retain paired third, fourth, and sixth arches and usually also a pair of fifth arches. Afferent and efferent branchial arteries are, of course, present

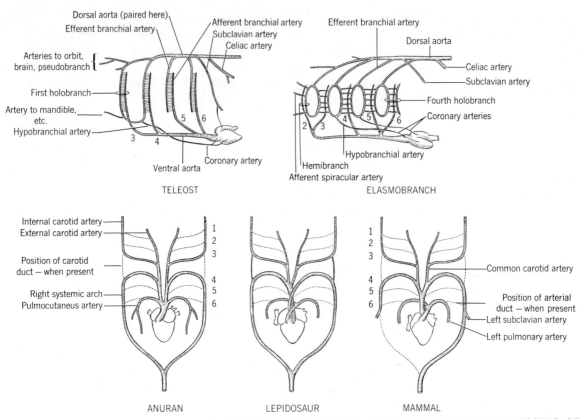

Figure 14-13 DERIVATIVES OF AORTIC ARCHES AND RELATED VESSELS OF REPRESENTATIVE VERTEBRATES. Semidiagrammatic. Left lateral views (above) and ventral views (below). Aortic arches are numbered.

in species retaining gills. Pulmonary arteries branch from the sixth arches. (There are variations on the typical pattern, and, unfortunately for students of morphology, the familiar *Necturus* is unusual in that common carotids are not identified and the proximal parts of the sixth arches are missing.)

ANURA have a "standard" carotid system to transmit blood to the head. The fourth arches are systemic arches transmitting blood to the posterior part of the body. The short segment of the paired dorsal aortae between the dorsal roots of the third and fourth arches (called the **carotid duct**) is thus a useless shunt between diverging streams of blood; although retained by most urodeles, it is lost in anurans. The fifth arch is always absent in the adult. The useless (indeed detrimental) part of the sixth arch that is distal to the pulmonary artery (called the **arterial duct**) is also lost. The proximal part of the sixth arch contributes to the base of the pulmonary artery.

The basic pattern of aortic arch derivatives in modern REPTILES is the same as for anurans, with the important difference that the conus has divided into three instead of two channels. The distribution of blood to these vessels was presented above with the heart. The carotid system (which is variously modified in the different groups) relates only to the right arch, which may have the more oxygenated blood. It is curious that the carotid duct still persists in some lepidosaurs, and the arterial duct in *Sphenondon* and several other genera.

BIRDS modify the carotid system so that internal carotids substitute for common carotids in the long neck (Figure 14-21). Right and left vessels lie side by side in the chick, and it is of interest to systematists that both or either may be retained in the different orders. The right, fourth aortic arch becomes the systemic arch. Characteristic brachiocephalic arteries, which branch from the systemic arch, give rise to carotid and subclavian arteries, the pattern of branching being various.

MAMMALS retain only the left systemic arch. The carotid system is less modified than in birds, but the proximal relations of the common carotids to the arch and the subclavian arteries are often asymmetrical.

EVOLUTION OF POSTERIOR ARTERIES   Posterior to the pharynx the **dorsal aorta** is the most conservative vessel of the body. It is always a large, median longitudinal artery lying ventral to the notochord or spine. It extends into the tail as the **caudal artery.** Moreover, there is little evolutionary significance to the variations of the branches of the aorta. **Ventral visceral branches** may be numerous (amphibians) but typically include only a **celiac artery** to stomach, duodenum, liver, and pancreas and one or several **mesenteric arteries** to the remainder of the gut.

Lateral visceral branches serve the urogenital organs. They are numerous if these organs are long (most anamniotes) and otherwise few. They take their names from the organs served—renal, ovarian, spermatic.

Dorsal somatic branches of the dorsal aorta serve spinal cord, muscles, and skin. They are segmental in fishes, but in tetrapods tend to be modified for functional reasons noted below. The large **subclavian arteries** extend into the pectoral appendage as **brachial arteries** (do not confuse with branchial arteries); the corresponding **iliac arteries** extend into the pelvic appendages as **femoral arteries.**

THE INITIAL PATTERN OF VEINS   The initial ontogenetic and phylogenetic pattern of veins comprises three somewhat independent systems. The **subintestinal-vitelline system** drains the tail, digestive tract, and yolk sac (Figure 14-14). A **caudal vein** runs forward to a **cloacal loop** that skirts the cloacal area. **Subintestinal veins** continue the drainage forward, receive tributaries from the digestive tract and, after penetrating the liver, empty into the common cardinal veins (see below) close to the heart. This completes the subintestinal component of the system as it may have occurred in ancestral vertebrates. A vitelline component consists of large **vitelline veins,** which come from the yolk sac and enter the subintestinals (Figure 14-12). Morphologists find it convenient to think of the vitelline veins as large precocious branches of the subintestinals. Embryologists

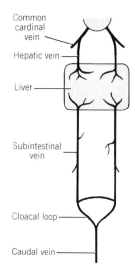

Figure 14-14   THE ANCESTRAL SUBINTESTINAL SYSTEM OF VEINS. Ventral view.

are more likely to regard the subintestinals as belated branches of the vitelline veins.

A second system of veins is the **cardinal system,** which drains the head, dorsal body wall, and kidneys (Figure 14-12). Its principal vessels are the **anterior cardinal veins,** which are lateral to the internal carotid arteries; **posterior cardinal veins,** which are located in or adjacent to the dorsal portion of the kidneys; and short **common cardinal veins,** which are formed by the confluence of the preceding vessels. After being joined by veins of other systems these return all blood to the heart. Several additional pairs of longitudinal veins are related to the embryonic kidneys. These anastomose freely with one another and drain into the posterior cardinal veins. Two of these that contribute to the story to follow are the **subcardinal** and **supracardinal veins.**

The third system of veins is the **abdominal system,** which drains the ventral body wall and paired appendages. It consists of paired **lateral abdominal veins,** which receive **iliac** and **subclavian branches** from the posterior and anterior appendages and drain into the common cardinal veins near the entrances of the subintestinal veins (Figures 14-12 and 14-15).

EVOLUTION OF ANTERIOR VEINS   Vessels draining the anterior part of the body have been conservative. They represent part of the ancestral cardinal system and associated veins. Anterior cardinal veins drain the brain and part of the head in cyclostomes and fishes (Figure 14-15). In tetrapods the same vessels are called **internal jugular veins** and are somewhat modified peripherally by various combinations of the parent vessels with their embryonic tributaries (Figures 14-16 to 14-18). The more ventral and external parts of the head are drained in fishes by veins of uncertain homologies usually called **inferior jugular veins.** The same regions are drained in tetrapods by **external jugular veins,** which are apparently enlarged tributaries of the primitive anterior cardinals. They join the internal jugulars in the neck. The common cardinal veins are paired in fishes; one or the other is lost in adult cyclostomes. Venous sinuses are common in the cardinal and jugular drainages—particularly in cyclostomes, cartilaginous fishes, and (adjacent to the brain) in mammals.

Between the heart and the confluence of internal and external jugulars the derivatives of the cardinals of tetrapods are called **precavae** (also anterior vena cavae). These vessels are paired all the way to the heart in tetrapods other than mammals, and also in some primitive mammals (most marsupials and rodents, insectivores, and some others). In other mammals a shunt called the **brachiocephalic** (or **innominate**) vein carries blood from the left jugulars to the right precava. The left precava is then lost except for a contribution to the coronary sinus (Figure 14-23).

The anterior appendages are drained by subclavian veins, which enter the anterior or common cardinals or their equivalents, the precavae or brachiocephalic.

EVOLUTION OF THE HEPATIC PORTAL SYSTEM   The evolution of the drainage of the digestive viscera has also been conservative. It relates to the subintestinal system of the ancestral circulation (Figure 14-14). The subintestinal veins of

Paired dorsal aorta

EFFERENT BRANCHIALS

SUBCLAVIAN

Lateral

Brachial

Ventrolateral = epigastric

Genital

CELIAC

DORSAL AORTA

Hepatic

Gastrohepatic

Gastric

Pancreaticomesenteric

Intraintestinal

Anterior intestinal

Posterior intestinal

Gastrosplenic

Posterior mesenteric
(to rectal gland)

ILIAC

Posterior epigastric

Femoral

Caudal

Inferior jugular

ANTERIOR CARDINAL

Sinus venosus

COMMON CARDINAL

SUBCLAVIAN

HEPATICS

Brachial

Posterior cardinal sinus

Genital sinus

LATERAL ABDOMINAL

POSTERIOR CARDINAL

HEPATIC PORTAL

Gastric

Pancreaticomesenteric

Anterior intestinal

Anterior splenic

Intraintestinal

Posterior splenic

Posterior intestinal

Kidney

RENAL PORTAL
(dorsal to kidney)

ILIAC

Cloacal

Caudal

ARTERIES   VEINS

Figure 14-15 CIRCULATORY SYSTEM OF THE SHARK *Squalus.* Ventral views. Principal vessels identified by capital letters. For the ventral aorta and branchial vessels see Figures 14-3 and 14-13.

embryos are paired and continue to the yolk sac as vitelline veins. Adults of all vertebrates lose the vitelline veins and establish a single large **hepatic portal vein** (hepatic = of the liver) by the selective retention of parts of the left and right subintestinals and of several anastomoses that occur between them within and just posterior to the liver. Inside the liver the system breaks up into sinusoids,

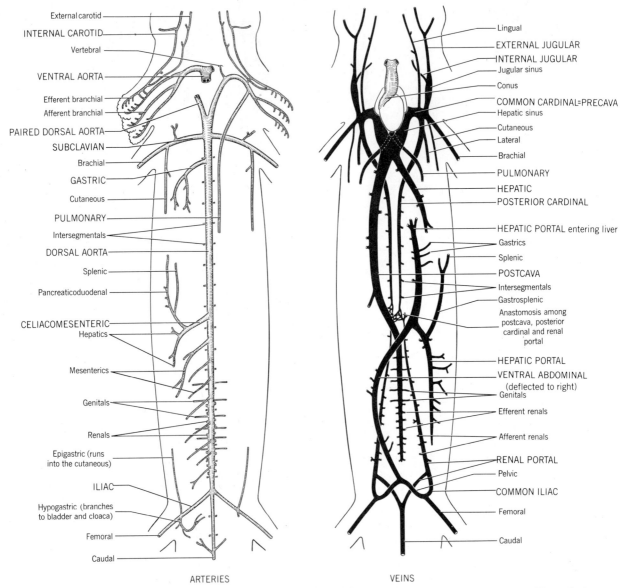

ARTERIES   VEINS

Figure 14-16 CIRCULATORY SYSTEM OF THE URODELE *Necturus*. Ventral view. Principal vessels identified by capital letters.

and anterior to the liver it continues toward the heart as one or more **hepatic veins** that drain the liver (Figure 14-23).

Primitively, the hepatic portal vein drains not only the gut but also the tail via the caudal vein and cloacal loop. This condition is retained by hagfishes and some teleosts. In other vertebrates the hepatic portal loses its connection to the cloacal loop. Consequently, the caudal blood must enter another system of veins

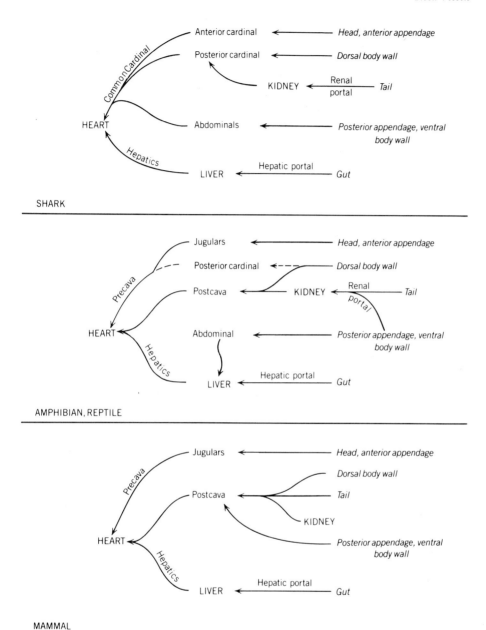

Figure 14-17   DIAGRAM OF THE EVOLUTION OF VENOUS BLOOD FLOW.

(see below). (As noted later, the liver of amphibians and reptiles may also receive portal blood by way of the abdominal system of veins.)

EVOLUTION OF THE RENAL PORTAL SYSTEM   The renal portal system evolved from the ancestral posterior cardinal veins and associated channels. Cy-

Figure 14-18 EVOLUTION OF THE CARDINAL SYSTEM OF VEINS (solid black) AND SOME RELATED VESSELS (open). Ventral views.

clostomes and larval fishes retain the ancestral condition: The posterior cardinal runs from the cloacal loop to the common cardinal, receiving tributaries from urogenital organs and body wall.

In FISHES an interruption develops in the posterior cardinals just anterior to the kidneys (Figure 14-18). Blood from the cloacal loop (which is now detached from the hepatic portal system) usually flows into the *posterior* segments of the posterior cardinals—now called **renal portal veins**—passes into the tissues of the kidneys, and emerges into subcardinal veins, which transmit it to the *anterior* segments of the posterior cardinals. The **renal portal system** is thus established (which, as we shall see later, is of great physiological importance).

DIPNOANS and URODELES retain the same pattern of vessels but add a new vessel, the postcava (see below), which receives most of the blood from the kidney. The reduced anterior segments of the posterior cardinal veins drain the body wall. ANURANS and REPTILES are further advanced in that the anterior segments of the posterior cardinals are usually represented only by variable **vertebral veins,** which drain the anterior part of the thorax. All blood in the renal portal veins now enters the kidneys, but in some species part of the blood is shunted through the organ to the postcava without entering a capillary bed.

The situation is the same in BIRDS except that nearly all blood shunts through the kidney to the postcava. It enters kidney tissues only if a valve closes and occludes the direct route. MAMMALS have no renal portal system at all. An **azygous vein** and, in some species, a hemiazygous vein, which drain part of the thorax, are the only derivatives of the anterior segments of the posterior cardinals.

EVOLUTION OF POSTERIOR SOMATIC AND PLACENTAL VEINS   The tail, posterior appendage, and body wall are drained by veins having a relatively diverse evolutionary history. We have already noted that blood from the tail and cloacal loop drained into the *subintestinal system* in hagfishes and some teleosts, but that this ancestral connection is otherwise lost. We have also seen that the *posterior cardinals* join the cloacal loop, and that posterior segments of them, as the renal portal veins, carry caudal blood to the kidneys in all vertebrates except mammals. The renal portal veins of tetrapods also receive at least part of the blood from the hind limb. Furthermore, anterior segments of the posterior cardinals (or derivatives of one or both of them) drain part of the body wall.

Another drainage, the *abdominal system* drains the pelvic fins and ventral body wall of sharks, and this seems to be the primitive condition (Figures 14-15 and 14-17). This system is curiously absent in cyclostomes and ray-finned fishes. In adult dipnoans, amphibians, and reptiles, abdominal veins (single or paired) are present, but with the difference that they join the cloacal loop posteriorly and (except in dipnoans) enter the liver anteriorly. Thus, blood from the hind limbs of amphibians and reptiles may be diverted *either* into the renal portal veins on its way to the kidneys, *or* into the abdominal veins whence it goes to the liver (Figures 14-17 and 14-20).

Adult birds and mammals lack the abdominal system, but embryonic amniotes have paired abdominals. Fetuses retain both (reptiles) or the left only as **allantoic** or **umbilical veins.** These vessels enter the liver as their antecedents did in ancestral amphibians. However, within the liver a **venous duct** usually connects them directly with hepatic veins, so the fetal circulation is not functionally portal.

Finally, and importantly, another major posterior somatic vein, the **postcava** (or posterior vena cava), is present in dipnoans, the living coelacanth, and tetrapods. This is not a new vessel in an evolutionary sense, but is instead derived from segments of preexisting systems. In dipnoans it forms from parts of the right posterior cardinal and the adjacent subcardinal. (The postcava is foreshadowed in other fishes by asymmetries in these channels.) In mammals, right hepatic and supracardinal veins, and anastomoses among the primordia, also contribute (Figure 14-24). In amphibians, reptiles, and birds, the postcava extends only to the posterior end of the kidney (Figures 14-16 and 14-20). In mammals, there being no renal portal system, the postcava collects directly from the tail and hind limb (Figure 14-23). Considering its complicated ontogeny it is not surprising that the postcava and its tributaries show much individual variation.

SOME FUNCTIONAL CONSIDERATIONS

It is important that the heart and blood vessels be studied not merely as a pattern of channels, but also as a functional system. One approach is to select any two organs of a vertebrate and name, in sequence, the vessels and heart chambers that would be traversed by blood in moving from one organ to the other. The exercise can be repeated for the equivalent organs of several other vertebrates.

It is also instructive to seek functional explanations for certain configurations of the system. As a first example, the conservative **circulation to the liver** always

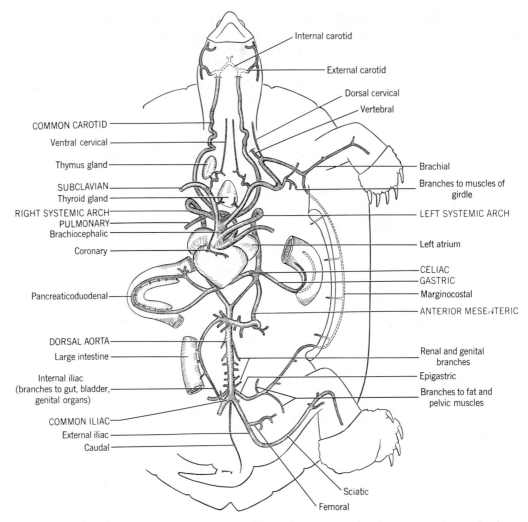

Figure 14-19   ARTERIES OF A TURTLE, family Testudinidae. Principal vessels identified by capital letters.

receives all of the blood from the digestive tract and may (in amphibians and reptiles) also receive venous blood returning from the tail and hind limbs. The supply of arterial blood, by contrast, is a trickle. Why so much blood, and why mostly venous? The many functions of the liver (see p. 221) require that it process quantities of blood. It receives the blood from the gut because some of the nutrients absorbed there would be not useful, or would even be detrimental in the general circulation if not first transformed by the liver. Another reason for furnishing the liver with portal blood is that since it requires little arterial blood to do its job, energy would be wasted if a large quantity of high pressure, oxygenated blood were provided.

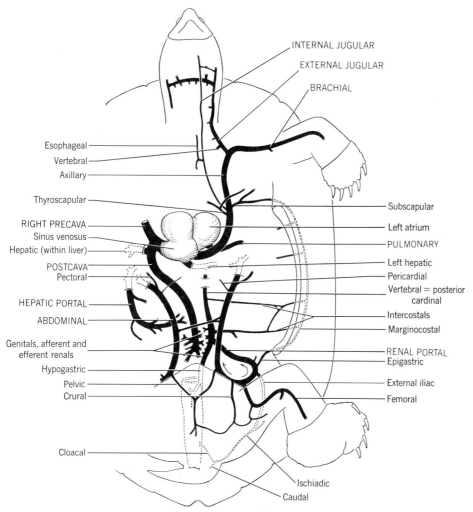

Figure 14-20   VEINS OF A TURTLE, family Testudinidae. Principal vessels identified by capital letters.

Turning to the **circulation of the kidney,** we note that most vertebrates have a functional renal portal system, but that mammals have none, and some others shunt part or all of their portal blood directly into the postcava. Furthermore, some vertebrates direct a large supply of arterial blood to the kidney, and some do not. How are these differences explained?

The vertebrate kidney usually has two capillary beds. One is organized as tiny knots called glomeruli, which lie within the renal capsules and are associated with the filtration phase of kidney function. The other is organized as nets that surround the nephric tubules and are associated with excretion and selective resorption (Figure 14-25). Animals that produce a large volume of urine (fresh-

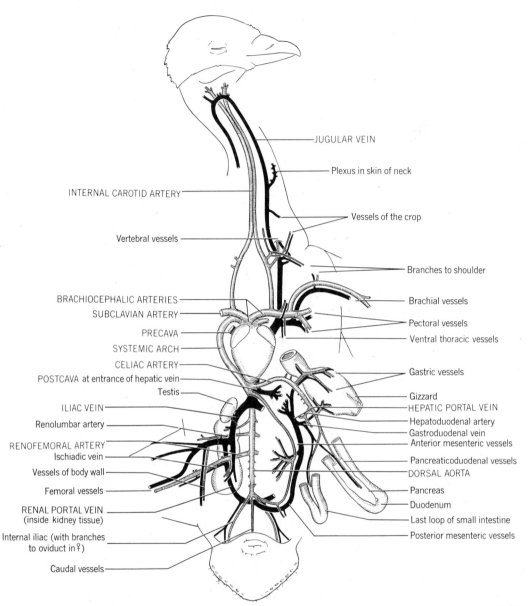

Figure 14-21   CIRCULATORY SYSTEM OF THE PIGEON. Principal vessels identified by capital letters.

water fishes, amphibians, mammals) must have a large high-pressure (arterial) flow of blood to the glomeruli, because it is this pressure that is the driving force of the filtration process. These animals can either reuse the same blood for the tubules (mammals) or use portal blood instead. The oxygen content of this blood is unimportant, it is low pressure that is required so reabsorption can take place.

Figure 14-22 ARTERIES OF THE CAT. Ventral view. Principal vessels identified by capital letters.

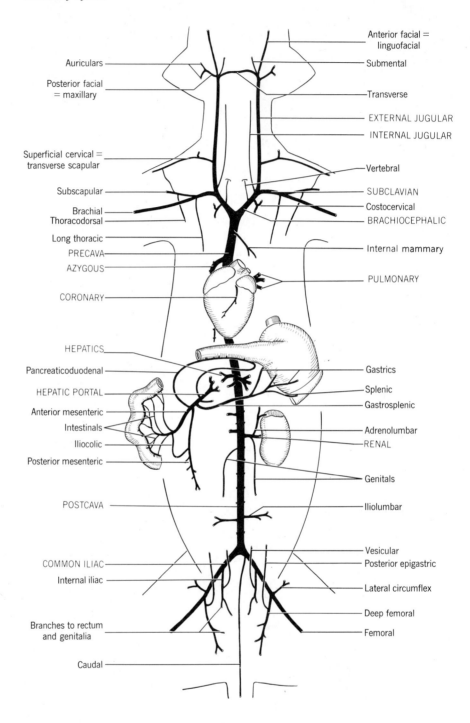

Figure 14-23 VEINS OF THE CAT. Ventral view. Principal vessels identified by capital letters.

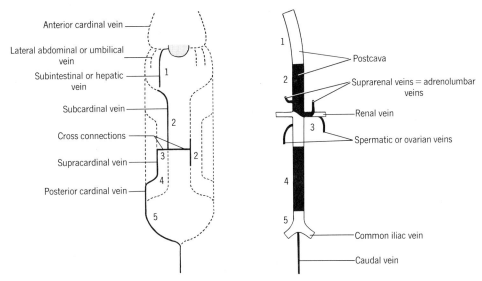

Figure 14-24 FORMATION OF THE POSTCAVA IN A MAMMAL. Left, embryonic and evolutionary primordia; right, adult derivatives. Ventral views. Contributions of the different primordia vary among tetrapods.

Figure 14-25 BLOOD CIRCULATION TO THE EXCRETORY UNITS OF THE KIDNEY IN RELATION TO FUNCTION.

Animals that conserve water and salts (marine fishes, reptiles, birds) have only a small high-pressure flow to the glomeruli (or none in some fishes) so filtration will be limited. Therefore, in order to supply enough low-pressure blood to the tubules a renal portal system is needed.

The function of the kidney in osmoregulation is discussed further on page 287.

The **coronary circulation** differs markedly between vertebrates using gills and those using lungs. The heart muscle of fishes cannot be supplied by blood from adjacent major vessels because that blood is low in oxygen and high in carbon dioxide. It is supplied instead by long coronary arteries that branch from branchial efferents, subclavians, or dorsal aorta (Figure 14-13). The coronary arteries of vertebrates returning oxygenated blood to the heart from lungs can, by contrast, be short vessels that branch from the base of the systemic arch (or right systemic arch if there are two).

In order for the **respiratory circulation** to be most effective, blood delivered to the respiratory organ should have minimum oxygen content and maximum carbon dioxide content, and blood leaving the organ should reach the metabolizing tissues without prior mixing with oxygen-poor blood. These conditions often pertain, but not always. Blood oxygenated in the gills of fishes and branchiate amphibians does continue directly to the tissues of the body. However, a consequence is that blood pressure in the dorsal aorta can be only one-half or one-fourth of the pressure in the ventral aorta. This is adequate for fishes because their hearts need not pump against gravity and their metabolism is low. Also, the swimming muscles or ventilation muscles of various fishes assist in moving the blood.

One might expect air-breathing fishes to supply blood to their lungs from the ventral aorta or branchial afferent arteries, those vessels having oxygen-poor blood. It is surprising that none does. Instead, they supply their lungs with blood coming either from the dorsal aorta or the efferents of the fifth and sixth aortic arches. Either arrangement can be effective only if the more posterior gills do not accomplish a significant degree of oxygenation of the blood before it is sent on to the lungs. This circumstance would pertain in stagnant water or if much blood were shunted past the gill capillaries. In any event, to take blood to the lungs from dorsal aorta or posterior branchial efferent arteries has what appears to be a serious weakness: Blood flowing to the posterior part of the body can be no richer in oxygen than blood that will go *to* the lungs. Also, these fishes mix oxygen-rich blood from the lungs with oxygen-poor blood from the tissues and then send the mixture to both lungs and tissues. However, this is adequate because it has met the needs of air-breathing fishes. (Air-breathing fishes eliminate most carbon dioxide by way of the gills, not the lung. Their need for gills is continuous; their need for lungs is intermittent.)

This general plan also serves the needs of the embryos and fetuses of amniotes. The respiratory organ is a yolk sac or allantois (or placenta, which is derived in part from the yolk sac or allantois). The vitelline arteries to yolk sac, and allantoic or umbilical arteries to allantois or placenta, are like the pulmonary arteries of some fishes in being branches of the dorsal aorta. For fetuses also, therefore, the posterior part of the body must get along with blood having the same oxygen content as that going *to* the respiratory organ.

Fetuses alter this situation by abandoning their fetal respiratory organs at birth or hatching. Adult fishes that supply blood to the lungs from the dorsal aorta cannot alter the situation and seemingly have no need to do so. Tetrapods do it by keeping oxygen-rich and oxygen-poor blood separate (or nearly so) in the heart and then sending the latter only to the lungs by way of the sixth aortic arch (Figure 14-13).

The evolution of an effective respiratory circulation required one further change for air-breathing vertebrates: Blood returning to the heart from the lungs had to be kept separate from systemic blood. Most fishes drain blood from the gas bladder or lung into the postcardinal or hepatic portal veins, which are far from the heart. The bowfin, a holostean, drains blood from the lung into the left common cardinal vein. Blood returning from the respiratory skin of amphibians also mixes with systemic blood. The pulmonary veins of dipnoans and tetrapods, at last, enter directly into the left, or "oxygenated side" of the heart.

Some features of the **flow of blood to and from the tissues** remain to be mentioned. Fishes maintain a nearly constant body configuration; vessels may be interrupted by injury but not occluded by accidents of posture. Furthermore, gravity does not affect blood circulation of fishes because of the buoyancy of the surrounding water. Tetrapods temporarily restrict or occlude vessels as posture is changed, and gravity does affect the distribution of blood. Several provisions are made to ensure that tissues will not be deprived. Arterial blood pressure tends to be higher than in fishes, and blood volume is relatively greater. Most parts of the body have a double blood supply by vessels that are parallel (e.g., ulnar, radial, and interosseus arteries of forearm) or that form **distribution loops** having inflow at each end (e.g., volar arches that deliver blood to digits, arches along curvatures of stomach and intestine, confluence of internal mammary and epigastric arteries in body wall, confluence of internal carotid and vertebral arteries in a common distribution loop ventral to the brain). Moreover, throughout the body, adjacent major arteries are joined by frequent small anastomosing vessels, and isolated major arteries have frequent small branches that form somewhat meandering and criss-crossing adjacent channels. These small vessels comprise the **collateral circulation.** They are standby channels in case of occlusion or injury to the major channels. Fishes also have a collateral circulation, but it is enhanced in tetrapods.

The venous drainage of the more active tissues is similarly primitive in fishes. The veins have valves only at the junctions of certain tributaries with major channels. This is apparently because the channels are relatively direct (there being no long neck or appendages) and the flow of blood is not complicated by gravity. Nevertheless, venous pressure is very low in fishes, and some of them have evolved accessory venous pumping mechanisms. Caudal musculature of elasmobranchs squeezes blood from segmental veins through valves into the caudal vein which, being in the rigid hemal canal, is not constricted by the muscles. In hagfishes (and perhaps some other fishes as well) respiratory movements hasten venous return in nearby vessels.

The veins of tetrapods differ in that they, like the arteries, tend to develop rich collateral circulations. Valves are usually present; the pull of gravity contributes to their necessity.

EVOLUTION OF THE LYMPHATIC SYSTEM

Amphioxus has no lymphatic system. CYCLOSTOMATA, CHONDRICHTHYES, and CHONDROSTEI have networks of fine vessels that resemble lymphatic vessels but differ in containing some red blood cells and in joining the veins and venous sinuses in many places. The term **hemolymphatic system** is appropriate for these vessels. The more superficial channels tend to be more like veins and the visceral channels more like lymphatics. **Hemolymph propulsors** are present in hagfishes but absent in these other fishes. These valved reservoirs passively propel hemolymph as extrinsic muscles impinge on them. The hemolymphatic system of these vertebrates almost certainly represents an early stage in the evolution of the true lymphatic system.

The lymphatic system is incomplete in Dipnoi. It is fully developed in HOLSOSTEI and TELEOSTEI (Figure 14-26). Four **subcutaneous ducts** are typical: one dorsal, one ventral, and two lateral. One or usually two **subvertebral ducts** (also called cardinal ducts) extend the length of the body cavity. Cranial and visceral networks complete the system. Small pairs of **lymph propulsors** are usually present in the tail and near the pharynx. There are no valves in the vessels. Lymph usually enters the venous system by a pair of openings in the anterior cardinal veins and another pair near the iliac veins.

AMPHIBIANS and REPTILES agree with teleosts in having well-developed lymphatic systems with subcutaneous, subvertebral, and visceral vessels. Valves are absent. Urodeles usually have numerous small pairs of segmentally arranged **lymph hearts** (with their own intrinsic musculature) and also a larger pair near the bases of each set of limbs. Anurans have fewer small hearts but retain the two larger pairs. They have large **lymph sinuses** under the skin. Reptiles have the posterior pair of hearts that drain into the renal portal veins. They are located at the ends of transverse processes of basal caudal vertebrae.

The subvertebral ducts of BIRDS are also called **thoracic ducts.** They drain into the precavae. Smaller ducts may enter the iliac veins. Valves may be present but are not numerous. Small lymph hearts appear in the fetus but are rarely present in the adult.

Most MAMMALS have a large single thoracic duct. It is derived from the subvertebral ducts and drains most of the body and enters the left jugular or subclavian vein. A small duct on the right drains the forelimb and shoulder on that side and enters the right subclavian vein. Numerous bicuspid valves are present in the vessels. Lymph hearts are absent. **Lymph nodes,** sparcely represented in the other tetrapod classes, are numerous in mammals—particularly among the viscera, in the neck, and at the bases of the limbs. The system has many anastomoses and parallel channels. Lymphatic vessels penetrate most tissues but are absent from the central nervous system, bone marrow, deeper parts of the liver and spleen, epithelium of the skin, cartilage, and placenta.

The **thymus gland** occurs in all vertebrates except possibly the cyclostomes. It is derived from the epithelium of one or (usually) several pharyngeal pouches and comes to lie in the gill area, neck, or anterior thorax. It may be diffuse or discrete. The thymus produces lymphocytes and also establishes the immunological potential of the young animal.

Figure 14-26 LYMPHATIC SYSTEM. Ventral views. The teleost and lizard are composite drawings. Modified from Kampmeier.

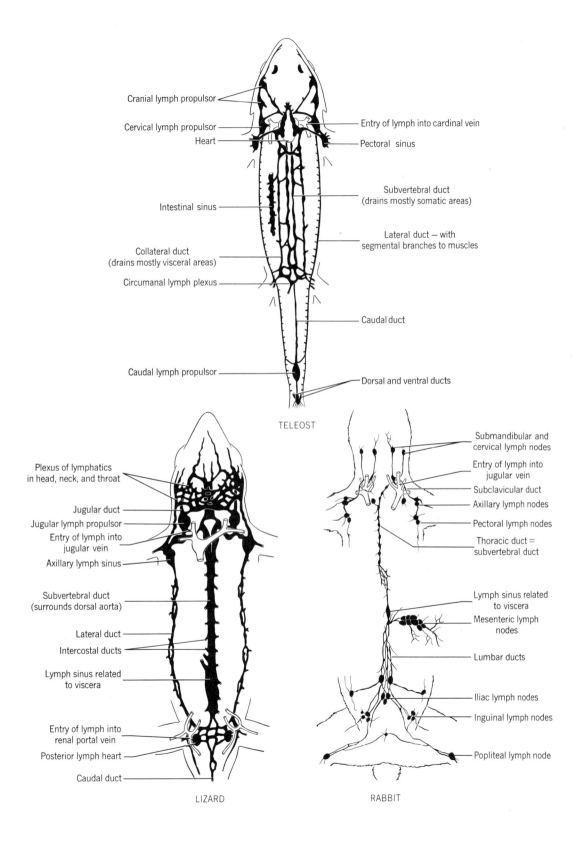

Cranial lymph propulsor

Cervical lymph propulsor

Heart

Intestinal sinus

Collateral duct
(drains mostly visceral areas)

Circumanal lymph plexus

Caudal lymph propulsor

Entry of lymph into cardinal vein

Pectoral sinus

Subvertebral duct
(drains mostly somatic areas)

Lateral duct — with
segmental branches to muscles

Caudal duct

Dorsal and ventral ducts

TELEOST

Plexus of lymphatics
in head, neck, and throat

Jugular duct

Jugular lymph propulsor

Entry of lymph into
jugular vein

Axillary lymph sinus

Subvertebral duct
(surrounds dorsal aorta)

Lateral duct

Intercostal ducts

Lymph sinus related
to viscera

Entry of lymph into
renal portal vein

Posterior lymph heart

Caudal duct

LIZARD

Submandibular and
cervical lymph nodes

Entry of lymph into
jugular vein

Subclavicular duct

Axillary lymph nodes

Pectoral lymph nodes

Thoracic duct =
subvertebral duct

Lymph sinus related
to viscera

Mesenteric lymph
nodes

Lumbar ducts

Iliac lymph nodes

Inguinal lymph nodes

Popliteal lymph node

RABBIT

**THE
CIRCULATORY
SYSTEM AND
THERMO-
REGULATION**

The circulatory system participates in most kinds of thermoregulation by selectively conserving, dissipating, and distributing heat in the body. Heat that enters or leaves the body by conduction (a snake resting on a warm rock), or convection (a whale fluke moving through cold water), or radiation (a basking lizard), or the combination provided by evaporative cooling (a panting bird or sweating mammal) alters the temperature of the skin, or of the epithelium of the tongue, mouth, pharynx, nasal chamber, respiratory tree, or gill. Also, much heat is produced by active muscles. The circulatory system can extend the exchange of heat to other parts of the body by the dilation of small vessels at the affected area and the transport of the blood elsewhere. It can limit the heat exchange to the local area by constriction of the local vessels. Several examples will illustrate the effectiveness of the mechanisms.

When subjected to heat stress, many kinds of cats and artiodactyls cool the nasal epithelium by rapid shallow breathing. The surface area of the vascular epithelium may be increased by the complexity of the turbinate bones. Blood from the nasal chamber goes to a **carotid rete** (= mesh). There the carotid artery, which carries blood to the brain, breaks up to vessels of 200 to 300 μm in diameter. These are bathed by a pool of blood from the nasal chamber. The two streams of blood flow in opposite directions, thus forming a countercurrent heat exchanger that can be very effective: The blood in the brain of a desert-living gazelle may be 2.9°C cooler than the blood leaving the heart.

The fins, flippers, and flukes of whales are potential heat exchangers having enormous surface. An active whale dissipates excess heat by deflecting blood into small surface vessels where it can be cooled by the water. Conversely, a less active animal in very cold water must prevent heat loss. Arteries entering the flukes divide into small arterioles each of which is surrounded by a rosette of venules (Figure 14-27). In this countercurrent exchanger blood entering the fin is cooled by the returning blood and, conversely, the returning blood is warmed by the outflowing blood. The consequence is that the fluke, but not the body, is cooled by the arctic sea. Similarly, there are countercurrent heat exchangers in the legs of penguins and many other birds, the tails of beavers, and the limbs of some primates. In each instance the exchanger can be used or bypassed according to need.

Certain fishes (e.g., bluefin tuna, mako shark) swim continuously using dark muscles deep in the body (see p. 519 regarding red muscle fibers). The metabolic heat produced by these muscles is retained by countercurrent exchangers. The central part of the body may be 10 or even 20°C warmer than the water. The elevated temperature enables the muscles to contract more rapidly and probably also more powerfully and efficiently. The fish can control the mechanism according to need.

Figure 14-27 One unit of the COUNTERCURRENT HEAT EXCHANGER in the dorsal fin of a dolphin as seen in cross section.

15

Excretory System and Osmoregulation

GENERAL NATURE AND DEVELOPMENT OF KIDNEYS

FUNCTION AND CHARACTERISTICS Kidneys are the primary adult excretory organs. Other organs that may contribute to the elimination of wastes from the body are the gills, lungs, skin, parts of the digestive system, and various salt glands. Together, these organs perform two related and essential functions: (1) They remove the nitrogenous waste products of protein metabolism and also many other harmful substances, and (2) they eliminate controlled amounts of water and salts, thus maintaining the internal environment within the narrow limits necessary for life.

Embryologists and morphologists have studied the urinary system for various reasons. Like the pharynx and circulatory system, it is outstanding for the recapitulation that takes place during its development: The system was much studied several generations ago in an effort to elucidate vertebrate origins. Conclusions were tentative, however, and most current research relates structure to the function of the system instead of to its evolution. Furthermore, the urinary system is of interest because it provides striking examples of gradients. Embryos and ancestors each form a series of excretory structures of increasing complexity that replace one another in orderly sequence in time and space.

The kidneys of vertebrates are compact organs derived from mesoderm and consisting of numerous **nephric tubules** (also called renal tubules—"nephros" and "ren" both mean kidney) that open either into the general coelom or into cup-shaped spaces called **renal capsules,** which are coelom derivatives (Figure 15-1). Such kidneys are called **metanephridia.** In the metanephridia of

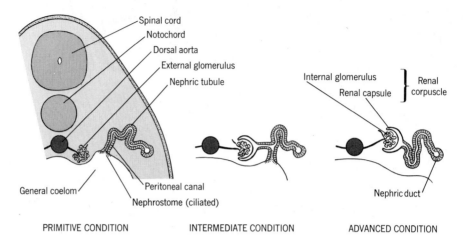

Figure 15-1 RELATIONS OF THE NEPHRIC TUBULE TO THE GLOMERULUS AND GENERAL COELOM. Cross sections.

vertebrates, quantities of water and other constituents of the plasma leave the blood by filtration from small knots of capillaries called **glomeruli** (= small balls). If the capillary bed is surrounded by a renal capsule, it is said to be an **internal glomerulus.** Glomerulus and capsule together form a **renal corpuscle.** If the capillary bed discharges instead into the general coelom, it is then called an **external glomerulus** and the filtrate reaches the tubules indirectly. A second capillary bed meshes around each tubule where many constituents of the filtrate are selectively returned to the bloodstream (see figure on p. 275). A tubule with related corpuscle is the functional unit of the vertebrate kidney and is called a **nephron.** There are thousands or even millions of nephrons in adult kidneys. Urine, the product of excretion by the kidneys, moves out of all the tubules into a long pair of **nephric ducts** that open into the cloaca or a derivative of the cloaca.

Amphioxus has short excretory tubules derived from ectoderm. They do not open into coelom derivatives but instead terminate blindly at their inner ends in flask-shaped cells that drain wastes from the body fluids surrounding them. At its other end each tubule drains into the atrium, which in a sense is outside the body. Such kidneys are called **protonephridia** and are found also in many invertebrates. Protonephridia have been considered ancestral to metanephridia, but if so, the changes that brought the tubules into relation with the coelom or its derivatives, that caused the tubules to share a duct system, that made the common ducts discharge into the cloaca instead of outside the body, and that caused them to develop from mesoderm instead of ectoderm are all lost in antiquity. In truth, the kidneys of amphioxus do not support close affinity of Cephalochordata and Vertebrata, and the origin of metanephridia is in doubt.

DEVELOPMENT   The embryonic mesoderm differentiates on each side of the body into a segmented dorsal epimere, a small lateral mesomere, and an unsegmented ventrolateral hypomere (see figure on p. 80). Earlier chapters have presented the derivatives of the epimere and hypomere; we now come to the relatively inconspicuous but important mesomere.

The mesomere is as long as the general coelom, or even a little longer. Gradually it pinches off from the overlying epimere. Anteriorly it remains relatively thin and becomes segmented into units called **nephrotomes.** Posteriorly it does not become segmented and is called the **nephrogenic cord.** Nephrotomes and nephrogenic cord merge into one another at a level that varies according to species and is unimportant to subsequent development. At mid and posterior levels the developing kidney bulges somewhat into the coelom to form a **nephric ridge.**

The differentiation of the mesomere is subject to control by induction: It will not form normal organs in the absence of either the adjacent epimere or hypomere, and undifferentiated mesoderm from elsewhere in the body will form urogenital organs if transplanted to the position of the mesomere.

The general coelom or splanchnocoel occurs in the hypomere. It is initially continuous with small coelomic spaces or **nephrocoels** in the nephrotomes. The narrow channels that join splanchnocoel to nephrocoels are called **peritoneal canals,** and their openings into the splanchnocoel are **nephrostomes** (Figure 15-1). Functional peritoneal canals and nephrostomes are ciliated. Posteriorly, at levels of the nephrogenic cord, peritoneal canals may be present, but in most amniotes they fail to form. In their absence, spaces regarded as nephrocoels form within the cord by a process of cavitation. Nephrocoels are important because they become the renal capsules of the mature kidney.

Nephric tubules develop either as outgrowths of the walls of nephrocoels (probably the primitive method) or from small solid masses of mesenchyme. Gradually they lengthen and become more or less convoluted. The anterior tubules are the first to mature, and their outer ends join to form a nephric duct on each side of the body, which then extends itself caudad by terminal proliferation until it has grown all the way to the cloaca (Figure 15-2). As the more posterior tubules mature they merely join the preexisting duct. Indeed, the duct induces their development.

(The developmental steps outlined above are typical, but not constant.)

HOLONEPHROS: ANCESTRAL KIDNEY   Although the mode of origin of vertebrate kidneys from invertebrate kidneys is highly speculative, the general nature of the early vertebrate kidney can be approximated from embryological and comparative morphological evidence. This somewhat hypothetical organ is called a holonephros. It was derived from the entire mesomere and hence was long. It was segmented throughout most or all of its length and had one pair of nephrocoels and one pair of tubules per pair of nephrotomes. Glomeruli were large. It is not certain if they were initially internal or external. In either event, nephrostomes were present.

A holonephros is found only in larvae of hagfishes and caecilians. Usually, the developing mesomere has a marked tendency to become subdivided into regions. Development always sweeps along the mesomere from anterior to posterior, but the wave is interrupted at two levels so that anterior, middle, and posterior kidneys form in sequence. The more anterior regions usually degenerate

EVOLUTION OF KIDNEYS

as posterior regions become functional. Furthermore, there is a tendency for the tissues to fail to mature between adjacent regions, thus establishing gaps. The more posterior break may be omitted, however, so two instead of three pairs of kidneys are formed in sequence. The various resultant kidneys will now be reviewed in order.

PRONEPHROS: LARVAL KIDNEY OR SPECIALIZED REMNANT   The most anterior kidney and the first to form is the **pronephros** (Figure 15-2). Relatively few nephrotomes are incorporated—one to twelve pairs according to species, but rarely more than four. All participating segments form discrete units in the early larvae of some species (notably in hagfishes and caecilians) but before becoming functional the pronephros is usually modified by the degeneration of the first and last pairs or so of tubules and by the partial or complete fusion of the remaining segments to form a giant corpuscle, or **glomus,** on each side of the body. Right and left primordia even fuse in the midline to form a single glomus in many Actinopterygii. Pronephric tubules are relatively simple unless related to a glomus, in which case they may become long and coiled. Glomeruli are internal (caecilians, most bony fishes), external (amphibians except caecilians), or even intermediate (some birds and reptiles). Ciliated nephrostomes are often present and may open into that portion of the anterior splanchnocoel that will become pericardial cavity in the adult.

The pronephros appears in at least rudimentary form in all vertebrates. It is functional in the free-living larvae of bony fishes and amphibians and possibly briefly in embryos of some reptiles. Renal corpuscles fail to form in cartilaginous fishes. A somewhat modified pronephros—usually compacted into a glomus— remains functional in adults of hagfishes and several bony fishes, where it may be called the **head kidney** because it is usually not the only functional kidney of these animals. Where not functional in the adult, pronephroi may vanish as development proceeds (cartilaginous fishes, most amphibians, birds) or may leave derivatives that are lymphatic or glandular in nature. The duct of the pronephros,

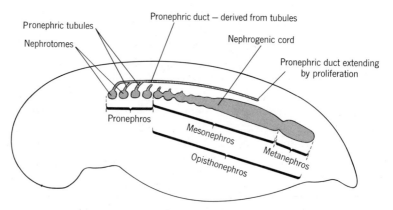

Figure 15-2 DEVELOPMENTAL ORIGIN OF THE ELEMENTS OF THE EXCRE- TORY SYSTEM. Left lateral view of a generalized vertebrate embryo. (Pronephric duct is actually lateral, not dorsal to the nephrogenic cord.)

which is identified as the **pronephric duct,** is the most constant part of the organ and is not lost even though the tubules degenerate.

It is seen that the pronephros is highly variable in development and structure. This is the result of ancient origin and reduced function. Few phylogenetic conclusions can be drawn, and the nonspecialist will not find it profitable to memorize individual peculiarities.

OPISTHONEPHROS: KIDNEY OF ANAMNIOTES   It was explained above that there is an anterior break in both time and space in the development of each mesomere and that the kidney that forms anterior to the break is the pronephros. If a second break forms in the mesomere, then middle and posterior kidneys develop in sequence. The middle kidney is called the **mesonephros** and is present only in fetuses of animals that retain the posterior kidney, or **metanephros,** and none other as adults. These are the amniotes. If all or most of the mesomere posterior to the pronephros instead forms one kidney, then that kidney is called an **opisthonephros.** Such a kidney is typical of late larval and adult anamniotes (see figures on p. 295 and 296).

[In spite of the distinction made above, the terms mesonephros and opisthonephros are used somewhat interchangeably and not entirely without reason. Some vertebrates complicate the concept of multiple kidneys by having an opisthonephros that excludes most of the posterior segment of the mesomere (anurans) or that is incompletely divided into anterior and posterior portions (dipnoans, many urodeles, some elasmobranchs, some teleosts). Some other teleosts have completely separate middle and posterior kidneys and retain each as adults. These organs are not identified as mesonephros and metanephros (as would be logical) but instead as an opisthonephros divided into a trunk kidney and a tail kidney. The more posterior kidneys tend to be larger and more complex than anterior kidneys, but there are no single structural distinctions between adjacent organs. Nephrons at opposite ends of an opisthonephros may be more dissimilar than one of the anterior end of the opistonephros and one of the pronephros. The paired organs of teleost embryos sometimes fuse to become a single median kidney. Also, the anterior end may be lymphoid and not excretory. Students should be cognizant of the concept of variable yet graded development along the length of the mesomere and not be unduly dismayed by the multiplicity of terms or difficulties in their definitions.]

The opisthonephros, then, develops later in time than the pronephros and forms from much or all of the long middle and posterior portions of the mesomere. It is initially segmental, at least anteriorly, and remains so in hagfishes. In other vertebrates, one to twenty generations of secondary tubules develop from the primary tubules, thus causing the organ to lose any segmental character. Opisthonephric tubules tend to be more coiled than unfused pronephric tubules. The entire organ bulges more into the coelom. Anteriorly, the glomeruli may be external but usually all are internal and relate to renal capsules. Nephrostomes are usually present in early development. Some of them may be retained (some sharks, several primitive bony fishes, some amphibians), but usually all are lost.

Opisthonephric tubules usually tap into the preexisting pronephric duct, which then can be called the **opisthonephric** (or in amniotes the mesonephric)

duct. In cyclostomes, some amphibians, and some bony fishes, however, the opisthonephric tubules contribute to the formation of the posterior portion of the duct, and in caecilians the duct sends buds to meet the growing tubules. Which mode of origin is the most primitive is in doubt.

The shorter but otherwise equivalent mesonephros is functional in fetuses of amniotes and even briefly after hatching or birth in some reptiles, monotremes, marsupials, and some ungulates. The organ varies from small (some rodents) to medium (carnivores, primates) to large (some ungulates), its development being inversely proportional to the excretory efficiency of the placenta.

METANEPHROS: KIDNEY OF AMNIOTES   The metanephros is the most posterior of the kidneys and is the last to develop in both ontogeny and phylogeny ("meta" = later in time). The part of the nephrogenic cord from which it develops lies opposite to only one or two body segments. The organ itself is never segmental, and hundreds of thousands of nephrons are formed by successive generations of budding. Glomeruli are always internal, and nephrostomes are absent. Metanephroi differ from other kidneys in being of dual origin. Early in development a **ureteric bud** grows out from each mesonephric duct near its entrance into the cloaca and approaches the differentiating nephrogenic cord (Figure 15-3). The lengthening neck of the bud becomes the duct of the metanephros, or **ureter,** and the end branches within the kidney to form the **collecting tubules** that become continuous with the nephric tubules derived from the nephrogenic cord. The entire organ migrates forward somewhat as it matures, particularly in mammals. It is of interest that the tail kidney of certain teleosts, and also the posterior part of the opisthonephros of some other anamniotes, are drained by independent ducts. These are called **accessory ducts** rather than ureters, but they are nonetheless probably homologous with the amniote ureter.

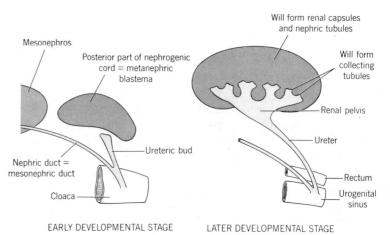

EARLY DEVELOPMENTAL STAGE   LATER DEVELOPMENTAL STAGE

Figure 15-3 STAGES IN THE DEVELOPMENT OF THE METANEPHROS AND URETER.

Figure 15-4 THE COMPOUND KIDNEY OF THE SEA OTTER, *Enhydra*.

The metanephroi of reptiles are of varying irregular shapes and may be crenulated by lobulation. The kidneys of birds fit into hollows in the synsacrum and are usually three-lobed (see figure on p. 302). Kidneys of mammals have a collecting basin called the **renal pelvis** into which urine oozes from one or more **renal papillae** that project into the pelvis. The organs are bean-shaped, simple, and smooth in primitive and small mammals and in primates and most carnivores (see figure on p. 303). In some other species they are smooth externally but have numerous lobes internally and multiple or otherwise complicated papillae. In marine mammals and some ungulates and other large species they are compound organs resembling clusters of rat kidneys (Figure 15-4). The development of compound kidneys in large animals seems to be a means of avoiding very long collecting tubules that would require excessive pressure for driving the contained fluid.

In summary, trends in the evolution of the vertebrate kidney include the loss of nephrostomes and external glomeruli, the occurrence of more numerous and nonsegmental nephrons, and development from a progressively more posterior, and ultimately shorter part of the nephrotome.

A more detailed account of kidney structure is now desirable to provide a basis for interpretation of anatomy in relation to function. The glomerulus is a tuft of capillary loops and anastomoses that hangs into the renal capsule. The capsule is thus cup-shaped, with an outer, or **parietal wall** and an inner, or **visceral wall** (Figure 15-5). The parietal wall is of squamous or cuboidal epithelium. In higher vertebrates the visceral wall enfolds the capillaries of the glomerulus. Its cells (podocytes) have numerous fingerlike projections called **pedicels** that reach to the basement membrane immediately under the endothelium of the capillaries.

KIDNEY STRUCTURE IN RELATION TO OSMOREG-ULATION

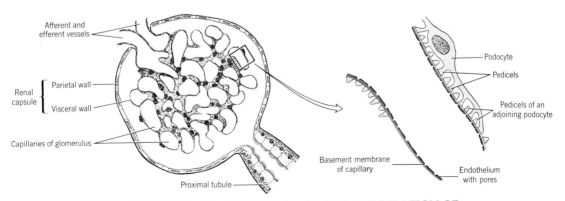

Figure 15-5 RENAL CORPUSCLE OF A MAMMAL (left) AND THE RELATION OF THE VISCERAL WALL OF THE RENAL CAPSULE TO A CAPILLARY (right).

The electron microscope reveals apparent pores in this endothelium, which thus create a filter. The renal corpuscle is, indeed, a device for filtering from the blood quantities of a fluid that is essentially a protein- and fat-free plasma. The driving force of the filter is the relatively high pressure of arterial blood (review p. 271). In mammals the total filtration surface is at least half the body surface, and the daily production of filtrate is equal to many times the volume of the extracellular fluids of the entire body.

Renal capsules of poikilothermous vertebrates are usually separated from their convoluted nephric tubules by narrow ciliated **neck segments.** Nephric tubules are characteristically divided into a **proximal tubule** and a **distal tubule** which are separated by a short yet sharply demarcated **intermediate segment.** The proximal tubule has cuboidal or low columnar cells with microvilli. The intermediate segment is thin and ciliated. Cells of the distal tubules are cuboidal and have few microvilli or none. The proximal tubule returns sugars, amino acids, vitamins, and various salts to the bloodstream and may secrete certain foreign materials into the filtrate. The distal tubule acidifies the filtrate and removes sodium and chloride ions. Water may be returned to the blood by both tubules. The details of tubule function are varied and complicated. Excretion is influenced by hormones in at least some classes.

Most marine invertebrates have body fluids that are isotonic (equal in osmotic concentration) to seawater. Little water enters or leaves their bodies, and their excretory organs lack filtering devices for removing water. Most vertebrates, regardless of habitat, have body fluids that are hypotonic (lower in osmotic concentration) to seawater but hypertonic (higher in concentration) to freshwater. Consider first the FRESHWATER FISHES and AMPHIBIANS. Water constantly enters their bodies from the environment because of the osmotic gradient (Figure 15-6). The skin may be moderately waterproof, but the gills and oral membranes admit much water. There is also water in their food. Consequently, even though these animals scarcely drink at all, they must eliminate copious quantities of urine to maintain their water balance. Accordingly, they have prominent renal corpuscles. Short proximal and distal tubules return solutes to the bloodstream, yet, although the distal tubules resorb some salt, the urine is more salty than the environment, and salt intake is essential. Some salt is taken with the food. Many amphibians can selectively absorb salt through the skin, and most of the fishes can extract salt from freshwater with the gills. Nitrogen is eliminated from the body by the gills as ammonia (fishes, larval amphibians) or by the kidneys as urea (adult amphibians).

MARINE BONY FISHES have the opposite problem with osmosis: Their body fluids constantly tend to leak away into the sea, particularly through the gills. In order to conserve water they must pass little urine and therefore have relatively small and poorly vascularized renal corpuscles. Some species have only rudimentary renal corpuscles or none at all. In order to replace water these fishes drink freely, which introduces an excess of salt. The gills excrete monovalent ions, and the proximal tubules of the kidneys excrete divalent ions. The gills also actively excrete urea.

Figure 15-6 DIAGRAM OF THE STRUCTURE AND FUNCTION OF THE NEPHRON IN RELATION TO EXCRETION AND OSMOREGULATION. (There is variation within the groups illustrated.)

CARTILAGINOUS FISHES are also marine but have solved the problem of water balance in a different way. Nitrogenous wastes of vertebrates are produced and excreted in several forms but in the bloodstream occur primarily as urea. Urea is eliminated by the kidneys of amphibians, many turtles, and mammals. These animals excrete urea because they do not need it, but it is scarcely toxic and may to a degree be eliminated because it diffuses readily and cannot be retained. The only vertebrates that can retain urea are the cartilaginous fishes (and marine toads and the modern crossopterygian, *Latimeria*). An unknown mechanism makes their gills impervious to the substance, and their large and distinctive nephric tubules are able to return urea to the blood from the filtrate.

Consequently, the blood contains enough urea to be slightly hyperosmotic to seawater. Some water enters the gills. Little water is drunk, and urine volume is moderate. Renal corpuscles are nevertheless large. Excess urea is excreted by the kidneys and gills. Excess salt is also excreted by the kidneys which, however, may fall behind in the elimination of the monovalent ions. These are then excreted by the rectal gland, which discharges into the hind gut.

CYCLOSTOMES are largely marine, but some lampreys migrate into freshwater. The structure and function of their kidneys vary among species and are not adequately known. Renal corpuscles are unaccountably the largest of all vertebrates. Hagfishes control water intake by having blood of uniquely high salt concentration. In freshwater, lampreys absorb some salt through the skin and excrete much urine.

AMNIOTES use water to carry out excretory wastes and to condition the surfaces of the lungs. They also lose water through the skin. Since drinking is the only source of water in a usually dry environment, it is desirable for the animal to conserve water. This is done in various ways, but most importantly by the kidneys. The kidneys of mammals do the job in one way whereas those of reptiles and birds do it in another.

Urea is highly soluble. It flushes from the body in water and escapes from aquatic embryos into the environment. (Some sharks must provide their embryos with waterproof cases to conserve urea for their special needs.) Most BIRDS and REPTILES have insufficient water available to pass dilute urine, and their terrestrial eggs have no way to dispose of soluble urea. They excrete, instead, uric acid, which is insoluble and can be passed from the body in a semisolid state with very little water. Since the formation of quantities of filtrate would be detrimental, these animals have small, poorly vascularized renal corpuscles. Nephric tubules are short in reptiles and of moderate length in birds. Most of these animals can eliminate excess salt in the urine. However, various marine species of both reptiles and birds take quantities of salt with the food and have only seawater to drink (if they drink at all). Many of these animals (sea snakes, marine iguanas, sea turtles, cormorants, albatrosses, petrels, gulls, terns, sea ducks, etc.) have salt glands that excrete a very concentrated brine. These glands may be derived from lacrimal glands or nasal or orbital glands and are variously located near the eyes, jaws, or tongue, or in the nasal chamber.

MAMMALS excrete urea, which is removed from the blood by prominent renal corpuscles. Quantities of dilute filtrate are thus produced. The mammalian kidney, however, is more effective than any other at returning water from the filtrate to the blood. Only about one one-hundredth of the filtrate is passed as urine. Concentration is dependent on the distinctive **loops of Henle** (Figure 15-7), which are derived from parts of the proximal and distal tubules and extend toward the pelvis of the kidney. (There is no intermediate segment.) Some loops are long and some of moderate length. The thin descending and thicker ascending limbs of a single loop are straight and immediately adjacent to one another. This is of functional significance because one of the means of modifying the urine is the cycling of sodium by countercurrent multiplication between the limbs. Mammals are the only vertebrates to pass urine that is more concentrated than the

Figure 15-7 STRUCTURE AND FUNCTION OF THE MAMMALIAN KIDNEY AND NEPHRON illustrated by the dog.

blood. Birds also have loops of Henle (apparently independently evolved), but their loops are short and are not borne by all nephrons.

Renal corpuscles and convoluted parts of the tubule are located in the **cortex,** or outer zone of the kidney ("cortic" = bark). The loops of Henle are located in the **medulla,** or inner zone ("medull" = pith). Desert-living mammals that must excrete a particularly concentrated urine accentuate medullary tissue to the extent that it commonly extends into the renal pelvis as one or more **papillae** (Figure 15-8).

The urinary bladder provides temporary storage for urine and, except in mammals, often modifies the concentration and composition of urine by selective absorption or secretion.

Most FISHES have urinary bladders, but even if much urine is passed, their bladders tend to be of small or moderate size because the aquatic environment makes long retention of urine unnecessary to accomplish sanitation. Enlargement of posterior segments of the nephric ducts provides temporary storage for dipnoans, primitive ray-finned fishes, and female elasmobranchs. The two ducts may fuse to form a single median bladder. The accessory ducts instead enlarge to form bladders in male elasmobranchs. Similar expansions of the posterior ends of the urinary ducts serve as bladders in lampreys and teleosts, but in this instance

URINARY BLADDERS

Figure 15-8 KIDNEY OF A SMALL MAMMAL FROM A DRY HABITAT showing the position of one nephron.

a pocket of the embryonic cloaca is considered to contribute to the adult bladder which is, therefore, largely a median structure.

AMPHIBIANS evolved a different kind of bladder of evolutionary importance. It is a large ventral outpocketing of the cloaca. Nephric ducts enter the cloaca dorsally so urine must cross the cloaca to enter the bladder. Almost certainly it was the enlargement and more precocious development of this bladder that produced the allantois of amniotes.

Some REPTILES (turtles, some lizards, *Sphenodon*) have an identical cloacal bladder, but this time it is formed by the retention in the adult of the base of the fetal allantoic stalk. Other reptiles and BIRDS understandably have no urinary bladder because they excrete a semisolid urine containing uric acid. Ureters enter the sides of the cloaca and urine mingles with fecal material before being discharged.

The ureters of MAMMALS join the ventrolateral surfaces of the embryonic cloaca and hence are associated with the urodeum after the cloaca is divided (see figure on p. 306). Parts of the urodeum and allantoic stalk together form the urinary bladder. The bladder is lined by a distinctive kind of epithelium that thins as it stretches. Smooth muscle in the walls of the bladder contracts to empty the organ.

Some of the ducts that serve the excretory system also serve the reproductive system. Accordingly, the urogenital ducts are discussed in the next chapter.

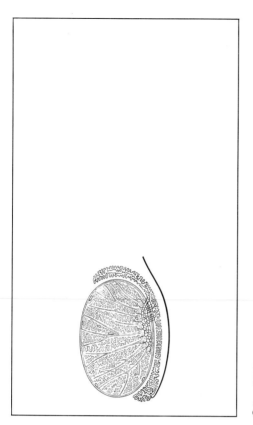

16

Reproductive System and Urogenital Ducts

The functions of the reproductive system are to produce the sex cells, or **gametes,** bring egg and sperm cells together, provide for the nourishment of the embryo or fetus until hatching or birth, and to release eggs or young from the maternal body. Structural adaptations of other parts of the body, and behavioral adaptations may subsequently furnish defense, care, nourishment, or warmth to the eggs or young.

EARLY DEVELOPMENT AND ANCESTRY OF GONADS

Embryonic primordia of the sex organs, or **gonads,** come from two diverse sources. The first is the mesomere. After opisthonephroi, or mesonephroi, as the case may be, are well established, **genital ridges** appear on their mesial surfaces. These ridges are shorter than the nephric ridges which preceded them, yet are longer than the definitive gonads—particularly in amniotes. Anterior and posterior parts of the genital ridges that do not form gonads form instead fat bodies and mesenteries that support the gonads. The latter are called **mesorchia** in males and **mesovaria** in females. Where reflected over the genital ridges, the lining of the coelom thickens to become the **embryonic germinal epithelium.** At the center of the developing organ is another type of tissue called **blastema** (Figure 16-1). The germinal epithelium soon subtends sheets of tissue called **primary sex cords** which penetrate into the blastema.

The second source of gonadal tissue is unexpected and provides one of embryology's more fascinating stories. Several thousand cells of the endoderm of the yolk sac become distinctively large. Some of these stay behind, but many

INDIFFERENT STAGE   LATER DEVELOPMENTAL STAGE

Figure 16-1   STAGES IN THE DEVELOPMENT OF THE AMNIOTE OVARY. Cross sections of the body.

migrate individually and actively into the splanchnic mesoderm of the gut, up the body stalk and mesenteries, and laterally into the genital ridges. These are the **primitive sex cells.** They divide in route, yet some are lost along the way. The remainder space themselves in the germinal epithelium and sex cords where they have important derivatives, as noted below. They have been much studied, yet the cause of their migration and the developmental and evolutionary significance of their remote origin remain unknown.

In summary, the early gonad consists of a superficial germinal epithelium and a deep blastema. Sex cords penetrate the blastema, and primitive sex cells are distributed within the epithelium and sex cords. The outer portion of the organ is called the **cortex;** the inner portion, including blastema and sex cords, is the **medulla.** Although the sex of an individual is determined at fertilization, development is identical in the two sexes to the point described, and this phase of development is called the **indifferent stage.** The indifferent stage may be completed early in embryonic development (mammals) or delayed until adult body size is approached (hagfishes).

Gonads of the ancestral vertebrate probably extended the entire length of the coelom. They may have had gonocoels, though these are absent from the adult organs of known vertebrates. Furthermore, ancestral gonads may have been at least partly segmental as they are in amphioxus and some embryonic urodeles and caecilians. Finally, the ancestral organs may have been bisexual. Dominance of the cortex produces a female and dominance of the medulla produces a male. At the end of the indifferent stage, cortex and medulla are in potential physiological competition. The genetic sex usually masks the other, but under abnormal or experimental conditions the nongenetic sex may prevail.

STRUCTURE OF GONADS

The structure of gonads varies within some taxa (Teleostei, Reptilia), but in general is uniform for each class and too conservative to provide evidence for evolutionary lineages.

OVARY   The germinal epithelium of the indifferent stage ovary becomes the thin peritoneal covering of the organ and the important adult **germinal epithelium.**

Testis

Efferent ductules

Oviduct

Ostium

Ovary

Shell gland is
behind ovary

Mesovarium

Oviduct

Opisthonephros   Opisthonephros

Opisthonephric duct

Deferent duct

Uterus

Seminal vesicle

Accessory excretory duct

Cloaca

Sperm sac

Cut edge of body wall

Urinary papilla

Cloaca

Cut edge of body wall

Urinary papilla

MALE   FEMALE

Figure 16-2   UROGENITAL SYSTEM OF THE ELASMOBRANCH *Squalus.* Ventral
views. In the male, sperm are transferred from the cloaca to grooves in the claspers
(see figure on page 163).

The latter is one cell layer in thickness and includes thousands of **oogonia,** some
of which sink below the surface in each consecutive breeding cycle to enlarge
and undergo maturation divisions and thus become **ova.** It is probable that the
primitive sex cells within the epithelium form all ultimate germ cells in some
species and only part of them in others. Even if they contribute only part of the
germ cells, their presence seems to be essential for the maturation of all. Primary
sex cords degenerate and are replaced by **secondary sex cords** (Figure 16-1).
These form **follicle cells,** which surround, nourish, and support the ripening
eggs, and **theca** or envelopes that enclose the follicles. Following ovulation, the
follicles of mammals (and probably of some elasmobranchs and birds) are temporarily converted to **corpora lutea.** The inner theca and corpora lutea have
endocrine functions (see p. 383 and 384). The blastema becomes only the
stroma, or matrix of connective tissue and vessels within the ovary.

Ovaries vary widely in character, being long or short, compact or flat,
smooth or lumpy, solid or flabby, according to species (Figures 16-2 and 16-3).
Eggs differ in size in proportion to the amount of yolk present. Relatively few

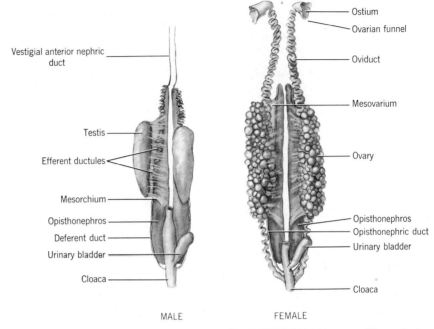

Vestigial anterior nephric duct

Testis

Efferent ductules

Mesorchium

Opisthonephros

Deferent duct

Urinary bladder

Cloaca

Ostium

Ovarian funnel

Oviduct

Mesovarium

Ovary

Opisthonephros

Opisthonephric duct

Urinary bladder

Cloaca

MALE   FEMALE

Figure 16-3   UROGENITAL SYSTEM OF THE AMPHIBIAN *Necturus*. Ventral views.

eggs ripen at one time if the eggs are large (cartilaginous fishes, reptiles, birds) or if the animal retains the eggs in the body until hatching (many fishes, some caecilians, and some reptiles) or gives birth to live young (mammals). The production of eggs is always cyclic, and the size of the ovaries of some vertebrates fluctuates markedly according to breeding condition. Long-bodied amphibians have long slender ovaries. Nevertheless, paired ovaries of large size are not easily accommodated in a streamlined body. Consequently, the two embryonic organs of some animals fuse more or less completely in the adult (lamprey, many teleosts) whereas other animals partially or completely suppress the left (many cartilaginous fishes) or right organ (hagfishes, most birds).

Ovaries of crocodilians, turtles, birds, and mammals are solid with relatively much stroma. Those of cyclostomes, cartilaginous fishes, dipnoans, and some primitive ray-finned fishes are also solid but less compact. In amphibians, regression of the blastema leaves one (urodeles) or several (caecilians and anurans) large lymph spaces within the ovary (Figure 16-4). Stroma is then virtually absent and the organ is soft and pleated. Ripening eggs hang into the central cavity but at ovulation rupture outward into the coelom. Smaller central lymph spaces occur also in ovaries of Lepidosauria.

Most teleosts have hollow ovaries, but the cavity is of different origin. A thin margin of the developing organ curls over and fuses either to the body wall or to the ovary itself, thus sealing off a bit of the general coelom. Ovulation is into the resultant cavity. In a restricted sense, the ovary only appears to be hollow.

Ovum that contacts surface of ovary at another plane
Follicle cells and smooth muscle
Maturing ovum with yolk and large nucleus
Lymph space
Ovulated follicle
Oogonium
Peritoneal membrane and germinal epithelium
Blood vessels

Mesovarium
Fat
Cavity of Graafian follicle = antrum
Stratum granulosum
Ovum
Theca
Primary follicle
Stroma
Oogonia
Young follicle
Blood vessels
Corpus luteum
Germinal epithelium
Corpus luteum forming from recently ovulated follicle

AMPHIBIAN   MAMMAL

Figure 16-4   SECTIONS OF CONTRASTING TYPES OF OVARIES. (It is unlikely that all developmental stages shown would be present at the same time in the reproductive cycle.)

The coelomic space that is adapted to receive the eggs is continuous with the special oviducts of these fishes, so eggs never reach the unmodified coelom.

Still another kind of cavity is present in ovaries of mammals. Other vertebrates have solid follicles, but in mammals a space appears within each maturing follicle, which is then called a **Graafian follicle.** Corpora lutea are more prominent in mammalian than in other ovaries, and in this class there is a tendency for the ovaries to migrate posteriorly as they mature.

TESTIS   Testes tend to develop a little earlier than ovaries. Their development emphasizes derivatives of the medulla of the indifferent stage gonad; the germinal epithelium of the embryonic organ forms only the peritoneal covering of the adult testis. Primary sex cords do not degenerate, as they do in females, and secondary sex cords are not formed. The primary cords separate from the overlying germinal epithelium and become hollow, thus producing slender, coiled **seminiferous tubules** (amniotes), sac-like **seminiferous ampullae** (cyclostomes, urodeles), or intermediate structures.

Seminiferous tubules usually branch, and may end blindly (reptiles) or anastomose peripherally (mammals). There are dozens, or even hundreds of tubules in each testis. Their walls, when in breeding condition, consist of stratified epithelium having germ cells at various stages of maturation (Figure 16-5). Undif-

Basement membrane

Primary spermatocyte

Supportive (or Sertoli) cell

Metamorphosing spermatids

Spermatogonia

Spermatids

Sperm

Lumen of seminiferous tubule

Interstitial cells

Blood vessel

Figure 16-5　CROSS SECTION OF SEMINIFEROUS TUBULE OF A MAMMAL.

ferentiated **spermatogonia** are peripheral. As the **sperm cells** are maturing they are held for a time by supportive cells scattered among the germ cells. Finally the mature sperm are released into the duct system of the reproductive tract.

Spermatogenesis within ampullae differs in that potential germ cells enter an ampulla from adjacent tissue and then divide many times to form clusters of cells all of which mature together (Figure 16-6). The result is masses of sperm that will all be evacuated at once, leaving each ampulla empty and collapsed.

As in the ovary, primitive sex cells are considered to play a critical role in the formation of germ cells. The blastema of the indifferent stage gonad forms supportive tissue and, in tetrapods, also cords that hollow out to become **rete tubules** (Figure 16-7). These come to be continuous with the seminiferous tubules. They form an anastomosis within the testis and conduct sperm to the margin of the organ.

Testes are usually smoother, firmer, and smaller than the ovaries of the same species. Usually they are paired, but the two organs may be partially fused in elasmobranchs and are completely fused in adult cyclostomes. Testes are elongate in slender vertebrates (cyclostomes, most fishes, caecilians, urodeles) but compact and ovoid in some cartilaginous fishes, anurans, and amniotes. Some asymmetry in the size and position of the two members of a pair is not unusual.

Mammalian testes are distinctive for the degree to which several features are expressed: The organ is enclosed in a tough envelope of connective tissue called the **tunica albuginea** (= tunic + white). Internal lobulation is more pronounced than for most other vertebrates, and **septa** separate adjacent lobules. **Interstitial cells,** which produce male hormones, are packed in the interstices among the seminiferous tubules. Lower vertebrates also have male hormones and doubtless also interstitial tissue. However, in them the tissue is never so prominent as it is in mammals and for many it has not been identified at all.

Figure 16-6 SECTION OF A SALAMANDER TESTIS SHOWING AMPULLAE with primary spermatocytes (small, spherical), secondary spermatocytes (large, spherical), and spermatids (linear).

Spermatogenesis ceases in testes that are warmer than about 36.5°C. Diurnal temperatures of birds usually exceed this figure by several degrees, but at night the body is cooler, and even during the day the testes are probably somewhat cooled by the abdominal air sacs. The testes of mammals are abdominal or pelvic in position if body temperature is relatively low (monotremes, edentates, whales, elephants). In many mammals (primates, most carnivores, most ungulates, and others) the testes descend at maturity out of the abdominal cavity into a cooler pouch of skin in the inguinal area called the **scrotum.** Small pockets of coelom extend into the scrotum to facilitate the descent of the testes into the organ and to give them some freedom of motion within their sac. The paired canals joining scrotal coelom to general coelom usually close, but in such seasonal breeders as rodents may remain open so the testes can descend when active and return to the abdominal cavity at other times. Some temperature control is provided by scrotal muscles, which raise or lower the testes, and by countercurrent exchange in the blood supply. In females the **large labia,** which flank the genital area in some species, are sexual homologs of the scrotum.

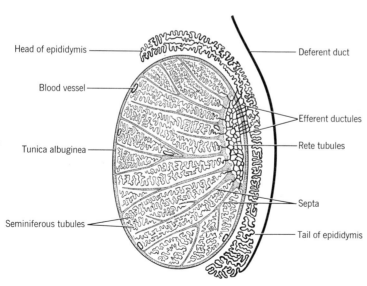

Figure 16-7 STRUCTURE OF THE MAMMALIAN TESTIS. Diagrammatic longitudinal section.

**UROGENITAL
DUCTS AND
ACCESSORY
ORGANS**

ORIGINS   Cyclostomes are unique in having nephric ducts but no genital ducts. Eggs and sperm are both released into the coelom whence they exit into the cloaca or urogenital sinus by way of a pair of **genital pores.** Similar openings, also called **abdominal pores,** are said to be present but not functional in various higher vertebrates: some sharks, dipnoans, and several primitive ray-finned fishes. The evolutionary origin of abdominal pores is obscure. Some anatomists believe they represent a posterior pair of ancient, segmentally arranged genital openings. If so, their presence is a primitive rather than a specialized condition, a view that is supported by their wide distribution.

All other vertebrates have two pairs of urogenital ducts that remain identical in the two sexes to the end of the indifferent stage of development and then are modified according to sex and taxon. One pair is the **nephric ducts,** which have already been described as the ducts of fetal kidneys other than metanephroi (Figure 16-8). The other pair is the **paramesonephric** ducts ("para" = beside). The phylogenetic origin of paramesonephric ducts is again obscure. In sharks and urodeles they form by a lengthwise splitting of the more precocious nephric ducts. In other vertebrates they either develop from long solid cords at the surface of the nephric ridges or from shorter anterior primordia which extend themselves back to the cloacal area. It is not possible to say what method of origin is primitive.

Other structures that contribute to the urogenital duct system of adults are the more anterior of the opisthonephric or mesonephric tubules, the ureter of amniotes, the accessory ducts of certain anamniotes, and, in teleosts, still other structures to be identified below.

The retention, loss, or modification of these fetal structures to produce the adult male and female patterns of ducts in the different taxa will now be summarized. To memorize the varied arrangements can be a bewildering and useless

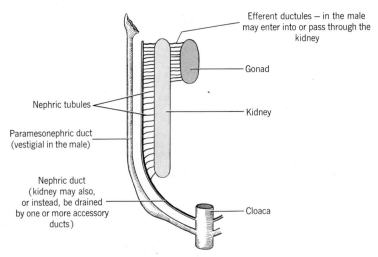

Efferent ductules — in the male
may enter into or pass through the
kidney

Gonad

Nephric tubules

Kidney

Paramesonephric duct
(vestigial in the male)

Nephric duct
(kidney may also,
or instead, be drained
by one or more accessory
ducts)

Cloaca

Figure 16-8   THE PRINCIPAL UROGENITAL DUCTS AS THEY OCCUR EARLY IN BOTH PHYLOGENY AND ONTOGENY. Ventral view of right side only.

exercise, but a review of the kinds of arrangements with attention to several principles that are illustrated is worthwhile. First, one must be impressed by nature's resourcefulness in adapting these "raw materials" to her needs. Evolution is opportunistic and in this instance has selected a variety of patterns of near functional equivalence. Second, the developmental mechanism of hormonal influence is well-illustrated by the maturation of the urogenital ducts: The indifferent stage ducts are modified under the influence of sex hormones elaborated in the fetal gonads. Male and female patterns can be altered or even reversed experimentally, and variation between the two norms is common. Finally, the concept of **sexual homology** is exemplified. Organs of the two sexes that are derived from identical primordia of the indifferent stage, yet become different in structure and function, have an equivalence which differs from phylogenetic homology.

MALE DUCTS   The fetal paramesonephric ducts of males always regress. Vestiges are left which may be prominent (elasmobranchs and amphibians—particularly caecilians) but usually are not.

Except in cyclostomes, sperm are not released into the coelom. Instead they are conveyed in a closed system of ducts which usually is appropriated, at least in part, from the urinary system. If so, sperm cells leaving each testis enter a dozen or so **efferent ductules** derived from anterior nephric tubules. These ciliated ducts are small and cross the mesorchium to enter the nephric duct. The nephric duct takes the new name **deferent duct** when it carries only sperm or both sperm and urine. It has ciliated cuboidal or columnar epithelium and walls of smooth muscle. In addition to conveying sperm it provides temporary storage and contracts during mating to ejaculate its contents. Various accessory organs are established by modification of parts of the deferent duct.

Typical CHONDRICHTHYES have paired deferent ducts that convey only sperm (Figure 16-2). The anterior end of each duct is convoluted to form an **epididymis** (= upon + testes) where sperm are stored and fluids augmented. The posterior end of each duct expands to become a **seminal vesicle.** The two vesicles may fuse before emptying into the cloaca. Urine is transported in several accessory ducts.

Various patterns serve the OSTEICHTHYES. Several primitive ray-finned fishes and one dipnoan pass sperm into the anterior end of the deferent duct, which conveys both sperm and urine. Other primitive ray-fins and dipnoans have short paired sperm ducts, without apparent counterpart in other vertebrates, which run to the posterior ends of the opisthonephroi. There, efferent ductules cross into the nephric ducts, which carry both urine and sperm for part of their length only. In teleosts the same distinctive sperm ducts, formed by folds of the peritoneum, extend posterior to the kidneys and either enter the nephric ducts just before their exit from the body or else exit independently.

All AMPHIBIA have deferent ducts, but in some species of each major group they only convey sperm and in others they also convey urine (Figure 16-3). Accessory ducts are present or absent as needed. When present they drain principally the posterior parts of the kidneys. Seminal vesicles (as defined above) are usually present; epididymides (plural of epididymis) may be present.

Deferent ducts of AMNIOTES carry only sperm (Figures 16-9 and 16-10). Each duct forms an epididymis, though this organ is reduced in birds. Seminal vesicles are present in reptiles and birds. Mammals have glandular outgrowths from the deferent ducts that augment seminal fluids but do not store sperm. They are commonly called seminal vesicles, but are better termed **vesicular glands.** They are variable in size and form, and sometimes are lacking. In mammals the **prostate gland,** which is an outgrowth from the urethra, also augments the seminal fluid.

FEMALE DUCTS   Variation is less extreme in the patterns of ducts of females, and urinary and genital systems are more independent of one another.

Opisthonephric ducts and ureters convey only urine in the female, but drainage of opisthonephroi is curiously supplemented by accessory ducts in some sharks and urodeles. The mesonephric ducts leave vestiges in adults of some amniotes.

It was noted above that the developing ovaries of TELEOSTEI fold to enclose pockets of coelom into which eggs rupture. Similar folding of the parts of the genital ridges posterior to the ovaries extends the same spaces as short oviducts, which are apparently without counterpart in other vertebrates. This arrangement correlates with the prodigious numbers of small eggs released by most of these fishes.

MALE REPTILE   FEMALE BIRD

Figure 16-9   UROGENITAL SYSTEM OF THE LEPIDOSAUR *Crotaphytus* AND THE BIRD *Gallus.* Ventral views.

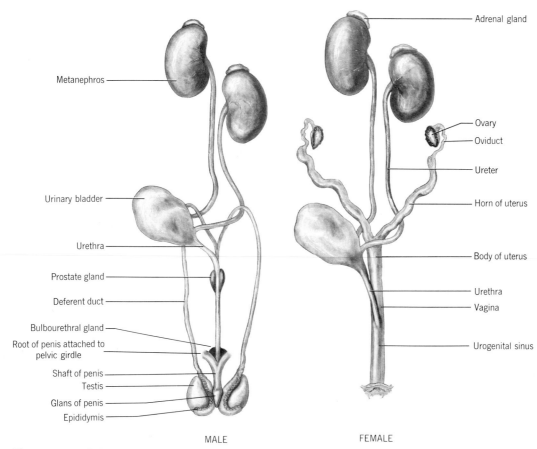

Metanephros

Urinary bladder

Urethra

Prostate gland

Deferent duct

Bulbourethral gland

Root of penis attached to pelvic girdle

Shaft of penis

Testis

Glans of penis

Epididymis

Adrenal gland

Ovary

Oviduct

Ureter

Horn of uterus

Body of uterus

Urethra

Vagina

Urogenital sinus

MALE   FEMALE

Figure 16-10   UROGENITAL SYSTEM OF THE CAT. Ventral views.

Eggs of other gnathostomes rupture into the coelom before entering a duct system derived from the paramesonephric ducts. The anterior ends of the passages enlarge as **ovarian funnels** that have openings called **ostia.** The funnels of at least some species are thought to represent an anterior pair of pronephric tubules, and the ostia the nephrostomes of those tubules. Much of the paramesonephric ducts, including all the anterior portions, become **oviducts.** These are lined by ciliated columnar epithelium with interspersed goblet cells. Eggs are moved along by peristaltic contractions of smooth muscle in the walls of the ducts. Posterior portions of the paramesonephric ducts usually become more muscular to expel eggs and more glandular to provide nutritive or protective coats to eggs or to nourish unborn young. The physical and physiological readiness of the female tract to perform its functions vary markedly with the breeding cycle.

The ovarian funnels of CHONDRICHYTHYES lie relatively far forward in the body and often fuse in the midline (Figure 16-2). Oviducts lead into **shell glands** which envelop eggs in albumen and, in some species, a horny shell. Some sharks retain their developing eggs in expanded **ovisacs** or **uteri.** Nourishment of the embryo by yolk may be supplemented by a simple yolk-sac placenta.

DIPNOI and some PRIMITIVE ACTINOPTERYGII have paired oviducts apparently derived from paramesonephric ducts, though some doubt exists about the homologies of the ducts of most bony fishes.

AMPHIBIA have glandular and somewhat coiled oviducts. The lining of the coelom is provided with cilia that beat toward the ostia. Posterior portions of the tract apply coats of jelly to the eggs, which are usually held temporarily in ovisacs, though many caecilians bear living young that develop in the oviduct.

The genital tracts of female REPTILES, BIRDS, and MONOTREMATA vary widely in detailed structure, but since all have large eggs they agree in general form. The right side of the tract tends to be larger in reptiles, whereas the same side is vestigial in birds (Figure 16-9). Ovarian funnels are large and pleated. The glandular upper part of the tract applies albumen to the eggs as they spiral along their course. Near the cloaca the tract enlarges as a combined shell gland and ovisac.

In female THERIAN MAMMALIA the genital tract is divided into three regions. First are the oviducts. Their ciliated epithelium is thrown into folds, but externally the ducts are relatively straight and slender in correlation with the small eggs, without albumen or shells, which they convey. The second region is the **uterus** (or uteri) that houses the fetus during pregnancy and provides the maternal contribution to the placenta. It is lined by a mucous membrane called the **endometrium.** When functionally active the endometrium is thick, soft, glandular, and highly vascular. Following each pregnancy or breeding cycle it regresses by sloughing or resorption. The uterus has a thick wall, the **myometrium,** composed of smooth muscle having circular, longitudinal, and oblique fibers. Posteriorly the uterus is closed by a muscular neck called the **cervix.** The third region of the tract is the **vagina** (= sheath) which receives the penis during copulation and serves as the birth canal. The vagina is soft and distensible. Its stratified epithelial lining may be glandular or cornified according to breeding condition and taxon.

The entire genital tract remains paired in monotremes. In marsupials the terminal part of the tract is fused to form a single urogenital canal. Anteriorly the two embryonic uteri and vaginae are retained, and in addition a third, or pseudovagina grows down from near the cervixes toward the urogenital canal (Figure 16-11). In other mammals fusion of the embryonic primordia includes all of the vagina and usually extends to the uterus. If the uterus is completely double (and there are two cervixes), it is said to be **duplex,** the condition found in monotremes, marsupials, elephants, many rodents, and some other groups. If the uterus is Y-shaped externally but nearly divided internally (except for the cervix), it is said to be **bipartite** (most ungulates, most carnivores, and others). If fusion is more nearly complete, yet does not include the anterior end of the organ, the uterus is **bicornuate** (some members of several orders). Finally, if there is a single

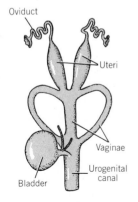

Oviduct

Uteri

Vaginae

Urogenital canal

Bladder

Figure 16-11 REPRODUCTIVE TRACT OF A FEMALE MARSUPIAL showing tripartite vagina. Ventral view.

uterine chamber, the organ is **simplex** (most primates, some edentates). Degree of fusion does not follow clear evolutionary lines, and departures from the norm are not unusual within species.

Seminal vesicles and vesicular glands have been mentioned as common male derivatives of the nephric ducts, and similarly, albumen and shell glands have been mentioned as adjuncts of the oviducts. Additional accessory glands of the reproductive system (which unfortunately have in some instances been given the same names) may be derived from the urethra, urogenital sinus, and adjacent tissues. Products of these glands augment the seminal fluids, soften and lubricate the genital organs to facilitate copulation, and provide scents that act as sexual attractants.

The embryonic hindgut forms before urogenital ducts have developed. When the pronephroi are established, the nephric ducts extend posteriorly to enter the hindgut, and later the paramesonephric ducts do likewise. The common passageway for products of the digestive, urinary, and reproductive systems is then called the **cloaca** (= sewer). This primitive arrangement is retained by adults of hagfishes, elasmobranchs, dipnoans, amphibians, reptiles, and birds (Figure 16-12). The posterior part of the gut of adult lampreys, chimaeras, and bony fishes, by contrast, is no longer joined by the urogenital ducts and is called a **rectum.** Their nephric and genital ducts either exit from the body independently or join each other to exit at a common papilla. In either instance, urine and gametes are discharged posterior to the anus.

CLOACA AND DERIVATIVES

A different evolutionary path has been followed by mammals. In monotremes the embryonic cloaca is partly divided by a septum that wedges between the gut and allantoic stalk. The result is a dorsal **coprodeum** (= dung + divide), a ventral **urodeum** ("ur" = urine) that is joined by the ureters and paramesonephric ducts, and a common posterior **proctodeum** ("proct" = anus). In therian mammals the embryonic septum continues to push back until a dorsal rectum is completely separated from the urogenital structures, which then exit anterior to the anus. In males, urine and sperm are discharged by a common **urethra.** In females of most mammals, urinary and genital tracts exit by a common **urogenital sinus.** In primates and some rodents, however, fetal eversion of the common passageway eliminates the urogenital sinus by separating an anterior urethral opening from a posterior vaginal opening.

CYCLOSTOMES and most BONY FISHES lay their eggs in water where the attendant male promptly discharges his sperm over them. Successful fertilization is dependent on behavioral rather than on structural adaptations. Copulatory organs are lacking, though the two sexes of some species entwine their bodies during egg laying and rarely some kind of holding organ is present. Teleosts that retain their eggs during development, or bear live young, however, require internal fertilization. Males of most such species have a margin of the anal fin modified as a copulatory organ called a **gonopodium.** It is enlarged, rigid, and movable. When inserted into the female tract it transmits sperm through a duct

COPULATORY ORGANS

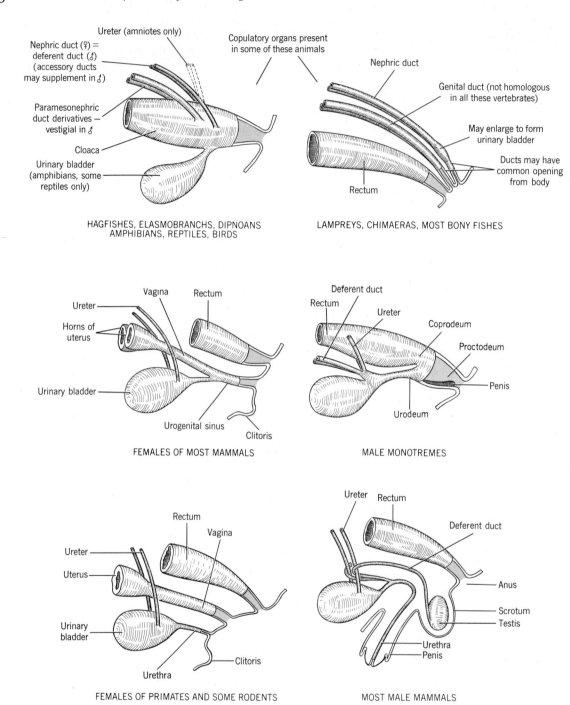

Figure 16-12 REPRESENTATIVE DIVISIONS OF THE CLOACA AND THEIR RE-LATIONS WITH UROGENITAL DUCTS AND THE URINARY BLADDER. Left lateral views.

or groove. Several cottid fishes have instead an enlarged genital papilla which serves as a penis.

CARTILAGINOUS FISHES also have internal fertilization. This time it is the pelvic fins of males that develop copulatory organs called **claspers** (see figure on p. 163). They are supported by the fin skeleton and may be of elaborate configuration. In some rays they include erectile tissue. Again, grooves convey sperm into the female cloaca.

Many AMPHIBIANS return to water to breed. Fertilization is usually external in anurans so copulatory organs are not needed. Fertilization is usually internal in urodeles but external genitalia are still lacking. The two sexes either press their cloacas together to transfer sperm or the male deposits packets of sperm, which are later picked up by the cloaca of the female. Male caecilians use an evertable extension of the cloaca for internal fertilization by means of copulation.

Internal fertilization is necessary if copulation takes place out of the water, if there is internal development of the young, or if eggs are provided with shells before deposition. With few exceptions amniotes must have internal fertilization for the first and either the second or third of these reasons. REPTILES evolved two kinds of copulatory organs. Male lepidosaurs have paired structures called **hemipenes,** which lie concealed in long sacs opening to the outside of the body on each side of the cloacal aperture (see figure on p. 195). During copulation the hemipenes are everted and introduced into the female cloaca. Most other reptiles have evolved the **penis,** which in this class is a grooved organ located internally on the floor of the cloaca. It is largely composed of two long, spongy, vascular bodies called **corpora cavernosa.** A **glans penis** caps the end of the organ. During sexual stimulation sphincters reduce the outflow of blood from these structures, thus causing engorgement and enlargement. The groove then closes over and the penis protrudes from the cloaca to serve as an intromittent organ. Female turtles and crocodilians have a small **clitoris,** which is the sexual homolog of the penis. The penis of MONOTREMES is like that of reptiles except that the sperm channel is permanently separated from the cloaca.

Most BIRDS copulate by pressing the cloacas together for the transfer of sperm. Avian ancestors doubtless had a penis, however, because the ostrich and its relatives and ducks and geese—all relatively primitive birds—have a small penis.

At the indifferent stage of development THERIAN MAMMALS have a genital tubercle anterior to the cloacal opening. When the cloaca becomes divided, its ventral portion contributes to the urinary bladder anteriorly and forms the female urogenital sinus or male pelvic urethra posteriorly. The genital tubercle of females becomes the clitoris and is not penetrated by the urethra. In males the pelvic part of the urethra is extended by a phallic part that grows into the tubercle. The tubercle enlarges to become the penis. Corpora cavernosa and glans are present (Figure 16-13). The organ may be hidden under the skin but is usually at least partly external even when not erect. The glans is variously shaped and is forked

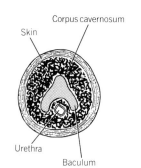

Figure 16-13   CROSS SECTION OF THE PENIS OF A DOG.

in monotremes and marsupials to correspond with the divided vagina of the female. The penis may be stiffened by a bone, the baculum (see figure on p. 176).

REPRODUCTIVE
STRATEGIES

The diverse reproductive strategies of vertebrates often relate directly or indirectly to structure.

Nearly all vertebrates are **dioecious** (having separate male and female individuals), but hagfishes and some teleosts are **hermaphroditic** (have both sexes functional in the same individual). Eggs and sperm are then produced in different parts of the gonads and usually at different times, thus assuring cross-fertilization. Several teleosts, however, are self-fertilizing. Some perches, darters, basses, and lizards are **parthenogenetic,** the eggs maturing without entry of sperm.

Most vertebrates are **oviparous;** that is, they lay eggs that mature outside the maternal body. Included are cyclostomes, most bony fishes, holocephalians, some elasmobranchs, most amphibians, most reptiles, birds, and monotremes. The number of eggs laid per season ranges from 1 to 20 (some birds), 100 (pythons), or even 28 million (several teleosts), the number relating to the hazards of development. They may be spherical, ovoid, teardrop-shaped, or tubular, and are variously provided with stalks, tendrils, filaments, floats, jelly coats, calcareous shells, or keratinized cases. Eggs may be broadcast singly, deposited in clumps, placed in nests, or carried about by a parent in the mouth (a catfish), in a pouch (sea horse, spiny anteater), on the back (some frogs), or elsewhere.

Brooding of eggs is common. It usually provides protection from predation, and according to species may also provide aeration (some fishes), humidity control (terrestrial amphibians, some squamates), protection from mold (some amphibians), temperature control (some reptiles, birds, platypus), and turning to prevent adhesions (birds). Most birds develop a naked, highly vascular **brood patch** on the breast or under the body.

Some species retain the eggs in the female reproductive tract until development is partly completed, thus reducing exposure to the vicissitudes of the environment. **Ovoviviparous** species retain their eggs in the body until hatching; embryos are nourished by egg yolk, yet eggs are not laid. Examples are found among fishes, amphibians of each order, and squamates. Since there is a tendency for such animals to reduce or eliminate the ancestral egg shells, they are not sharply demarcated from **viviparous** species, which nourish the embryo by physiological exchange with maternal tissues, and also give birth to "live" young.

Viviparity is found among teleosts (rockfishes, sea perches, blennies, cyprinodonts, etc.), elasmobranchs, amphibians (many caecilians), snakes, lizards, and all therian mammals. This reproductive strategy is a drain on the female, but provides maximum protection for larva or fetus.

Viviparous teleosts either nourish the fetus within the hollow ovary or within the ovarian follicle. Other viviparous vertebrates retain their fetuses in the oviducts or uterus. Respiratory gases and water are always exchanged between fetus and mother, and nourishment is usually provided from "milk" secreted by the female reproductive tract (some fishes), juices released by the lysis of maternal epithelia or blood (early stages in mammalian development), or physiological exchange between fetal and maternal blood streams. The latter mechanism is facilitated by a **placenta,** which is an intimate juxtaposition of fetal and maternal tissues in such a way as to assure a large area of contact. A wide variety of pleats, folds, filaments, and villi have evolved for the purpose. The yolk sac is commonly involved on the fetal side, though the allantois supplements or substitutes in most mammals.

Various fishes and mammals have evolved provisions for the storage of sperm cells in the male body long after the seasonal loss of testicular function, or within the female body for as long as 10 months after copulation. Some mammals in at least five orders postpone development instead by delayed implantation of the blastocyst (corresponding to the blastula of other vertebrates). These mechanisms time birth with seasonal food, dormancy, or return to breeding grounds.

Vertebrates with short brooding periods (11 days for several birds) or gestation periods (13 days for some marsupials) tend to have small, dependent hatchlings or young. Other animals have long brooding periods (about 79 days for one albatross) or gestation periods (22 months for the elephant) and usually have larger, more active young (ducks, grouse, ungulates, cetaceans). Sexual maturity is reached by several fishes by the time of hatching, and in humans and some large vertebrates only after many years.

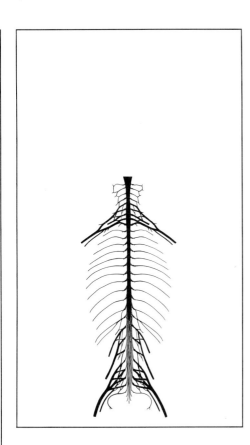

17

Nervous System: General, Spinal Cord, and Peripheral Nerves

Nerve cells are uniquely adapted for conducting stimuli from one place to another. They may be activated by sensory cells or by other nerve cells. A stimulus is usually carried to numerous other nerve cells before reaching the muscle or gland that will respond. Messages may be diminished, reinforced, selected, stored, blocked, or integrated within the system. Thus, the nervous system (with help from the endocrine glands) determines responses of the body to changes in the internal and external environments. It is the body's messenger and coordination system for most activities—for all activities that are rapid or complex.

The nervous system is of particular interest to morphologists because it is the most complex of the body, yet it is conservative in terms of change. Although it is too soft to be directly preserved in the fossil record, the size and shape of the brain, and the distribution of cranial nerves are remarkably well-recorded by fossil skulls. This record, comparative study of surviving vertebrates, and, to a lesser extent, developmental studies, are enabling morphologists to construct the outlines of the phylogeny of the system. The general habits of an animal can be accurately determined (where information is adequate) from the nervous system. It is by adding to knowledge of the fine structure and ultrastructure of this system that morphology can contribute the most to physiology, psychology, and behavioral sciences. The nervous system is an active field of research.

ELEMENTS OF THE NERVOUS SYSTEM

NEURONS AND NEUROGLIA Nerve cells, or **neurons,** range from short to uniquely long, yet are always small in bulk. Each has a **nerve cell body** that may be oval in outline or irregularly star-shaped (Figure 17-1). Within the nerve

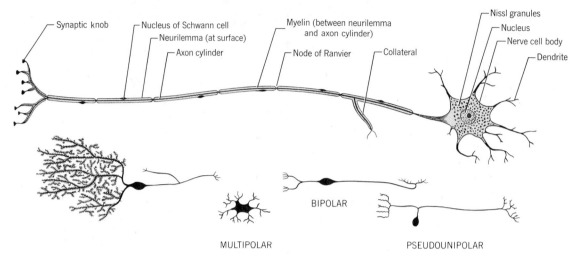

Figure 17-1 STRUCTURE AND SHAPES OF REPRESENTATIVE NEURONS.

cell body is the nucleus and many distinctive granules that contribute to the very high rate of protein synthesis characteristic of nervous tissue. Most nerve cell bodies, particularly of the brain and spinal cord, support many filamentous processes and are termed multipolar. Neurons related to the nose, eye, ear, and lateral line usually have only two processes (i.e., are bipolar), and most related to spinal nerves have two processes which, however, exit from the nerve cell body at the same place and course together for a short distance before separating (pseudounipolar).

Each neuron has one to many processes called **dendrites.** Traditionally these have been thought to carry impulses only toward the nerve cell body, but it is now known that microcircuits occur between dendrite and dendrite. Dendrites of motor and association neurons are up to 1 mm long; those of sensory neurons may be much longer. They branch in the direction away from the nerve cell body, sometimes achieving an exceedingly diffuse and complex pattern. There is a single process, the **axon,** which transmits impulses away from the nerve cell body. Axons may be short or, in large animals, 1 m or more in length. They tend to branch less than dendrites, yet in some regions they give off collaterals and usually have twigs at their far ends where they communicate with other cells.

Between their short initial segment and their terminal twigs, most axons (and only axons) are sheathed. **Schwann cells** arrange themselves along developing axons located outside the brain and spinal cord. A fold of each cell then wraps around the axon much as a window shade wraps around its wooden core. Thus, the axon is ensheathed in lamellae derived from the cellular membranes of Schwann cells (Figure 17-2). These enclose lipids and proteins. The entire coating is called **myelin.** Myelin is essential for rapid conduction. It is interrupted at intervals along the enclosed **axon cylinder** to form **nodes of Ranvier.** Outside the brain and spinal cord, axons are also covered by a second sheath that is external to the myelin. This delicate transparent membrane is also formed from

Figure 17-2 SCHWANN CELL ENSHEATHING AN AXON WITH MYELIN.

Schwann cells and is the **neurilemma.** (There are some long unmyelinated fibers in the peripheral nervous system—particularly in cutaneous nerves.)

Within the brain and spinal cord, both myelinated and unmyelinated parts of neurons may be in contact with cells called **neuroglia** (= nerve + glue). Three kinds of neuroglia (astrocytes, oligodendroglia, microglia) are distinguished histologically. They fill interstices among neurons, bind fiber to fiber, and the oligodendroglia provide myelin to axons. Neuroglia cells also contribute to the energetics of neurons in various ways, including potassium transport, and possibly nutrition, excretion, regeneration, and repair. About half of the bulk of the brain is neuroglia.

NERVE IMPULSE AND SYNAPSE A nerve impulse is an electrical phenomenon that passes as a wave along the surface membrane of a nerve fiber. The fluids bathing the inside and outside of the membrane differ chemically. The crucial difference is that there is about thirty times more potassium on the inside of the resting membrane and about ten times more sodium on the outside. The potassium leaks out through the membrane, but in some unknown way the membrane resists the entrance of sodium. The consequence is a difference in electric potential of about 60 μV across the membrane, with the inside negative. When the fiber is locally excited, the membrane suddenly, but briefly, allows sodium to rush in, and the inside of the membrane changes to a positive potential of about 50 μV. The excited part of the membrane now differs from adjacent parts, and a tiny eddy of current is set up between the excited portion and adjacent portions of the membrane. This, in turn, depolarizes the resting membrane and in this manner the impulse is propagated along the fiber, switching charges ahead and restoring them behind as it travels. An excised fiber can transmit thousands of impulses without using energy. In time, however, enough sodium would pass into the fiber to destroy the mechanism. To avoid this, the membrane pumps sodium out, in a way that is not yet fully understood, and this does use energy.

Nerve fibers of vertebrates range in diameter from about 0.5 to about 22 μm. The rate of travel of an impulse increases with fiber size and temperature. In mammals the rate ranges from about 1.0 to 120 m/sec, values below 1.5 being characteristic of unmyelinated fibers.

An impulse is propagated without decrement or is not propagated at all, and all conducted impulses are alike. It is by the frequency of impulses in each fiber, number of active fibers, and connections made by the neurons that the system decodes the messages. Other electrical charges are not conducted, especially in dendrites in the brain, but instead affect excitation. These may have graded responses.

There is no cytoplasmic continuity between neurons. The functional union of an axon of one neuron with a dendrite or nerve cell body of another neuron is called a **synapse.** Some neurons have few synapses, but others have thousands. The synapse transfers impulses only in the direction away from the axon, and only if a threshold level of excitation is achieved. At its terminus, the axon

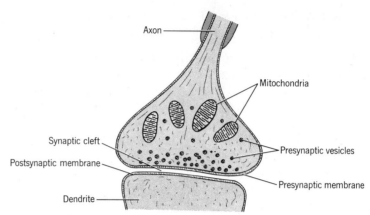

Figure 17-3 SYNAPTIC KNOB AND SYNAPSE.

enlarges to form a **synaptic knob** (Figure 17-3). The knob is separated from the adjacent dendrite or nerve cell body by a narrow cleft (10–30 nm wide). In the knob are mitochondria and **presynaptic vesicles** that contain a chemical known as a **neurotransmitter.** When the transmitter is released into the cleft by the arrival of a nerve impulse at the synaptic knob, it alters the permeability of the postsynaptic membrane to the ions bathing it, and thus causes propagation of the nerve impulse in the connecting neuron. At present, several transmitters are known including acetylcholine, noradrenalin, serotonin, and dopamine. These are found at different synapses, and their exact distributions and functions are under study.

Junctions between axons and muscles or glands tend to be broader but apparently function in a similar manner.

TRACTS, NERVES, AND GANGLIA   Nerve cell bodies of functionally related neurons tend to mass together, and their fibers, if long, tend to run parallel to one another in bundles. Within the cord such bundles are called **tracts.** Myelinated tracts are whitish and together make up the **white matter,** whereas nerve cell bodies and associated unmyelinated fibers, being of darker color, together make up the **gray matter.**

Outside the brain and cord, bundles of fibers are termed **nerves.** Aggregates of nerve cell bodies cause marked swellings on nerves and are termed **ganglia.** Opposite the appendages, adjacent nerves usually exchange bundles of fibers, thus weaving to form a **plexus.**

SOME DIVISIONS OF THE SYSTEM   The parts of the nervous system are functionally interrelated to a remarkable degree. Nevertheless, it is convenient to recognize structural and functional divisions. Brain and spinal cord comprise the **central nervous system,** leaving nerves and ganglia to the **peripheral nervous system. Afferent,** or **sensory fibers** of the peripheral nervous system carry impulses from receptor organs to the central nervous system; **efferent,** or **motor fibers** carry impulses from the central nervous system to effector organs. Nerves may be entirely sensory or motor, or may be mixed, having each kind of fiber. Within the central nervous system there are also **association neurons** (also called **interneurons**) that make up local circuits and are not themselves afferent or efferent.

Somatic fibers (sensory and motor) relate to the skin and its derivatives and to voluntary muscles. **Visceral fibers** (again sensory and motor) relate to involuntary muscles and glands of the various organ systems. It is useful to designate fibers as somatic sensory, visceral sensory, somatic motor, or visceral motor, because these types of fibers tend to be structurally independent.

Most organs innervated by visceral nerves receive two complete sets of fibers that elicit opposed responses. The sets have a measure of structural and functional independence from the remainder of the peripheral nervous system and are together called the **autonomic system.**

Neurulation, or the process that establishes the central nervous system, was described on page 82. Each step occurs first at the anterior end of the embryo and progressively later and later in moving toward the posterior end. This is one of many developmental gradients of the body. In the head of the embryo, the neural tube becomes relatively large and soon is compartmentalized into vesicles that foreshadow the regions of the brain. Further discussion of the development of the brain will be deferred to the next chapter.

As seen in cross section, the embryonic neural tube forms three layers having conspicuously different staining properties. In order, from neurocoel outward, these are the **ependymal, mantle,** and **marginal layers** (Figure 17-4). Some cells of the ependymal layer remain in place to become the thin, ciliated lining of the adult central canal. Most ependymal cells migrate outward to join mantle cells in forming both neurons and (a little later) neuroglia. These will be the gray matter of the adult cord. The marginal layer forms neuroglia and is penetrated by nerve fibers growing out of the deeper layers. It becomes the white matter of the cord.

As the cord grows in diameter, it enlarges every place except at the thin roof and floor of the neurocoel. Growth, therefore, establishes longitudinal surface grooves at these places known in the adult as the **dorsal median sulcus** (= furrow) and **ventral median fissure** (= a split).

DEVELOPMENT OF SPINAL CORD AND PERIPHERAL NERVES

Spinal ganglion — derived from neural crest

Dorsal root

Ventral root

Spinal nerve

Neurocoel
Ependymal layer
Mantle layer
Marginal layer

Figure 17-4 STAGE IN THE DEVELOPMENT OF THE MAMMALIAN SPINAL CORD, SPINAL NERVES, AND GANGLIA. Cross section. Arrows show directions of fiber growth.

Nerve cell bodies of the sensory fibers of spinal nerves are located in spinal ganglia located near the cord. These are derived from neural crest cells. Fibers from the developing ganglia grow outward to receptor organs and inward to penetrate the cord. Sensory fibers of cranial nerves and their ganglia develop in the same way except that there are contributions from the ectodermal placodes that will form certain sense organs of the head (see p. 360). The autonomic system also has ganglia. These are derived in part from neural crests and probably in part from cells that migrate out of the cord. Fibers of motor nerves grow out of the mantle layer of brain and cord.

Cells from the versatile neural crests also migrate to contribute to the formation of Schwann cells, myelin, and the neurilemma.

Throughout life, the nerve cell body actively produces cytoplasm that slowly flows outward along axon and dendrites. Neurons do not divide further after infancy; if one is lost, it is not replaced. If a nerve is cut, fibers severed from their nerve cell bodies degenerate whereas fibers that are still nourished slowly regenerate. They tend to feel their way along the empty sheath tubes and may reach their accustomed destinations. However, if the tubes are also destroyed, the regenerating fibers do not find their original end organs. In some instances (but seemingly not in others) anamniotes can then retrain the central nervous system to respond correctly to the new stimuli, but amniotes cannot.

SPINAL CORD   **FUNCTION AND STRUCTURE**   In the simplest possible reflex arc, messages from receptor organs are transferred within the cord directly from afferent fibers to efferent fibers, which then send appropriate messages to effector organs. Nearly always, however, one or more association neurons are interposed between afferent and efferent neurons. Since each afferent fiber synapses with many association neurons, possible pathways run to any of very many effector fibers. These may be on the same side of the cord as the afferent fiber or on the opposite side, at the same level of the cord or at a different level, or in the brain. Thus the function of the cord is to receive incoming impulses, integrate and coordinate them, transmit them wherever they should go within the central nervous system, and send responses to the peripheral nervous system as appropriate.

The general structure of the spinal cord is best exemplified by a cross section of the cord of an amniote. The gray matter is internal and has an irregular shape resembling the letter H (Figure 17-5). The upper arms of the H are the **dorsal gray columns** (or horns) and the shorter, broader, lower arms, the **ventral gray columns** (or horns). Nerve cell bodies of association neurons that synapse with somatic sensory fibers are on the medial side of a dorsal column. Cell bodies of association neurons synapsing with visceral sensory fibers are in a smaller, lateral, and slightly more ventral part of the dorsal column. Nerve cell bodies of somatic motor neurons fill the ventral columns. Nerve cell bodies of visceral motor neurons are in a small, intermediate, and lateral position. The **gray commissure,** just above and below the central canal, makes up the cross arm of the H and transmits fibers from one side of the cord to the other.

The external white matter is divided into right and left sides by the dorsal median sulcus and ventral fissure of the cord. Each half is further divided by the

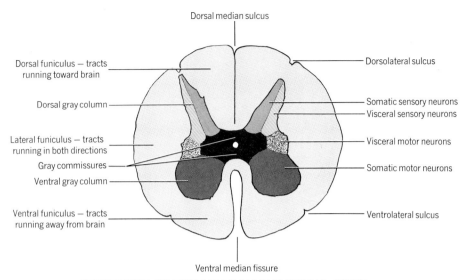

Figure 17-5   STRUCTURE OF THE MAMMALIAN SPINAL CORD.

gray columns into three funiculi. The **dorsal funiculus** is between the dorsal column and the dorsal median sulcus. It transmits axons toward the brain. The **ventral funiculus** is between the ventral fissure and the ventral column. It transmits axons away from the brain. The **lateral funiculus** is between the dorsal and ventral columns and transmits fibers in both directions; those going toward the brain tend to be more superficial. The positions of the specific pathways vary among the vertebrates. Many of the fibers to and from the brain cross from one side to the other; the change is sometimes in the cord and sometimes in the brainstem. The reason for this crossing is not known. One author thinks it relates to an ancient coiling reflex in the direction away from a stimulus.

The spinal cord is supported and protected by one or more layers of tissue called **meninges.** These are similar to, and continuous with, meninges of the brain and will be described in the next chapter.

EVOLUTION OF THE SPINAL CORD   The spinal cord of AMPHIOXUS is seemingly primitive in that the embryonic neural folds do not completely fuse (Figure 17-6). Accordingly, the adult neural canal communicates with the space around the cord by a slitlike groove. Gray and white matter cannot be distinguished because nerve fibers of amphioxus are not myelinated. There being little cephalic neural specialization, there are no long spinal tracts.

CYCLOSTOMES, like other vertebrates, complete the neurulation process to enclose a central canal. The boundary between gray and white matter remains indistinct. The cord is wide and its ventral surface is concave where it fits against the notochord.

The cord of FISHES and AMPHIBIANS is nearly circular in cross section. Gray and white matter have become distinct. The configuration of the gray matter is various; ventral gray columns are usually evident. Dorsal median sulcus and

Figure 17-6　COMPARATIVE ANATOMY OF THE SPINAL CORD.

ventral median fissure are making their appearance among these animals as increasing complexity of nuclei and tracts cause the cord to enlarge. The diameter of the cord enlarges moderately opposite the appendages where there is greater need for nervous integration.

AMNIOTES have a deep sulcus and fissure. The cord again enlarges opposite the appendages, the **cervical enlargement** being the more pronounced if the pectoral appendages are emphasized (bats, apes) and the **lumbar enlargement** more pronounced if the pelvic appendages are emphasized (ostrich, bipedal dinosaurs). In cross section, the gray matter now is shaped like an H with thick arms.

A distinctive feature of BIRDS is the **glycogen body.** This is a conspicuous glandlike mass of tissue wedged into the cord of the lumbar area where the dorsal median sulcus opens to receive it. Its function has been debated but is not surely known.

Spinal cords of MAMMALS frequently have **dorsolateral** and **ventrolateral sulci** (Figure 17-5). The dorsal and ventral spinal nerve roots, respectively, join the cord along these grooves. The cords of anurans and some fishes are shorter than the canal within the vertebral column. This condition is not seen in reptiles and birds, but appears again in mammals (except monotremes). The embryonic cord fills its bony housing, but grows more slowly than the spine so that the adult

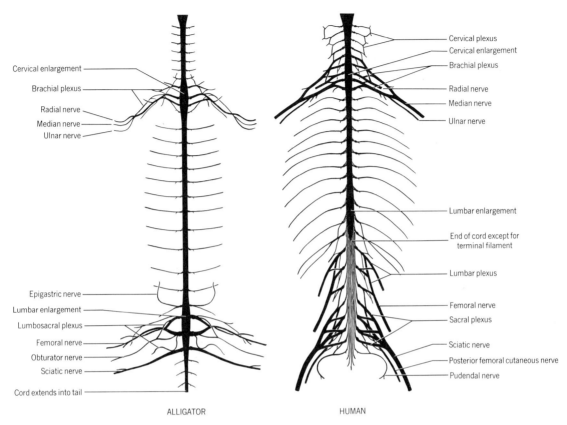

Figure 17-7   SPINAL CORD, PRINCIPAL NERVES, AND PLEXUSES OF A REPTILE
AND A MAMMAL. Dorsal views.

cord ends in the lumbar region (except for a terminal filament), and the more
posterior nerves must angle back to reach their destinations (Figure 17-7).

AMPHIOXUS has a series of paired "dorsal" spinal nerves that contain three
kinds of fibers: somatic sensory from skin and muscle, visceral sensory from
internal organs, and visceral motor. These nerves are intersegmental and run with
the myosepta between muscle segments. All nerve cell bodies of sensory neurons
are located within the cord. Consequently, there are no ganglia on the spinal
nerves. The structures formerly called ventral spinal nerves are really segmentally
arranged specialized muscle fibers that run to the surface of the spinal cord where
their motor end plates are located. In this amphioxus is specialized.

   The further evolution of spinal nerves follows a relatively simple and straight-
forward progression. LAMPREYS have intersegmental dorsal spinal nerves that
are like the spinal nerves of amphioxus except that some of the sensory neurons
have cell bodies outside the cord (Figure 17-8). Lampreys also have segmental
ventral spinal nerves that contain only somatic motor fibers. Each ventral nerve

**EVOLUTION OF
SPINAL NERVES**

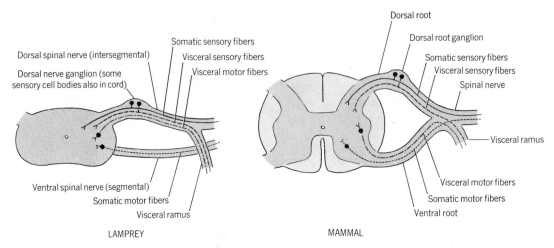

Figure 17-8   DISTRIBUTION OF NERVE COMPONENTS IN SPINAL CORD AND NERVES AT TWO EVOLUTIONARY LEVELS. More detail for the branching of spinal nerves and the distribution of visceral motor fibers is shown in Figure 17-11. Size of the nerves is exaggerated.

joins the cord anterior to the corresponding dorsal nerve. Dorsal and ventral spinal nerves do not join in lampreys (or cephalaspids) but run independently to their destinations.

The HAGFISH, FISHES, and AMPHIBIANS illustrate the next structural level. The dorsal and ventral nerves of each body segment join outside the vertebral column. This establishes a single spinal nerve per segment on each side of the body. The nerve joins the cord by separate **dorsal** and **ventral roots.** Each root tends to emerge from the cord as a row of adjacent twigs. Close beyond the union of the roots, the spinal nerve divides into a **dorsal ramus** that goes to structures of epaxial origin, a relatively large **ventral ramus** that goes to the appendages and structures of hypaxial origin, and a **visceral ramus** that goes to structures derived from the hypomere (visceral and branchial muscles and glands). At the levels of the paired appendages the ventral rami form plexuses of varying complexity.

Nerve cell bodies of sensory neurons are now all located in the **dorsal root ganglion** of each nerve. The distribution of fibers in the dorsal and ventral roots is the same as for the dorsal and ventral nerves of the lamprey except that some visceral motor fibers exit in each root.

Several further advances are found in AMNIOTES. Dorsal and ventral roots of spinal nerves join inside the vertebral column (some salamanders excepted). Each dorsal root joins the cord at the same level as the corresponding ventral root rather than posterior to it. Usually all visceral motor fibers exit from the cord in the ventral root. This completes the gradual shift of these fibers from the dorsal spinal nerve and leaves the dorsal root with only sensory neurons. **Brachial** and **lumbosacral plexuses** tend to be more complex than for anamniotes, but the patterns of interweaving are various (Figure 17-7).

ORIGIN AND NATURE   Spinal nerves are conveniently uniform in regard to occurrence, configuration of roots and branches, nerve fiber components, and relation to the central nervous system. This is not true of cranial nerves: A cranial nerve may be present in some vertebrates and missing in others (e.g., terminal nerve). A nerve may split in the course of evolution to become two (spinal accessory nerve from vagus nerve). Conversely, two nerves may fuse to become one (evolution of amniote trigeminal nerve). The same nerve that is a cervical spinal nerve of one vertebrate may be a cranial nerve of another (hypoglossal nerve). Also, fiber components believed to have been present in an ancestral nerve may become lost.

A further difference between spinal and cranial nerves stems from the nature of the segmentation of the central nervous system. The somites are already segmented when they first appear in the embryo, but the central nervous system is not. Motor nerves grow out of the cord at intervals to penetrate segmented somite derivatives. Sensory nerves and ganglia are derived from neural crests that become intersegmental as they are squeezed between the bulging somites. Thus, the segmentation of spinal nerves is regular but seems to be secondary. The head, by contrast, may in part never have been segmented, was in part segmented like the body but only in remote ancestors (derivatives of head myotomes and sclerotomes), and is in part segmented in an independent series (derivatives of visceral arches). In consequence, serial homology is less evident in cranial than in spinal nerves. Cranial nerves are numbered in spatial sequence, but it should be understood that in terms of fundamental nature the assignment of numbers is rather arbitrary.

These difficulties were a welcome challenge to morphologists at the turn of the century, and much attention was given to the analysis of cranial nerves. Various kinds of clues were used. Developmental evidence shows that sensory fibers have been lost from the accessory nerve: ganglia form in the embryo and then regress. Comparative anatomical evidence shows (for instance) that the trigeminal nerve of amniotes is a composite nerve: one of its principal branches joins the brain independently in some fishes. Physiological evidence shows the unique nature of some nerves (only aquatic vertebrates have a lateral line system and it is served only by cranial nerves).

Putting together many clues of these kinds it is evident that cranial nerves are of three general categories. First, there are seven nerves (numbers 0, V in two parts, VII, IX, X, XI) that are in series with dorsal roots of spinal nerves, or better, with the dorsal spinal nerves of lampreys which, as described in the previous section, do not join their respective ventral spinal nerves (Figure 17-9). These nerves all join the brainstem at a lateral (not ventral) level. It is postulated that in the ancestral condition, each nerve carried the same components as dorsal spinal nerves: somatic sensory, visceral sensory, and visceral motor. A sensory ganglion was present close to the entrance of each nerve into the brain. Furthermore, it is postulated that in the remote ancestral vertebrate, these nerves served branchial and pharyngeal areas: Each nerve had a major branch to each hemibranch bordering a gill slit, and minor branches to the adjacent pharyngeal wall and skin.

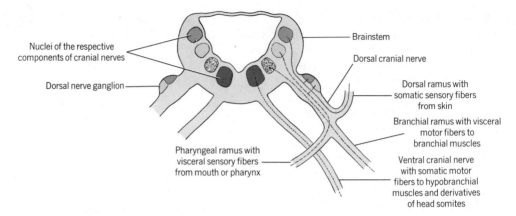

Figure 17-9 POSTULATED ANCESTRAL DISTRIBUTION OF NERVE COMPONENTS IN THE BRAINSTEM AND IN CRANIAL NERVES OF THE DORSAL AND VENTRAL SERIES.

The second category includes four cranial nerves (III, IV, VI, XII) that are in series with ventral spinal nerves. All but one join the brainstem at the expected ventral level. (The exception is only partial; the nucleus of the trochlear nerve is ventral in the brainstem, but the fibers arch over the brain to emerge high on the opposite side.) These nerves carry somatic motor fibers. Ganglia are, of course, absent. The nerves innervate hypobranchial muscles and derivatives of head somites.

The third category has no counterpart in the spinal series because its nerves serve structures that are peculiar to, or centered on, the head: the nose, eye, ear, and lateral line system. Nerves I, II, VII, VIII, IX, and X are in this category. These nerves are sensory. Their ganglia, unlike those of other sensory nerves are derived, at least in part, from ectodermal placodes. These nerves are usually considered to be somatic, though the designation is somewhat arbitrary and perhaps superfluous. Common practice is to call them "special" sensory nerves in recognition of their distinctive nature. To a degree, this is a category of leftovers because the nerves of the nose and eye are not serially homologous either with one another or with the other nerves of the category.

There is one further complication. The visceral motor nerves of the body are so distinctive in various ways that together they make up the autonomic system, which is described at the end of this chapter. Even at levels of the cord, it was the visceral motor fibers that were least stable, emerging sometimes in the dorsal root and sometimes in the ventral root. In the head, visceral motor fibers of the autonomic system join four cranial nerves (III, VII, IX, X). These include nerves of both the dorsal and ventral categories.

Counting seven nerves in the dorsal series, four in the ventral series, six in the special series, and four with autonomic fibers, and knowing that there are thirteen cranial nerves in all, it is evident that there is some doubling up. Thus, the oculomotor nerve is in the ventral series but has autonomic fibers tagging

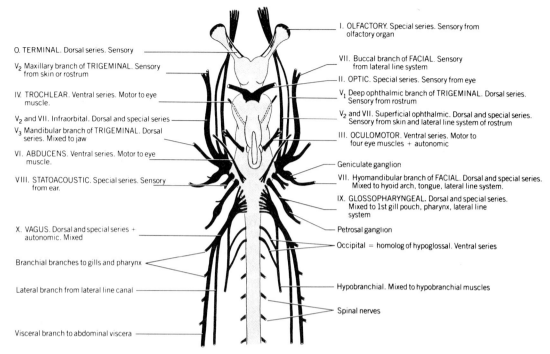

Figure 17-10   CRANIAL NERVES OF THE SHARK *Squalus*. Ventral view. (Trigeminal and facial nerves are more independent in most other vertebrates.)

along; the facial nerve is in the dorsal series but may be augmented by both special and autonomic components.

This general information provides the basis for a more specific account of the evolution of each cranial nerve. The following account is clarified by Figure 17-10 and also figures on p. 145 and 335.

STRUCTURE AND EVOLUTION OF CRANIAL NERVES   The TERMINAL nerve was not discovered until the other cranial nerves had been given numbers from I through XII. Hence, it has the number 0. Regarded as being in the dorsal series, it originates in nasal epithelium and sometimes in the vomeronasal organ, yet seems not to be olfactory in function. Having one or more ganglia, it is classed as a somatic sensory nerve, any ancestral visceral fibers having been lost. The nerve is present in all vertebrates except cyclostomes, birds, and some mammals (including man). It is largest in elasmobranchs.

The OLFACTORY nerve (number I) is in the special series. It runs from the olfactory epithelium and vomeronasal organ (if present) to the olfactory bulb of the brain. It is unique in that its fibers are extensions of the receptor cells. Consequently it has no ganglion even though it is sensory. The nerve is present in all vertebrates, its size relating to the excellence of the olfactory sense. It is long if the rostrum is long and the olfactory tracts of the brain are short. Otherwise the nerve is short. Often it is divided into many twigs. It is paired in cyclostomes in spite of the unpaired nature of that animal's nasal pouch.

The OPTIC nerve (II) is again in the special series. It runs from the eye to the brain and is unique in that it develops as a tract of the embryonic brain (see p. 372). The associated ganglion cells are in the retina. The nerve is constant in all vertebrates. The two optic nerves may completely cross under the brain (teleosts, birds, and some other vertebrates), but often some fibers cross and some do not. In mammals, half the fibers of each nerve cross to the other side, an arrangement that in this class probably contributes to the coordination of eye movements.

The OCULOMOTOR (III), TROCHLEAR (IV), and ABDUCENS (VI) nerves are in the ventral series. They innervate the extrinsic muscles of the eye. All are somatic motor nerves, though the oculomotor is joined by autonomic fibers passing to muscles of the iris and ciliary apparatus of the eye. These nerves are constant in virtually all vertebrates.

The DEEP OPHTHALMIC (or profundus) nerve (V_1) is the second nerve of the dorsal series, though like the terminal nerve it does not relate to a gill slit in any surviving vertebrate. It runs to the skin of the rostrum. It is exclusively somatic sensory; any visceral fibers it may once have had were lost with the associated gill slit. The nerve is represented in all vertebrates, but is an independent nerve with its own ganglion only in ostracoderms, placoderms, and some primitive bony fishes. In other vertebrates it becomes a branch of the next nerve to be described.

The large TRIGEMINAL nerve (V) has two branches in those vertebrates having an independent deep ophthalmic nerve. In other vertebrates the deep ophthalmic (in mammals simply called ophthalmic) (V_1), maxillary (V_2), and mandibular (V_3) nerves are the three branches that give the trigeminal nerve its name. (A superficial ophthalmic branch of the trigeminal may also be present, and unfortunately the facial nerve has a branch by the same name.) The maxillary and mandibular branches apparently represent a branchial nerve that once served the premandibular gill slit. The maxillary branch retains only somatic sensory fibers from the teeth, gums, and skin of the upper jaw. The mandibular branch similarly serves the lower jaw. It also has motor fibers to the various jaw muscles derived from the mandibular arch. These are striated, voluntary muscles, but since they relate phylogenetically to the pharynx (see p. 187), the associated nerves are designated as visceral. The large **semilunar** (or Gasserian) **ganglion** is located where these branches all merge before entering the brain.

The FACIAL nerve (VII) is in part the nerve in the dorsal series that is associated with the spiracular cleft and derivatives of the hyoidean arch. It is the first nerve of this series to retain all the components of the ancestral dorsal nerves: somatic sensory to related areas of the skin, visceral sensory to much of the mouth and taste buds, and visceral motor to all muscles derived from the hyoidean arch. Included among the latter are the muscles of facial expression of man—hence the name of the nerve. The facial nerve of fishes also has a large component from the special series. This consists of the sensory fibers from the cranial part of the lateral line system. Finally, there is a component of the autonomic system serving tear glands and several salivary glands. The composite nerve is large. Its ganglion is the **geniculate ganglion.**

The STATOACOUSTIC nerve (VIII) (also called vestibulocochlear and auditory) serves the inner ear. It is, therefore, in the special series, and relates, developmentally and phylogenetically, to the lateral line components of nerves VII, IX, and X. It always has two principal branches. In most vertebrates the more anterior branch serves most of the organ of equilibrium, and the posterior branch serves both the organs of equilibrium and hearing. In eutherian mammals the posterior branch functions only in hearing.

The GLOSSOPHARYNGEAL nerve (IX) is in part the nerve of the dorsal series associated with the first branchial gill slit and arch of fishes. Somatic sensory fibers are present in some vertebrates. Visceral sensory fibers run to part of the pharynx and some taste buds. Visceral motor fibers serve some small muscles. As for the previous nerve, there is a component from the special series of nerves innervating the lateral line system, this time at the back of the head, and another from the autonomic system, this time to a salivary gland. The composite nerve is usually small. Any somatic sensory fibers present have a **superior ganglion;** visceral sensory fibers have a larger **petrosal ganglion.**

The VAGUS (X) and ACCESSORY (XI) nerves can best be listed together because the latter is present only in amniotes and is derived by splitting away from the original vagus. The vagus spans the levels of several ancestral head somites and joins the brainstem in a linear series of twigs. These nerves are in part the last of the dorsal series, and as such are associated with all remaining branchial structures. A small somatic sensory branch of the vagus serves skin in the gill and ear region. Visceral sensory fibers come from posterior taste buds and the pharynx. Sensory fibers of the accessory nerve do not mature. Visceral motor branches serve muscles of the branchial arches and their derivatives (including some muscles of the shoulder). A large component of the vagus nerve of fishes innervates the part of the lateral line system that is on the body, and hence is from the special series of nerves. Finally, the large, long, and important **visceral branch** of the vagus is the autonomic component of that nerve. Its fibers serve the heart, lungs (if present), and gut. Somatic sensory fibers have a **jugular ganglion;** visceral sensory fibers have a large **nodose ganglion.**

The HYPOGLOSSAL nerve (XII) is in the ventral series of nerves. It is an exclusively somatic motor nerve and innervates the hypobranchial muscles of throat and tongue. It is a cranial nerve in amniotes and some labyrinthodonts and a cervical nerve (called the hypobranchial nerve) in cyclostomes and fishes. It is derived from several postotic somites (the same for which the vagus was the dorsal nerve) and joins the central nervous system by a linear series of twigs.

AUTONOMIC SYSTEM

The autonomic system is not isolated, structurally or functionally, from either the central or peripheral nervous systems. Hence, it is difficult to set its limits. It relates exclusively to involuntary functions of the body. Accordingly, the system includes only visceral fibers, and the visceral fibers serving the striated branchial muscles are excluded. Structures innervated by autonomic nerves are, therefore, the heart and vessels, some respiratory organs, glands, gut tube, urogenital organs, pigment cells, and intrinsic muscles of eye and skin. These structures are, of course, served by both sensory and motor fibers. There is nothing very remarkable, however, about the sensory neurons: like somatic sensory neurons they have their nerve cell bodies in dorsal root ganglia and certain cranial ganglia. Sensory fibers are, therefore, omitted from descriptions of the autonomic system, if not from its definition.

What, then, is distinctive about the visceral motor neurons that innervate involuntary organs? First, nearly all such organs are served by two sets of autonomic nerves—and the sets elicit antagonistic responses. These sets comprise the **sympathetic** and **parasympathetic divisions** of the autonomic system. Second, every pathway includes a neuron having its cell body inside the central nervous system, and in addition (except in the adrenal medulla) a neuron (or in some instances several neurons) having its cell body outside the central nervous system (Figure 17-11). Cell bodies of the latter are in motor ganglia. Fibers between ganglia and the central nervous system are designated **preganglionic;** those between ganglia and end organs are designated **postganglionic.** Preganglionic fibers are myelinated; postganglionic fibers have little or no myelin.

The following descriptions of the two divisions of the autonomic system are based on amniotes (particularly mammals). Comments on the system in lower vertebrates will conclude the chapter.

The SYMPATHETIC division of the system is said to have a **thoracolumbar outflow** because that term indicates the levels at which its fibers emerge from the central nervous system in spinal nerves (Figure 17-12). Preganglionic fibers are relatively short. They reach the sympathetic ganglia by way of the visceral rami of spinal nerves (or the white part of the connection—see Figure 17-11). Most of the small but tough ganglia are **chain ganglia** arranged like two parallel strings of beads on the **sympathetic trunks** that are just ventral to the vertebral

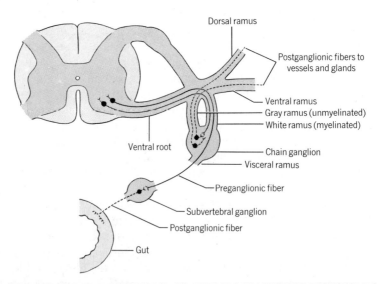

Figure 17-11 TYPICAL PATHWAYS OF THE SYMPATHETIC DIVISION OF THE AUTONOMIC SYSTEM OF AMNIOTES. Cross section at posterior thoracic level. Other fibers run lengthwise in the body from the chain ganglia. Size of the nerves is exaggerated.

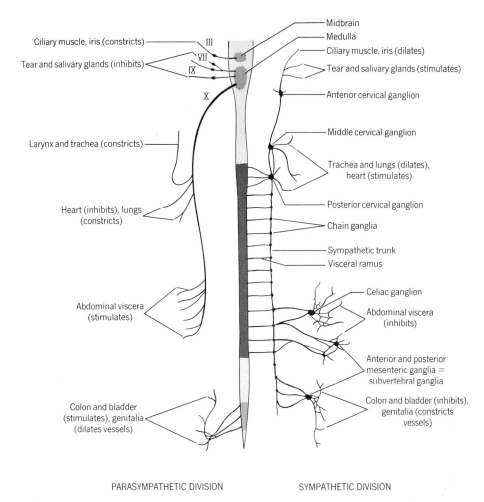

Figure 17-12 DIAGRAM OF THE MAMMALIAN AUTONOMIC NERVOUS SYSTEM. General distribution of fibers to vessels and skin is not shown.

column. There are also three pairs of **cervical ganglia** against the carotid arteries in the neck, and about three unpaired **subvertebral ganglia** at the bases of the major arteries into the viscera.

A preganglionic fiber may synapse with a postganglionic fiber in the chain ganglion closest to its exit from a spinal nerve, or may go through that ganglion to a chain ganglion on the other side of the body, a chain ganglion at another level of the spine, or a cervical or subvertebral ganglion. Postganglionic fibers are relatively long. Most of them that emerge from chain ganglia reenter a spinal nerve and run with it and its branches to their destinations. Most postganglionic fibers that emerge from cervical or subvertebral ganglia run parallel with blood vessels to their destinations.

Postganglionic fibers, like the medulla of the adrenal gland, release **noradrenalin.** There are fewer terminal fibers than there are muscle and gland cells to be influenced. These nerve junctions, therefore, release relatively large quantities of noradrenalin, which then spreads to adjacent effector cells.

In general, the sympathetic division of the system elicits responses of alertness, excitement, alarm, and the expenditure of energy as necessary to meet emergencies. Vegetative functions tend to be inhibited.

The PARASYMPATHETIC division of the autonomic system is said to have a **craniosacral outflow** because it emerges from the central nervous system with cranial nerves III, VII, IX, and X and with about three sacral spinal nerves. Preganglionic fibers are relatively long. There are about four pairs of small parasympathetic ganglia in the head. These are located near the organs served (eye, salivary glands, tear glands). Elsewhere, ganglion cells are dispersed in the tissues of the viscera and are not found by dissection. Postganglionic fibers are quite short.

Postganglionic fibers of parasympathetic neurons are like most somatic fibers and preganglionic fibers of both divisions of the autonomic system in releasing acetylcholine, though again, the quantity is large.

The parasympathetic division of the system elicits responses appropriate to quiet, vegetative activities such as digestion and maintenance of resting levels of blood sugar.

The above account probably fits most AMNIOTES, though the autonomic system is well-known only in mammals. The system was late to evolve. AMPHIOXUS has visceral motor fibers running from each spinal nerve to the gut, but there are no ganglia outside the viscera. CYCLOSTOMES have autonomic fibers in the vagus nerve, but otherwise the system is rudimentary. CARTILAGINOUS FISHES send visceral motor fibers to vessels as well as the gut, and have some delicate autonomic ganglia under the spine. However, sympathetic and parasympathetic divisions of the system are scarcely, if at all, distinguished. In BONY FISHES and AMPHIBIANS the two divisions of the system are established in much of the body, though usually not in the more posterior organs of fishes and urodeles.

18

Nervous System: Brain

The brain is the most complicated, and for many persons the most amazing organ of the body. It is under study by many researchers in various fields. Since understanding of its function must ultimately be based largely on its wonderfully intricate structure, the morphologist has a major role to play in the analysis of this master organ.

In order to gain an adequate background in brain structure for either teaching or advanced study it is desirable to study the brain in enough depth to acquire basic vocabulary, learn general spatial relationships, and identify the major functional components.

HOW THE BRAIN IS STUDIED

The brain is studied in many ways. Descriptions are made of gross structure using entire or dissected brains, of fine structure using serial sections, and of ultrastructure using electron micrographs. Thick slices of the brain are stained to distinguish, in striking color contrast, the myelinated tracts from the unmyelinated cortex and nuclei. Histological techniques reveal individual neurons in detail.

A device called a stereotaxic instrument is used to position the head and brain in a precise standard position. The location of every minute part of the brain can then be expressed in terms of the three coordinates of space. Using the stereotaxic atlases that are now available for many experimental animals, the investigator can introduce microinstruments into any desired part of the living brain. Having done this, various techniques are used: Restricted areas are destroyed with electricity, freezing, or chemical agent, or very small bits of tissue are removed by suction. Impairment of function is then noted, and after a period of time, the animal is killed, and suitable staining techniques are used to reveal on sections of the brain the paths of degenerating nerve fibers. Alternatively, normal stimuli, microelectrodes, or chemicals are used to activate specific parts of the brain, and resultant behavior is observed. Animals may be taught to stimulate their own brains for the reward of pleasant sensations, and conscious human subjects have reported sensations and thoughts accompanying local stimulation of the brain during surgery. Following stimulation of one part of the brain, electric phenomena elicited in other parts are recorded using an oscillograph, multigraph pen recorder, or counters that translate data to tape for analysis by computer.

The impairment of function that follows accident or pathology is studied. Parts of the brains of experimental animals are severed and changes in behavior analyzed. Differences in the brains of different vertebrates are related to their widely different adaptations. Progressive maturation of the fetal brain is correlated with the onset of function. Natural electrical discharges of the organ (brain waves) are recorded during various activities. Culture techniques are used to study the growth and physiology of isolated neurons. The response of the brain to both the general and local administration of drugs is revealing. Finally, the science of cybernetics adapts mathematical models to brain function in an effort to elucidate mechanisms.

From all this come several overlapping levels of advancement in the study of this challenging organ. The first level is descriptive. Though not finished for any species and not even begun for many, this necessary level of investigation has produced hundreds of technical terms and shelves of much-labeled drawings. Study of the brain can be tedious if the emphasis is on description. Nevertheless, description is an essential first step. A second level of study relates structure and function. A third level identifies electrical circuits and produces complicated wiring diagrams. These last two levels have dominated brain research in recent years. Now it is gradually becoming apparent that the wiring diagram, however important, is conceptually inadequate. For some of the best-known circuits (e.g., that of vision), it seems that no part of the system is the exclusive site of any one function. Concepts of scanning mechanisms, holograms, and interference effects may prove useful for understanding memory. Such "mass action' of the brain points to further levels of analysis. The brain does not merely transmit, reject, or store the information in the 3 billion impulses that reach its 10^{10} cells in every waking second. It transforms the information, adapts it, and chooses among alternative responses in ways that surpass present comprehension.

DEVELOPMENT OF THE BRAIN

By the time the neural folds close over the neurocoel in the later stages of neurulation, the future brain is already of greater diameter than the spinal cord. As soon as the tube is formed, the developing brain expands at three levels to form vesicles separated from each other by constrictions. These **primary vesicles** are the **forebrain** or **prosencephalon, midbrain** or **mesencephalon,** and **hindbrain** or **rhombencephalon** (Figure 18-1). The prosencephalon lies anterior to the notochord; the other vesicles are dorsal to the notochord.

There is reason to believe that the unknown ancestors of vertebrates had, as

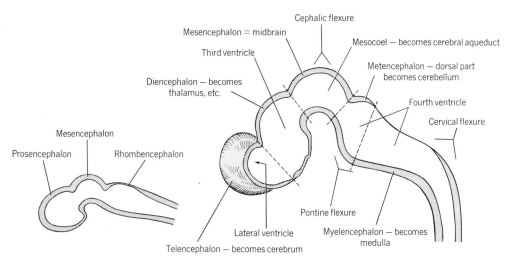

Figure 18-1   DEVELOPMENT OF THE MAMMALIAN BRAIN: stage of primary vesicles on left; stage of secondary vesicles on right. Brains cut in the sagittal plane.

adults, brains similar to this present-day embryonic stage. At that time in evolution the brain may have been similar to the cord except for its response to sense organs peculiar to the head. It was formerly believed that the prosencephalon was related to olfaction, the mesencephalon to vision, and the rhombencephalon to taste, equilibrium, and the lateral line system, but recent research indicates that this view is too simplistic.

At the next stage of development, additional constrictions divide the brain further into five **secondary vesicles.** The anterior part of the prosencephalon becomes the **telencephalon,** largely through expansion of its lateral walls. These expansions will form the **cerebral hemispheres** of the adult. The posterior part of the prosencephalon becomes the **diencephalon.** The mesencephalon remains undivided. The rhombencephalon forms an anterior **metencephalon,** which forms the adult **cerebellum,** and a posterior **myelencephalon.** ("encephal" = brain; most of the prefixes indicate position.)

Divisions between certain of these secondary vesicles are slight, and there is little functional and evolutionary basis for recognizing them. Nevertheless, it is a convenience to divide the brain into these parts for purposes of instruction.

The embryonic neurocoel is larger in the brain than in the cord. Within the vesicles it forms expansions called **ventricles.** The **lateral ventricles** occupy the cerebral hemispheres. If, as in fishes, the hemispheres are partly joined, then they share a common ventricle. The **third ventricle** is in the diencephalon. The neural canal expands within the mesencephalon of most vertebrates, but it is relatively restricted and tubelike in mammals and is then called the **cerebral aqueduct.** The **fourth ventricle** is located in both the metencephalon and the myelencephalon.

Most brains have nearly straight axes. Brains of birds and mammals, however, acquire three **flexures** in the embryo. The sharpness of each flexure depends on posture and cranial architecture, and does not relate directly to function. The **cephalic flexure** is in the mesencephalon and is concave ventrally. The **pontine flexure** is in the part of the metencephalon called the pons and is concave dorsally. The **cervical flexure** is within the posterior part of the myelencephalon and again opens ventrally. This flexure is most evident in bipeds that hold the head at an angle to the neck.

It is useful to divide cross sections of the tubelike embryonic brain (and cord) into quadrants. The dorsolateral quadrants are called **alar plates** and the ventrolateral quadrants **basal plates.** In the cord, the dorsal gray columns (see figure on p. 317), with their association neurons and axons of sensory neurons, develop from the alar plates, whereas the ventral gray columns, with their motor neurons, develop from the basal plates. In the brain, alar and basal plates become discontinuous, yet (as explained further below) they form neurons corresponding to their counterparts in the cord. The basal plate terminates at the diencephalon.

The three tissue layers of the embryonic spinal cord (ependymal, mantle, and marginal) are also present in the brain. The mantle layer is thick, and in the cerebral hemispheres and cerebellum of amniotes most of its inner cells migrate peripherally into the marginal layer. The vacated region will thus become white matter, and the invaded surface of the brain will become gray matter. This is the reverse of the spatial relationship elsewhere in the central nervous system where gray matter is internal.

Myelination is a slow process. It does not start until the fiber tracts are well-established and is not finished until well after birth.

MORE ABOUT THE ORGANIZATION OF THE BRAIN

In the previous section the brain was divided into five regions on the basis of development. Another division into three regions—**brainstem, cerebellum,** and **cerebrum**—is also useful. The central axis of the brain is the brainstem. It is the first region to form in ontogeny, the least variable, and the most like the spinal cord in structure. It receives all cranial nerves except the atypical terminal and olfactory nerves; relays impulses to the other two regions; and independently controls various vegetative functions of the body. Part of the adult metencephalon and all of the diencephalon, mesencephalon, and myelencephalon are included in the brainstem. The adult mesencephalon is usually called the **midbrain** and the adult myelencephalon the **medulla.**

The cerebellum and related pons (if present) are the principal adult derivatives of the metencephalon. They contribute to the coordination of motor functions. The cerebellum of amniotes and of some fishes is a conspicuous appendage of the brainstem covering much of the posterior part of its dorsal surface.

The cerebrum is the adult derivative of the telencephalon. Gradually this region of the brain enlarged and added new parts and functions until, in mammals, it dominates the brain both in size and control.

Within the spinal cord, nerve cell bodies mass in the dorsal and ventral gray columns, and these are continuous throughout the length of the cord. In the brain, by contrast, functionally related nerve cell bodies either mass at the surface of the cerebrum or cerebellum, where they make up the **cortex** of those organs, or cluster in discontinuous masses within the brain. Such a cluster is usually called a **nucleus,** but may also be termed a center, ganglion, or body.

Similarly, within the cord, bundles of functionally related nerve fibers are usually called tracts. In the brain, such a bundle is also commonly called a tract, but unfortunately may have another name (fascicle, capsule, brachium, peduncle, lemniscus) depending on size, shape, and relationships.

Sensory fibers enter the brain from the cord and cranial nerves, and terminate in a nucleus of the brainstem or in the cortex of the cerebellum. Incoming impulses are usually passed from a first nucleus in the brainstem to one or more other nuclei or to the cortical areas (including the cerebral cortex). Each incoming impulse "comes to the attention of" many (usually thousands) of neurons in the gray matter of the brain. In this process of "considering" messages received, the nuclei are not mere relay stations for transmitting impulses; each is also an association center that processes, alters, and redistributes the messages in some way. The more complicated and the less automatic the ultimate response, the more likely that relevant impulses will reach the anterior part of the brainstem and cerebrum. Messages passing posteriorly may also be processed in "lower" nuclei before exiting from the brain in the spinal cord or cranial nerves. Furthermore, tracts called **commissures** pass from one side of the brain to corresponding parts on the other side. These permit the integration of sensations and learning experiences from the two sides of the body. Crossings of sensory or motor fibers from one side of the brain or cord to different parts on the other side are termed **decussations.**

NUCLEI OF CRANIAL NERVES   It is again useful to refer to structure of the spinal cord as a frame of reference. There, somatic motor neurons are derived from the embryonic basal plates, whereas other neurons are from the alar plates. The same is true in the brainstem. In the cord, there is a more or less dorsal to ventral linear arrangement in the gray columns of somatic sensory, visceral sensory, visceral motor, and somatic motor neurons. This linear arrangement is roughly preserved in the brainstem. However, spreading of the dorsal parts of the alar plates of the medulla to accommodate the fourth ventricle causes the orientation of the linear series to change from dorsal–ventral to somewhat lateral–medial (see figure on p. 322). Also, these neurons are pushed up close to the floor of the fourth ventricle as other structures come to occupy the ventral part of the medulla. Furthermore, these components are continuous in the gray columns of the spinal cord, but are broken up into discontinuous nuclei in the brain. Finally, nuclei without counterpart in the cord are present in the brainstem to relate to the "special" senses of the head. These shoulder their way into position near their nonspecial equivalents.

It follows from this general plan that each cranial nerve tends to have a nucleus in the brainstem for each kind of component fiber it carries and that although corresponding nuclei of different nerves are at different anterior–posterior positions in the brainstem, they tend to be at roughly equivalent lateral–medial and dorsal–ventral positions.

These general relationships are useful for interpreting the anatomy of the nuclei of cranial nerves in specific animals, but nature contrives to complicate matters somewhat: One nucleus may serve two or more adjacent nerves and one ancestral nucleus may split into several nuclei. Some nuclei are long and some

POSTERIOR BRAINSTEM: MEDULLA THROUGH MIDBRAIN

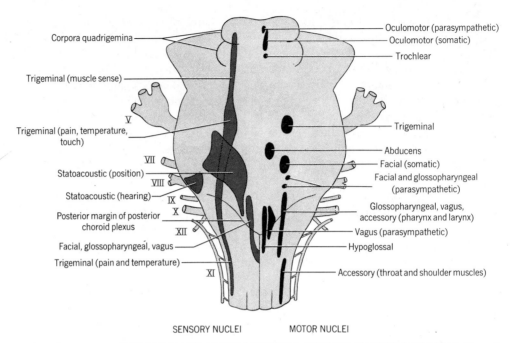

Corpora quadrigemina

Oculomotor (parasympathetic)
Oculomotor (somatic)
Trochlear

Trigeminal (muscle sense)

V

Trigeminal (pain, temperature, touch)

Trigeminal

VII

Abducens
Facial (somatic)
Facial and glossopharyngeal (parasympathetic)

Statoacoustic (position)
VIII

Statoacoustic (hearing)
IX

Glossopharyngeal, vagus, accessory (pharynx and larynx)

Posterior margin of posterior choroid plexus
X

Vagus (parasympathetic)

XII

Hypoglossal

Facial, glossopharyngeal, vagus

Trigeminal (pain and temperature)

XI

Accessory (throat and shoulder muscles)

SENSORY NUCLEI   MOTOR NUCLEI

Figure 18-2   NUCLEI OF CRANIAL NERVES IN THE BRAINSTEM OF MAN. Dorsal view. Labels show nerve affiliations of nuclei, not their technical names.

short; some large and some small. Jostled by neighboring nuclei and tracts, a nucleus may stray somewhat from its postulated ancestral position.

The oculomotor and trochlear nerves join the midbrain, and their nuclei are confined there. Nerves V–XII join the medulla, and most of them have nuclei there. Nuclei of the more anterior of these nerves commonly also (or instead) are located in the metencephalon, and a nucleus of the trigeminal nerve reaches forward to the midbrain.

To name the various nuclei of cranial nerves and to compare their numbers and positions in different vertebrates would not serve the objectives of this book. However, several representative nerve–nucleus relationships will be noted to illustrate the structural plan already presented: The oculomotor nerve has a medially situated nucleus for its somatic motor neurons and another smaller nucleus for its autonomic (visceral motor) fibers (Figure 18-2). The single nucleus of the trochlear nerve is found, as expected, just posterior to the somatic motor nucleus of the oculomotor nerve, even though the nerve itself exits from the brain at an unaccountably dorsolateral position. The large trigeminal nerve has a long somatic sensory nucleus in the expected dorsolateral part of the medulla. The visceral motor fibers of its mandibular branch (which innervate jaw muscles of branchial origin) have a midbrain nucleus which is relatively large in animals that chew their food. The facial nerve, having several kinds of component fibers, has several nuclei in the brainstem. The statoacoustic nerve has separate "special" nuclei (or clusters of nuclei) for its vestibular and acoustic branches. The spinal accessory nerve shares one of its nuclei with the vagus from which it is derived.

Figure 18-3 BRAINSTEM OF THE COW in ventral (left) and dorsal (right) views. Cerebrum and cerebellum removed by dissection. Letters show the levels of the cross sections in Figure 18-4. (The brain of the sheep, commonly studied in the laboratory, is similar to the brain of the cow.)

RETICULAR FORMATION The reticular formation occurs in all vertebrates. It is derived from the basal plates and is located in the central part of the brainstem from midbrain to medulla and, in reduced form, also in the anterior part of the cord. It consists of a diffuse, interlocking mass of nerve cell bodies and fibers; hence it is a mixture of gray and white matter. Its boundaries are indistinct (Figure 18-4). More or less sharply defined clusters of nuclei tend to form in its substance at several levels of the brainstem.

The reticular formation has sensory input from virtually all parts of the body and all senses. It in turn projects to the cerebrum, cerebellum, various cranial nuclei, and the cord. It is essential for consciousness. Stimulation awakens a sleeping animal and makes an awakened animal more alert. It contributes to the activities of both voluntary and involuntary muscles by facilitating, inhibiting, screening out noise, and coordinating stimuli. It also contributes to control of the cardiovascular and respiratory systems.

OTHER NUCLEI OF THE POSTERIOR BRAINSTEM There are numerous paired nuclei in the posterior brainstem other than those relating directly to cranial nerves. Several of the more prominent ones will be mentioned. The **olivary nuclei** arise from the alar plates, yet during development migrate down to the ventrolateral wall of the medulla. This complex of nuclei is present in all vertebrates. It is best developed in the most active representatives—mammals, birds, and some fishes. The largest component is the **inferior olive,** which may form a low bulge on the sidewall of the medulla (Figure 18-3). In man the inferior olive, which is a motor coordination center, has the size and crumpled shape of a raisin, receives association fibers from some other nuclei of the brainstem and

Anterior colliculus

Central gray substance

Tract between midbrain and cord

Spinothalamic tract

Ruber nucleus

Substantia nigra

Medial geniculate body

Cerebral aqueduct

Cerebral peduncle

SECTION AT A

Middle cerebellar peduncle

Tract of facial nerve

Nucleus of trigeminal nerve

Trapezoid body

Tract of abducens nerve

Vestibular nuclei

Cochlear nuclei

Statoacoustic nerve

Superior olivary nucleus

Pyramidal tract

SECTION AT B

Posterior cerebellar peduncle

Reticular formation

Cerebrospinal tract (continues into pyramidal tract)

Nuclei of glossopharyngeal and vagus nerves

Nucleus of hypoglossal nerve

Complex of olivary nuclei

SECTION AT C

Figure 18-4　CROSS SECTION OF THE BRAINSTEM OF THE COW at the levels shown by letters in Figure 18-3. The contrast between nuclei and tracts has been enhanced by staining.

proprioceptive (muscle sense) impulses from the spinal cord (and some other input), and projects dorsally to the cerebellum.

The **ruber** (or red) **nucleus** and **substantia nigra** (which is pigmented in some mammals) are located deep in the mesencephalon (and may extend into the diencephalon). The ruber nucleus is present in all vertebrates but is most developed in mammals, whereas the substantia nigra first appears in reptiles and is best developed in primates. Each nucleus may have evolved from the reticular formation. In man the ruber nucleus resembles a pea and the substantia nigra a lima bean. These are relay stations between the forebrain, on one hand, and posterior brainstem and cord, on the other. The ruber nucleus of carnivores (at least) plays a role in coordination of motor functions, particularly of flexor muscles. The substantia nigra is involved with the memory of learned tasks.

The roof of the midbrain is called the **tectum** (= covering). In all vertebrates except mammals it consists primarily of a bilateral pair of conspicuous hemispherical eminences termed **optic lobes** (Figure 18-10). With mammals again excepted, the optic lobes are the primary center for the perception of vision (though even in fishes, the retina, diencephalon, and elsewhere in the mesencephalon are also involved in this complicated sense). The optic lobes are distinctly striated internally, and the visual image is projected onto the lobes point for point. The lobes are the most prominent feature of the brains of fishes that locate food by sight. They remain conspicuous in reptiles and birds.

The perception of vision is largely transferred by mammals to the cerebrum, though the midbrain tectum still has an important function: It tells the mammal where in space a visual object is, whereas the cerebrum tells what the object is. The much smaller optic lobes are here called **anterior colliculi** (Figure 18-3). Behind these are the **posterior colliculi,** which are involved possibly in coordination of auditory reflexes. Taken together, these four bumps are the **corpora quadrigemina.** There are commissures between the lobes of a pair.

Some of these brainstem nuclei, and others not named, are sufficient to support various vegetative functions in normal circumstances. Even with other parts of the brain removed, experimental animals maintain heartbeat, respiration, swallowing, and digestion.

ADDITIONAL FEATURES OF THE POSTERIOR BRAINSTEM   The brainstem transmits all impulses moving to or from other parts of the brain. Obviously, much of its substance must consist of tracts of nerve fibers. The pathways are complex, yet are well-known for many vertebrates. Several tracts that are large and can usually be seen on the surface of the brain will be identified here.

Particularly large paired tracts of motor fibers run directly from the cerebral cortex of mammals to the spinal cord without interruption. Similar, but less prominent tracts are found in other tetrapods and sharks. In mammals they emerge at the surface of the brain on the ventrolateral wall of the midbrain where they converge on each side to form the **cerebral peduncles** (= small feet, hence supports). The fibers are lost from surface view in the metencephalon but continue in most mammals as the prominent **pyramidal tracts** on the ventral surface of the medulla as they move to the lateral and ventral funiculi of the cord.

The **trapezoid body,** which may be evident externally in mammals, transmits fibers from the statoacoustic nerve to the superior olive and elsewhere. Large paired columns of sensory fibers enter the medulla from the dorsal funiculi of the spinal cord. These **gracile** and (more lateral) **cuneate fascicles** flank the dorsal median sulcus of cord and posterior medulla until they terminate in the **gracile and cuneate nuclei.**

The cerebellum, which arches over the fourth ventricle, joins the brainstem by three pairs of peduncles (or brachia) that flank the ventricle. If the cerebellum is large, the peduncles are then prominent surface features. The **posterior peduncle** carries fibers between the cerebellum and spinal cord, inferior olive, and vestibular nuclei. The **middle peduncle** of mammals is the largest and most lateral peduncle, partly hiding the others (Figure 18-3). It transmits fibers between the two sides of the cerebellum and from cerebrum to cerebellum. The **anterior peduncle** has fibers joining the cerebellum to the ruber nucleus and diencephalon.

The roof of the fourth ventricle is covered by the **posterior choroid plexus.** This is a delicate, much pleated, highly vascular membrane to which the brain contributes only its thin ependymal layer. The function of choroid plexuses is described at the end of this chapter.

ANTERIOR BRAINSTEM: DIENCEPHALON

The anterior part of the brainstem differs from the posterior part in being derived entirely from the embryonic alar plates, in having no nuclei for cranial nerves and no reticular formation, and in relating to functions that are more highly evolved. The embryonic telencephalon forms the cerebrum which, in most vertebrates, is bilaterally paired and has no structures in the midline except for several commissures and the thin anterior wall of the ventricle. Consequently, the diencephalon forms the most anterior part of the brainstem. It is convenient to divide the diencephalon into three parts.

The narrow dorsal part of the diencephalon is the **epithalamus,** much of which is nonnervous in function. Most anterior is the **anterior choroid plexus** of the third ventricle (Figure 18-8). Posterior to the plexus is an evagination of similar structure (the paraphysis) which, however, does not mature in man. Next in line are the **habenular nuclei** or habenulae (which are often of different size on right and left!). They are present in all vertebrates and contribute to the coordination of olfactory reflexes. Behind the habenula are two evaginations, the **parietal organ** and **pineal body.** These structures do not lend themselves to the organization of chapters by organ system. They are mentioned here because of their origin and relationships, in Chapter 19 because each may function as a sense organ, and in Chapter 20 because one may function as an endocrine gland. The habenula has a small commissure, and a larger **posterior commissure** marks the posterior boundary of the epithalamus. Its fibers join some nuclei of the two sides of the diencephalon and perhaps of the mesencephalon as well.

Each thick lateral wall of the diencephalon is called a **thalamus.** This is the largest part of the anterior brainstem, being 4 cm long in man. The two thalami are bordered in part by the lateral ventricles and are separated from one another by the third ventricle, except where the large **middle commissure** or intermediate mass spans that cavity (Figure 18-8). Each thalamus is a compact oblong mass of many nuclei—30 or more are recognized. A ventral cluster of nuclei (**ventral thalamus**) is represented in all vertebrates and projects motor fibers posteriorly in the brain. A dorsal cluster (**dorsal thalamus**) projects sensory pathways (except olfactory pathways) to the cerebrum (to both its striatal and cortical parts, as defined below). The dorsal nuclei are most developed in tetrapods—particularly in mammals.

Two of the dorsal nuclei are of particular importance. These are located at the posterior end of the thalamus close to the tectum of the midbrain. One is the **medial geniculate body,** which forms part of the auditory pathway and is joined by a pathway to the posterior colliculi of the tectum (Figures 18-3 and 18-4). The other is the **lateral geniculate body,** which is a relay station in the primary visual pathway. Seen in cross section, the lateral geniculate bodies of some mammals (like the optic lobes of many vertebrates) are striated into about six conspicuous layers. The visual field is projected onto these layers.

The thalamus is a relay center to the cerebrum and also is the last center short of the cerebrum where bodily functions are modulated. There is evidence that in several classes of vertebrates a level of awareness, including perception of both pain and pleasure, is located in the thalamus.

The ventral part of the diencephalon is the **hypothalamus.** It contains about a dozen pairs of nuclei that together integrate and largely control the autonomic functions of the body including water balance, temperature regulation, appetite and digestion, blood pressure, sleep and waking, sexual behavior, and emotions. Consistent with the dual nature of the autonomic system, each function has two centers in the hypothalamus—one for facilitation and one for inhibition.

On the ventral surface of the hypothalamus is the **optic chiasma** where the optic nerves converge and cross (usually with partial decussation of their fibers) before continuing up the sides of the brain as the **optic tracts.** The optic tracts terminate in the lateral geniculate bodies. Just posterior to the optic chiasma is an area called the **tuber cinereum,** which encloses several nuclei of the para-sympathetic system and subtends the **hypophysis.** The important hypophysis is in part of nervous origin but functions (partly under the influence of nuclei in the hypothalamus) as an endocrine gland. Accordingly, it is described in Chapter 20. Posterior to the hypophysis is a pair of small but evident lumps called the **mammillary bodies.** Within are nuclei of the same name that function in olfaction. These are present in all tetrapods and may be represented in fishes.

CEREBELLUM AND PONS

The cerebellum is an ancient part of the brain, yet has enlarged and changed in the course of evolution much more than has the brainstem. It follows that it is a variable part of the brain and that portions of the cerebellum of mammals are relatively "new." The cerebellum develops from the dorsal part of the metencephalon and lags a little behind the development of the brainstem.

The cerebellum of cyclostomes and amphibians is small and smooth. Because it is merely a thickening of the wall of the brain tube, it has no cavity. The cerebellum is usually still smooth in fishes and reptiles, yet in them is prominent and encloses part of the fourth ventricle. In birds and mammals the organ is very large, lobed, and convoluted into tight **gyri** (convex folds) and **sulci** (concave grooves) (Figure 18-5). Its solid walls then nearly exclude the fourth ventricle.

Unlike the spinal cord and brainstem, the gray matter of the cerebellum is in a thin superficial cortex. The central white matter branches to each lobe and gyrus. Because of its appearance in longitudinal section, this branching white matter is called the **arbor vitae** (= tree of life).

The cerebellar cortex of all vertebrates is divided histologically into three regions: a deep **granular layer,** which has several types of cells, a middle **Purkinje cell layer,** and a superficial **molecular layer,** which has scattered cells but consists mostly of synapses. Afferent impulses are relayed twice within the cortex before reaching the large and exceedingly branched Purkinje cells (see the most branched neuron of the figure on p. 312). The arrangement of cells in the cortex assures that impulses will spread widely.

The cerebellum functions to control motor coordination and to maintain equilibration. It does not initiate motor activities, but processes those initiated

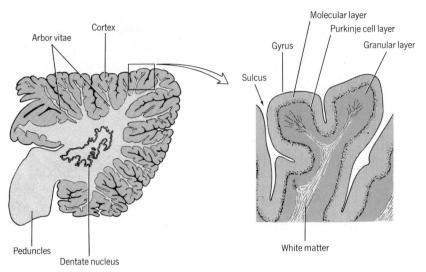

Figure 18-5   PARASAGITTAL SECTION OF THE HUMAN CEREBELLUM.

elsewhere. It provides unconscious timing and integration of muscles that must contract together and of antagonistic muscles that must simultaneously relax. If the cerebellum is damaged, muscle activity loses control and precision.

In order to do this job the cerebellum needs extensive input from the sensory system. In primitive vertebrates the cerebellum has two parts. One is the **archicerebellum.** Its principal input comes from the labyrinth of the ear and the lateral line system by way of associated nuclei in the brainstem, though pathways from the cord may also exist. The archicerebellum persists in amniotes as the auricular, or flocculonodular lobes of the more evolved organ (Figure 18-12). The other part is the **paleocerebellum,** which is located in the midline of the organ. Its input is largely from proprioceptor (muscle sense) organs of the trunk, and is relayed by the spinal cord and inferior olivary nucleus. This part of the cerebellum is important for the control of swimming in fishes. Finally, when appendicular muscles became more prominent with the evolution of amniotes, and the cerebral cortex became larger and more dominating, the cerebellum responded by evolving an associated **neocerebellum.** Though not sharply set off, the neocerebellum forms the hemispheres of the organ. There is also input to the cerebellum from the senses of touch, sight, and hearing, and from the reticular formation.

The cerebellum has fewer efferent than afferent fibers. Outgoing impulses originate in the Purkinje cells and are relayed by **cerebellar nuclei.** The location of these nuclei is varied and complex in anamniotes. They are located in the base of the larger cerebellum of amniotes, and in mammals split to become three or (in primates) four separate pairs of nuclei. The largest and most distinctive is the **dentate nucleus,** which (like the inferior olive of the medulla) has a crumpled contour. There is a topographic relationship between the cerebellar nuclei and the regions of the cerebellar cortex. Efferent pathways run from these nuclei to the vestibular nuclei, reticular formation, ruber nucleus, and dorsal thalamus.

Afferent and efferent pathways between the cerebellum and cerebral cortex evolve as the newer, more dominant parts of the cerebral cortex evolve. Associated with this change is the appearance in some birds, and the prominence in mammals, of the **pons.** This addition to the ventral part of the brainstem at the level of the cerebellum receives in its **pontine nuclei** fibers from the cerebral cortex and relays impulses to the cerebellum through the middle peduncles. (The peduncles supporting the cerebellum were described above with the posterior brainstem.) The pons also transmits impulses from one side of the cerebellum to the other.

GENERAL STRUCTURE   The cerebrum is the adult telencephalon. The most anterior part of the cerebrum is always bilaterally divided. Nearly all of the organ is divided in amniotes and each half is then called a cerebral hemisphere. **CEREBRUM**

The cerebrum is like the cerebellum in that it is of ancient origin yet enlarges and changes in the course of evolution. However, the changes are here more pronounced and do not form a single progression but instead follow separate trends leading to different end points in teleosts, birds, and mammals. The phylogeny of the cerebrum is one of the most striking in vertebrate evolution.

At the anterior end of each hemisphere is the **olfactory bulb,** which receives the olfactory nerves. The remainder of each hemisphere is divided into two principal parts: The **corpus striatum** is in a ventral position. It is composed of various prominent nuclei and often swells to become large and globular. The **cortex,** or pallium, forms the roof and side walls of the cerebrum. Its association centers tend to be spread out into a sheet. In vertebrates other than actinopterygians, the corpus striatum and cortex are in part separated by the lateral ventricles. Where cortex and corpus striatum merge, the boundary may be indistinct. Homologies of the various nuclei are sought using structural, comparative, functional, and histochemical evidence. The cerebrum is completed by various tracts and commissures. These parts will be discussed in order.

OLFACTORY BULB AND TRACT   Axons of the conductive receptor cells of the nasal epithelium enter the olfactory bulbs (Figure 18-8). There they converge to synapse with two kinds of neurons within ball-like tangles of nerve endings called **glomeruli.** Axons of these second-level neurons enter the olfactory tracts, which conduct impulses out of the bulbs. Another kind of neuron provides local feedback circuits within the bulbs. There are fewer fibers in the outgoing tracts than in the incoming nerves, so the bulbs accomplish a summation of impulses. Third- and fourth-level neurons elsewhere in the brain are also involved in the complicated processing of olfactory input.

The size of these structures relative to the remainder of the brain varies over a wide range according to the importance of olfaction in the life of the animal (Figure 18-12).

CORPUS STRIATUM AND BASAL NUCLEI   The corpus striatum (= body + striped) is named for the appearance of part of its substance in some vertebrates as seen in section. For mammals, **basal nuclei** is an approximate synonym. The term **basal ganglia** is also used. The corpus striatum has three principal parts, which are named according to function. It should be recognized, however, that historical misconceptions about sequence in their evolution have given emphasis to their distinctions. Also, the homologies assigned to the parts among the different classes of vertebrates are in some instances tentative.

One part of the striatum is called the **archistriatum** (Figure 18-6). It integrates olfactory and general somatic senses. The archistriatum of fishes consists of several indistinctly segregated nuclei called the **amygdaloid** (= almond shaped) **complex.** Tetrapods retain the structure, and in mammals the corresponding amygdala is a globular mass that tends to be ventral to the other basal nuclei. Even in mammals it remains in part an association center for olfactory input, but also contributes to arousal, food intake, and emotions.

The **paleostriatum** is represented in all vertebrates, with various names assigned to its parts. The homologous basal nucleus of primates is the **globus pallidus** (= ball + pale), the function of which is noted below.

The name **neostriatum** reflects the long-held belief that this part of the striatum evolved only in amniotes—a view that is no longer held. Some structures to which this name has been assigned may prove not to be homologs (particularly in birds), but the derivative mammalian basal nuclei are the long **caudate** (= tailed) **nucleus,** which arches over the others, and the **putamen** (= shell).

The corpus striatum is the highest integrative center of fishes (though, as we shall see, the cortex should not be discounted). The striatum appears to be enormous in reptiles and is usually considered to be their highest integrative center, though it now seems probable that part of the cortex is in fact incorporated into the "striatum." Paleostriatum and neostriatum are also large in birds and over these is a unique, thick, four-layered **hyperstriatum.** Recent evidence indicates that this is again derived from the cortex.

Considering their prominence, the function of the basal nuclei is poorly known for mammals. The caudate nucleus, putamen, and globus pallidus are here dominated by the cerebral cortex. They receive fibers from the cortex and thalamus. They project fibers to each other and to the thalamus and substantia nigra. Together they are thought to contribute to the function of muscle masses, as opposed to delicate control of individual muscles. Lesions of one or more of the basal ganglia result in specific motor disturbances, recognizable as different from those resulting from lesions of cerebellum or cerebral cortex.

(Two additional nuclei are commonly named in articles about the mammalian corpus striatum. Although phylogenetically and functionally more closely related to the caudate nucleus, the putamen may merge structurally with the globus pallidus to form the lentiform (= lentil-shaped) nucleus. Also classed as a basal nucleus is the flat, laterally placed claustrum. It may in fact be derived from the pallium rather than the corpus striatum. Although present in reptiles as well as mammals, its function is not clear.)

CORTEX Three principal parts of the cortex are recognized: **paleocortex, archicortex,** and **neocortex.** Except in actinopterygians and mammals, the paleocortex is lateral to the ventricle and the archicortex is dorsal or median to the ventricle. The neocortex may be between the other parts or, apparently, in reptiles and birds, ventral or lateral to the ventricle in association with the striatum.

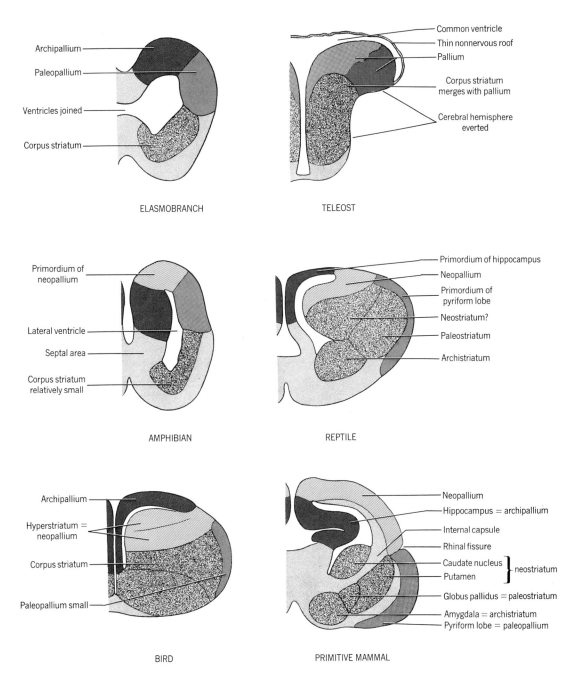

Figure 18-6   DIAGRAMMATIC CROSS SECTIONS OF CEREBRAL HEMISPHERES SHOWING THE COMPARATIVE STRUCTURE OF THE CORPUS STRIATUM AND CORTEX.

Cut surface of corona radiata where it emerges from the internal capsule

Body of caudate nucleus (bordering on the lateral ventricle)

Fornix

Hippocampus

Cerebral hemispheres removed, but outline shown in frontal section at level of corpus callosum

A — — A

B — — B

Interventricular foramen

Putamen = dorsal part of lentiform nucleus (exposed by dissection)

Body of corpus callosum

Cut surface of corpus callosum

Corona radiata

Body of corpus callosum
Caudate nucleus
External capsule
Internal capsule
Anterior commissure

Cerebral cortex

Lateral ventricle
Septum pellucidum
Claustrum
Septal nuclei
Lentiform nucleus

SECTION AT A

Median longitudinal fissure

Fornix

Tail of caudate nucleus
Internal capsule
External capsule
Lentiform nucleus
Hippocampus
Intermediate mass of thalamus
Hypothalamus

Lateral ventricle (choroid plexus removed)

Third ventricle
Thalamic nuclei
Amygdala
Rhinal fissure
Optic tract
Third ventricle
Cerebral peduncle

SECTION AT B

Figure 18-7 FOREBRAIN AND THALAMUS OF THE COW shown by a dissection seen in dorsal view (above) and by cross sections (below) made at the levels indicated by letters. On the sections, contrast between nuclei and tracts has been enhanced by staining.

Figure 18-8 SAGITTAL SECTION OF THE BRAIN OF THE COW.

It was long believed that the telencephalon was initially associated only with olfaction. The paleocortex and archicortex were thought to relate only to olfaction and to be the only parts of the cortex of fishes. It is now known that much of the cortex of many fishes is *not* olfactory in function. Experiments have shown that parts of the cortex contribute to schooling, aggressive, and reproductive behavior and make learned responses possible. Accordingly, the neocortex is probably also ancient, though its limits are not clear at the fish level of evolution.

The architecture of the cerebrum of actinopterygians is unique. Cortex and corpus striatum are thick, merge together, and both lie lateral and ventral to a common ventricle. Paleocortex and archicortex are certainly present, and probably also the neocortex, but their spatial relationships are atypical, and thus it is difficult to homologize their nuclei with those of other vertebrates.

In reptiles and birds a thin cortex is stretched wide to arch over the enlarged corpus striatum. As noted above, however, it now seems likely that part of the large striatum (the hyperstriatum portion in birds) is really neocortex. Birds do not seem to be seriously inconvenienced by the surgical removal of their thin superficial cortex, though there are indications that some capacity for memory is sacrificed. Contrary to common belief, some birds are more intelligent than many nonprimate mammals, and it has been proven that their ability to solve problems, and to remember how to solve new problems of a related kind, resides in the hyperstriatum.

In mammals the paleocortex and archicortex are wedged apart as the evolving neocortex enlarges between them. Anteriorly, the paleocortex includes the olfactory tracts. Posteriorly it is pushed down around the sidewall of the hemisphere to a position flanking the anterior brainstem. The resulting **pyriform** (= pear-shaped) **lobe** is prominent in mammals having a keen sense of smell. It is separated from the neocortex above by the **rhinal fissure.** The pyriform lobes are the olfactory cortex.

The archicortex is pushed by the neocortex in the other direction to the crown of the brain near the midline (reptiles, monotremes, marsupials) or over the edge onto the medial wall of the hemisphere (other mammals). As it moves, it rolls on itself lengthwise and sinks largely below the surface, thus forming a long arching band that impinges on the lateral ventricle. Its name, **hippocampus** (= sea horse), is suggested by its rolled appearance as seen in cross section (Figures 18-6 and 18-7). The hippocampus of mammals relates to memory of spatial relationships. Together with the hypothalamus, amygdala, habenula, and mammillary bodies it contributes to the **limbic system,** which functions in aspects of sexual and emotional behavior, memory, learning, and motivation. It seems to regulate more primitive parts of the brain by inhibiting stereotyped behavior, thus permitting accommodation to new circumstances.

All large mammals and some smallish ones have a convoluted neocortex (or cerebral cortex, or simply cortex); the gyri and deep sulci greatly increase total surface area. Intelligence relates to absolute brain size, relative brain size, and convolutions, but only in a general way; man's brain is not the largest, nor the largest in relation to body weight, nor even the most convoluted.

Gray matter is internal in the cortex of anamniotes but external in that of amniotes. The various parts of the cortex can be distinguished histologically. Characteristic cell layers differ in the density, size, configuration, connections, and staining properties of their neurons. Cells in the six successive layers of the cortex of primates tend to be arranged in vertical columns, each having as many as 100 cells. Small mammals have more cells per unit volume of brain than large mammals.

The primate cortex is divided into regions for ease of reference (i.e., frontal, temporal, parietal, and occipital), and specialists have a standardized way of indicating subregions. In a band across the crown of the human brain, at the posterior margin of the frontal region, is the primary somatic motor cortex (Figure 18-9). Ultimate control of voluntary motor activity rests here, though this part of the cortex is "coached" and "advised" by other parts of the cortex and by the cerebellum, reticular formation, basal nuclei, ruber nucleus, and other centers.

Just behind the somatic motor cortex, at the anterior margin of the parietal region, is the primary somatic sensory cortex. Similarly, the primary auditory cortex is at the top of the temporal lobe, and the primary visual cortex (which develops in mammals as the optic lobes of the midbrain relinquish most of their function) is in the occipital region. These areas have a point-for-point relationship with function (though in some mammals there is overlap among adjacent points): Motor control of the thumb is at a specific place just below the forefinger point and above the neck point; tones are perceived at given points in treble to bass sequence; sight points are arranged in relationship to the visual field. If the face is relatively sensitive (as in man), the corresponding areas of the cortex are rel-

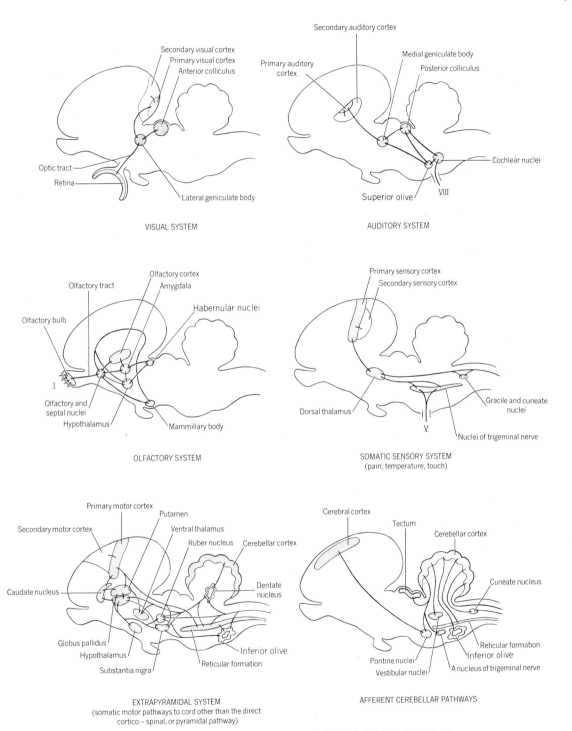

Figure 18-9   DIAGRAMS OF SOME PRINCIPAL COMPONENTS OF SOME FUNC-
TIONAL SYSTEMS OF THE MAMMALIAN BRAIN.

atively large; if the hands are relatively sensitive (raccoon), the hand areas are large.

Although the above is true, it does not go far enough to indicate the complexity of cortical function. For unknown reasons, the two hemispheres of man are not symmetrical, either in structure or in their control of various functions. For example, speech, reading, writing, and rational and practical behavior are vested in the left hemisphere (of right-handed persons), whereas artistic ability, intuition, spatial recognition, and body awareness are dominated by the right hemisphere. The primary cortical areas noted above account for most of the cortex of lower mammals (marsupials, insectivores) but for only one-quarter of the cortex of man. In the parietal and occipital regions there are secondary and tertiary projection areas. These are not related to function on a point-to-point basis and have fuzzy boundaries. They code messages, store information, combine input, and provide spatial orientation. If they are impaired, a man might see well, yet confuse right and left; might walk well, yet get lost in a familiar place. Finally, the frontal region of the cortex relates to programs, intentions, orientation to goals, and sequence in the performance of activities. The entire cortex of experimental animals becomes relatively heavy if the animals live in a relatively diverse, stimulating, and "enriched" environment.

There is constant interaction among these parts of the cortex and constant interaction (sometimes both ways) between sensory input and motor output. Short-term memory, whatever it is, requires the frontal cortex and probably the thalamus. Long-term memory is different and seems to be stored by virtually all of the brain as a diffuse code. No one structure of the brain is the exclusive site of any one function, and alternative circuits seem to be available for controlling any function.

All in all, several things remain to be learned about this organ!

SOME OTHER FEATURES OF THE CEREBRUM   The tracts of the cerebrum are at least as constant as the nuclei. Association fibers, short and long, loop between the different parts of the cortex. The olfactory tracts divide into several paths as they merge into the brain: One crosses to the other olfactory bulb. Another stops in **olfactory nuclei,** which send third-level neurons to the habenula and mammillary bodies of the diencephalon. Still another runs to the amygdaloid nuclei and pyriform lobe. All these structures function in the perception and integration of olfactory stimuli (Figure 18-9).

The conspicuous motor tract that exits from the primary motor cortex runs right through the mammalian corpus striatum as a broad white band called the **internal capsule** (Figures 18-6 and 18-7). It separates the caudate nucleus from the functionally related putamen. It is this band that continues, as the cerebral peduncles and subsequently as the pyramidal tracts, directly into the spinal cord. Collectively this is the **pyramidal system,** as contrasted to the **extrapyramidal system,** which includes the same cortical areas but also other parts of the cortex and efferent relay centers in the corpus striatum, thalamus, and elsewhere.

The cerebrum has several commissures. The small **habenular** and larger **anterior commissures** are derived from the archicortex and appear in the lower classes. Therian mammals have evolved a commissure, the **corpus callosum,** which in all but marsupials is the largest of all (Figures 18-7 and 18-8). It joins one neocortex to the other and is seen in the dorsal midline as a horizontal shelf when the hemispheres are pulled apart. If the corpus callosum is cut, something learned or experienced on one side of the body cannot be adequately used or acted upon by the other. The **fornix** is a conspicuous arching pathway positioned ventral to the corpus callosum anteriorly and lateral to the hippocampus posteriorly. It transmits fibers between the hippocampus and the hypothalamus.

Each lateral ventricle encloses a large choroid plexus that is rooted at the roof of the diencephalon with the smaller plexus of the third ventricle.

EVOLUTION OF THE BRAIN

The brain of CYCLOSTOMES is primitive, but also in some ways specialized and degenerate. The anterior part of the brain is foreshortened in response to crowding by the terminal mouth and dorsal nasal chamber (Figure 18-10). The large olfactory bulbs are separated by only a shallow constriction from thick cerebral hemispheres. Corpus striatum, paleocortex, and archicortex are somewhat vaguely delimited. Parietal organ, pineal body, and large habenula are all visible on the

LAMPREY

SALMON

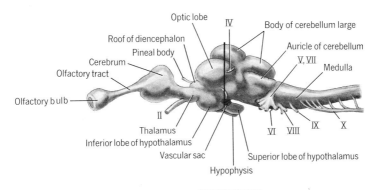

DOGFISH SHARK

Figure 18-10   BRAINS OF REPRESENTATIVES OF THREE CLASSES OF FISHES.

roof of the diencephalon. Optic lobes are evident in lampreys and small in blind hagfishes. The medulla is relatively large and supports a large everted posterior choroid plexus. The cerebellum is rudimentary, as would be expected for these sluggish, parasitic creatures.

The brain of ELASMOBRANCHS is much advanced over that of cyclostomes, yet is also somewhat specialized according to species. Olfactory bulbs are large, widely separated, and enclose extensions of the lateral ventricles. Olfactory tracts are in evidence. Cerebral hemispheres are broadly joined in the midline and share a common ventricle. Corpus striatum and cortical areas are well-established. Present in the diencephalon are pineal body (but not parietal organ), choroid plexus, habenula, lateral geniculate body, thalamus, and large hypothalamus. The hypothalamus has paired **inferior lobes** flanking the hypophysis, and a thin **vascular sac** that functions in depth perception and correlated behavior. These two structures are characteristic of fishes in general, particularly of deep sea fishes, and are absent in tetrapods. Optic lobes are usually prominent and surround an expansion of the midbrain ventricle. The ruber nucleus has evolved. The cerebellum is large in active species, median in position, somewhat fissured in large sharks, and contains an expansion of the fourth ventricle. Archicerebellum and paleocerebellum are distinguished, and the cortex has differentiated into the three layers characteristic of higher vertebrates. Olivary nuclei and reticular formation are present in the medulla.

Brains of BONY FISHES are diverse. Those of Dipnoi resemble those of Elasmobranchii, whereas those of Actinopterygii have forebrain architecture shared by no other vertebrates. The brain of the surviving representative of the Crossopterygii is intermediate.

The cerebral hemispheres of dipnoans evaginate in the usual manner, each becoming convex on its outer surface. Those of actinopterygians evert instead: The dorsal lip of the hemisphere curls outward so the hemisphere becomes concave on its outer surface. Cortex and corpus striatum are contiguous and thick. The homologies of the parts of the cortex are uncertain.

Conspicuous inferior lobes and vascular sac are present under the diencephalon as for cartilaginous fishes. Optic nerves pass one another at the optic chiasma or weave through one another in various patterns. Bulging optic lobes are usually the most prominent part of the brain of bony fishes. The cerebellum is smooth but usually large. It is more solid than in cartilaginous fishes. The cerebellum of actinopterygians has a distinctive anterior projection called the **valvula,** which pushes into the ventricle of the midbrain, thus contributing to the separation of the optic lobes.

Large pathways relating to taste are present in the medulla. If food is tasted primarily on barbels and lips, a median **facial lobe** forms behind the cerebellum to house nuclei of the facial nerves. If food is tasted primarily within the mouth and pharynx, lateral **vagal lobes** form to house nuclei of the glossopharyngeal and vagus nerves. In the medulla of actinopterygians (and urodeles) there is a single pair of giant neurons called **Mauthner cells** that relate to the ear and lateral line system. They mediate escape reflexes to muscles used in swimming. The choroid plexuses of fishes are everted.

FROG

GOOSE

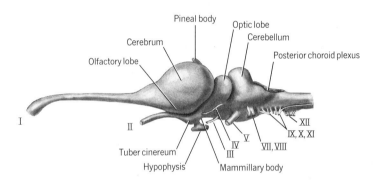

SEA TURTLE

Figure 18-11   BRAINS OF REPRESENTATIVES OF THREE CLASSES OF TETRA-PODS.

The brain of AMPHIBIANS is remarkably unspecialized (particularly in urodeles) and is scarcely more advanced than that of cartilaginous fishes and dipnoans (Figure 18-11). The cerebral hemispheres are more separate from one another than in fishes, so they share little common ventricle. Primitive hippocampal and pyriform areas have formed, respectively, from the archicortex and paleocortex. The corpus striatum is small. The pineal body is well-developed in anurans. The dorsal thalamus is beginning to enlarge. Incipient mammillary bodies are present in the hypothalmus. Optic lobes are of moderate (anurans) or small (urodeles) size. The cerebellum is rudimentary.

The brain of REPTILES is narrow, elongate, and nearly straight. Olfactory bulbs tend to be smaller than for fishes. Olfactory tracts are long. The cerebrum is large because of the expansion of the corpus striatum and associated neocortex. The superficial parts of the cortex are thin, and gray matter has become external. Relative sizes and positions of the divisions of cortex and corpus striatum indicate that there have been two trends in the evolution of the reptilian forebrain: one line (represented by turtles) in the direction of mammals and another (represented by crocodilians) in the direction of birds.

The parietal organ is functional in lizards (see Chapter 19). The dorsal thalamus is larger and more complex than in the lower classes, and the ventral thalamus has all the nuclei regularly recognized in mammals. Since most reptiles have excellent vision, the optic lobes are conspicuous. The midbrain still encloses an expanded ventricle as in anamniotes.

The reptilian cerebellum is smooth. It is largest in swimmers and rudimentary in snakes. Cerebellar nuclei are now within, rather than below the organ. Choroid plexuses are inverted.

The brains of BIRDS are relatively large, uniform, and distinctive. The organ is short and broad. There are marked cranial, pontine, and cervical flexures. Olfactory bulbs and tracts are evident in the scavengers, but in general are smaller than in other vertebrates. The avian cerebral hemisphere is surpassed in size only by that of some mammals. This is because of the enormous development of the corpus striatum with its associated ''hyperstriatum'' or neocortex. The parts of the cerebrum have a different configuration in birds that emphasize hearing (owl) than in birds that emphasize touch and manipulation with the beak (duck, snipe, parrot). The superficial parts of the cortex are exceptionally thin and have little function.

The dorsal thalamus is even more developed than in reptiles. Optic nerves, chiasma, and tracts are large. Optic lobes are particularly large and are layered within. They have connections from all sense organs and with the cerebrum. Squeezed between the cerebrum and cerebellum, the optic lobes have a uniquely lateral position.

The cerebellum is larger than in other vertebrates except some mammals. It is tightly convoluted, and the organ is high and narrow. Related to the marked development of the cerebellum are the appearance of the pons under the brainstem, and enlargement of the olivary nuclei within the broad medulla.

The olfactory bulbs and tracts of MAMMALS range from huge (aardvark, armadillo, anteater) to very small (primates) (Figure 18-12). Although relatively smaller than in reptiles and birds, the corpus striatum is prominent. It is represented by the basal nuclei, of which the caudate and putamen, derived from the neocortex, are relatively large. The extensive neocortex is the hallmark of the class. It dominates the entire brain both structurally and functionally. The hemispheres are smooth in most small mammals and convoluted in most large mammals. A new commissure, the corpus callosum, joins the hemispheres of therian mammals. The archicortex is represented by the large hippocampus. Pyriform lobes are lateral or ventral in position. They are extensive if olfaction is acute; otherwise they are restricted.

Thalamus and hypothalamus are highly differentiated. The midbrain is exposed in only a few mammals. Optic lobes, now called anterior colliculi, are small because the cerebral cortex has taken over much of their function. Posterior colliculi are present to complete the corpora quadrigemina of the midbrain tectum. The ventricle of the midbrain is restricted to a narrow cerebral aqueduct.

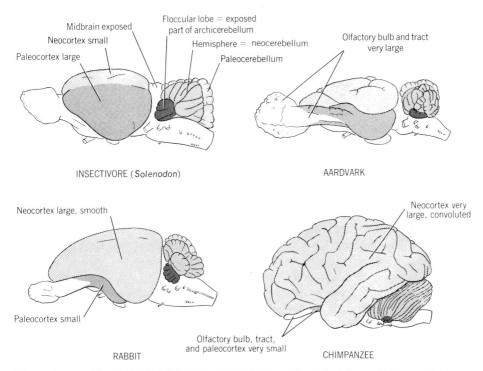

Figure 18-12 VARIATION IN THE STRUCTURE OF THE MAMMALIAN BRAIN. (Derivatives of the archicortex are hidden from lateral view.)

The mammalian cerebellum is large, much convoluted, and relatively broad. The cerebellar nuclear complex has differentiated into three or four distinct pairs of nuclei. A pons is prominent.

The central nervous system of cyclostomes and fishes is loosely surrounded by a protective fibrous envelope called the **primitive meninx.** Amphibians and reptiles have two envelopes (or meninges), an outer tougher **dura** and an inner **pia-arachnoid.** Birds are structurally intermediate between reptiles and mammals in having the beginnings of a third layer.

Mammals have three meninges (Figure 18-13). The strong **dura mater** (= hard + mother) is outermost. It is of mesodermal origin. In the vertebral column it is separated from the vertebrae by fat; in the cranium it adheres directly to the bones. The innermost meninx is the thin vascular **pia mater** (= tender + mother) that adheres to the nervous tissue, following contours into every fissure and sulcus. Between the dura and pia is the nonvascular **arachnoid** (= spiderlike), which is delicately fibrous and sends strands to the pia mater. Pia and arachnoid are both derived in part from neural crests and in part from mesenchyme.

Between the dura and arachnoid is a shallow, fluid-filled, **subdural space.** Between arachnoid and pia is a deeper **subarachnoid space.** The central canal

SUPPORT AND NOURISHMENT OF THE CENTRAL NERVOUS SYSTEM

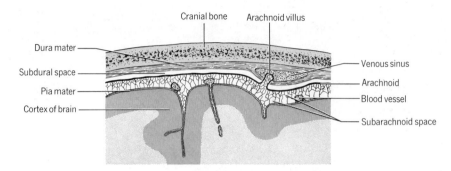

Figure 18-13   MENINGES OF THE MAMMALIAN BRAIN.

of the spinal cord, the ventricles of the brain, and (in mammals) the subarchnoid space are filled with a considerable quantity of **cerebrospinal fluid.** This fluid is produced primarily by the vascular choroid plexuses of the brain, but also by other tissues—probably including the ependymal cells of the spinal cord. Each choroid plexus consists of ependymal layer joined to pia mater. Cerebrospinal fluid is clear, colorless, and similar to blood plasma except that it contains more chloride and virtually no protein. It flows slowly by secretion pressure and also by action of ciliated ependymal cells. Motion is toward the fourth ventricle from both the anterior ventricles and the central canal of the cord. The fluid escapes through holes in the roof of the medulla to the subarchnoid space. From there it reenters the bloodstream in various places, particularly through venous sinuses near the brain. Some cerebrospinal fluid also drains into lymphatics.

The pattern and extent of this circulation must be somewhat different in nonmammals (there being no subarachnoid space), but such vertebrates also have some cerebrospinal fluid outside the central nervous system: The choroid plexuses are usually everted into thin sacs of the fluid instead of inverted into their respective ventricles.

Such cerebrospinal fluid as bathes the outside of the central nervous system protects the system from damage when there is a blow to the head or back. The fluid is also said to nourish the system by way of the ependymal cells—particularly in lower vertebrates in which the blood supply to the system is less developed.

The structure and function of the circulatory system assure a rich blood supply to the central nervous system. Associated vessels are usually large and numerous, collateral systems are present, and in time of stress this system is given priority over others. Brain and cord need much blood for two reasons. First their metabolic rate is high and their need for oxygen is constant and great. Second, the exchange between blood and tissues may be less efficient than elsewhere. Nervous tissue is dense, interstitial spaces are minute, and there are no lymphatic vessels. The blood-brain barrier is distinctive in that respiratory gases and water pass quickly, but larger molecules leave the capillaries less readily than elsewhere.

19

Sense Organs

All cells of the body are responsive to their environments. Certain cells and organs, however, are specialized for monitoring the environment, and these comprise the sensory system. They stimulate the central nervous system, and it is there that sensation is perceived and integrated and where action, if any, is initiated.

Receptors range from mere nerve endings, through simple microscopic capsules, to the large and complex eye. Various classifications of these diverse organs are used: (1) general (widely distributed like pressure receptors) versus special (localized like the ear); (2) somatic (conscious reception of relatively superficial stimuli) versus visceral (unconscious reception of deep stimuli); (3) stimulated from internal sources (muscle tone, balance) versus external sources (cold, light); and (4) responsive to mechanical stimuli (touch, sound) versus electromagnetic stimuli (heat, light) versus chemical stimuli (taste, smell). None of these schemes is entirely satisfactory for our purpose of analyzing the major advances of vertebrate structure. The more complicated sense organs contribute interesting histories to the story of evolution and will be presented in approximate order of increasing complexity after giving brief attention to the relatively simple receptors.

SOME RELATIVELY SIMPLE RECEPTORS

The structural bases for the perception of such human sensations as hunger, fatigue, sex drive, and anxiety are unknown, and many vertebrates have sense organs, not shared by humans, for which functions can only be approximated.

Figure 19-1 PACINIAN CORPUSCLE.

We can never know what it feels like to be a fish, and much remains to be learned about sense reception—particularly in lower vertebrates.

Various sensations, including touch, pressure, stretch, heat, cold, and general chemical sense, are at least sometimes received by naked nerve endings. There can be little to say about the phylogeny of these. Numerous kinds of **sense capsules** are found in the epithelial and connective tissues of tetrapods. These consist of variously modified nerve endings surrounded by small simple capsules having many configurations. They appear to monitor heat, cold, touch, pressure, and pain. The best-known is the **pacinian corpuscle,** which is a marvelously effective tiny pressure transducer (Figure 19-1). Slight distortion of the 30–50 onionlike layers of the capsule sets up a **generator potential** in the nerve ending within the organ. If the potential reaches threshold intensity, the capsule greatly steps up the impulse at the first node of Ranvier. The resulting **action potential** is then transmitted along an axon.

It is difficult to identify the function of each kind of sense capsule, and it is possible that some kinds respond to more than one stimulus. Virtually nothing is known of their phylogeny, and homologies can rarely be made.

The **proprioceptive senses** apprise an animal of the relative positions of the parts of its body. Half awaking from deep sleep one may not know for an instant how the legs are flexed or the arms placed, but the slightest tensing of the muscles makes one's position known. Three kinds of proprioceptors are involved, all of which respond to tension and contribute to postural reflexes. They appear to be present in all tetrapods. The **muscle spindle** is a special muscle fiber wound up by a net of nerve endings and enclosed in connective tissue (Figure 19-2). It discharges when it is stretched. These organs are most abundant where muscle is merging into tendon. **Joint receptors** have complex nerve endings within the connective tissue of joint capsules. Similar **tendon organs** are located where tendons join bone. They discharge when muscles contract.

Several dissimilar kinds of sense organs are macroscopic yet relatively simple in structure. The **carotid body** is a small mass of cells of disputed origin that lie in the fork of the internal and external carotid arteries of tetrapods. Richly supplied with both nerves and blood, it detects fluctuations in the concentrations of carbon dioxide and oxygen in the bloodstream and sends impulses via the ninth cranial nerve to the parts of the brain controlling circulation and respiration. **Otoliths** are hard objects within the internal ear of some vertebrates. In addition to their function in equilibrium (see below) they probably act as accelerometers. Also, the large otoliths of several teleost fishes have been shown to have peizo-electric properties (i.e., converting force to electric potential); theoretically they could function as depth registers. **Pit organs** give their name to the American snakes called pit vipers. The conspicuous pits are located between the nostril and the eye. The floor of the blind pit is vascular and rich in nerve endings. The

Figure 19-2 MUSCLE SPINDLE.

organ is a remarkably sensitive thermoreceptor. The background discharge is modified if the temperature increases or decreases by as little as 0.002°C. This enables the snake to detect within half a second the presence and position of warm-blooded prey that comes within striking distance. Pythons and boas either have series of similar, but smaller and less sensitive, pits in the scales surrounding the mouth, or they have thermal-sensitive skin areas. Pigeons have in their heads, between the brain and the skull, an innervated, magnetite-rich, naturally **magnetic organ** that may serve as the compass needed for homing on cloudy days.

OLFACTORY ORGANS Seemingly unspecialized nerve endings at various places on the surface of the body of many vertebrates are sensitive to the chemical environment. Response of the human eye to smog and onions is an example. The principal chemical senses are olfaction and taste. Olfaction is a primitive sense. The earliest vertebrates appear to have had a keen sense of smell, and most of the forebrain was then devoted to olfactory signals.

ORGANS OF OLFACTION AND TASTE

Olfactory epithelium is localized in the nasal pits of fishes and in protected outpocketings of the respiratory passages in air breathers. Only part, and usually a small internal part, of the epithelium lining the nasal pit or passage contains olfactory cells. These are columnar cells each of which has about eight hairlike filaments on its free surface (Figure 19-3). These cells are unique in that they continue as axons into the olfactory bulb of the central nervous system. Supporting cells are dispersed among the thousands or millions of olfactory cells.

The combined area of all the filaments of the olfactory epithelium may exceed the surface area of the body, and it is on these filaments that dissolved chemicals are detected. The sense of smell is so wonderfully acute that the theoretical limit of sensitivity is reached by many animals—each filament can respond to a single molecule of certain odoriferous materials. Each kind of animal has its own spectrum of sensitivity, being very sensitive to some odors and little sensitive to others. The exact mechanism of response is not known, but it appears that the filaments can distinguish molecules having about seven different shapes,

Mucus bathing olfactory filaments

Olfactory cell

Supporting cells

Neck of mucous gland

Axon running to olfactory bulb

Figure 19-3 SECTION OF THE OLFACTORY EPITHELIUM OF A TETRAPOD.

Figure 19-4   DISSECTION OF THE NASAL PIT OF AN EEL, *Anguilla*.

possibly by a submicroscopic lock-and-key mechanism. The almost infinite number of odors that can be distinguished would then be combinations of these in various relative concentrations.

Olfaction has not been detected with certainty in amphioxus. CYCLOSTOMES have a median olfactory sac that is ventilated by water passing in and out of the single nasal pouch. However, it is probable that the olfactory structures of the ancestors of these animals, and doubtless also of anaspids and cephalaspids, were paired as for other vertebrates. This is indicated by the bilobed nature of the sac, its innervation, and its ontogeny.

FISHES admit water into each nasal pit by one compressed opening (Selachii) or by two openings (most Teleostei) constructed so that a continuous stream passes over the olfactory epithelium. The nasal pit of cartilaginous fishes lies in the respiratory current, thus assuring ventilation even when the fish is stationary. The internal nostrils of Sarcopterygii may have evolved partly to benefit smelling in water (through increased ventilation) rather than breathing in air. The lining of each pit is pleated (Figure 19-4). The number and shape of the pleats relates to both species and age. These folds are assumed to control water flow in a beneficial way. The olfactory epithelium is largely confined to the clefts between the pleats.

AIR-BREATHING VERTEBRATES add mucous cells to the olfactory epithelium. This is necessary both to dissolve the particles to be smelled and to wash away material that has already been detected so that fresh samples of air can be examined.

The size of the nasal chamber increases, and its structure becomes more complex, as the secondary palate evolves. Most amniotes increase the surface area of the nasal epithelium by folding it over several pairs of turbinates (= conchae), or scrolls of bone, which curl into the nasal chamber from its lateral walls. Turbinates are relatively simple in most vertebrates but exceedingly complex in some mammals. The area of the olfactory epithelium may be increased in this way, but the primary function of the folded nasal epithelium is to clean, moisten, and sometimes to alter the temperature of respired air before it reaches the lungs. The olfactory sense is well developed in only few birds (pelagic scavengers, vultures, and several other groups) and is weak in aquatic mammals which, unlike fishes, cannot ventilate the olfactory epithelium while submerged.

Tetrapods also depart from their ancestors among fishes in having a second and independent olfactory system consisting of a **vomeronasal organ** with its nerve and accessory olfactory bulb in the brain. The organ opens into the nasal chamber of amphibians. Some salamanders have paired external nasolabial grooves that conduct fluids from the rostrum through the nostrils and into the vomeronasal sacs. In some reptiles, the vomeronasal organ connects with the mouth instead of the nasal chamber. Squamate lepidosaurs protrude the forked tongue and then insert the tips of the tongue into the paired vomeronasal organ to sense the quality of adherent particles. In some other reptiles, and most mammals, the organ communicates instead with the small incisive foramen in the front of the palate. In many mammals it is particularly sensitive to sexual odors. The organ is rudimentary or absent in turtles, crocodilians, birds, aquatic mammals, and some primates.

TASTE ORGANS   Taste organs are similar to olfactory organs in being chemoreceptors having epithelial hair cells. They differ from olfactory organs in being less sensitive by about four orders of magnitude, having more restricted response (there are fewer different tastes than odors), relating to different parts of the brain, synapsing with sensory nerve fibers rather than running directly to the brain, being of endodermal instead of ectodermal origin, and having the receptor cells aggregated into groups, or **taste buds.**

Each taste bud consists of supportive cells and 30–40 columnar taste cells all arranged in a barrel-shaped cluster (Figure 19-5). The buds may be dispersed or grouped on little hummocks of the epithelium called papillae. Papillae are of various sizes and shapes and may be arranged in various patterns—all of unknown functional significance. Taste buds, unlike olfactory epithelium, are exposed and subject to wear. Accordingly, their cells are short-lived and are constantly replaced. Each bud is most sensitive to one of the four basic tastes: salt, sour, sweet, or bitter. Nevertheless, all taste buds look alike, and the different vertebrates are not equally sensitive to these tastes. Being derived from epithelium at the levels of several visceral arches, taste buds are innervated by several cranial nerves: the seventh, ninth, and tenth.

Taste buds are abundant in the mouth and pharynx of cyclostomes and fishes, but they also may be located on the surface of the body, particularly on the heads and oral feelers of fishes that find food in sand, mud, or murky water. Amphibians have taste buds on the tongue, pharynx, and skin. Frogs have taste buds positioned to taste bits of tissue abraded by the palatal teeth and dissolved by enzymatic action of an oral secretion. Reptiles and birds, having dry skins and usually somewhat keratinous tongues, distribute most of their taste buds in the

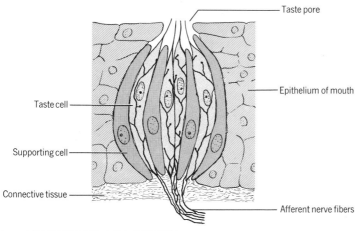

Figure 19-5   SECTION OF A TASTE BUD.

pharynx. These animals, especially birds, have a relatively poor sense of taste. Mammals also have taste buds in the mouth generally and on the pharynx, but concentrate them on the fleshy tongue.

LATERAL LINE SYSTEM

The lateral line system and internal ear are so related by structure, function, and ontogeny that together they are called the acousticolateralis system. The lateral line system will be considered first because it is less complex than the ear. The system is present in fishes and in both larval and aquatic adult amphibians. It consists of thousands of microscopic organs called **neuromasts.** These are always freely distributed at the surface of the skin and usually are also located in shallow surface pits, along canals in the skin (Figure 19-6), or along horizontal tubes tunneling under the skin and even through dermal bones. Such tunnels always communicate with the surface by pores spaced along their length. The longest and most constant tube or canal is the **lateral line canal**—hence the name of the system. These ways of distributing and protecting neuromasts intergrade. Somewhat set apart are neuromasts located at the bottoms of **ampullary organs,** which are deep, tubelike pits located in the skin and subcutaneous tissue of the head.

Early in development, several ectodermal placodes form on each side of the head and neck area of the embryo. It is probable that cells from neural crests play a part in their induction. One placode on each side forms the internal ear. Cells move away from the other placodes to become the lateral line system and the ganglion cells of the nerves that will innervate that system. These are branches of the seventh cranial nerve, which serves most neuromasts on the head, the ninth nerve serving a small area at the back of the head, and the tenth nerve serving the remainder of the system.

A neuromast looks somewhat like a taste bud. It is a clump of sensory hair cells and supportive cells. Each sensory cell has on its exposed surface a bundle of 20–50 sensory hairs called **stereocilia** and usually also one larger **kinocilium,** which stands at one edge of the stereocilia. Each neuromast that is free at the surface of the skin is capped by a tall, fragile, and transparent dome of extracel-

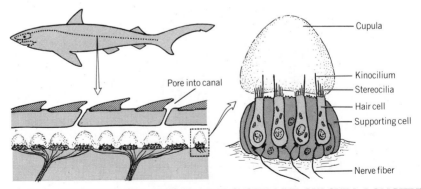

Figure 19-6 POSITION OF LATERAL LINE SYSTEM BELOW SKIN, LONGITUDINAL SECTION OF A CANAL, AND STRUCTURE OF A NEUROMAST.

lular material that protrudes into the water. This structure is the **cupula.** All sensory hairs of the neuromast are imbedded in the base of the cupula. Neuromasts opening into subsurface canals usually have cupulas also, but some of these, and all neuromasts within ampullary organs, are instead covered by a jelly that fills the surrounding space.

Free neuromasts, and those in canals and pits, differ somewhat in structure and function from neuromasts within ampullary organs. We shall consider the former first. Any motion of the surrounding water causes a cupula to pivot, thus tilting the imbedded sensory hairs in the direction of motion. Each hair cell discharges electrical signals at a basic frequency that increases when displacement of the hairs is in one direction and decreases when displacement is in the opposite direction. There are two sets of hair cells, one positively activated by motion of fluid along the canal in one direction and the other by motion in the opposite direction.

Collectively, free neuromasts and those of canals and pits form a system of distant touch that is so well-coordinated that it gives both spatial and temporal information. Even when blinded, fishes can determine the position and motions of nearby prey and of other fishes. They can also sense the relative motion of their own bodies and the water. Sound waves from a distant source are not sensed. However, low-frequency sound waves from a near source cause relatively great displacement of the water and hence are heard (whatever they may "sound" like to a fish). The lateral line system contributes to obstacle avoidance and orientation of the body relative to currents.

Cyclostomes and amphibians have no sensory canals on the head or body. Their neuromasts are free or in pits that tend to be linearly arranged. The lateral line system is well-known from the armor of ostracoderms and placoderms. Fishes have lateral line canals (sometimes doubled or branched) and a complicated but rather stable system of head canals. The canals are best developed in active fishes and least protected from the surface in fishes inhabiting quiet water. Most of the canals of chimaeras are open to the surface (see figure on p. 51).

Ampullary organs are found in elasmobranchs, chimaeras, and several groups of bony fishes. These organs are primarily detectors of weak electric stimuli (see p. 201 and 202). They sense the muscle potentials of nearby fishes. They also respond to thermal and mechanical stimuli. Thus, they are also organs of distant touch, though the principal stimulus is different from that of the remainder of the lateralis system.

ORGAN OF EQUILIBRIUM   Details of the origin of the ear are lost in antiquity, **EAR** but the story can be surmised with reasonable confidence. It is probable that the lateral line system evolved before the ear and that of the two basic functions of the ear, equilibration and hearing, the former was perfected before the latter was well-initiated. It is desirable for an organ of equilibration to be on the head where it is close to the feeding apparatus and other special senses, and away from the oscillating locomotor mechanism. Furthermore, it is desirable for the organ to be away from the surface of the body where signals of the type monitored by the lateral line system would be distracting. Evidently a portion of the lateral line

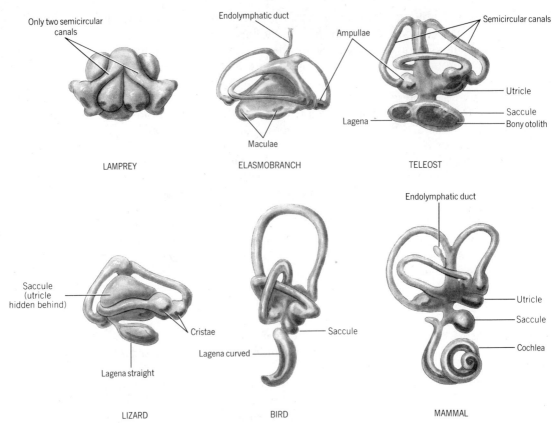

Only two semicircular canals

LAMPREY

Endolymphatic duct

Ampullae

Maculae

ELASMOBRANCH

Semicircular canals

Utricle

Saccule
Bony otolith

Lagena

TELEOST

Saccule
(utricle hidden behind)

Cristae

Lagena straight

LIZARD

Saccule

Lagena curved

BIRD

Endolymphatic duct

Utricle

Saccule

Cochlea

MAMMAL

Figure 19-7   COMPARATIVE ANATOMY OF THE LABYRINTH. Lateral view of the right organ.

system at the side of the head sank below the skin and was modified to become the new organ.

This probable phylogeny appears to be recapitulated by embryos in that the internal ear forms from an ectodermal placode which, in fishes, is in the middle of the series of placodes that form the lateral line system. This otic (or auditory) placode develops under the inductive influence of the hindbrain, thus assuring its position at the back of the head. The placode sinks in, forming a pit, that then pinches away from the skin ectoderm to become the hollow **otic vesicle.** Gradually, the vesicle assumes the complicated shape that makes appropriate its adult name of **labyrinth** (Figure 19-7). Ganglion cells of the associated eighth cranial nerve are also derived from the vesicle.

The body of the labyrinth becomes more or less divided into a dorsal chamber, or **utricle,** and a ventral chamber, or **saccule.** A diverticulum from the utricle is called the **endolymphatic sac,** and another from the saccule is called the **lagena.** The labyrinth is completed by three looping **semicircular canals,** each of which joins the utricle at both ends. The labyrinth and its canals are filled with the fluid **endolymph.**

In both utricle and saccule there is at least one patch of sensory and supportive cells. Each patch, called a **macula,** resembles a neuromast of the lateral line system, but is somewhat larger. Each sensory cell has the familiar clump of stereocilia and single, asymmetrically placed kinocilium. The hairs of each macula are imbedded in a modified cupula that is made heavy by deposits of calcium. These are in the form of crystals in some groups and in the form of solid masses called **otoliths** in others. The size and shape of otoliths is distinctive for many vertebrates.

When the head of the animal moves to a new position, the cupulas or otoliths tend to slide over the maculae in response to gravity, thus establishing the shearing force to which hair cells are sensitive. This makes the labyrinth an excellent position register.

The labyrinth evolved a related function, which is the detection of angular acceleration, or rotation of the head. To understand the mechanism involved, imagine a straight garden hose filled with water. If one jerked the hose in the direction of its length, the inertia of the long column of contained water would cause it to tend to remain stationary while the hose itself moved. As a result, pressure would increase at the end of the hose away from the motion and water would tend to escape there. (If the hose were jerked sideways, there would be little pressure on any area equal to that of the end of the hose because there would be no long column of water in that plane.) Similarly, when a semicircular canal is rotated in its own plane, the canal moves slightly past the contained endolymph. At one end of each canal is a swelling, the **ampulla,** within which is a patch of sensory cells, here called a **crista,** which responds to the motion. Cristae have high, domed cupulas that extend across the canals, completely occluding the lumens. The hair cells are all oriented so as to respond to motion in the appropriate direction. (Hair cells of maculae are oriented in various directions.) Since one of the three semicircular canals lies in each plane of space, every rotational motion is registered.

In order to protect the delicate labyrinth, and to mask out distortions from irrelevant sources, the entire structure is surrounded by cartilage or bone. However, the labyrinth does not fit quite snugly inside its bony housing. The thin space between labyrinth and skeleton is filled (at least in places) with the fluid **perilymph,** which cushions the organ against harm and plays a role in the hearing function of the ear of terrestrial vertebrates.

ORGAN OF HEARING In order for far-field sound to be heard underwater, it is necessary that (*1*) pressure waves be received and amplified by a confined compressible medium (that is, a gas), and (*2*) pressure changes in the gas be translated to motion changes in the endolymph bathing hair cells. The gas bladder of bony fishes was a ready-made resonator, and two groups of fishes independently evolved mechanisms for translating pressure waves from bladder to labyrinth, as described further in the next section. These fishes are known to hear well at frequencies below 1000 Hz (cycles per second).

Most airborne sound waves do not strike the body with sufficient force to carry through soft tissues to a deep resonator such as the gas bladder (much of the force is simply reflected away). Consequently, in order for an animal to hear when out of water, it is desirable that it have a thin surface membrane, the **tympanum,** or eardrum, that is free to vibrate against the low resistance of an air chamber on its inner side. It has been postulated that the rhipidistians among crossopterygian fishes learned to hold a bubble of air in the upper wing of the first, or spiracular pharyngeal pouch. If so, this may have improved hearing right away, but gradually the pouch became adapted as a true middle ear. The tympanum probably evolved from part of the overlying operculum, but this portion of the story is obscure and it is not certain that the tympanum is homologous in all vertebrates.

This evolutionary sequence is recapitulated by embryos of tetrapods: The extremity of the first pharyngeal pouch becomes the middle ear, its neck becomes the eustachian tube joining middle ear to pharynx, and the closing plate of the second pouch becomes the tympanum.

These advances accomplish the first requirement for hearing in air; sound waves are received by a tympanum vibrating freely against a middle ear cavity. There remains the second requirement, which is to translate this motion to part of the labyrinth. Merely to have the cavity of the middle ear impinge on the labyrinth would not be adequate, because sound waves (*1*) would be damped by the soft walls of the eustachian tube, and (*2*) would lose greatly in force in being translated from the gas of the middle ear to the fluid within the labyrinth. The first of these problems is solved by mechanically transmitting sound waves across the cavity. One or more bony ear ossicles accomplish this function. (The interesting origin of these ossicles from leftovers when the tetrapod jaw evolved was told in Chapter 12.) The second problem is solved in one, or both, of two ways. The footplate of the ear ossicle (or innermost ossicle, if there are three) vibrates against the **oval window,** which is an opening in the bony housing of the labyrinth (Figure 19-8). The area of the oval window is always much smaller than the area of the tympanum, thus increasing the force of vibration per unit area. Furthermore, the outermost ossicles of mammals (incus and malleus) are constructed so as to mechanically decrease amplitude but increase force as they pivot. In all, force is increased about 14–20 times.

What remains is to translate motion of the innermost ossicle to shearing force over hair cells. When the footplate vibrates against the oval window it creates compressional waves in the perilymph beyond. This perilymph is in a tube, called the **scala vestibuli,** which extends away from the oval window and then turns sharply on itself to return as a parallel tube, the **scala tympani.** Between these two tubes is an extension of the lagena called the **scala media.** The three adherent channels lengthen and coil—slightly in birds and markedly in mammals—and are then called the **cochlea** (= snail shell) (Figures 19-7 and 19-9). Within the scala media is a special auditory macula called the **basilar papilla** or, if much lengthened in a cochlea, its derivative, the **organ of Corti.**

The basilar papilla or organ of Corti rests on the **basilar membrane** separating scala media from scala tympani. This organ has several to many rows of

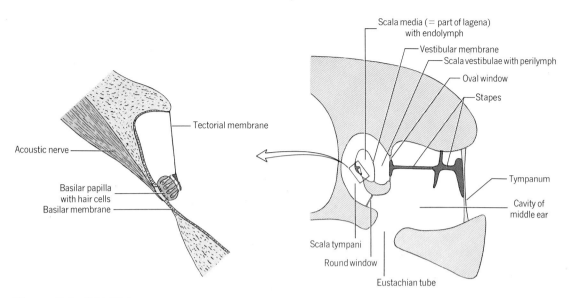

Figure 19-8 DIAGRAMMATIC SECTION THROUGH THE AUDITORY APPARATUS OF A LIZARD.

hair cells which, however, lack kinocilia in the adult. The stereocilia are embedded in a derivative of the ancestral cupula now called the **tectorial membrane.** Vibrations at the oval window cause traveling waves in the scala vestibuli that are translated across the thin vestibular membrane to the endolymph of the scala media, and thence across the basilar membrane to the scala tympani. These incompressible fluids can move within their bony walls because the scala tympani meets the cavity of the middle ear at a thin membrane covering the **round window.** Thus, when the tympanum and oval window move inward, the round window moves outward.

The basilar membrane and tectorial membrane pivot at different points, so

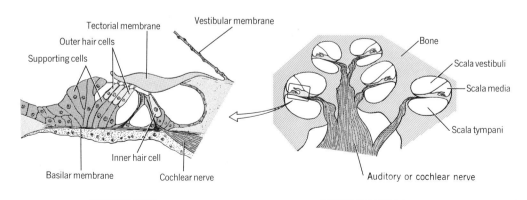

ORGAN OF CORTI   SECTION OF COCHLEA

Figure 19-9 STRUCTURE OF THE MAMMALIAN INNER EAR.

their motions create shearing force between them, which activates the hair cells. There is a systematic change along the length of the organ of Corti in pattern of hair cells, arrangement of stereocilia, and breadth of the organ. This is the structural basis for differential response at different levels along the organ; high frequencies are sensed at the base of the cochlea and low frequencies toward the tip. All in all, this is an amazingly complex but effective structure.

EVOLUTION OF THE EAR   The labyrinth, or inner ear, evolved very early in vertebrate history and, with many variations in configuration but none of basic design and function, has been retained by all vertebrates. The middle ear evolved as tetrapods evolved, and the external ear is scarcely found except in mammals.

CYCLOSTOMES, and at least some ostracoderms, have fewer than three semicircular canals (Figure 19-7). It is not known if this is a primitive or degenerate condition. Utricle and saccule are not set apart, and the organ is compact. Some ELASMOBRANCHS retain a slender canal, the endolymphatic duct, where the embryonic otic vesicle pulled away from the skin ectoderm. Fine sand grains enter the saccule through this canal and clump to form an otolith. Utricle and saccule are broadly joined.

The labyrinth of BONY FISHES is only partly surrounded by bone. Hard calcareous otoliths rest on the maculae, of which there may be three or four. The otolith of the saccule ranges from small to so large that it nearly fills the chamber. Otoliths of bony fishes are useful to systematists because they have distinctive shapes and annular growth rings. Two groups of fishes have achieved far-field hearing of low-frequency sounds by adapting the gas bladder as a resonator. One group (cods, herrings) have a pair of long extensions of the gas bladder that penetrate the brain cavity to impinge on extensions of the endolymphatic sacs of the labyrinths. The other group (minnows, catfishes, goldfishes) has modified processes of adjacent vertebrae into paired chains of three or four ossicles that pivot on the spine to convey vibrations from gas bladder to endolymphatic sac.

Ears of ancestral LABYRINTHODONTS probably were similar to those of such REPTILES as have not developed degenerate or highly modified ears. A large tympanum is either flush with the surface of the head or protected by a flap of skin (Figure 19-8). The eustachian tube may be broadly open to the pharynx. The single ossicle, or stapes, is commonly a slender but complicated structure having cartilaginous arms at its outer end. The lagena is somewhat lengthened but coils only in mammal-like reptiles. The basilar papilla is elongate and, like the tectorial membrane, is constructed in a variety of ways often departing considerably from the mammalian condition. The number of hair cells along the papilla gives some indication of the range of frequencies that can be detected and of the general sensitivity of the ear.

The ears of LIVING AMPHIBIANS and SOME REPTILIA are off the main line of descent in having specialized ears. Most anurans have short heavy stapes and also another ossicle unique to themselves. These animals are most responsive to low-frequency sounds and do not hear sounds of more than 3000–5000 Hz. Caecilians, urodeles, some anurans, snakes, and some other reptiles lack a middle ear cavity and tympanum, but usually retain the stapes, which is attached

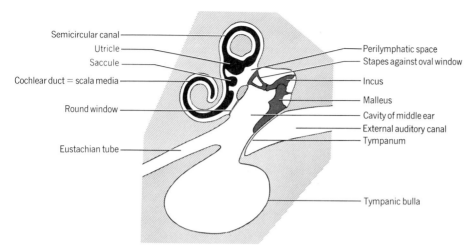

Figure 19-10 DIAGRAMMATIC SECTION THROUGH THE AUDITORY APPARATUS OF A MAMMAL. Somewhat distorted for two-dimensional representation. Bulla may be relatively much larger.

(usually by means of muscle, ligament, or cartilage) to the skin, squamosal, quadrate, hyoid, or even scapula. These ears may be sensitive to airborne vibrations and also are sensitive to vibrations conducted from the ground to the labyrinth by way of the bones of limbs or jaws.

The vestibular membrane is delicate in MAMMALS, but heavy in BIRDS. The basilar membrane of birds is much shorter and broader than that of mammals. The organ of Corti has one inner, and three or four outer rows of hair cells in mammals, but about ten times as many in birds. A coiled (hence lengthened) cochlea is present in each class (except in monotremes), making as many as five turns in mammals. Ears of mammals are distinctive in several ways: An external ear, or pinna, is usually present. This structure helps to funnel sound waves toward the tympanum, but may have other functions (cooling, recognition) and can be an inefficient funnel. The middle ear is housed in a bony tympanic bulla, and extensions of the middle ear cavity may penetrate adjacent bones as mastoid sinuses. There are, of course, three ossicles (Figure 19-10). Mammals commonly hear sound over the wide range of 20–20,000 Hz. Dogs hear at 40,000 Hz, and some volant and aquatic species emit and receive sound at much higher frequencies as a means of echolocation. At the other extreme, elephants can hear sounds as low as 16 Hz. Loudness of sound is detected over about ten orders of magnitude. Very strong vibrations are damped to prevent injury by tiny muscles and by buckling of the chain of ossicles.

EYES

All vertebrates have a pair of lateral eyes, and some also have an unpaired dorsal eye. The lateral eye is a complex precision instrument. It is unmatched for range of adaptation, being modified to function in water or air, by day or night, at short range or far, and in habitats ranging from the sky to the depths of the ocean.

From the eye alone, morphologists can determine much about the habits of any vertebrate. Evolutionary lineages are less easily established from eye structure, though some trends are evident. In several instances, prior knowledge of phylogeny helps to explain otherwise puzzling characteristics of the eye.

STRUCTURE AND FUNCTION   Structure and function can profitably be studied in terms of the requirements of the eye as an optical instrument. A first requirement is for a firm housing—sharp images would not be possible on the filmplate of a camera having soft walls. The capsule of the eye is kept rigid by its outermost envelope, the **sclera,** which is stiffened by cartilage, or bone, or a tough mesh of collagenous fibers, or by a combination of these (Figure 19-11). Pressure of the fluid within the eye also contributes to keeping the envelope distended and firm.

Another requirement is for control of the amount of light entering the eye. This is accomplished by a curtain, called the **iris,** which surrounds the light window, or **pupil.** Delicate muscles within the iris cause it to pull back and enlarge the pupil when light is dim, and extend to narrow the pupil to a small circle or slit when light is bright.

There must be no random reflections of light within the eye, so the capsule is usually lined by a nonreflective black pigment. This **pigment layer** is just behind the light-sensitive cells where it can, in some vertebrates, temporarily surround the ends of those cells to screen them from excessively bright light. (There is directed reflection of light by some eyes—see below.)

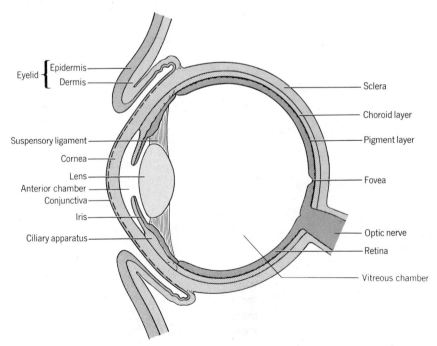

Figure 19-11   GENERAL STRUCTURE OF THE EYE OF A DIURNAL MAMMAL.

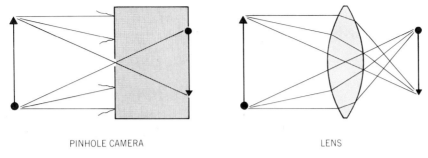

PINHOLE CAMERA   LENS

Figure 19-12   TWO METHODS OF FORMING AN IMAGE.

An important requirement is for the formation of an image. When light enters a box through a small hole, a dim inverted image is formed, as shown by Figure 19-12. This is the principle of the pinhole camera, and it is actually used by several vertebrates (camel, some geckos) in bright light when the pupil is made very small. Usually, however, another principle must be followed. When a ray of light passes from one material into another having a different refractive index, it changes direction at the interface. The direction and amount of the change depend on the materials and on the angle the ray of light makes with the interface. A **lens** is a transparent object shaped so as to bend light into an undistorted image. The image is "upside down" both on the filmplate and the retina. The brain somehow turns the image over, or so it seems.

Over the front of the eye, the sclera merges into the transparent **cornea.** The cornea has about the same refractive index as water. Accordingly, corneas of eyes that function underwater do not bend light and can be of any shape. They are usually flat or streamlined. Rays of light entering the cornea from air are bent sharply. Hence, the cornea of terrestrial animals must be accurately curved to avoid the distortion of the image called astigmatism. The lens then completes the image formation initiated by the cornea.

When passing through an ordinary lens, central and marginal rays have slightly different focal points, thus causing spherical aberration. The lensmaker overcomes this problem by building fine lenses from several apposed elements. The eye overcomes it by having the cornea curve slightly less at its margins, and by having the lens more dense at its core. Furthermore, light rays of different wavelengths bend slightly differently when passing through an ordinary lens, thus causing chromatic aberration. The maker of fine lenses combines crown and flint glass, which have compensating properties. The eye that requires a sharp color image screens out light having the shortest wavelengths (violet and blue) by using a color filter. Yellow pigment is added to lens or retina, or oil droplets of yellow or red color are distributed in the photoreceptor cells.

Pinhole cameras select light rays rather than bend them; every object is in focus regardless of its distance from the camera. When images are made by lenses, however, provision must be made for focusing the image where it will be recorded. In the eye, this is called **accommodation.** Some vertebrates (like the photographer) move the lens outward from a farsighted resting position to focus on near objects. Others move it inward from a nearsighted resting position to

Sclera

Choroid layer

Pigment layer

Rod

Cone

Direction of light

Outer plexiform layer

Bipolar cells

Inner plexiform layer

Ganglion cells

Fibers of optic nerve

Figure 19-13   HISTOLOGY OF A MAMMALIAN RETINA seen in cross section.

focus on far objects. Amniotes instead focus by changing the shape of a stationary lens, causing it to bulge for near vision and flatten for far vision. This is done by changing the tension in the **ciliary apparatus** that suspends the lens. Different amniotes use different mechanisms as told in a later section. The very small lenses of small vertebrates naturally have relatively great depth of focus and close near points, so little accommodation is needed.

The next requirement is for perception of the image. Photoreceptor cells are of two kinds, **rods** and **cones.** Each has a nervelike base, an inner segment with nucleus and mitochondria, and an outer segment which is long and cylindrical for rods but shorter and conical for cones (Figure 19-13). The outer segment, which is narrowly connected to the nourishing inner segment, is intricately folded into a stack of hundreds of lamellae.

On these lamellae is **visual pigment,** which is instantly altered chemically in the presence of light. This change is translated to nervous impulses in a manner as yet not understood. Rods are all alike. Their pigment is **rhodopsin** (or, in freshwater fishes and tadpoles, a related pigment), each molecule of which responds to a single quantum of light energy—the theoretical limit. Cones all look alike, yet are of three classes having different pigments and responding to light of different wavelengths. Thus, cones are the basis for color vision. Rods are more sensitive than cones by about two orders of magnitude but do not record color. Nocturnal animals have only rods.

Beyond the rods and cones (in the direction of the inside of the eyeball) are **bipolar cells,** and beyond these, **ganglion cells,** whose axons extend through the optic nerve. There are also two plexiform layers of nerve cells, one meshed into the system at each end of the bipolar cells. This complex of nerve cells accomplishes summation of stimuli from different photoreceptors (particularly rods) and in other ways performs the first steps of forming the image perceived by the brain.

In order for the eye to have acuity of vision, as is usual for diurnal animals that rely on vision in feeding, moving about, or avoiding danger, the retina must record the image with a fine grain; photoreceptors must be tightly packed so different cells can respond differently to adjacent parts of the small image. In most diurnal vertebrates having color vision (most teleosts, frogs, most reptiles and birds, some mammals), the sharpest image is perceived at a place on the retina called the **area centralis** where the cells are most slender and closely packed (to as many as $10^6/mm^2$), where there are only cone cells, and where many animals have a pit called the **fovea,** which bends light rays to enlarge the image by as much as 30%. Man has a "good" area centralis (we move our eyes when we read to keep the print focused there), but not the best: He needs a low-power binocular to gain the acuity of vision of a hawk!

If the eyes move independently in their orbits, their fields of vision may be independent or may at least partly overlap. If they have overlapping fields of vision, depth perception, or stereoscopic vision, is achieved. In mammals, nervous coordination prevents independent movement of the eyes.

Finally, the eye requires some structures that are accessory to its optical functions. A vascular **choroid layer** nourishes the eye. Extrinsic muscles turn the eyeball in its socket (see figure on p. 191). Glands and lids (sometimes including a third eyelid, or **nictitating membrane**) moisten and protect the eye in air. The eye appears to look through a hole in the skin, but the edges of the lids are in fact folds in the skin, not breaks, and the skin continues over the cornea as the thin, transparent **conjunctiva.**

This book cannot cover many of the specializations of the vertebrate eye, but two principal adaptations will be noted: The general structure of eyes adapted for vision underwater is presented in Chapter 26, and adaptations for vision when there is little light are presented here.

The many amphibians and mammals, and few reptiles and birds that are active at night, some fishes that frequent murky water, and many of the fishes that live deep in the ocean (where sunlight does not penetrate but where animals are luminescent) must be able to see in dim light. Their eyes are very large in front to gather in as much light as possible (Figure 19-14). The pupil opens very wide (though it may also close to a narrow slit by day). The lens is large, spherical, and placed well back in the eye to be close to the retina. The retina is relatively small, so the entire eye is deep in the optical axis, or even tubular. This arrange-

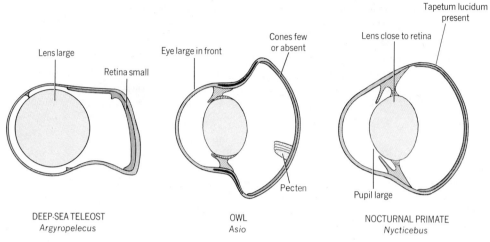

Figure 19-14 ADAPTATIONS OF THE EYE FOR VISION IN DIM LIGHT.

ment mimics the slide projectionist who gains a bright image by placing his projector close to a small screen.

Cone cells are few or absent. The more sensitive rod cells are slender and closely packed. There is so much summation by the ganglion cells that light striking a thousand or more rods may trigger an impulse in a single nerve fiber. This enormously increases sensitivity to dim light, but slightly blurs the image. An area centralis is often lacking so the eye, which may be too large to turn much in its socket, need not be oriented exactly toward the objects perceived. The entire head is turned if need be—some owls can rotate the head 270°!

Finally, most nocturnal vertebrates have a mirror, or **tapetum lucidum,** behind the rods that reflects light back out of the eye, so light passes through the rods twice instead of once. Analogous mirrors have evolved several times; reflection may be from crystals of guanine, lipid particles, or other pigmented layer, or supportive tissues.

DEVELOPMENT AND ORIGIN   Developmentally, the eye has three principal parts: retina and pigment layer, lens, and supportive tissues. First to appear is the primordium of the retina, which is an evagination of the sidewall of the diencephalon. This expansion, or **optic vesicle,** pushes toward the skin ectoderm, trailing an **optic stalk** behind (Figure 19-15). Next, the lateral surface of the vesicle sinks into the cavity of the vesicle, thus forming a double-walled cup resembling the bulb of a rubber syringe when indented on one side by the thumb. The outer wall of the cup becomes the pigment layer and the inner wall the retina. The lip of the cup forms the iris. The optic stalk becomes the optic nerve as axons from ganglion cells in the retina extend down its length. In a sense, the retina is a nucleus of the brain and the optic nerve a tract of the brain.

The lens forms from a placode of the skin ectoderm that develops only under the inductive influence of the underlying optic vesicle. The **lens placode** be-

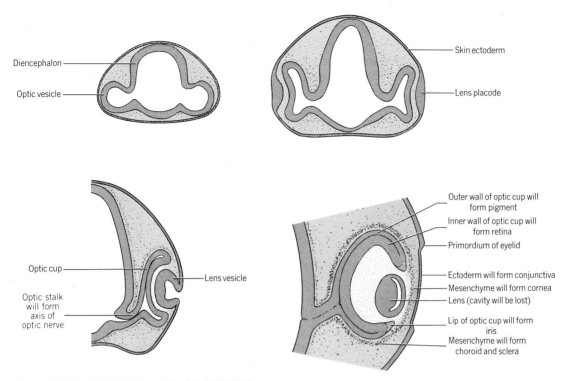

Figure 19-15   DEVELOPMENT OF THE EYE.

comes a hollow **lens vesicle,** but the cavity is soon diminished and then obliterated by the thickening of the inner wall of the vesicle.

Mesenchyme surrounding the optic cup differentiates into choroid, sclera, cornea, and ciliary apparatus. Developmentally, these are continuous with, and apparently also homologous with, the meninges of the brain.

The phylogenetic origin of the vertebrate eye is unknown; the first known vertebrates had already perfected the organ. This has not discouraged morphologists from speculating, however, and nearly a dozen theories have been proposed since the 1870s. The most plausible theory is strongly supported by developmental, and weakly supported by histological and comparative anatomical clues. The neurocoel and ventricles of the brain are lined with ependymal cells that are usually ciliated. In the head region of many chordates, from amphioxus to birds, these cells are light-sensitive. If phylogeny is repeated by ontogeny, then rods and cones evolved from ependymal cells of a part of the forebrain, and the eye could have been functional throughout its evolution. The outer segments of rods and cones are considered to be modified cilia.

(A dermal light sense is common in cyclostomes, fishes, and amphibians. in some instances the receptor is unknown, but light-sensitive cells are identified in the skin of some species. This is probably an ancient condition.)

EVOLUTION OF LATERAL EYES   The hagfish, a scavenger in deep water, has degenerate eyes and is blind. The well-developed eyes of the LAMPREY are primitive in two respects: The conjuctiva is not fused to the cornea, and the ependymal layer is retained in the core of the optic nerve. Other distinctive features may or may not be primitive: The size of the pupil is fixed, the sclera is not stiffened by cartilage or bone, accommodation results when an extrinsic muscle pulls against the cornea, thus pressing the lens inward, and the lens is held in position by pressure only, not by a suspensory apparatus. As for primary swimmers in general, the eye is large and shallow along its optical axis, the lens is large and spherical, eyelids and glands are absent, and the extrinsic oblique muscles rotate the eyeball around its optical axis.

There is a wide range of eye structure within the large assemblage of fishes, but some generalizations can be made. ELASMOBRANCHS stiffen the sclera with cartilage (Figure 19-16). They accommodate somewhat by pulling the lens forward from its resting position with a small intrinsic muscle of ectodermal origin. A unique cartilaginous pedicel props the eyeball away from the back of the orbit;

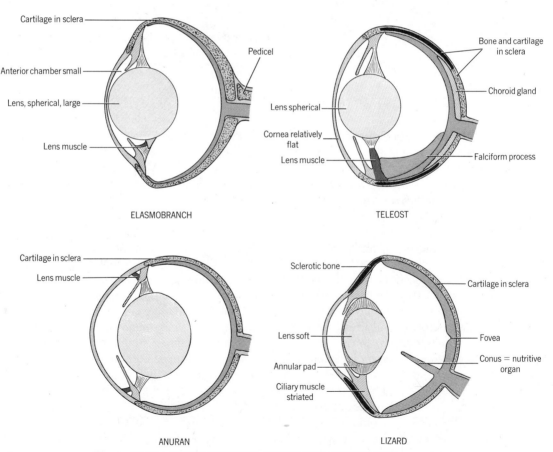

ELASMOBRANCH

Cartilage in sclera
Anterior chamber small
Lens, spherical, large
Lens muscle
Pedicel

TELEOST

Bone and cartilage in sclera
Choroid gland
Lens spherical
Cornea relatively flat
Lens muscle
Falciform process

ANURAN

Cartilage in sclera
Lens muscle

LIZARD

Sclerotic bone
Cartilage in sclera
Lens soft
Annular pad
Ciliary muscle striated
Fovea
Conus = nutritive organ

Figure 19-16   COMPARATIVE ANATOMY OF THE EYE.

its function is not clear. Cones are few or absent. An area centralis is present. Crystals of guanine in the choroid cause it to reflect light as a tapetum lucidum. The chamber between lens and cornea is small.

BONY FISHES stiffen the sclera with cartilage and, in most teleosts, by several bony plates. The cornea is flat or streamlined. Cones are usually present, so color vision is typical. Usually, an intrinsic mesodermal muscle pulls the lens inward to accommodate for far vision (chondrosteans, and some other fishes, have no accommodation). Teleosts have a nutritive structure derived from the choroid, called the **falciform process,** which projects into the cavity of the eyeball. Many teleosts have an area centralis, and some have a fovea.

As larvae, AMPHIBIANS have fishlike eyes. At metamorphosis they develop eyelids and glands and an evenly curved cornea. A small mesodermal muscle within the eyeball moves the lens forward for near vision. The eyes of anurans are better developed than those of urodeles. Some have color vision as, in all probability, did many labyrinthodonts. (Caecilians and some salamanders are blind.)

REPTILES and BIRDS, most of which are diurnal, have the finest visual acuity and accommodation of all vertebrates. The eye is large, particularly in birds; eyes of hawks are larger than those of humans. The sclera is stiffened by a cartilaginous cup behind and by a ring of about 15 small, overlapping bones on the forward wall, where the eyeball might otherwise be distorted by the ciliary muscles. These muscles are striated (though of ectodermal origin) and, hence, faster in action than those of other vertebrates. The small, soft, somewhat flattened lens has a peripheral **annular pad** that makes firm contact with the ciliary apparatus. Accommodation is active and instant; muscular contraction causes the lens to bulge in front. Cones are usually numerous and color vision excellent (except in snakes and crocodilians). An area centralis is present and there is one or, in many birds and some lizards, two foveas. There is a vascular projection into the cavity of the eyeball. This is particularly large and complexly folded in birds, where it is called the **pecten** (= comb.) (Figure 19-14). Of the many functions postulated for this structure, that of nutrition is the most probable. Lacrimal gland and nictitating membrane are present. Iris musculature is striated. A tapetum lucidum is rare. Some binocular vision is frequent, particularly in predatory birds. (The eyes of snakes are in some ways atypical because of derivation from burrowing ancestors for which vision was unimportant. Amphisbaenians, some lizards, and some snakes are blind.)

MAMMALS first evolved as small nocturnal creatures, and at that stage lost the perfection of eye structure of their reptilian ancestors. Some of the loss was later regained, but not all. There is no cartilage or bone in the sclera. Muscles of the ciliary apparatus and iris are smooth and, therefore, relatively slow. The shape of the lens is adjusted to focus the image, but the mechanism is inferior to that used by reptiles and birds; contraction of the ciliary muscles relieves tension on the suspensory apparatus that then allows the lens to bulge of its inherent elasticity. This process is relatively slow, particularly in old age, when it may also become incomplete, making near vision impossible (without eyeglasses). There is no pecten or corresponding nutritive organ. A tapetum lucidum

Figure labels: Paraphysis, Dorsal sac, Habenula, Posterior commissure, Pineal body, Optic lobe

BRAIN IN SAGITTAL SECTION

Figure labels: Cornea, Lens, Skin, Retina, Parietal nerve

PARIETAL EYE

Figure 19-17   DORSAL EYE OF A LIZARD.

is confined to several orders; a nictitating membrane is rare. Color vision appears to be partially regained in several orders but is complete only in primates and some rodents. An area centralis is present in some orders; a fovea is present only in higher primates. Binocular vision is frequent and coordination of the eyes is superior. (The eyes of monotremes are in some respects atypical of the class.)

DORSAL EYES   On top of the diencephalon there are, in the midline, two small evaginations: an anterior **parietal organ** (or parapineal) and a posterior **pineal body** (or epiphysis). One or both of these structures may be photoreceptive and is then termed the third eye.

Asymmetry of these structures in adult lampreys and embryo lizards, and the conformation of ostracoderm head armor, indicate that parietal and pineal organs may be derived phylogenetically from a bilateral pair of organs, the left member of which shifted forward to the midline while the right member slipped behind.

The pineal organ is present in nearly all vertebrates. It has photoreceptive cells in lampreys, and endocrine properties are known for various vertebrates. Initially, at least, the endocrine functions seem to have been light-related (see p. 385). The parietal organ also has photoreceptor cells in lampreys, but in these animals is subordinate to the pineal eye. The parietal eye is functional in tadpoles and salamander larvae and in many lizards, where it may have a lens, retina, and tiny nerve (Figure 19-17). Histologically, its photoreceptive cells are closely similar to the cones of lateral eyes. Such light-associated behavior as sunning and activity rhythms is influenced by this organ in some lizards.

It is evident from the widespread presence of a parietal (or pineal) foramen in skulls of ostracoderms, placoderms, and early representatives of bony fishes, labyrinthodonts, and reptiles, that a third eye has been persistent in the evolution of vertebrates.

20

Endocrine Glands

**GENERAL
NATURE OF
ENDOCRINE
GLANDS**

The endocrine glands are ductless glands having secretions called **hormones** that are discharged into the circulatory system for distribution. Only certain tissues of the body are responsive to each hormone. Response may be morphologic, as for sex hormones that influence the development of secondary sexual characteristics, or physiologic, as for an adrenal hormone that influences kidney function. Many hormones act directly on tissues that are receptive to their "messages." Generally, however, the action is indirect. Thus, a hormone secreted by the hypothalamus of the brain and released by the hypophysis causes the ovary to secrete a hormone to which the lining of the uterus responds.

Taken together, the endocrine glands function like the nervous system in several respects: Each controls and integrates bodily functions, each mediates control through the release of chemicals (the same chemicals in several cases), and each may accomplish interaction within its own system to coordinate its activities. Endocrine control differs from nervous control in tending to be slower and more sustained. The functions of the two systems merge in several instances.

All endocrine glands are small and highly vascular. They are often diffuse in lower vertebrates, but tend to be discrete in tetrapods. Most are constructed of cords of more or less cuboidal cells arranged among sinusoids and supported by a matrix of connective tissue (Figure 20-1). Several endocrine glands (neurohypophysis, urophysis) are instead constructed of thin attenuated cells, and one (thyroid) is constructed of follicles. The hormones of glands of mesodermal origin (gonads, adrenal cortex, placenta) are steroids, whereas hormones of glands of

Figure 20-1   TWO HISTOLOGICAL TYPES OF ENDOCRINE GLAND.

ectodermal or endodermal origin are proteins, peptides, or other derivatives of amino acids.

Since endocrine glands utilize the circulatory system to transmit signals, their shape is of little importance, and a gland may usually be single, multiple, or diffuse without relation to function. Only the aggregate volume of cells is critical. Even the secretions have been remarkably constant over the ages, though there are some exceptions, and responses to the secretions are various. Endocrine glands have diverse developmental origins, are usually unrelated to one another in space, and are incompletely related in function. They do not comprise an organ system in the usual sense. Endocrine function is still inadequately known for most lower vertebrates.

STRUCTURE, FUNCTION AND EVOLUTION OF THE GLANDS

HYPOPHYSIS.   The **hypophysis,** or **pituitary gland,** is located under the brain (see figures on p. 345, 349, and 351). In mammals it is housed in a bony pocket of the basisphenoid bone. Although it is small, this gland is, in both structure and function, one of the most complicated organs of the body. Developmentally it has a surprising dual origin: The adult portion called the **neurohypophysis** forms from the part of the floor of the embryonic diencephalon termed the **infundibulum.** This structure may evaginate (most tetrapods) or remain nearly unfolded (amphibians, some fishes). The remainder of the gland, or **adenohypophysis,** forms instead from an evagination of the ectodermal part of the embryonic mouth cavity (the stomodaeum) called the **hypophyseal pouch,** or Rathke's pouch (Figure 20-2). This pouch and its derivatives are variously lobed in the different vertebrates. Its connection to the mouth is usually lost during maturation.

In general, the neurohypophysis has an anterior subdivision, the **median eminence** (not identified in forms below lungfishes), and a posterior or ventral

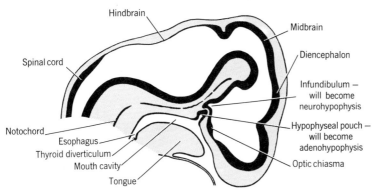

Figure 20-2   EMBRYONIC ORIGIN OF THE HYPOPHYSIS shown by a sagittal section of the head of a mammalian embryo.

subdivision, the **pars nervosa,** or posterior lobe of the gland (Figure 20-3). The adenohypophysis has several parts: The largest, most constant, and most active is the **pars distalis** (or anterior lobe). A **pars intermedia** is usually present but is lacking in birds. Other parts are of variable occurrence and doubtful function. They differ widely among the vertebrates, and homologies can be made, if at all, only by combining clues from embryology and histochemistry. The median eminence and pars distalis have a common blood supply; the pars nervosa has an independent blood supply.

The neurohypophysis is atypical of endocrine glands in that it is constructed largely of long parallel nerve fibers originating in the hypothalamus of the brain (Figure 20-4). Indeed, this part of the hypophysis functions by storing and releasing into the bloodstream hormones elaborated in the hypothalamus and transferred to the neurohypophysis by neurosecretion (of which, more below). Very different is the adenohypophysis, which has cords of secretory cells of two kinds that branch without pattern among sinusoids.

The pars intermedia secretes one hormone, **melanocyte-stimulating hormone,** which influences the pigment of the skin. The pars nervosa releases two hormones, **vasopressin** (which causes the kidney to hold back fluid) and **oxytocin** (which in mammals causes the letdown of milk and contraction of the uterus). The pars distalis usually secretes six hormones: **growth, thyrotropic, adrenotropic, follicle-stimulating** (which is active during the ripening of eggs), **luteinizing** (active during the formation of the corpus leuteum in the ovary), and **lactogenic** (which is poorly named because though it is needed for lactation it has other functions in nonmammals). The hypophysis is, in turn, influenced by the hormones of other glands, notably the ovary, thyroid, and adrenal.

It has been postulated that any of several structures of amphioxus may be homologous with the vertebrate hypophysis, but no conclusions can be drawn. In cyclostomes and fishes other than Sarcopterygii, the neurohypophysis is the more or less flat floor of the brain above other parts of the gland; a pars nervosa cannot be clearly distinguished. The atypical adenohypophysis of hagfishes is scattered as islets among other tissues. The adenohypophysis of Selachii is unique

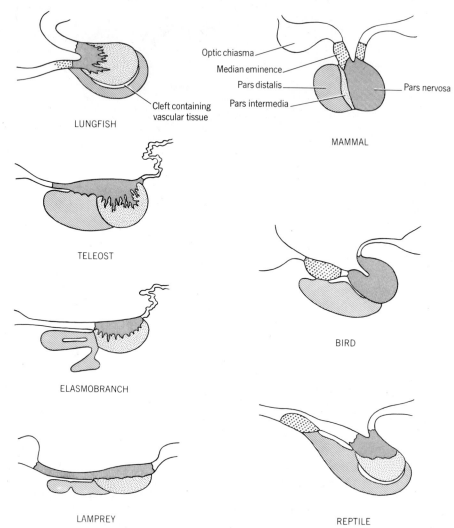

Figure 20-3 COMPARATIVE ANATOMY OF THE HYPOPHYSIS as seen in sagittal section. Anterior is to the left.

in having a ventral lobe of unknown homology. Fishes, and particularly ray-finned fishes, are distinctive for the way that the adenohypophysis and neurophypophysis interdigitate over a broad area.

In dipnoans and tetrapods, interdigitation between the neurohypophysis and adenohypophysis is much reduced, or, more often entirely lost. A pars nervosa forms from all or part of the infundibulum, which is now usually evaginated from the floor of the brain. This causes the gland to subtend more from the brain. The pars intermedia is large in reptiles, small or absent in mammals, and absent in birds. When absent, its hormone, melanocyte-stimulating hormone, may be produced by the pars distalis.

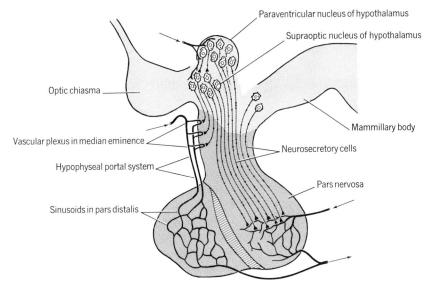

Figure 20-4   RELATIONSHIPS OF THE NEUROSECRETORY CELLS OF THE HYPOPHYSIS. Anterior is to the left.

THYROID   The secretory cells of the thyroid are derived from a midventral evagination of the endoderm of the embryonic pharynx at about the level of the second pharyngeal pouch (Figure 20-2). Surrounding mesenchyme contributes supportive tissues. The gland always consists of a cluster of rounded follicles, and each follicle is lined by a single layer of cells that are usually cuboidal (sometimes columnar when highly active) and have microvilli on their free surfaces (Figure 20-1). The thyroid has an exceedingly rich blood supply for its size. It is the only endocrine gland to have extracellular storage of its secretion. This secretion, called **colloid,** fills the follicles and contains **thyroglobulin.** This iodine-rich protein is converted by hydrolysis to either of two hormones, **thyroxine** and (in much lesser quantity) **triiodothyronine.** These are amino acids that are linked to blood proteins for transport.

The gland begins to function early in ontogeny, contributing to the control of differentiation, growth, metamorphosis, the distribution of pigment, and sexual development. It has a profound affect on metabolic rate and may influence molt (amphibians and reptiles), feather shape, body temperature, and functions of the nervous, digestive, and excretory systems. The thyroid interacts with the hypophysis and in at least some instances (anurans) with the hypothalamus. The gland enlarges when diseased.

All vertebrates have a thyroid, and its origin traces back to cephalochordates. The endostyle on the floor of the pharynx of amphioxus, and also of the larval lamprey, ammocoetes (see figure on p. 46), functions in the production and movement of mucus for filter-feeding. Nevertheless, there is evidence that part of the endostyle is the phylogenetic precursor of the thyroid: (*1*) The endostyle, like the thyroid, forms from a midventral evagination of the pharynx. (2) Part of it, like the thyroid, concentrates iodine from the blood. (3) At metamorphosis the endostyle of ammocoetes is partly converted to adult thyroid.

In cyclostomes and many teleosts the thyroid is relatively diffuse and is variously distributed near the ventral aorta, branchial afferent arteries, heart, gills, head kidney, spleen, brain, or eye. It is more discrete in other vertebrates but may be paired (amphibians, lizards, birds), bilobed (dipnoans, many mammals), or single (cartilaginous fishes, most reptiles) (see figures on p. 190 and 270). The thyroid of tetrapods is usually near the larynx, trachea, or bronchi.

PARATHYROID The secretory portion of the parathyroid glands differentiates from the epithelium of the third and fourth (and in reptiles also the second) pharyngeal pouches. Curiously it is the dorsal wings of the pouches that contribute in mammals, but the ventral wings in other vertebrates. It is clear from this and other evidence that the pharyngeal pouches share the potential for forming glandular tissue, yet the kind of gland formed by a specific region is not constant.

The gland consists of densely packed cells arranged in cords and clumps. The parathyroid hormone is a polypeptide called **parathormone.** It affects the level of calcium, and less directly the level of phosphorus, in the blood. In the absence of the hormone, calcium disappears from the blood in a matter of hours, tetanus occurs in muscles, and death follows. Deficiency of the hormone leads to abnormalities of bones and teeth.

Glandular tissue of unknown function has been identified in cyclostomes and fishes that may be homologous with the parathyroid. However, the gland is known with certainty only in tetrapods, and hence evolved relatively late. It is usually divided into one or two pairs of small glands. These may be linear (birds), but are usually more or less globular. They are located in the throat area, commonly near, or even imbedded in, the thymus or thyroid (hence the name, parathyroid).

ULTIMOBRANCHIAL BODIES AND PARAFOLLICULAR CELLS **Ultimobranchial bodies** occur in all vertebrates except mammals and perhaps cyclostomes. They may be single or paired, and are located near the esophagus or thyroid gland. They develop from the most posterior pair of pharyngeal pouches. Their hormone is **calcitonin,** which affects calcium metabolism in a manner similar to, but more rapid than, that of parathormone. In mammals, **parafollicular cells** within the thyroid are homologous with ultimobranchial bodies, and also secrete calcitonin.

INTERRENAL ORGAN AND ADRENAL CORTEX The adrenal glands of amniotes are located adjacent to the kidneys. There are two kinds of adrenal tissues, though only in mammals are they separated into distinct regions, the **cortex** and the **medulla.** The two tissues are different in function and embryonic origin. The cortical type of tissue is discussed here, and the medullary tissue is discussed in the next section. In bony fishes the tissue that is equivalent to the adrenal cortex is called interrenal tissue.

These structures are similar to the gonads in the steroid nature of their hormones and also in their embryonic origin. Cortical tissue is derived from mesoderm lining the coelomic cavity close to the place of origin of the genital ridges.

The secretory cells of cortical and interrenal tissues form cords that are arranged in three layers in mammals but have little or no organization in other vertebrates. Many cortical hormones are known in mammals, but some of these are also produced elsewhere in the body, some are readily converted to others, and only about a dozen are known to be physiologically active. Together these hormones are called **adrenocorticosteroids.** They are classed according to chemical structure and general function into four groups. One group (including cortisone, corticosterone, and cortisol) functions in carbohydrate and protein metabolism. A second group (including deoxycorticosterone) affects salt and water metabolism. The third group (including aldosterone) is typical only of mammals and relates to sodium and potassium metabolism. The last group (including adrenosterone) resembles the sex hormones. Most of these hormones interact with the pituitary, and many are involved in responses to stress.

Interrenal tissue of cyclostomes is scattered along the posterior cardinal veins and other vessels. In teleost fishes, interrenal tissue may be diffuse or discrete, but usually forms numerous small flecks located near or within the head kidneys. The gland is characteristically elongate and between the kidneys in cartilaginous fishes, elongate and adherent to the kidneys in anurans, and diffuse and adherent to the kidneys in urodeles. The cortical tissue of amniotes forms a pair of compact bodies located on or near the anterior ends of the kidneys (see figures on p. 302 and 303).

CHROMAFFIN BODIES AND ADRENAL MEDULLA   Tissue corresponding to the adrenal medulla tends to be much scattered in some vertebrates and is then termed **chromaffin tissue** because of its staining properties. Chromaffin and medullary tissue is innervated by preganglionic fibers of the autonomic nervous system. These nerves and glands are all derived from the ectodermal neural crests of the embryo, and all secrete **adrenalin** and **noradrenalin,** though the glands produce much more (particularly of adrenalin) than does the nervous system. The body responds to these hormones in many ways that better enable it to meet sudden emergencies (e.g., increased blood sugar and blood pressure, inhibition of smooth muscles).

The distribution of chromaffin tissue in the body corresponds to that of interrenal tissue but tends to be even more diffuse, particularly in fishes where it may occur along the postcardinal veins as well as near, on, or in the kidneys. Chromaffin tissue may lie near, but separate from, interrenal tissue (some fishes and some lepidosaurs), may be intermingled with interrenal or cortical tissue (some fishes, most amphibians and reptiles, birds), or may lie as a medulla within a covering of cortical tissue (most mammals). Even mammals, however, have chromaffin bodies or **paraganglia** associated with some sympathetic ganglia.

GONADS AND PLACENTA   The development and structure of the gonads were presented in Chapter 16, but these organs should be mentioned again as endocrine glands. The ovary produces several **estrogens** (the principal ones are estradiol and estrone), **progesterones,** and in mammals, **relaxin.** Estrogens control the growth and development of the female genital duct system and are essential for reproduction. They also initiate and maintain secondary sexual characteristics, which are marked in some vertebrates and inconspicuous in others. The site of estrogen secretion appears to be the internal theca in mammals, and probably also the stratum granulosum (see figure on p. 297) but is in doubt for vertebrates having virtually no theca.

Following ovulation, the ruptured mammalian follicle is transformed into a temporary but pronounced gland termed the **corpus luteum.** This structure then secretes progesterone, a hormone that is essential for the final differentiation of the female reproductive tract in preparation for fertilization and pregnancy, and also for maintaining pregnancy. Structures similar to the mammalian corpus luteum form in the ovaries of various other vertebrates (sharks, teleosts, urodeles, birds, some reptiles) but seemingly not in others. It is therefore puzzling that all vertebrates have one or more progesterones.

Relaxin is the only gonadal hormone that is not a steroid. In some mammals, at least, it acts on the pelvic symphysis, mammary glands, and genitalia, readying them for their functions at delivery and thereafter.

The interstitial cells of the testis (see figure on p. 298) produce male hormones collectively called **androgens.** The principal androgens are **testosterone** and **androstenedione.** Androgens are required for the growth, differentiation, and function of the male genital ducts and copulatory organ (if present) and for control of secondary sexual characteristics and sexual behavior. All vertebrates have androgens. Interstitial tissue has not been identified in all, but homologous cells may occur in the seminiferous tubules or bounding the seminiferous ampullae. To complicate the situation, males produce estrogens and females produce androgens, all from inadequately known sources, though the adrenals are probably involved. Supportive (or Sertoli) cells within seminiferous tubules may produce a hormone necessary for sperm maturation.

The mammalian **placenta** is a rich source not only of estrogen and progesterone, but also of a gonadotropin and of prolactin, which is otherwise produced by the adenohypophysis. Prolactin is needed for milk production in mammals but is also present in lower vertebrates where it has varied functions.

All the gonadal hormones have complex interactions with the hypophysis, some relate to interrenal or cortical function or to the activities of the thyroid or pineal. Gonadal function is usually seasonal, the ultimate control often being photoperiod as mediated by the hypothalamus. The pathways are poorly known for fishes.

MISCELLANEOUS, POSSIBLE, AND NEAR ENDOCRINE GLANDS   The origin and exocrine nature of the pancreas were noted in Chapter 12. Its thousands of islets (see figure on p. 225) secrete **insulin** which, in mammals, controls the deposit of glycogen in the tissues, and **glucagon** which controls its release. Among fishes there may be fewer, larger islets located within the pancreas or along the bile duct. There is great variation in response to these hormones—protein metabolism and solutes of the blood may be involved.

It is clear that hormones are secreted by the **stomach** and **small intestine** (e.g., gastrin, secretin, pancreozymin, cholecystokinin). These act on other parts of the digestive system to coordinate the digestive process. Although apparently present in all vertebrates, these hormones are virtually unknown except in mammals, and even in that class the cellular source of the secretion is unknown.

The tiny **pineal** organ atop the brain (see figure on p. 376) is a photoreceptor in the lower classes. In tetrapods it assumes endocrine functions that seem usually to involve the relations between homeostasis or reproduction and illumination. The pineal may influence activities of the hypothalamus and hypophysis. In most tetrapods the gland secretes **melatonin** under chemical signals from nerves. This is a hormone that affects melanophores of frogs and suppresses the gonads of mammals.

Fishes have what appear to be neurosecretory cells in the caudal part of the spinal cord. In teleosts these relate to a vascular, ventral swelling termed the **urophysis.** It is suspected to have endocrine function, possibly affecting osmo-regulation, ion regulation, or renal blood pressure. Near the afferent arterioles of the glomeruli of the kidney are **juxtaglomerular cells,** which contain granules and are believed to be the origin of the renal hormone **renin.** This poorly under-stood hormone seems to influence salt balance and blood pressure. Most fishes also have in their kidneys **corpuscles of Stannius.** These bodies are derived from the excretory ducts. They usually number two to six, but some fishes have dozens. A hormone with osmotic affects may be produced.

The **thymus gland** has been suspected of endocrine function, but none has been certainly established.

Not only are there these apparent endocrine glands of unknown function, but there are apparent "hormones" known not to be produced by glands. Thus, urea and carbon dioxide, derived from nonendocrine tissues, are distributed by the blood and convey "messages" to organs distant from their place of release. In order not to confuse the definition of hormone, these and similar substances are called **parahormones.**

NEURO-SECRETION

Emphasis should be given to the close relationship noted above between certain parts of the nervous system and certain endocrine organs. Neurons in two or more nuclei of the hypothalamus actually extend into and constitute the neuro-hypophysis. Those reaching the pars nervosa elaborate the hormones of that part of the gland. These hormones are released into the general circulation. Those neurons reaching the median eminence release hormones that enter a miniature portal system running from the median eminence to the adenohypophysis. This portal system is lacking in fishes, but the interdigitation of neurohypophysis and adenohypophysis in many fishes may accomplish the same interaction.

The structure of the urophysis suggests that its nerve cells may also be se-cretory and that the swelling of the cord is in part the result of temporary storage of the product. The adrenal medulla and other chromaffin tissues do not resemble nervous tissue histologically, yet they function as though they had evolved from postganglionic fibers of the autonomic nervous system. Nervous stimulation causes the medulla to secrete. The pineal gland of at least some mammals secretes its hormone on stimulation of the autonomic nervous system. Neurosecretion is also known in crustacea, insects, and other invertebrates.

These observations make it clear that the distinction between the nervous system and endocrine glands is not sharp—and that much remains to be learned about the relationship.

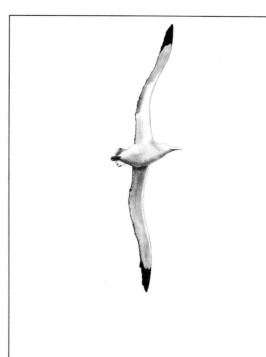

Structural Adaptation:
Evolution in Relation to Habit and Habitat

21

Structural Elements
of the Body

ANIMALS AS SPECIALISTS

Part II of this book includes much functional interpretation of structure, particularly for the respiratory, circulatory, and excretory systems and for the eye and brain. Emphasis, however, is on analysis of structure in relation to the long sweep of phylogeny; on conservative evolutionary changes common to all animals in such large taxa as classes and subclasses. Primitive and unspecialized characters are featured. Structures that are useful to animals of varied habits are stressed: Jaws are generally advantageous; two pairs of appendages proved to be a good general plan; a circulation divided into pulmonary and systemic circuits is superior for tetrapods.

Part III deals with the parallel or convergent influence of functional adaptation on different vertebrates; particular attention is paid to locomotor mechanisms in which parallels are seen most clearly. Animals with similar specialties are found scattered among the systematic categories, but are here brought together, their common problems are identified, and their various adaptations are interpreted from morphological clues.

Analysis is complicated by several factors. Different animals may do similar things in different ways: Squirrels and pottos both climb, but pottos slowly grasp the branches whereas squirrels run on the limbs or cling to them with sharp claws. No one kind of animal has *all* the structural modifications that are associated with its general habit. Furthermore, one kind of animal may have several specialties: Frogs jump and swim, flying squirrels climb and glide, cormorants fly, swim, and dive. Also, the activities of animals are determined not only by structural adaptations but also by behavioral factors. Thus, gray foxes climb trees whereas red foxes do not. One cannot learn by dissection which climbs, or, indeed, that either climbs. Even without special structural adaptation, many animals can run, swim, climb, and dig somewhat. Conversely, animals may fail to move in ways for which they seem, on the basis of morphology, to be adapted. Thus, the adult gorilla has the structure that correlates with climbing by arm-swinging under the branches, but because it is very large, it seldom does so.

In spite of these complications, however, it is rarely difficult to determine the principal habits of a vertebrate animal from its structure. Some clues are subtle and some obvious, but all make sense. Their identification and interpretation can be very engaging. Animals are so good at their specialties!

This part of the book begins with two chapters presenting basic mechanical principles that relate to feeding, posture, and locomotion in general. Subsequent chapters analyze specific adaptations.

PROPERTIES OF SUPPORTIVE MATERIALS

The materials of the body that accomplish support and movement are bone, cartilage, muscle, tendon, and ligament. (Soft organs are further supported by meshworks of collagenous fibers.) The suitability of these materials for the various requirements of the body depends on their properties.

Three important properties of living supportive tissues are not shared by any material available to architect or engineer: First, all display **growth** without interruption of function. They have remarkable capacity for repair of both major breaks and minor damage. This property protects these tissues from fatigue, or loss of strength with repeated loading, which is characteristic of nonliving supportive materials. The rate of repair is faster for muscle and bone than for cartilage, and in all the capacity diminishes with age. Second, all have amazing **capacity to adjust to circumstance,** slowly altering their substance and configuration in response to demand. It is common knowledge that muscular strength increases with exercise; the adaptability of other supportive tissues is exemplified later in this chapter. Finally, these properties taken together assure adequate **durability** for a lifetime of constant use. No man-made apparatus having even remotely comparable complexity of moving parts approaches the body in this regard.

The strengthening of supportive materials that results from their heterogeneity and the presence within them of microscopic lacunae was noted on pages 118 and 119. Before considering further the strength of the structural elements of the body it is necessary to review or introduce several concepts and terms. The important concept of **force** was presented on page 184, and should be reviewed there now. The weight of an animal pressing on the ground is a force, the pull of a muscle on its insertion is a force, and the push of a fish tail against the water is a force. Since forces of the body are concentrated at such places as insertions of tendons and contacts between bones, it is useful to consider **pressure,** or force per unit area. Pressure may be expressed as kilograms per square centimeter (or more precisely as dynes or newtons per square centimeter).

Load is a general term referring to any force that is applied to a solid object. Adjacent bones of the legs and spine load each other; active muscles load related bones. In order for loaded objects to remain in equilibrium, equal forces must operate in opposite directions (this is an application of Newton's Third Law). Thus, as a tetrapod stands at rest, the downward force of its weight is opposed by an equal upward force by the ground. The weight is transmitted through the bones of the legs to the ground. When objects transmit loads, there are internal forces of one part of the object acting on adjacent parts. The internal condition that results from transmission of a load is called **stress.** External loads cause internal stresses. Stress is expressed in units of pressure.

When any load is applied to any object, **deformation,** or **strain,** results. Strain is expressed as the relative deformation or change of shape that occurs when an object is stressed. Strain may be expressed as change in length, in volume, or in angle. Deformation may be permanent or temporary. Even moderate loads cause permanent deformation of modeling clay. Such materials could hardly support the animal body. If deformation is temporary, recovery may be almost immediate, as for bone, or somewhat slower, as for cartilage, tendon, and ligament. The capacity of a material to return completely to its original shape after a load is removed is called **elasticity.** The structural materials of the body have virtually perfect elasticity within usual load limits. Some elastic materials, like rubber, are much deformed by moderate force, whereas others, like steel, are only slightly deformed by great force. This property is expressed by the **modulus of elasticity,** which is stress divided by strain. If the modulus is high, the material is rigid; if it is low, the material is easily deformed.

The important property of **strength,** as applied to supportive materials, is the capacity to resist force without breakage or permanent deformation. Strength varies, of course, with material, and is proportional to the cross-sectional area of the object (i.e., the more bone the greater the strength). As noted on page 119, strength of heterogeneous materials varies with relative orientation of force to grain. Strength also varies importantly according to the direction of the applied force in relation to a surface; that is, the interaction between adjacent objects differs depending on the direction of the forces acting between them. This is true both for a load applied to an actual external surface of an object and for a stress applied to an imaginary internal surface. Forces are of only two kinds: perpendicular to a surface or parallel to a surface. They can, of course, be applied at intermediate angles, but analysis of a kind presented in Chapter 22 shows that such forces can always be broken down into a perpendicular component and a horizontal component.

Perpendicular forces, in turn, are of two kinds: **Compression** results from force directed *toward* an object. It tends to make the object shorter in the direction of the applied force (strain is then said to be negative) (Figure 21-1). **Tension** results from a force directed *away* from an object. It tends to make the object longer (strain is positive). Columns and pillars withstand compressive force; guy wires and cords that suspend objects withstand tensile forces. Compression and tension occur together, but at right angles to one another.

Forces applied parallel to a surface but in opposite directions cause **shear.** Shear slides one part of a material crosswise to adjacent parts. Scissors cut by shearing. If a closed book is held between the palms of the hands and one cover is pushed or twisted relative to the other, the book is distorted by shear as the pages slip over one another.

COMPRESSION   TENSION   SHEAR

Figure 21-1   THE THREE PRINCIPAL KINDS OF FORCES and the distortions they tend to cause in solid objects.

Equipped with these concepts, let us now consider the strength of the supportive materials of the body. Fresh compact bone (not dry or embalmed or cancellous bone) loaded parallel to its grain has a compressive strength of 1330–2100 kg/cm² (19,000–30,000 lb/in.²). About 170 students would somehow have to stand on a single 1 in. cube of compact bone in order to crush it! This is roughly four times the compressive strength of concrete. Values for cartilage vary, but are lower than those for bone. Tendons and ligaments, like string, merely crumple when compressed lengthwise.

Fresh compact bone loaded parallel to its grain has a tensile strength of 620–1050 kg/cm², or about half its compressive strength. The tensile strength of cartilage is again less than that of bone. Tendon and ligament, however, although softer and lighter materials, have about the same tensile strength as bone.

The resistance of compact bone to shear may be as low as 500 kg/cm² if stressed parallel to the grain, and as high as 1176 kg/cm² if stressed crosswise to the grain. Cartilage, tendon, and ligament have less resistance to shear.

It might seem that resistance of tendon and ligament to tension, and of bone to all forces, are far in excess of demand. So they are for maintaining posture and engaging in moderate activity. In strenuous activity, however, forces exerted on the skeleton by individual tendons may reach several hundred kilograms, and excessive loads, as from a fall, occasionally cause tearing or breakage. It is clear from the above figures that shearing forces would be limiting in the body if they approximated usual compressive and tensile forces. Actually, pure shear is unusual, but bones may be sheared by twisting (i.e., rotating) at the same time they are compressed, and bending (i.e., bowing, or curving) forces, which are very common in the skeleton, combine shear, compression, and tension. The relative magnitudes of the kinds of stresses in the skeleton seem usually to be in proportion to the capacity of bone to withstand them: Compressive forces are largest and shearing forces are smallest. When bones do fail, any of the types of forces may have been responsible, though compressive fractures are least common.

STRESS AND STRESS LINES   It will be easier to understand how the structural elements of the body are constructed for maximum effectiveness after considering the transmission of forces

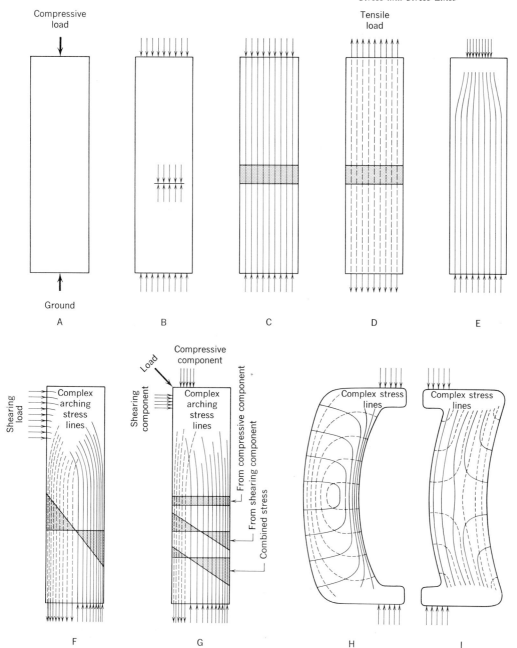

Figure 21-2 DIAGRAMS OF STRESS LINES WITHIN CYLINDRICAL OBJECTS.

within homogeneous objects. When a solid cylinder resting on the ground is compressed by a uniform load, the downward force of the load is opposed by an equal and opposite upward force at the ground. These forces are represented by large arrows in Figure 21-2A. If the opposed forces are depicted instead by many arrows, each representing a unit of force, more information is included because even spacing of the arrows then shows that pressure is uniform over the ends of the cylinder (part B). Within the cylinder, units of stress have the same magnitude and direction as the external pressure, so at any arbitrary plane they can also be represented by arrows, the number of arrows being proportional to the area taken. The lines we draw to represent the paths followed by units of force as they pass through an object are called **stress lines.** In this example the lines are straight and evenly spaced because loading is uniform (part C). They represent only compression and can also be called **compression lines.** The magnitude of compressive stress in the plane of the illustration is proportional to the height of the shaded rectangle (the units being arbitrary). At surfaces of the cylinder, and at vertical planes within, there is no tension or shear. (In this, and the following examples, stress resulting from weight of the object itself is ignored for the sake of simplicity.)

Tensile force applied uniformly to one end of a cylinder or rod, and that which opposes the load at the other end, can also be depicted by arrows representing units of force. Again, straight lines show the paths of the forces within the object, but this time they are **tension lines** (represented by dashed lines in part D). The magnitude of tensile stress is again proportional to the height of the shaded rectangle. This model is closely approximated by stressed tendons and ligaments.

Bones are never cylinders evenly compressed over their ends in the direction of their long axes. If the end of a cylinder is compressed over a restricted area, then adjacent stress is great and is represented by compression lines that are close together (part E). As the lines pass away from the point of application of the load, however, they spread out until they are evenly distributed. The upper, outer parts of the cylinder are devoid of stress, but at the boundaries between stressed and unstressed areas, and also immediately under the load, stresses are complicated (and not figured). This model was approximated by the nearly solid, cylindrical, 2 m long femur of the great 54,500 kg (120,000 lb) dinosaur *Brontosaurus*, but for reasons given below, the long bones of tetrapods are rarely solid cylinders. We progress, therefore, to other applications of the concept of stress lines.

If a load is not applied perpendicularly to the end of a cylinder (as in the above examples) but is instead applied along an upper edge perpendicular to the axis of the cylinder, then the resultant stress must resist bending. Compression lines arch away from the load and come to run lengthwise in the opposite side of the cylinder at some distance away from the load (part F). Tension lines run lengthwise in the side of the cylinder near the load. The configurations of compression and tension lines near the load are complicated and somewhat dependent on the material, and hence are not illustrated. Shear is also present. Note that compression and tension are each greatest at their respective edges of the cylinder (stress lines are closest together there) and each diminishes to zero at the central axis of the cylinder. The stress at any intermediate point between edge and center is proportional to the height at that point of the relevant triangle, as shown in part F. If a tension load replaces the compression load, the pattern of stress remains the same, but the kinds of stresses are reversed. These models are approximated by the force of food against the jaw (compressive load) and by the forces of muscles on opposite edges of the summits of vertical neural spines (tensile load), though for reasons explained in the next section these bones are not cylindrical.

Most loads applied to long bones are neither parallel nor perpendicular to their long axes, but instead are at an intermediate angle, as when one bone loads another across a flexed joint, or a tendon inserts obliquely onto a bone. Such loads can be converted, however, to longitudinal (compressive or tensile) and transverse (shearing) stresses (see p. 414). Part G represents this more general situation when both compression and shear are present. The magnitudes of these stresses in the plane illustrated can be independently represented, respectively, by the heights of a rectangle and a pair of congruous triangles as before. Total stress can be represented by combining these figures as shown. It is seen that compression exceeds tension and that the axis of zero stress is no longer the central axis of the cylinder.

Finally, although the cylinder has provided a conveniently simple model thus far, few bones closely approach cylindrical shape with straight parallel sides. Most large bones have somewhat enlarged ends and curved shafts. Solid models of this general shape can be loaded either by compression or tension tending to bend them further (see part H) or straighten them (part I). The patterns of stress lines in the curved shafts are shown; those near the applications of the loads are intricate and are omitted. Note that stress is distributed throughout each shaft (except for one central focal point in each), but is greatest near the convex and concave edges of the shafts.

Small cavities, notches, and channels all weaken materials by causing local concentrations of stress. It has been shown by Currey that bones are constructed to minimize such loss of strength. It is the shafts, not the ends, that have the greatest strain, and these tend to be very smooth. Canals for blood vessels usually run at an angle to the long axis of a bone, and the lacunae housing osteocytes have their shortest axes at right angles to the long axis of the bone. These configurations reduce the concentration of stress. (The presence of lacunae in cellular bone is probably also advantageous for stopping the spread of microfractures, and microfractures are removed by a constant turnover of bone tissue.) Furthermore, bones tend to have at least small elevations or crests where tendons join them. This causes less concentration of stress within the body of the bone than would otherwise occur.

There are various ways of determining patterns of stress in simple objects such as the kinds described. When models (usually two-dimensional) are cut from photoelastic plastic, like Plexiglas, and then loaded as desired and photographed with polarized light, they show light and dark bands from which stress lines can be calculated. Models of bones have been tested in this way, but results are subject to the criticism that bones are neither two-dimensional nor homogeneous. Strain gauges affixed to objects show the magnitude of local stress. When a hard brittle lacquer is painted onto an object which is then loaded, the lacquer develops microcracks in a pattern that indicates distribution of stress at the surface.

Application of these methods is not easy. Engineers have difficulty determining the patterns of stress in solid, homogeneous objects of irregular but relatively simple shape when single loads are applied. The patterns in living bones are much more complicated. Thus, long bones are hollow, quite irregular in both internal and external contour, and are stressed simultaneously by many external forces that change constantly. It is impossible to accurately calculate detailed patterns of stress in living bones. Diagrams such as the often-published figures of stress lines in the proximal end of the human femur are misleading if not qualified as approximations. Nevertheless, as will be shown, the use and design of the structural elements of the body *do* correlate well with expectation according to their properties, including the general pattern of stress lines under usual loads.

USE AND DESIGN OF STRUCTURAL ELEMENTS

TENDONS, LIGAMENTS, AND CARTILAGES   Tendons transmit the pull of muscles to bones, a function for which their flexibility and tremendous resistance to tension well suit them (Figure 21-3). They consist of tightly packed parallel bundles of collagenous fibers. Tendons that must move appreciably relative to adjacent tissues have sheaths, and in some instances slip through lubricated channels resembling the spaces around movable joints of the skeleton.

The force of contraction of even a large muscle is usually concentrated by its tendon on a small area of the skeleton. This contributes to precision of movement and allows several muscles to act in different ways at about the same place. It is important that tendons enable muscles to move skeletal parts at a distance

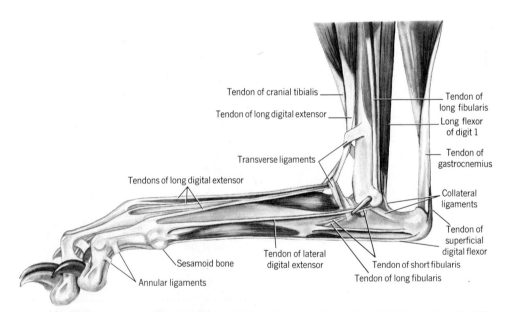

Tendon of cranial tibialis
Tendon of long digital extensor
Transverse ligaments
Tendons of long digital extensor
Sesamoid bone
Annular ligaments
Tendon of lateral digital extensor
Tendon of short fibularis
Tendon of long fibularis
Tendon of long fibularis
Long flexor of digit 1
Tendon of gastrocnemius
Collateral ligaments
Tendon of superficial digital flexor

Figure 21-3   EXAMPLE OF RELATIONSHIPS AMONG TENDONS, LIGAMENTS, AND BONES shown by a lateral view of the left ankle and foot of the gray fox, *Urocyon*. (Drawn from a freeze-dried dissection and hence slightly shrunken.)

from their own positions. Weight distribution and body contours are thus controlled in ways that contribute to speed, endurance, and agility (of which more in subsequent chapters). Human fingers would be useless if encumbered by all of their own muscles.

Some tendons transmit tension around corners at movable joints in the manner of cords passing through pulleys. The living pulley may be a tunnel of bone (as where tendons of digital flexors pass around the proximal end of the tarsometatarsus of some birds), a bony projection forming a channel (as where tendons of abductors of the foot angle at the outside of the ankle of many mammals), or a ligamentous loop (as where tendons of digital extensors turn in front of the ankle). Loaded tendons that are straight sustain only tensile forces, but where a tendon bends, shearing forces also occur. Accordingly, tendons compensate by becoming thicker at such places. If forces other than tension are particularly severe (as where the tendon of the quadriceps muscle passes in front of the knee), small bones, which are better able to withstand such forces, interrupt the tendons. These are called **sesamoid bones.**

Some ligaments have virtually the same structure and properties as tendon, whereas others have less regularly oriented collagenous fibers and contain elastic fibers in various proportions. Because tendons always relate to muscles, they function only when muscles function. Ligaments, by contrast, function passively and are therefore superior where constant tension is required. In the manner of lashings, ties, and elastics, they bind the skeleton together, limit the motion of some joints, and contribute to antigravity mechanisms.

Some ligaments are merely thickened portions of the capsules around movable joints. These merge with adjacent connective tissue and have indistinct margins. Other ligaments that also bind movable joints are tough and prominent. Ligamentous loops and sleeves guide tendons, particularly at joints. Some of these merge into the sheaths of the tendons.

The **nuchal ligament** is an example of an antigravity mechanism. This strong and extensible ligament (i.e., having a low modulus of elasticity) is prominent in large mammals with heavy heads and long necks (Figure 21-4). It extends from the summits of anterior thoracic neural spines to the back of the skull and neural spines of anterior cervical vertebrae. The head and neck are held in normal resting posture without muscular effort. A small muscular tug depresses the head to the ground, simultaneously stretching the ligament. When the muscles relax, the ligament shortens, thus elevating the head.

Ungulates have a suspensory mechanism that cushions their footfalls. Each foot is supported by an elastic ligamentous sling. This mechanism is described on page 452. The claws of cats are passively retracted by elastic ligaments.

Cartilage is found where moderate resistance to compression, tension, and shear (in different combinations according to circumstance) must be combined with firmness and some flexibility. It has the advantage over bone that it is lighter. The various types of cartilage (see p. 118) can be distorted in varying degrees,

Figure 21-4   NUCHAL LIGAMENT OF THE HORSE, an antigravity mechanism.

but all return to their original form when released. Elastic cartilage, the most flexible, supports the external ear, nose, and epiglottis of mammals. Fibrous cartilage provides a tough, but somewhat flexible cushion. It forms intervertebral disks and the pelvic symphysis of some tetrapods. It is also found at the insertions of some tendons. Hyaline cartilage forms the skeletons of embryos and elasmobranchs and parts of the skeletons of adults of many vertebrates. In tetrapods, it may substitute for bone where the greater strength of bone is not needed (e.g., in the carpus and tarsus of salamanders). It also stiffens the trachea and covers the articular surfaces of movable joints where its hardness, smoothness, and slight accommodation under pressure are advantageous. The principal kinds of cartilage may intergrade. Varying degrees of calcification may harden the hyaline cartilage of the sternal ribs of mammals, the epiphyses of amphibians, and the skeletons of elasmobranchs.

BONES THAT RESIST COMPRESSION OR TENSION   In order to save on weight, bulk, and metabolic requirements, the supportive elements of the body are designed to provide adequate strength with minimum material. This principle is important for analysis of the skeleton.

Adequate strength to resist all stresses is provided to a mouse by even a slender skeleton. For reasons explained in Chapter 22, however, very large tetrapods would be unable to sustain the resultant loads if they were proportioned

Figure 21-5 LONGITUDINAL SECTION OF A CENTRUM SHOWING ORIENTA-
TION OF TRABECULAE PARALLEL TO COMPRESSIVE FORCES. The specimen is
a lumbar vertebra of a caribou, *Rangifer*.

like enormous mice. Elephants, various extinct mammalian giants, and many
dinosaurs are (or were) obliged to modify structure, posture, and behavior to
minimize stresses on the skeleton and to resist the remaining stresses effectively.
Figures given above show that bone can sustain more compressive force per unit
of cross section than any other kind of force. An animal could support itself with
the least bone (and weight and bulk) if it could limit all loads to compression.
This is not even possible for a static table, let alone a moving animal, yet the
largest land animals do minimize stresses other than compression in ways that
can be predicted from study of Figure 21-2. The columnlike limbs of such animals
have bones that are relatively cylindrical with heads nearly in line with their
shafts, thus reducing bending forces (see figure on p. 429).

In Chapter 22 it is explained that the vertebral centra of most land tetrapods
are subjected primarily to compressive forces acting on their opposing ends. If
large enough for the necessary muscle attachments and leverages and also solid,
they would be much stronger (and heavier) than needed. They are not solid, but
neither are they hollow. Spicules of bone called **trabeculae** brace the inside of
each centrum, and these are largely oriented lengthwise in the direction of the
predicted compression lines (see Figure 21-5).

The arm bones of gibbons are stressed primarily by tensile forces as the
animals swing under tree branches. Few bones of few animals, however, are
loaded primarily by tension; if constant or frequent tension must be withstood,
a ligament, being as strong for its weight and less likely than a thin bone to
fracture if bent sideways, is substituted. An example is the sacrotuberous ligament
shown in Figure 21-6. Bones are heavily stressed locally by tensile forces where
tendons insert on them. This most often occurs near the ends of bones where
bending forces are also common.

BONES THAT RESIST BENDING IN ONE PLANE   When a solid cylinder resists
a bending force applied in one plane (e.g., the plane of the paper in parts F and
G of Figure 21-2), the resultant stresses are concentrated in that plane and are
greatest at the surface of the cylinder. Accordingly, although a cylinder is useful
for sustaining compression alone, or tension alone, it is not economical of material for resisting bending in one plane: Too much of the material is not stressed
and hence is wasted. The engineer uses instead an "I-beam" as a girder. This
beam has upper and lower bars of steel that are stressed when loaded, and a wall
to hold the bars apart. (Which bar is compressed and which is tensed depends

Figure 21-6 AN EXAMPLE OF A LIGAMENT THAT WITHSTANDS FREQUENT TENSION is the sacrotuberous ligament of the dog, which resists the tendency of the innominate bone to rotate on the sacrum in the direction shown by the arrow when muscles that swing the leg to the rear pull on the ischium.

on the relation of the load to the support (see Figure 21-7). Bones are never designed as simple I-beams, yet the same principle of construction explains the dumbbell-like distribution of material sometimes seen in cross sections of bones.

The carpenter's beam and joist provide another useful analogy. Lumber is used that has rectangular, but not square, cross section, and is always oriented so that the longer dimension is parallel to the load (i.e., usually is vertical). The reason for this is that resistance to bending is equal to a constant times the width of the beam (dimension transverse to the load) times the square of its height (dimension in line with the load): $R = cwh^2$. If one dimension is twice the other, then the beam is about twice as strong on edge as it is flat; if one dimension is three times the other, the beam is three times stronger when on edge. Clearly, the animal body should "know about" this, and it does. The zygomatic arch is a bony beam turned on edge to the muscles acting on it. The same is true of most of the neural spines. The pygostyle of birds is a blade of bone oriented parallel to the air resistance transmitted to it by the tail feathers. The lower jaw at the level of the teeth is a modified beam of expected orientation in relation to loads imparted by teeth and muscles.

Resistance of a beam to bending also varies inversely as the square of its length. For this reason, bony beams are not long; other kinds of construction resist loads that must be held at some distance from the support of the mechanism. The nuchal ligament is one example; others will be described later.

BONES THAT RESIST BENDING IN SEVERAL PLANES We have seen that a flat beam effectively resists bending when loaded edge on, but is weak when loaded flat-side on. The long bones of the appendages of tetrapods must resist

Figure 21-7   BONE STRUCTURE ANALOGOUS TO THE I-BEAM (left) AND JOIST (right).

bending in many directions; hence, they cannot be flat. The cylinder, discarded as wasteful of material when bending forces are in one plane, is here an effective model because compressive and tensile forces can concentrate at opposite edges of the cylinder no matter what the direction of loading. It is still true, however, that stresses are least toward the central axis of the cylinder (see again parts F-I of Figure 21-2). Therefore, the most strength for the least material is achieved by a hollow cylinder. This explains the hollow shafts of long bones.

Controlled bending of the vertebral column of a fish occurs at the intervertebral joints. The centra themselves must resist the bending forces produced by axial muscles. The spine as a whole functions as a somewhat flexible bony tube, the discontinuous cavity of which is the spaces between the markedly amphicoelous centra.

Returning to tetrapods, since resistance of a tube to bending varies inversely

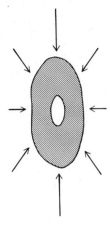

Figure 21-8   CROSS SECTION OF A PHALANX OF A BAT SHOWING STRUCTURE IN RELATION TO USUAL FORCES.

as the square of its length, and since the long bones are stabilized by muscles acting at their ends, they would be most subject to fracture at the centers of their shafts if there were no provision to the contrary. The shafts of such bones compensate by being a little thicker at midlength, by increasing their diameter, or both (see figure on p. 169).

Although long bones are subject to bending forces in various directions, force may be greatest in one plane. The bone then compromises between the beam and tube by becoming oval in cross section, with the long axis of the oval in the direction of the dominant load and with thicker walls on the sides toward and away from the load. The phalanges of bats tend to have this configuration (Figure 21-8). It sometimes happens that stress is greatest along the concave side of a curved bone (Figure 21-2, part H), and the wall of the bone is then thickest there. However, with numerous forces acting on a bone, one cannot assume that the shaft is asymmetrically stressed just because it curves.

It would be impossible for a morphologist to accurately determine the directions and relative magnitudes of the forces acting on a bone during the daily activities of an animal. The amazingly adaptable living skeleton *does* determine the magnitudes and directions of its predominant loads. Bones seem to have built-in sensors that monitor strain and "report" to the mechanisms of bone destruction and growth. When dominant strain is moderate to severe, the bone is slowly remodeled in such a way as to reduce the strain. The nature of the sensing device is not yet clearly understood but apparently involves bioelectric phenomena (piezoelectric effect and semiconductor effect) within the bone.

It is evident from Figure 21-2, parts F-1, that since most muscles insert near the ends of the long bones, and since forces transmitted by one bone to another across a flexed joint are rarely parallel to the shaft of either, stress lines arc across the ends of those bones. It follows that the ends of long bones should not be tubes. Since solid ends would be stronger (and heavier) than needed, the most economical design is a network of interconnecting trabeculae and thin sheets of bone that follow the stress lines, and this is what we find. Stress lines change somewhat as loads change so the body adopts lines that are a compromise of the more usual loads. (The arching of the trabeculae is usually evident, as for the lower trabeculae in Figure 21-9, but rarely are as regular as in the head of the human femur, which is so commonly figured. The same is true of the rather stylized Figure 21-11). Adaptability is again evident. The trabeculae and sheets become oriented as the young animal begins to move about, and their orientation changes if an injury alters the usual loads. The spongy nature of the ends of long bones also provides that they can function as shock absorbers.

Where several bones function as a unit in sustaining usual loads, stress lines, and hence trabeculae, also traverse those bones as a unit. The human tarsus is an example. It functions as a beam loaded in the middle by the tibia (though we shall see later that this is not all of the story).

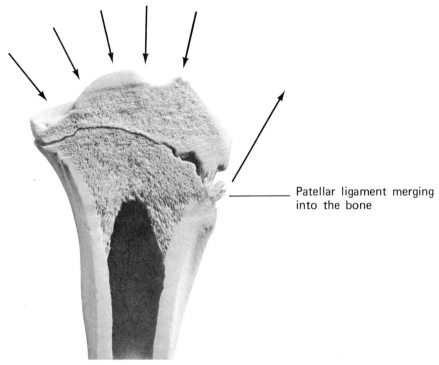

Patellar ligament merging
into the bone

Figure 21-9   LONGITUDINAL SECTION OF THE PROXIMAL END OF A COW
TIBIA SHOWING ORIENTATION OF TRABECULAE IN RELATION TO SOME OF
THE FORCES ACTING ON THE BONE.

TENDON TO MUSCLE; TENDON AND LIGAMENT TO BONE   The tensile
strength of tendons is roughly four times the maximum loads delivered to them
by their respective muscles. The union of tendon to muscle sometimes is a little
less strong than the muscle, but only a little. The tendon may appear to end
where the muscle begins, but it branches and pervades the muscle, its fibers
merging with those of the perimysium and endomysium. In pulling on its own
fibrous framework the muscle also pulls on its tendon.

Muscles that take origin from large areas of bone (e.g., supraspinatus) may
gain sufficiently firm attachment by merging their connective tissue with the
periosteum of the bone. Tendons and ligaments, however, concentrate so much
force on such small areas that a stronger attachment is needed. Imagine the
difficulty of joining with glue the end of a flexible cord to a hard smooth material
with enough strength to sustain loads of 900 kg/cm^2 even though the angle of
attachment changes! Insertions of ligaments and tendons do tend to be weak
points in the bone-muscle system, yet the body surpasses human technology in
solving the problem.

The collagenous fibers of tendons are not attached to bone; they merge into
it (Figure 21-9). Fibrous bone forms all of the skeleton of small animals. Large
animals have osteons throughout most of the skeleton, where compressive forces

**UNIONS OF
STRUCTURAL
ELEMENTS**

dominate, but retain fibrous bone at the insertions of tendons. Fibers of tendons penetrate the bone and there become indistinguishable from its fibers. (Fibrous cartilage may intervene. Also, calcification may merge into a tendon at the tendon–bone junction.)

There remains an apparent source of weakness. Consider a large tendon of circular cross section that inserts at right angles to the surface of a bone. If all bundles of the tendon are of equal length, they share the load equally when tension occurs. If the angle of insertion changes, as would be expected if the tension causes motion, the relative distance from muscle to bone decreases on the side of the tendon now forming an acute angle with the bone and increases on the side forming an obtuse angle. One might expect that fibers on the long side of the tendon would carry all the stress and would give way one by one. This does not happen, primarily because the collagenous fibers of a tendon, although parallel in the body of the tendon, weave at its insertion, thus distributing the load throughout the insertion. Furthermore, the strength of a tendon usually has a large safety factor; not all of it need support the pull of its associated muscle. Fibers of a resting tendon are a little wavy; a pull that straightens them on the long side transmits some tension also to the short side if the difference in length is slight. Moreover, the angle of insertion rarely changes very much.

The elastic fibers of ligaments are largely replaced by collagenous fibers at their insertions. Thus, the insertions of ligaments are like those of tendons, though details for each differ according to size of animal, general angle of insertion, and specific location.

KINDS AND FUNCTIONS OF JOINTS   Joints between bones are classified on the basis of both structure and function, though the two are, of course, related. A first structural category is the immovable joint or **synarthrosis** (= together + joint). The bones may be joined only by connective tissue, which is the rule for membrane bones (e.g., on the roof of the skull), or only by cartilage, which is usual for replacement bones (e.g., at the base of the skull and between shafts and epiphyses of long bones). The cracks between bones joined by synarthroses are called **sutures.** Although "immovable," some sutures are flexible enough to provide some shock-absorption, particularly in response to tension. These joints are places of growth; sutures must remain open for growth to occur. When the growth period terminates, synarthroses of birds and mammals tend to ossify and thus become obliterated one by one on a schedule characteristic of each species. Most sutures of marsupials, however, and some sutures of many other mammals, remain open for life.

Synarthroses are further characterized by the configuration of the suture, and this relates to function. If the suture is approximately straight and the bones have nearly squared-off edges, then a **butt joint** is formed, as between the two nasal bones, and between bones of the basicranium of most mammals (Figure 21-10). Butt joints can withstand compression but little shearing or bending.

If the same square-edged bones were instead joined by overlapping, the union, a **lap joint,** might be somewhat stronger. However, as human builders have learned, when a lap joint is compressed or tensed, the area of contact is not evenly stressed. When glue is used it tends to give way at the leading edges

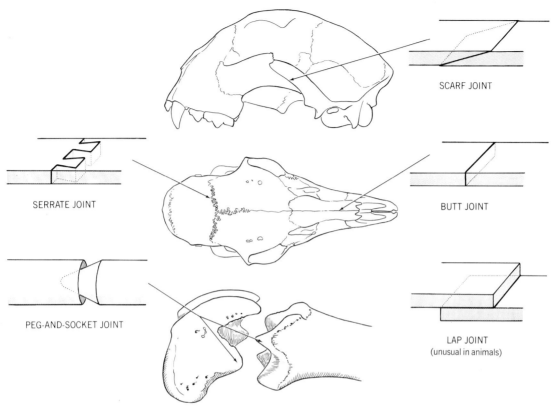

Figure 21-10 SOME KINDS OF SYNARTHROSES. Above, skull of a cheetah, *Acinonyx;* middle, skull of a deer, *Odocoileus;* below, distal epiphysis and shaft of the femur of a young wolf, *Canis.*

and to break toward the middle. If the overlapping edges taper instead so that the two members remain in line, the union is called a **scarf joint.** Here the entire area of contact is evenly stressed by most loads and strength is much improved. Scarf joints (also called squamous joints) often join thin flat bones. They occur between some bones of the mandible of reptiles, and in most vertebrates join various of the cranial bones.

A synarthrosis that is very effective for withstanding compression and shear between hard structures that are not thin and flat is the **peg-and-socket** (also called gomphosis). Such joints join thecodont teeth to the jaw bones and often the jugal to the maxilla. Most epiphyses of long bones join their shafts by complex joints including several pegs and sockets, sometimes relatively deep (distal end of femur of mammals) and sometimes shallow (proximal end of tibia). Another synarthrosis is the **serrate joint,** which has such an irregular suture that the adjoining bones interlock repeatedly throughout the union. This firm type of joint is found between roofing bones of the cranium of some tetrapods, particularly of amphisbaenians (which dig with the head) and artiodactyls (which support horns or antlers). The configuration of the interlocking surfaces relates to the type of force that is resisted.

These various synarthroses may occur in combination and may intergrade.

A second, and intermediate, structural category is the **amphiarthrosis** (= both + joint), which allows some motion in response to compression, tension, or twisting, yet is tough. The surfaces of the adjoining bones may be covered by hyaline cartilages which, in turn, are joined by a pad of collagenous fibers or by fibrous cartilage. The union between the bones of such a joint is called a **symphysis** instead of a suture. Examples are the mandibular symphysis of many vertebrates, the pelvic symphysis, which allows more motion in females toward the end of pregnancy than in males, and the joints between most vertebral centra, which provide for motions of the spine. The function of joints between centra is conditioned by configuration (for terminology, see p. 148 and 149). Joints between procoelous and between opisthocoelous centra allow adequate motion in any direction, withstand compression, and resist dislocation better than platyan centra. Procoelous and opisthocoelous vertebrae are, therefore, common in the necks of tetrapods and in their tails if the tail is strong. Platyan centra are usually restricted to the trunk, where shearing is minimal.

In another kind of amphiarthrosis called a **syndesmosis,** the bones are joined by moderately thick zones of collagenous fibers or by ligaments, and somewhat greater motion is allowed. Examples are the unions of radius to ulna and of fibula to tibia in certain mammals having slight play between these pairs of bones. However, syndesmoses are more characteristic of various other classes. Thus, such joints are common among bones of the protrusible upper jaws and movable opercula of bony fishes.

The last general structural category of joints is the freely movable joint or **diarthrosis** (= two + joint). The articulating surfaces of the bones are covered by extremely smooth hyaline cartilage (Figure 21-11). In fetal life the **joint cavity** develops, which is necessary for movement of one bone on the other. Where not bordered by the cartilage-covered bones the cavity is enclosed by a **joint capsule.** The capsule may be thin and membranous but is usually at least partly tough and fibrous, containing both collagenous and elastic fibers. The capsule is lined by a cellular **synovial membrane,** which is more or less folded. It contains fat cells and, in some joints, fat pads that encroach on the cavity and help cushion its changing configuration as the joint moves. Ligaments binding a diarthrosis may be within or partly outside the capsule, or may be inside the joint cavity, as at the hip and knee of mammals. Tough pads of fibrous cartilage called **menisci** (singular, meniscus) occur inside the cavities of several joints (knee and jaw articulation of mammals) where they seem to firm and guide the turning bones, though their association with these joints and not with others remains a puzzle. Bathing joint cavities is a small quantity of **synovial fluid,** apparently produced by the synovial membrane. This clear or yellowish fluid is similar to tissue fluids and contains mucin. It is more or less viscid according to the joint. Its function is to nourish the hyaline cartilage (which is devoid of blood vessels) and, importantly, to lubricate the joint.

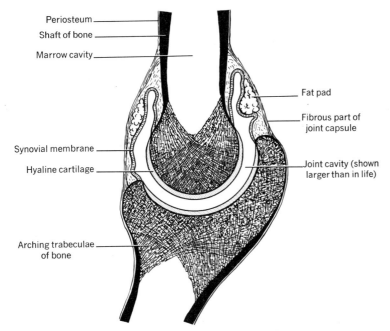

Periosteum
Shaft of bone
Marrow cavity
Fat pad
Fibrous part of
joint capsule
Synovial membrane
Hyaline cartilage
Joint cavity (shown
larger than in life)
Arching trabeculae
of bone

Figure 21-11　STRUCTURE OF A DIARTHROSIS.

No matter how congruously two surfaces may be shaped, and how carefully polished (e.g., two flat pieces of glass), when they rest together they touch only at microscopic elevations. When the dry surfaces slide over one another the microelevations grate, thus producing heat, friction, and wear. If a lubricant is introduced between the surfaces, fewer solid-to-solid contacts are made. Much of the shearing force takes place between molecules of the lubricant, and friction and wear are reduced. This kind of lubrication is used in man-made machinery and is called **boundary lubrication.** It occurs in diarthroses when there is no motion and just as motion starts or stops. Some wear takes place; white cells in the synovial fluid remove from the joint capsule the microscopic fragments of cartilage that result.

When a lubricant can be kept thick enough (about 50 μm) to hold two polished surfaces completely apart, even though they are under considerable pressure, then as they slide over one another the only shearing is within the lubricant—and there is no wear. This is **fluid film lubrication.** To achieve it, the surfaces must not be quite congruous so the lubricant is shaped like a wedge and pressure must force the lubricant into the joint. Engineers have learned to lubricate critical bearings in this way, and diarthroses probably have fluid film lubrication during motion. The moving bones do not quite touch any place and come very close to it at only two or three places at a time. Consequently, healthy joints have negligible friction and remarkably little wear.

Diarthroses are classified according to both function and shapes of articulating surfaces. Function and shape are correlated, but several shapes may serve

Figure 21-12   MECHANICS OF A SNAP JOINT. As the upper bone rotates, hinge-like, from position 1 to position 2 on the lower bone, which is fixed, insertion B of ligament AB moves to B′ in an arc around the pivot of motion, P. Since distance AB′ is shorter than AB, the joint is unstable in position 1.

similar functions, and one general shape may serve several functions. Thus, terminologies tend to be either inadequate or inconsistent. Furthermore, there are intergrades and combinations. Nomenclature does not substitute for interpretation.

A **hinge joint** has a more or less cylindrical head which rotates in a corresponding socket. Motion is primarily around one axis, as for a door hinge. The articulation of the mandible to the skull of carnivores is a simple hinge joint. The head of a hinge joint may have splines, flanges, or bulges that mesh with corresponding grooves or waists in the socket (see figure on p. 454). These devices further limit motion to one plane and resist dislocation. Examples are the elbow joint, ankle joint of many mammals, and joints between phalanges, particularly of runners.

Some hinge joints can further be designated **snap joints** (Figure 21-12). These are stabilized by their ligaments in the open and closed positions and are unstable in intermediate positions. Any hinge joint revolves around an axis which lies within the convex member and is transverse to the plane of motion. Ligaments holding the joint together are lateral in position. If one end of such a ligament inserts exactly at the pivot, then its length remains constant as the joint moves and tension does not change. If, however, a ligament crosses over the pivot to insert slightly beyond, then its length and tension are reduced as the joint moves in either direction from an intermediate position. The joint then snaps into the open or closed positions. Snap joints are found at the elbow and ankle (hock) of various large mammals. The mechanism provides some passive support in the standing posture. In order to better secure the union, snap joints usually have at each side of the hinge either two ligaments that cross or a single broad ligament that twists. (Can you say how the ligaments of a hinge joint could be arranged to snap the joint into a single given position, open, closed, or intermediate?)

A **ball-and-socket joint** has a hemispherical head that turns in a nearly congruous socket. A wide range of motions, including rotation, is implied, and the shoulder and hip joints are the usual examples. The union of the occipital condyle to the atlas of archosaurs can also be cited. A shallow socket (shoulder) allows more excursion than a deep socket (hip). Joints between procoelous and between opisthocoelous centra, and unions of the heads of ribs with their facets, are ball-and-socket in structure, but they have less range of motion and are usually not so designated, even if the joint is a synarthrosis. A modification of the ball-and-socket joint is the peg-and-socket diarthrosis that allows only rotation around the long axis of the peg. This unusual kind of joint joins the avian quadratojugal to the quadrate and functions with the mechanism that moves the upper part of the bill (see p. 562).

Somewhat similar to the last named joint in function is the **pivot joint,** which allows rotation of one bone around its own long axis. During pronation and supination of the manus the proximal end of the radius pivots on the ulna; its disk-shaped head revolves in the radial notch. Likewise, the manus pivots on the styloid process of the ulna.

If the convex head of a bone is biaxial instead of hemispherical, and fits into a biconcave socket, an **elipsoid joint** results. Motion is around two axes (e.g., flexion–extension and adduction–abduction), and motion around the third axis is prevented—very different from a pivot joint. The human radius-to-carpus joint is an example. The same in function, though different in structure, is the **saddle joint** found between the heterocoelous cervical vertebrae of birds (see the pelican in the figure on p. 149). The articulatory surface of the anterior vertebra (i.e., at the posterior end of the centrum) is convex horizontally and concave vertically, whereas that of the posterior vertebra is concave horizontally and convex vertically.

Another family of joints has the name **plane joint.** The articulating surfaces are more or less flat and permit various motions depending largely on the nature of associated ligaments. Contact between the bones may be maintained if they glide over one another as do the pre-and postzygapophyes of vertebrae. Usually the articulating surfaces are not quite flat, so that gliding motions force them apart. Some motions separate flat-ended bones to a surprising degree; the joints between the carpal bones of large mammals are a striking example (see the vicuna in the figure on p. 455).

The patella has a curved surface that slides in the patellar groove of the femur. Lumbar vertebrae of artiodactyls have postzygapophyses that are rolled into scrolls, and prezygapophyses that are trough-shaped. These joints limit some kinds of motions. They are no longer so ''plain,'' yet have no special name.

Still other kinds of joints defy current terminology yet invite attention. Thus, nature has designed the mammalian knee joint without regard for orderly classification: The femoral condyles largely rotate on the platform of the head of the tibia, but they also roll over it. Motion is mostly around one axis, as for a hinge joint, but not entirely so, and there is also slight rotation of the tibia around its axis. Many birds can move the upper bill on the braincase (of which more in Chapter 28). The joint usually consists merely of a zone of thin flexible bone—a type not named in human anatomy texts. Anurans have either a syndesmosis or diarthrosis between the sacral vertebra and the arms of the pelvic girdle. The joint may function as a hinge in the vertical plane or may allow side to side bending. In some frogs these motions are possible to a degree, and the vertebral column can also telescope forward and backward on the pelvic girdle under the control of apposed sets of muscles.

22

Mechanics of Support and Movement

FORCE VECTORS AND THEIR RESOLUTION   Force was defined on page 184. Forces are **vector quantities.** That is, they have both magnitude and direction. Each property is important to the analysis of bone-muscle systems and each can be represented graphically by an arrow called a **vector.** The arrow is usually placed so that its tail is at the point of application of the force, for example, the insertion of the tendon (alternatively, the head of the arrow could be placed at the insertion). The orientation of the arrow represents the direction of the force, and its length represents the magnitude of the force according to any arbitrary scale (e.g., 1 cm = 10 N). In Figure 22-1, part A shows the long head of the triceps muscle inserting on the olecranon process of the mammalian ulna. If the force of contraction is approximated from the cross-sectional area of the muscle, or better, by direct measurement from the live muscle, then the force of contraction can be represented by the vector $\mathbf{F}_1$ in part B.

The medial head of the triceps (part C) inserts at the same place by the same tendon and can similarly be represented by the vector $\mathbf{F}_2$ of part D. (If the second muscle inserted near the first but not by a common tendon, the forces could still be considered to act at a common point if extensions of their lines of action intersected.)

Having two muscles pulling on the same point but in different directions and with different tensions, it becomes important to learn the magnitude and direction of their common, or net effect. What, we ask, is the vector of the single hypothetical muscle which, acting alone, would load its insertion in just the same

Figure 22-1 FORCE VECTORS AND THE ADDING OF FORCES.

way as the actual muscles acting together? The desired force is called the **resultant** of the given forces and its derivation is called the adding of forces. The resultant is usually determined graphically by drawing a **parallelogram of forces** as shown in part E: F_1 and F_2 form two sides of a parallelogram that is completed by drawing dotted lines. The diagonal, R_1, is the desired vector of the resultant force.

Since opposite sides of a parallelogram are equal, an alternative graphical solution is to place F_2 so that its tail is at the head of F_1 (or vice versa), and then to draw the line that makes the third side of a triangle (i.e., the triangle that is half of the parallelogram constructed above). That third side is again the desired vector, R_1 (see part F).

A third muscle, the anconeus, also inserts at the same place (part G), its vector being F_3. To determine the resultant force, R_2, of all three muscles contracting simultaneously, one can either draw the diagonal of the parallelogram having vectors R_1 and F_3 as sides, or one can close the triangle having R_1 and F_3 as sides (parts H and I). Either method is a two-step solution, since R_1 had first to be determined from F_1 and F_2. The problem can also be done in one step by

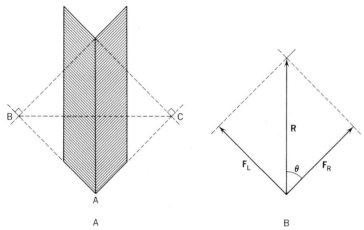

A   B

Figure 22-2 APPLICATION OF THE BREAKING DOWN OF FORCES TO A STYL-
IZED, FLAT, PINNATE MUSCLE.

joining vectors $\mathbf{F}_1$, $\mathbf{F}_2$, and $\mathbf{F}_3$ tail to head (in any sequence) and then closing the
polygon (part J). The closing line is again $\mathbf{R}_2$, or the vector of the resultant force
of all three muscles acting together.

The magnitude and direction of a resultant force can be calculated somewhat
more precisely, if need be, by trigonometric methods.

The determination of the magnitude and direction of the force of contraction
of a pinnate muscle provides an application of the adding of forces. Consider the
stylized, flat pinnate muscle shown in Figure 22-2 A. It can be regarded as two
muscles pulling in different directions on a common central tendon. The force
of contraction of each side taken alone is in the direction of the fibers of that
side and has a magnitude that is roughly proportional to the cross-sectional area
of all the fibers of the side (or to the distances AB or AC if the muscle is of
uniform thickness). Vectors $\mathbf{F}_R$ and $\mathbf{F}_L$ of part B represent the forces of the right
and left sides of the muscle. Completing the parallelogram of forces, $\mathbf{R}$ is found
to be the vector of the resultant force. (Alternatively, $\mathbf{R} = 2\mathbf{F} \cos \Theta$.) The value
of $\mathbf{R}$ is greatest when the muscle fibers of each side insert on the central tendon
at an angle of about 45°. Note, however, that the angle of insertion must change
as the muscle shortens.

In order to have the same force of contraction as this pinnate muscle, a
straplike muscle with parallel fibers would need to have the much greater width
BC (Figure 22-2A). Such a muscle having the same overall length as the pinnate
muscle would have much longer fibers, however, and could move its insertion
farther. Clearly, a pinnate muscle can develop great force for its overall width,
but has a short contraction distance. Pinnate muscles, unlike muscles with fibers
parallel to their tendons, do not become wider during contraction—an advantage
if the muscle must function in a confined space. Furthermore, pinnation allows
a muscle to have an irregular shape.

The central tendons of some pinnate muscles are ossified; examples are
found in the "drumstick" of the turkey. Similar splints of bone are found in the

neck and jaw muscles of certain birds and were present in the backs of some
large dinosaurs. Among mammals, pinnation is seen in most of the flexors of the
limbs and in the mylohyoid. The human deltoid is a complexly pinnate muscle
as is the subscapularis of some mammals (see figure on p. 475).

The reader may care to work out the direction of the resultant force when
a fan-shaped muscle, such as the pectoralis, trapezius, or latissimus dorsi, contracts as a unit (which it does not necessarily do). If a magnitude is assigned to
the force of each of several parts of such a muscle, then the magnitude of the
resultant force can also be determined.

COMPONENTS OF FORCES   Just as two or more forces can be combined into
one resultant force, so a given force can be broken down into two or more
components. And just as an infinite series of pairs of forces can have the same
resultant force (Figure 22-3), so conversely a given force can have an infinite
number of pairs of components. However, it is usually desirable to specify the
direction of each desired component, and there is then only one solution.

Consider the vector of the mammalian triceps muscle (one head or all heads
in combination) as shown in Figure 22-4A. When the muscle contracts, the ulna
turns counterclockwise on the humerus. The insertion of the muscle swings in
an arc around the pivot of motion. The radius of the arc is the distance from the
pivot to the insertion. At any instant in time the direction of motion of the insertion
is in the direction of the tangent to the arc that passes through the point of
insertion. (A tangent of a circle touches the circumference of the circle at one
point and is perpendicular to a radius drawn to that point. At successive instants
in time the tangent will change as the joint moves because the insertion will
move along the arc.)

Since the insertion usually does not move exactly in the direction of the pull
of the muscle, it is important to learn the magnitude of the part, or component
of the pull that *is* in the direction of motion. Two components must be selected
such that their resultant is the given force of the triceps. One component will be

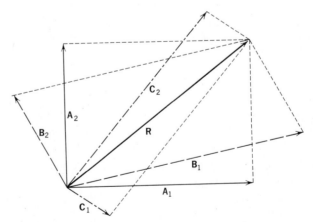

Figure 22-3   THREE PAIRS OF VECTORS, A_1 and A_2, B_1 and B_2, C_1, and C_2, ALL
HAVING THE COMMON RESULTANT, R.

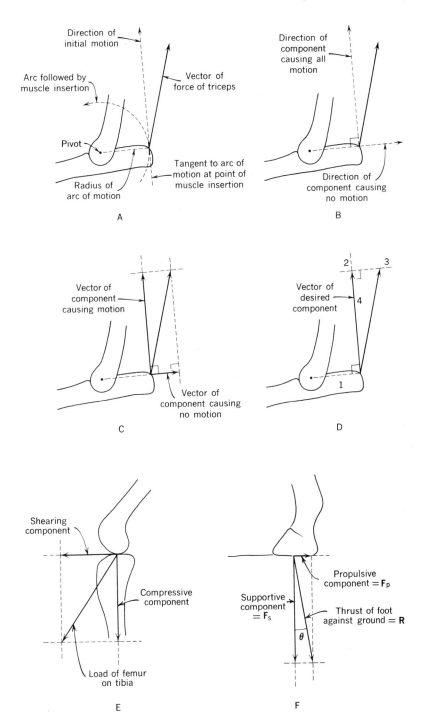

Figure 22-4 DETERMINATION OF A DESIRED COMPONENT OF A GIVEN FORCE.

selected to include all the force in the direction of motion, and, as just explained, this must be along the tangent passing through the insertion. The other component must be selected so as to cause no motion at all, either clockwise or counter-clockwise. Only one direction fits this requirement and that is normal (or perpendicular) to the arc of motion at the point of insertion (see part B). (The vector of this component is an extension of the radius of the arc at the insertion.) Having the directions of the two components, it is simple to derive their magnitudes graphically by completing the parallelogram of forces that has the vector of the force of the triceps as a diagonal. Since the components have been selected to be perpendicular to one another, the parallelogram is a rectangle (part C).

Because the component in the direction of motion is the only one of concern in this instance, and because the desired parallelogram has only right angles, the problem can be solved more directly as follows (part D): (*1*) Draw a line from the estimated pivot of motion to the insertion of the muscle. This is the radius of the arc of motion. (*2*) Draw the line perpendicular to this radius that passes through the point of insertion. This is the tangent giving the direction of the desired component. (*3*) Draw the line perpendicular to the tangent that extends from the tangent to the tip of the vector of the given force. This line is the side of the rectangle of forces that is opposite to (and therefore equal to) the component of force selected to cause no rotation of the ulna on the humerus. (*4*) The vector of the desired component in the direction of motion is now that part of the tangent between the point of insertion of the muscle and the intersection with the perpendicular drawn in the previous step. It only sounds complicated; one can draw the diagram in less time than it takes to read about it.

In Chapter 21, it was noted that a load applied diagonally to the end of a long bone can be converted to a longitudinal compressive component and a transverse shearing component that tends to bend the bone. The directions of the desired components are determined by the nature of the problem and are at right angles to one another. Completion of a rectangle of forces establishes their magnitude (part E).

When the foot of a running animal thrusts against the ground, it both supports and propels the animal. The direction of the supportive component is vertical (in opposition to the pull of gravity) and the direction of the propulsive component is horizontal in the line of travel. If the thrust of the foot at a given instant can be determined, then it is easy to solve for the magnitudes of the components (part F). Similarly, one can calculate the components of an applied force that are necessary for analyzing centrifugal force (see figure on p. 456), lateral undulation of snakes (see figure on p. 481), friction (see figure on p. 491), and the forward and lateral components of the diagonal thrust of the tail of a fish against the water (see figure on p. 515).

Components of forces can be determined trigonometrically as well as graphically. In this instance the calculations are more simple and direct than for the resolution of forces because only right triangles need be used. Thus, in Figure 22-4F, $\mathbf{F}_p = \mathbf{R} \sin\Theta$ and $\mathbf{F}_s = \mathbf{R} \cos\Theta$.

Note that as the angle between the given force and a component increases from 0 to 90°, the magnitude of the component decreases from that of the given force to zero. In terms of mechanics, muscles are most effective for pivoting bones when they pull in the direction of motion. By inserting onto its central tendon at an angle, a pinnate muscle increases the number of its fibers and hence its force of contraction. However, as the angle of insertion increases, the effective component of the force decreases. The optimum angle (usually a little less than 45°) represents a compromise between these opposing factors.

A **machine** is a mechanism that transmits force from one place to another, usually also changing its magnitude. Thus, when a screwdriver is used to pry the lid off a can, moderate downward force applied to the handle produces great upward force at the tip of the tool against the lid. Similarly, when the triceps muscle pulls up on the olecranon process, a downward force is produced at the forefoot (Figure 22-5, parts A and B). All bone–muscle systems are machines. It is useful to designate any input force applied to a machine as an **in-force** (F_i) and any output force derived from a machine as an **out-force** (F_o). In the body, in-forces are applied by the pull of tendons or tensed ligaments, by gravity, and by external loads; useful out-forces are ultimately derived at the teeth, feet, digits, and else-

BONE-MUSCLE SYSTEMS AS MACHINES

Figure 22-5 PRINCIPLES OF IN-FORCES AND OUT-FORCES, LEVER ARMS, AND TORQUE. (In- and out-torques are not adjusted to be in equilibrium.)

where. For now, we will consider only simple machines having one in-force and one out-force.

LEVER ARMS AND TORQUES   An in-force may be transmitted to an out-force by a crankshaft, hydraulic device, pulley, lever, or other mechanism. Most feeding and locomotor systems of the body transmit forces by levers, and only these will be considered here. A **lever** is a rigid structure, such as a crowbar or bone, that transmits forces by turning (or tending to turn) at a pivot. Each force is spaced from the pivot by a segment of the lever called a **lever arm;** the **in-lever arm** (I_i) (or power arm) extends from the in-force to the pivot, and the **out-lever arm** (I_o) (or load arm) extends from the pivot to the out-force. In the example of the screwdriver used to pry open a lid, I_i extends from the hand on the handle to the lip of the can, and I_o extends from the lip to the tip of the tool pressing on the lid. In the other example, I_i is the olecranon process and I_o the forearm from elbow joint to forefoot.

The product of a force times its lever arm is called a turning force, or moment, or **torque** (τ). Every functioning lever includes at least two torques, one for the in-system and one for the out-system. Thus (as qualified below), $\tau_i = F_i I_i$ and $\tau_o = F_o I_o$. Torques are expressed in dyne-centimeters, or equivalent units (or in gram-centimeters if one substitutes weight units for the more precise force units). When $F_i I_i > F_o I_o$, the level rotates in the direction of F_i; when $F_i I_i < F_o I_o$, the lever rotates in the direction of F_o; when $F_i I_i = F_o I_o$, the system is in equilibrium and there is no motion.

When a lever system is in equilibrium, any of the variables can, of course, be easily obtained if the other variables are known. This is the principle of a beam balance: The product of a known weight times its lever arm (the calibrated scale) is adjusted until it equals the product of the unknown weight times a fixed lever arm (Figure 22-5C). Likewise, when tetrapods stand, the forces of all postural muscles are adjusted so that in-torques equal out-torques and support is maintained without motion. Other examples are given later in this chapter.

It is important that the student of body mechanics be able to solve the equation $F_i I_i = F_o I_o$ for any of the variables and to understand the relation of each to the others. Thus, if it is desirable for a mammal that digs to produce a large out-force at the forefoot when the triceps contracts, then since $F_o = F_i I_i / I_o$, it is seen that the animal can increase the out-force by increasing F_i or I_i, or by decreasing I_o. The adaptations of many diggers include all of these (compare B with D in Figure 22-5).

ACTUAL VERSUS EFFECTIVE FORCES AND LEVER ARMS   There is one further qualification. The product of force times lever arm is only equal to the torque of the system when the force and lever arm are at right angles to one another. There are two ways to ensure that this condition is met. Each is simple and either may be the easier to apply to a dissection or experimental animal, so each will be presented.

The actual length of the lever arm of an in-force is the straight-line distance from the insertion of the relevant muscle to the pivot of the motion caused by contraction of the muscle. This is called the **actual lever arm** (I_a). As noted in

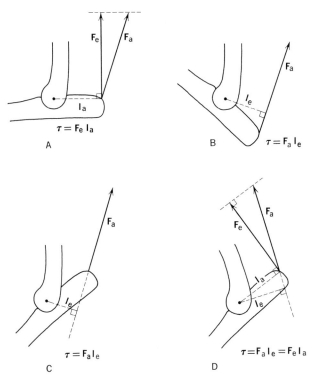

Figure 22-6 USE OF ACTUAL AND EFFECTIVE FORCES AND LEVER ARMS IN CALCULATING TORQUE.

the section on components of forces, this line (a radius of the arc traveled by the insertion as it moves) is at right angles to the effective component of force that is in the direction of motion. If we call the force of the muscle the **actual force** (F_a) and the force of the effective component of the actual force the **effective force** (F_e), then the required condition is met when $\tau = F_e l_a$ (Figure 22-6A). (To indicate that these variables relate to the in-torque, we can write $\tau_i = F_{ie} l_{ia}$.) Using this method, one must always calculate the effective force before calculating the torque.

The alternative method uses instead the actual force, regardless of the relation of its line of action to that of its effective component. The appropriate, or **effective lever arm** (l_e), is then the perpendicular extending from the line of action of the muscle to the pivot. This perpendicular strikes F_a short of the insertion when the angle between F_a and l_a is acute; it strikes a projection of F_a beyond the insertion when the angle between F_a and l_a is obtuse (contrast B and C of Figure 22-6). Now $\tau = F_a l_e$, and also for any given position of the joint, $F_e l_a = F_a l_e$. The two methods are compared in part D. Can you position the forces so $F_a = F_e$ and $l_a = l_e$?

RELATIONS OF IN-FORCE TO OUT-FORCE Thus far my examples have shown the pivot to be located between the in-force and out-force. The torques are then

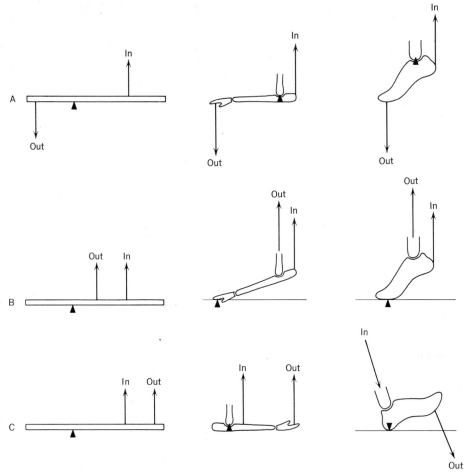

Figure 22-7   THREE WAYS THAT IN-FORCES AND OUT-FORCES CAN BE AR-RANGED IN LEVER SYSTEMS.

in different directions, one clock-wise and one counter-clockwise around the pivot. This relationship is common in the body and is usual with extensor muscles (see Figure 22-7A). Two other arrangements are possible, each having the in- and out-force on the same side of the pivot and turning in the same direction: either the out-force can be closer to the pivot than the in-force (a less common arrange-ment, but two examples are figured), or it can be farther from the pivot than the in-force (a usual arrangement with flexor muscles as shown by part C). Note that in the first arrangement (or first-order lever), the lever arms are independent and that either can be the longer, though in the body, the out-lever arm is usually much longer. In the other arrangements the longer lever arm (which is the in-lever arm for second-order levers and the out-lever for third-order levers) includes all of the shorter lever arm—they "share" part of the lever.

Figure 22-7 shows that one joint and one muscle can function as different kinds of levers. When one walks in sand, the foot (to one's distress) even functions as two kinds of levers at the same time. The important thing is not to memorize diagrams but to learn to identify the pivot, in- and out-forces, and in- and out-levers in any specific bone-muscle system.

SUMMATION OF TORQUES; TWO-JOINT SYSTEMS   More than one in-force may tend to turn the same lever. In order to determine the net effect, one determines the torque of each independently, adds all that tend to turn the lever in one direction and subtracts the sum of all (if there are such) that tend to turn it in the other direction. Multiple out-forces, as when several teeth simultaneously crush food, are treated in the same way. In strychnine poisoning all muscles contract, but the limbs are held in rigid extension because in general, extension torques exceed flexion torques.

If the weight of a part of a living machine is itself to be considered, as it should be in making an accurate calculation, then in addition to torques resulting from the action of muscles, one must allow for the torque from the pull of gravity on the lever (e.g., forearm or thigh). The lever responds to gravity as though all its weight were concentrated at its center of gravity. The weight of the lever times its effective lever arm, calculated using the center of gravity, is the desired torque. The center of gravity of an irregularly shaped object is difficult to locate by calculation but is easy to locate experimentally if the object can be isolated: The object is suspended (e.g., by a string) successively from two points on its surface. The point where extensions of the lines of support intersect is the center of gravity.

Two-joint muscles, which pass over two diarthroses, are common (e.g., gastrocnemius, gracilis, biceps, long head of triceps). Each can move either joint, both, or neither, depending on the actions of other muscles and loads. In any event, contraction moves, or tends to move, both joints, so the muscle simultaneously provides in-force to two lever arms. Tendons of some digital flexors pass over three or more joints. Such systems are virtually impossible to analyze in detail, either theoretically or experimentally, and even for one instant in time. However, practical approximations of the mechanics of a unit of activity (a digging stroke, a running stride) have been undertaken by monitoring the principal out-forces, learning the approximate strengths and activities of relevant muscles (by electromyography, by preparing length–tension curves, and in other ways), determining usual postures (by various methods including x-ray cinematography), and measuring lever arms. We should not expect anything so wonderfully complex as the moving body to lend itself to simple analysis.

BALANCE AND COUNTERBALANCE   The mechanics of the support of the trunk of tetrapods has long been a subject for speculation and analysis. Various kinds of bridges have been proposed as analogs. The curved spines of many small mammals resemble the arch bridge (see Figure 22-8A). This analogy is near the truth (see below) but fails because the spines of many large animals do not arch. The truss bridge, which has a rigid superstructure (part B), fails as an analog if the spine is likened to the roadbed; the spine is in fact under longitudinal

MECHANICS OF BODY SUPPORT

Figure 22-8 TYPES OF BRIDGES PROPOSED AS ANALOGS FOR THE SUPPORT OF THE BODY.

compression, whereas such a roadbed is under tension. If instead, the spine is likened to the top profile of a truss bridge and the abdominal musculature is likened to the roadbed, then the distribution of stress would be correct, but the rigid framework of braces would be missing. This suggests the analogy of a closed oblong bag (the trunk with its enclosed coelomic cavity) filled with fluid. Such a bag could be supported at its ends and would have peripheral stresses like those of the truss bridge. However, if pressures were sufficient to prevent sagging, the living bag would not be adequately flexible.

The analogy of the cantilever bridge (part C) has often been cited. The roadbed, which is this time under compression, is equated to the spine, and the top profile of the bridge is equated to the long epaxial muscles. The criss crossing braces of the bridge are equated to the neural spines and short epaxial muscles. It is here, as the Dutch morphologist Slijper and the German morphologist Kummer have pointed out, that the analogy partly fails, because the appropriate structures are not constructed to provide enough rigidity for constant support. It does not follow that no support can be provided in this way when muscular activity contributes tension.

Another characteristic of the cantilever bridge does apply to the support of the body of many tetrapods. The functional unit of a cantilever bridge is from midspan to midspan. During construction, the overhanging cantilevers are built out equally on each side of each tower. (A cantilever is a projecting member that is supported at only one end). Thus, the load on one side is counterbalanced by the load on the other. In different terms, the mass of one cantilever times the distance of its center of mass from the tower is a torque that is equal in magnitude, but opposite in direction, to the similar torque of the cantilever on the other side of the tower, hence there is equilibrium and the tower does not fall. The principle of balance and counterbalance is widely used by tetrapods (Figure 22-9).

Figure 22-9 THE PRINCIPLE OF BALANCE AND COUNTERBALANCE IN THE SUPPORT OF THE BODY.

BOWS AND BOWSTRINGS FROM BONES AND FIBERS Another analogy for the support of the body, proposed in 1798 but neglected until presented by Slijper in 1946, is more nearly valid than any of the bridge analogies. It equates the trunk vertebrae to a bow, which may be either curved like the archer's bow (many small mammals) or nearly straight with the ends bent down like the violinist's bow (salamanders, crocodilians, lizards, many large mammals) (see Figure 22-10, parts A and B). The principal string of this living bow is the ventral abdominal musculature, scalenus and related muscles, and intervening sternum. The psoas and quadratus muscles under the lumbar vertebrae form a shorter secondary bowstring. The spine cannot sag if tension is maintained in these bowstrings.

The cervical vertebrae of the larger mammals form another bow, this time inverted. The strands of the nuchal ligament form multiple bowstrings. The ligament can be stretched by muscles, thus decreasing the arch of the bow to depress the head.

All the analogs proposed have the shortcoming that they are passive, static systems, whereas the body is active and dynamic. The trunk cannot remain balanced on the limbs as a bridge is balanced on piers; the length and orientation of animal cantilevers are subject to change; the loads on animal bows and bowstrings shift in magnitude and direction.

Furthermore, there has been too much tendency for morphologists to choose

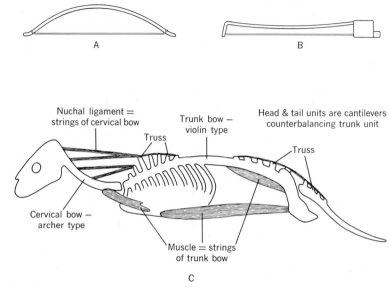

Figure 22-10   BOW ANALOGS AND COMBINED FACTORS IN THE SUPPORT OF THE BODY.

among the analogs instead of combining them. The trunk can be regarded as having a bow, the spine, which is prevented from collapsing by muscular bow-strings which, in turn, may be relieved in part by the counterbalance of a trussed cantilever (a heavy, extended tail) and in part by the counterbalance of an in-verted-bow cantilever (the neck) loaded at its extremity by the head (Figure 22-10C). We may assume that when a tetrapod stands at rest, this complicated system remains in equilibrium with only slight sustained tension in a few muscles and intermittent corrective tension in some others.

On page 402 the human foot was likened to a beam loaded at the center. So it is, but it is also a bow, touching the ground at the ball of the foot and the heel, that is prevented from collapsing by muscles and by a bowstring (the plantar aponeurosis) that can be tightened by lifting the toes. On page 400 the zygomatic arch was likened to a beam, and so it is, but it is often also an arch, firmly anchored at each end, and strengthened by a bony string, the basicranial axis, and sometimes also by the orbital ligament that suspends the arch from the postorbital process above.

STOPS, SLINGS, AND LOCKS   The factors contributing to support that have thus far been presented relate primarily to the axial part of the body. Since the legs are not rigid vertical columns, but instead are hinged, and usually bent struts, it is important to learn how they can support the body. During periods of activity, muscles provide support as well as motion. During periods of inactivity the an-imal may avoid a support role for the legs by resting sprawled on its ventral surface (amphibians, reptiles), by crouching (rodents, rabbits), by sitting (primates), or by lying down (carnivores, many artiodactyls). Nevertheless, some large tetrapods stand for long periods; their limbs must then support them without excessive muscular effort. Various factors contribute.

As described in Chapter 21, joints, together with their ligaments, are constructed to limit both the kind and extent of motion that can occur. In large mammals, no joint distal to shoulder and hip (and these to only a limited degree) can flex in the transverse plane; passive support is provided against lateral bending. Also, hyperextension in the sagittal plane is prevented by either ligamentous or bony stops: Elbows, knees, hocks, and digits cannot collapse by "bending backwards."

The tendency to collapse by flexion may be reduced or avoided in any of several ways other than muscular contraction. The largest tetrapods (living and extinct) stand (or stood) with limb segments vertically aligned, or nearly so, thus balancing one bone on the next and reducing bending torques (of which more in the next section). Ungulates have more or less cuboidal podials and flat-ended metapodials, which balance over one another in rows like a child's building blocks. During flexion of the "knee" (corresponding to the human wrist) the squared-off carpals separate by pivoting, hingelike, on their posterior margins (see vicuna in figure on p. 455). When weighted, this may lift the center of gravity very slightly. If so, a little work must be done to initiate flexion, which is thus resisted when the animal stands at rest (this analysis needs verification).

The human knee (and probably some upright limb joints of other animals) can be extended just beyond vertical. When weighted, there is then a slight torque in a direction in which the joint cannot bend further. This prevents bending in the opposite direction and provides the standing body with a lock against flexion. Some resistance to flexion is also provided by snap joints (see p. 408 for a description). This resistance may help to prevent collapse of a joint that is otherwise nearly in equilibrium.

Ungulates also have a sling mechanism that prevents collapse of the fetlock joint between metapodial (or cannon bone) and proximal phalanx. This joint is always bent when weighted; the angulation depending on the load. It is thus an important shock absorber. Because it is bent, it must be supported. This is accomplished by a remarkable ligamentous sling that extends from the posterior surface of the proximal end of the metapodial, down under the fetlock joint (where it is anchored to sesamoid bones), and then around the proximal phalanx to insert on the anterior surfaces of the distal phalanges (Figure 22-11). As the foot is heavily loaded by shifting weight when at ease, or by impact when moving, the ligament stretches, thus allowing the joint to flex sharply. In doing so, the ligament stores potential energy that is released by returning the joint to a neutral angulation when the load is diminished (see also figure on p. 456).

Ungulates, unlike elephants, usually have flexed stifle joints (corresponding to the human knee) and moderately flexed hock (or heel) joints. Passive support to relieve extensor muscles is provided to perissodactyls and some artiodactyls by a mechanism that is effective, simple, and subject to voluntary control. The tibia on one side, and the almost completely tendinous superficial flexor muscle of the digit (or digits) on the other side, form the long arms of a parallelogram that is completed above by the distal end of the femur and below by the calca-

Attachment of suspensory ligament to cannon bone

Check ligament — prevents hyperextension of flexor tendon

Suspensory ligament

Tendon of superficial digital flexor

Tendon of deep digital flexor

Anchor of suspensory ligament to sesamoid bone

Tendon of lateral digital extensor

Annular ligament

Tendon of deep digital flexor

Cruciate ligament

Figure 22-11 STRUCTURE OF THE RIGHT FOREFOOT OF THE COW, INCLUDING THE SUSPENSORY MECHANISM. (Drawn from a freeze-dried dissection and hence slightly shrunken.)

neum and other tarsal bones (Figure 22-12). All angles of this parallelogram must change simultaneously. The stifle and hock cannot flex independently, and if any of the angles is prevented from changing, the entire system remains rigid. This is accomplished by a locking device at the stifle. The ridge flanking the patellar groove on the medial side is enlarged and ends proximally in an eminence. The patella is anchored to the tibia by several strong ligaments that just permit the patella to be pulled (by the quadriceps muscle) behind this eminence when the joint is extended. When the patella is thus seated, the ligaments prevent flexion of the joint; the patella must be pulled sideways and down into the patellar groove before the limb can be flexed.

These support mechanisms are representative. Numerous others are known (some that relate to specific adaptations will be described in subsequent chapters) and doubtless still more await identification.

VERTEBRATE GIANTS   An evolutionary trend toward large body size has been common within vertebrate lineages, and as the body weight of a terrestrial animal

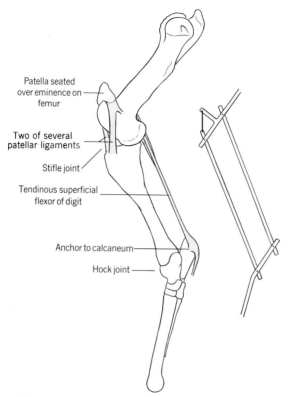

Patella seated
over eminence on
femur

Two of several
patellar ligaments

Stifle joint

Tendinous superficial
flexor of digit

Anchor to calcaneum

Hock joint

Figure 22-12 SUPPORT MECHANISM OF THE HIND LEG OF MANY UNGULATES shown by a medial view of the right leg of a horse.

approaches 900 kg (2000 lb), marked adaptations for support become necessary. Beasts having such adaptations are said to be **graviportal** (= heavy + to carry).

The selective advantages of large size include (*1*) virtual freedom from predation (except, alas, by man); (*2*) the ability to roam over large areas in search of food, drink, shelter, or breeding grounds; (*3*) the capacity to use and produce energy more slowly than small animals, so that relatively little food is required per unit of body weight and so food of low nutritive quality may be sufficient; and (*4*) a low surface-to-volume ratio and a high capacity for heat production, enabling the larger animals to heat up and cool off slowly.

The largest arthrodire was 9 m (30 ft) long. The largest living fish is the tropical whale shark, which reaches about 18 m in length. A Siberian freshwater sturgeon is said to have reached 1360 kg (3000 lb).

Gigantism has not characterized the amphibians, but the largest labyrinthodont was thickset and about 4.5 m long. There have been giants in many reptilian lineages: An extinct marine lizard (a mosasaur) was 10 m long. Plesiosaurs reached 9 m in length and were quite bulky. The longest snakes probably reached 12 m. The largest reptiles were archosaurs. A crocodile attained about 13 m, and many dinosaurs of the order Saurischia were gigantic: *Diplodocus,* with its long neck and tail, was probably the longest land animal at 26.6 m (87.5 ft); *Brachiosaurus,* which was almost as long but stouter, may have been the heaviest at an estimated 78,000 kg (86 tons). The largest bird was the extinct elephant bird, which is estimated to have weighed 436 kg—more than three times as much as the ostrich.

Bull elephant seals have reached 3500 kg, and bull walruses weigh as much as 1260 kg. Four extinct mammalian orders (Amblypoda, Embrithopoda, Notoungulata, Astrapotheria) included bulky land herbivores having the proportions of rhinoceroses or elephants. Steller's sea cow, exterminated by man, is said to have reached 7.5 m in length and 4000 kg in weight.

The hippopotamus rarely attains 4.5 m and 4500 kg. A bull giraffe may weigh 1900 kg. The living Indian rhinoceros reaches 4000 kg, and the extinct, hornless *Baluchitherium* was the largest of all land mammals—it stood about 5.3 m (17–18 ft) high at the shoulder. The bull African elephant, largest of living land animals, stands 4 m high and weighs as much as 7000 kg (7.7 tons). Human greed has nearly exterminated the largest of all vertebrates, the blue whale, and few large specimens remain. This magnificent creature reached 23–30 m (75–100 ft) A 27 m specimen weighed 136,400 kg (150 tons).

Aquatic giants support their bodies effortlessly by flotation. Terrestrial giants reduce their requirements for support by avoiding unnecessary oscillations (the spine is usually stiff), jolting (they rarely jump or gallop), and exertion (the limbs are little flexed in locomotion). Furthermore, they reduce demands on their skeletons by modifying structure and posture so that compressive forces, which can be withstood with minimal tissue, are increased and bending forces are decreased. Limb bones are oriented vertically (Figure 22-13). Their columnar shafts are straight, and articulatory surfaces are in line with the shafts. The olecranon process is deflected backwards so that the elbow joint can open completely. The heads of the humerus and femur face more nearly upward than for smaller tetrapods, and the acetabulum faces more nearly downward. Even the scapula and ilium tend to be broad and vertically oriented. Proximal limb segments are long; distal segments are short. Radius and fibula are large and free. The feet are broad. Usually, all five toes are retained, are about equal in length, and radiate over a pad that cushions and evenly distributes the great weight.

These modifications shift part of the support role from muscle to bone. Ligamentous slings and guys, and modifications of muscle mechanics to favor force and reduce velocity, contribute further to economy of effort.

MECHANICS OF MOTION

VELOCITIES AND LEVER ARMS Movement is change of position. Speed is rate of change of position. **Velocity** (**v**) is speed in a given direction. An animal may be said to be able to swim or run at a stated speed because it can move in any direction; muscles and bones, however, are said to attain certain velocities because their directions of motion relative to associated structures are always restricted. Velocity is expressed as centimeters per second, or equivalent units, and direction is stated or implied. Velocity, like force, is therefore a vector quantity and can be represented by an arrow (see p. 411).

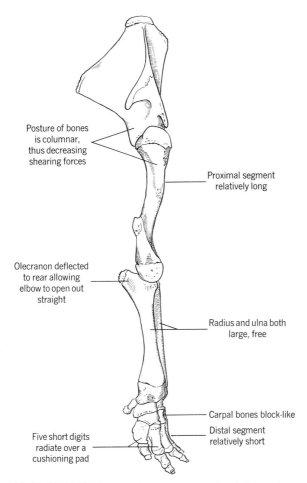

Posture of bones
is columnar,
thus decreasing
shearing forces

Proximal segment
relatively long

Olecranon deflected
to rear allowing
elbow to open out
straight

Radius and ulna both
large, free

Carpal bones block-like

Distal segment
relatively short

Five short digits
radiate over a
cushioning pad

Figure 22-13　SOME GRAVIPORTAL ADAPTATIONS of the right foreleg of a 12-year-old Indian elephant, *Elephas.*

Just as parallelograms can be drawn to add several forces acting together, so they can be drawn to add several constant or instantaneous velocities acting together. When a man swims straight out from the bank of a river, his velocity relative to the water and the velocity of the current relative to the bank combine to move him diagonally downstream at a rate exceeding either velocity taken alone. Similarly, if the distal end of the humerus of a running animal is extended on the scapula at the instantaneous velocity $\mathbf{v}_1$ (Figure 22-14A) as the body is moving relative to the ground at velocity $\mathbf{v}_2$, then $\mathbf{R},$ the diagonal of the parallelogram, is the resultant velocity of the distal end of the humerus relative to the ground.

Likewise, a velocity can be broken down into components. As for forces, a useful application establishes components at right angles to one another by constructing a rectangle of velocities as in part B, which shows the glide of a pigmy scaly-tailed squirrel.

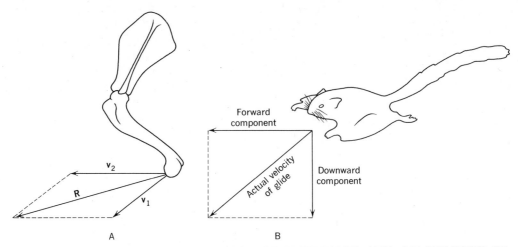

Figure 22-14　THE CALCULATION OF RESULTANTS AND COMPONENTS OF VELOCITIES.

　　It is often desirable to combine forces (or torques) that act independently on the same lever; two muscles can pull harder than one. Velocities, by contrast, cannot be applied independently to different parts of the same lever; disregarding the inertia of the system (which can be an oversimplification), two muscles cannot pull faster than one. The velocity of any point on a lever is determined only by its distance from the pivot and the angular velocity or rate of turning of the lever as a unit. Thus, when loads are small the relative velocities of different insertions acting together on one bony lever are determined primarily by their positions and not by inherent differences in rate of shortening of associated muscles.

　　We have learned that the relation of in- and out-forces to their respective lever arms is expressed by the formula $F_i l_i = F_o l_o$. Therefore, $F_i = F_o l_o / l_i$ and $F_o = F_i l_i / l_o$. Out-force increases with the length of the in-lever and decreases with the length of the out-lever: The digger needs a long olecranon and short forearm. It follows from the previous paragraph that **in-** and **out-velocities** (v_i and v_o) are also related to their respective lever arms, but in the reverse way: $v_i l_o = v_o l_i$, so $v_i = v_o l_i / l_o$ and $v_o = v_i l_o / l_i$. For high velocity at the foot, the olecranon should be short and the forearm long. The same lever cannot enhance both the force and the velocity of the same muscle. The problem of providing an animal with both low and high "gear systems" is discussed on page 446. (Optimum structure is determined not only by these mechanical factors but also by related physiological factors.)

　　It is often desirable to know the speed of an animal as a whole. It may also be useful to know the maximum velocity of a part of the body (as of the center of gravity of a jumper at takeoff) or the velocity at a given instant (as of the foot of a runner when it strikes the ground), but the parts of moving animals seldom attain constant velocities and average velocities are not often useful. Therefore, other variables must also be considered.

MASS AND ACCELERATION　Objects remain at rest or in uniform motion unless acted upon by external forces. (This is Newton's First Law.) They have the inherent tendency to remain at rest or in motion at constant velocity. This tendency is called **inertia.** It is not expressed in numerical terms, but experience tells us that heavy objects have more inertia than light objects. Actually, the critical factor is not weight, which is determined by the pull of gravity, but **mass** (*m*), which represents quantity of matter and remains the same in space as on earth. The vertebrates are on the earth where mass is measured by weighing, so the difference is not of practical importance; nevertheless, it is preferable to use the concept of mass.

The capacity of an object moving in a straight line to overcome resistance is called **linear momentum** (*M*) and is equal to *mv*. Momentum is conserved: When one system loses momentum, another system gains an equal amount. Thus, when a bird flies to a perch, the momentum lost by the bird is gained by the perch, which may cause it to sway.

The relation of mass (*m*) to force (**F**) and change in velocity per unit time, or acceleration (*a*), is stated by Newton's Second Law and is simply $F = ma$, when appropriate units are selected. Acceleration is expressed as centimeters per second per second, or equivalent units, and obviously relates to velocity (**v**), time (*t*), and distance (*s*). If an object is at rest when acted on, then the basic relationships are $v = at$, $s = vt$, and $Ft = mv$. Other equations can be derived by substitution. If an object is already in motion when acted on, then the equations must be modified somewhat. Derivations and interpretations are found in texts on elementary mechanics.

These relationships show that if an object is to be accelerated or decelerated rapidly (e.g., the body of a bird at takeoff or the tongue of a feeding chameleon), it should be light in order to avoid excessive forces; if an object is heavy (body of an elephant), relatively more time is needed to achieve maximum velocity; greater velocity is attained when a force acts on an object for a relatively long time (a reason for the very long hind legs of the jumping frog and tarsier).

These formulas and examples apply only to rectilinear motion. Since animals and their moving parts often do not travel in straight lines, other factors must now be considered.

CURVILINEAR AND ROTATIONAL MOTION   When a tetrapod jumps, or momentarily lifts all feet from the ground while running, or folds its wings for an instant in flight, then (disregarding wind resistance) its center of gravity tends to continue to move with the initial direction and velocity (according to Newton's First Law), but is simultaneously accelerated in a different direction by gravity. One component of its motion, the rate of fall, is not constant. Hence, the resultant motion is not rectilinear. While unsupported, the jumping body as a whole moves in a parabolic curve regardless of any motions of its appendages.

When a running animal turns sharply, its velocity also changes even though its speed may remain constant (because velocity has both magnitude and direction). There is then a sideways, or centrifugal, force out of the curve of the turn that must be balanced by an inward centripetal force in order to prevent skidding.

These are examples of curvilinear motion. Such motion relates particularly to jumping, dodging, turning in flight and throwing and will be noted further in subsequent chapters.

Man-made machines that move on land are supported and propelled by wheels. Vertebrates, instead, use moving parts (legs, fins, jaws, spines) that swing back and forth. Such parts may be thought of as segments of wheels that constantly reverse their direction of rotation. This is inefficient because the appendages must constantly be accelerated and decelerated. However, oscillating parts do have the advantage of versatility: the functional length and period of oscillation of a limb can be modified at will to adapt to rough ground or angle of terrain.

Turning wheels and the oscillating levers of moving animals rotate around axes; the mechanics of rotational motion apply. Consider an extended limb that is rapidly pivoting at the hip or shoulder during the supportive phase of a running stride. Its **angular velocity** (ω) is the rate of rotation, or degrees turned in a unit of time. Its **angular acceleration** (α) is the rate of change in angular velocity. Its **moment of inertia** (I), which corresponds to mass in rectilinear motion, is an index of the resistance of the limb to acceleration. This equals the mass (m) of the limb times the square of a constant (k) called the radius of gyration, the value of which depends on the distribution of the mass of the limb. The value is greater for long limbs than short and greater for limbs that are heavy distally than for limbs that are heavy proximally. The turning force, or torque (τ) of the limb is $I\alpha$. **Angular momentum** (L) equals $I\omega$, and like linear momentum it is conserved. By flexing a limb, or spine, or tail so as to concentrate its mass closer to its axis of rotation, k, and hence I, are decreased. Because L does not change, ω increases. Human divers and gymnasts learn to control the spin of their bodies in this way; cats, jerboas, and gibbons do likewise.

Complicating factors make it extremely difficult to apply these, and related formulas, to the quantitative analysis of actual animal movements: Articulated segments of the body commonly rotate around independent axes at the same time that they turn as a unit. Each segment then has its own values for each variable, and the values of the variables for the segments taken together change constantly. Furthermore, one segment may rotate simultaneously around more than one axis (e.g., extension with supination, or flexion with adduction).

In general, however, study of the formulas $I = mk^2$, $\tau = I\alpha$, and $L = I\omega$ makes it easy to interpret many morphological adaptations. It is seen that the mass and distribution of mass of an oscillating structure are critical to its function unless the part has little mass (legs and jaws of very small tetrapods), or moves slowly (legs of the sloth), or overcomes external resistance far in excess of resistance offered by its own mass (foreleg of a burrowing mammal). If an oscillating structure moves fast, then the muscles that move it, being heavy, should not be located in the fastest moving part of the structure (distal part of leg or wing) but instead should transmit their forces by levers or long tendons from regions of relatively slow motion (shoulder of antelope, breast of bird). Forces can be reduced and angular velocities increased if the distal parts of oscillating structures are light and slender. Very large animals eliminate such oscillations as can be avoided (e.g., flexion and extension of the spine).

FREE-BODY DIAGRAMS   In solving problems in functional morphology it is often useful to use a **free-body diagram.** One isolates the system under study, locates its center of mass,

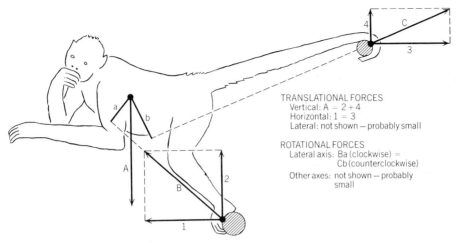

TRANSLATIONAL FORCES
Vertical: A = 2 + 4
Horizontal: 1 = 3
Lateral: not shown — probably small

ROTATIONAL FORCES
Lateral axis: Ba (clockwise) =
                     Cb (counterclockwise)
Other axes: not shown — probably
                     small

Figure 22-15   FREE-BODY DIAGRAM OF A MONKEY IN EQUILIBRIUM. A, B, and C are the external forces acting on the animal; 1 to 4 are components of those forces; a and b are the lever arms of the forces tending to rotate the body around its center of mass.

measures all external forces acting on the system, then sketches the system and represents all of the forces by vectors (Figure 22-15). There can be a maximum of six external forces: a translational force in each plane of space (tending to shift the entire system as a unit), and a rotational force around each axis in space (tending to revolve the system around its center of mass). When the system is at rest, the sum of all translational forces must be zero, and likewise the sum of all rotational forces must be zero. This is a powerful tool for learning if all forces have been identified and none overrepresented.

The isolated mechanical unit need not be an entire animal, as in the example figured, but can instead be a limb or limb segment, a jaw, a tooth, or other structural complex. It is also possible (when the variables can be identified and measured) to construct a dynamic free-body diagram for a moving system at an instant in time (as of a human diver). However, this is a more difficult task.

In order for an animal to produce relatively large forces, torques, velocities, momentums, or moments of inertia, it must pay the price of a relatively large expenditure of effort. The effort required of the body is proportional to the sum of the energies produced in its various mechanical systems. Some kinds of energy (electromagnetic, chemical) do not directly concern us here even if they are involved in muscle physiology. The production of heat energy is a desirable consequence of the locomotion of small homeothermous vertebrates in cold weather, but in general represents wasted energy. Of direct relevance to locomotion are potential energy (E_p) and kinetic energy (E_k).

Gravitational potential energy is energy of position. In climbing a tree, a flying squirrel does work against gravity, thus gaining potential energy for its glide. Birds gain potential energy as they fly high above the ground. The limbs

EFFORT, ENERGY, AND EFFICIENCY

of running animals gain potential energy when lifted. Quantitatively, $E_p = mgh$, where m is mass, g is the pull of gravity, and h is the height.

Elastic potential energy is exemplified by a loaded spring, a drawn archer's bow, a stretched tendon or ligament, or an inflated lung. It equals $\frac{1}{2} ks^2$, where k is the spring constant, or restoring force per unit displacement, of the particular material stretched, and s is the displacement. Some of the supportive mechanisms described earlier in this chapter rely on elastic potential energy to restore equilibrium after a system has been disturbed.

The body expends no effort when potential energy is released. Sometimes the body expends no effort to store potential energy: The soaring bird is lifted by updrafts; the suspensory mechanism of an ungulate's foot may be stretched by an external load. Often, however, the body must use muscular effort to derive needed potential energy.

Kinetic energy is energy of motion. In rectilinear motion (the charging rhinoceros and striking falcon), $E_k = \frac{1}{2}m\mathbf{v}^2$. In rotational motion (the beating wing, snapping jaw, and swinging limb), $E_k = \frac{1}{2}I\omega^2$. Since in each instance velocity is squared, it is clear that high speed in animals (as in automobiles) is costly. Animal swimmers and runners do not sprint for long.

Energy is never destroyed: The E_p of the flying squirrel at the top of a tree becomes E_k as it glides down; the E_k of a jumping salmon becomes E_p as the fish comes to rest in a higher pool. When a pendulum swings, it changes E_p to E_k on the downstroke and E_k to E_p on the upstroke. Disregarding friction and air resistance, no new energy need be introduced to keep it going. Oscillating appendages of moving tetrapods do the same to a degree—the E_p of lifted limbs and elevated wings is recovered in each cycle of motion—yet much energy is lost to the environment (largely as frictional heat), so the muscles of the body must constantly work to maintain the motion. The rate at which a pendulum naturally swings, and the rate at which it can be forced, both relate to its length (though the formulas are different): The greater its length the longer its period. This is a reason that each man can walk most comfortably at a certain rate, and why long-legged animals tend to walk with long, slow strides, whereas short-legged animals tend to walk with short, fast strides. However, the parallel between the pendulum and the animal limb can be overrated: Limbs are usually moved faster than their natural periods. Their functional lengths, and hence natural periods, are shorter on the recovery stroke than on the support stroke; the relative durations of these strokes must relate to rate of travel, not to the physics of pendulums.

The **efficiency** of a mechanical device is the ratio of the useful energy taken out (usually as force or work that produces potential or kinetic energy) to the energy put in (usually as fuel). The efficiency of an inorganic machine can be exactly measured: For a hydraulic turbine it might be 90%, for a steam engine 20%, and for many engines less. With practice, rhythmic human muscular effort, as measured by an ergometer such as a bicycle-pedaling apparatus, has an efficiency of 25% or a little more; the most efficient speed is always less than the maximum speed. However, when considering normal behavior of animals, efficiency is a less exact, though equally important concept. Thus, the cheetah's way of running is more extravagant of energy than that of the blackbuck, yet the cheetah overtakes the blackbuck often enough to survive, and in that (somewhat different) sense is the better runner. The weight of the hydraulic turbine is unimportant, but for flying machines, animal or man-made, the ratio of useful energy output to weight is more important than its ratio to fuel input. The ratio of thrust to weight is greater for the flight muscles of birds than for aircraft engines, which in turn, rate far ahead of steam engines. The skill of the piano player is limited by neuromuscular control, not the conversion of food to mechanical energy. The efficient animal performs the needed activity, however demanding or exacting, with sufficient economy of effort to make the activity advantageous.

23

Running and Jumping

Animals that travel far, fast, or easily on the ground are said to be **cursorial.** Quadrupedal cursors evolved from walkers, and in general are either predators or medium-to-large-sized herbivores. Animals that jump or hop are said to be **saltatorial.** Saltators are often bipedal, and if the hind legs are used in unison for a succession of jumps, kangaroo-fashion, the gait is called a **ricochet.** Saltators have evolved many times among small vegetarian tetrapods living in relatively open habitats, and among small arboreal vertebrates. Most cursors are at least fair jumpers. Saltators either move easily on the ground or are excellent climbers.

ADVANTAGES OF SPEED AND ENDURANCE

Cursorial and saltatorial animals have a number of selective advantages:

1. Cursors are able to forage over large areas. A pack of African hunting dogs may range over 3800 km^2 (1500 mile^2), and a mountain lion works a circuit some 160 km long.

2. Cursors can seek new sources of food and water when familiar supplies fail. Africa's big game animals may travel great distances, and individual arctic foxes have wandered 1300 km.

3. Cursors can take advantage of seasonal variation of climate and food sources. Some herds of caribou migrate 2500 km each year.

4. Predators run to overtake prey, exploiting superior speed or relay tactics according to their habits.

5.   Prey species run to escape predators. They are commonly about as swift as their pursuers (the latter rely partly on surprise) and may have superior endurance. Small prey species may be master dodgers; when chased, a kangaroo rat bounces in a different direction on nearly every hop. When startled on land, a frog can reach the safety of the water with several jumps.

6.   Animals may leap to clear obstacles or to see over obstructions. The long springy legs of the Peruvian maned wolf are said to enable it to keep mice in view in tall pampas grass. Rabbits make "spy-hops" to check up on pursuers.

7.   Arboreal saltators jump to climb and are able to move from branch to branch with great agility.

CURSORS, SALTATORS, AND THEIR SKILLS

SPEED   All cursors and many saltators are swift runners. More than two dozen species of lizards are bipedal when running fast, and some can attain 25 km/hr. Many thecodont dinosaurs were excellent cursors. Pheasants and the road-runner likewise run about 25 km/hr. The ostrich has been credited with 80 km/hr (50 mile/hr), which may be as fast as any biped has ever run.

The maximum speed of the kangaroo is about 65 km/hr. The marsupial wolf and rabbitlike bandicoots are also quite fast. Man is probably the fastest of primates, attaining about 37 km/hr (23 mile/hr) for 200 m. Jack rabbits are able to run 64–72 km/hr, and some of the larger rodents (agouti, cavy) are good runners. Kangaroo rats and mice, jerboas, spring-hares, and some other rodents are ricochetal except when moving slowly; the behavior evolved four or five times within the order.

The cats and dogs are cursorial, and many other carnivores moderately so: the whippet runs 55 km/hr, the coyote 69 km/hr, and the red fox 72 km/hr. The cheetah seldom runs more than $\frac{1}{2}$ km, but for short distances is the fastest of animals, probably attaining 110 km/hr (nearly 70 mile/hr). The horse has sprinted to about 70 km/hr, several antelopes run at 85–95 km/hr, and the pronghorn has been paced with a car at 98 km/hr.

ENDURANCE   A second skill of many cursors, endurance, is less well documented, but also impressive. Man runs 30 km at 19.5 km/hr. One fox, running before hounds, covered 240 km in $1\frac{1}{2}$ days. For rewards of food, a terrier climbed, on a treadmile, 1500 vertical m/hr for 17 hr. The horse has run 80 km at 18.2 km/hr. One Mongolian wild ass is reported to have run 26 km at 48 km/hr, and the camel has traveled 186 km in 12 hr.

LEAPING ABILITY   There is a report of a South African frog that averaged 3.28 m for three consecutive jumps. One kangaroo cleared a 2.7 m fence. The tarsier, which weighs only about 120 g, is a prodigious leaper, and the little galago can leap vertically more than 2 m. A jump of one rabbit measured more than 7 m, and the impala jumps 2.4 m high.

ACCELERATION AND MANEUVERABILITY   The ability of cursors and saltators to start fast, to speed over uneven terrain, and to dodge has not been quantified, yet is known to be outstanding for many small cursors and ricochetal mammals.

GREEN FROG
Rana

OSTRICH
Struthio

BASILISK
Basiliscus

SAND LIZARD
Uma

Figure 23-1 EXAMPLES OF A SALTATOR (above), TWO BIPEDAL CURSORS (center), AND A QUADRUPEDAL CURSOR (below) showing long hind legs and, for the bipedal cursors, balance of the body over the legs.

KANGAROO
Macropus

SPRINGHARE
Pedetes

KANGAROO RAT
Dipodomys

Figure 23-2   EXAMPLES OF RICOCHETAL MAMMALS shown at takeoff and landing and midway in a very high jump. Note body and leg proportions and use of tail
for balance.

**GENERAL
REQUIREMENTS
OF CURSORS**

Saltators and bipedal cursors tend to be more specialized than quadrupedal cursors.
Accordingly, we shall start with an analysis of adaptations for running on four
legs. In order to run well an animal must (*1*) overcome the inertia of its body to
attain speed; (*2*) overcome the movement and inertia of the legs, and any other
oscillating parts, with every reversal in the direction of their motion; (*3*) support
the body without benefit of wheels; (*4*) compensate for forces of deceleration,
including the action of the ground against the feet as they come down; (*5*) control
its course; and (*6*) maintain these functions as long as required.

A full cycle of motion of a running or walking animal is called a stride.
Speed equals length of stride times rate of stride. The giraffe emphasizes length,
and the wart hog emphasizes rate. Great speed requires both. Endurance is speed
sustained through economy of effort. Economy of effort is dependent on body
shape, the ways that parts of the body are moved, and on the magnitude and
distribution of masses. For convenience we shall consider these related factors
one at a time.

**LENGTH OF
STRIDE**

A galloping race horse covers about 7 m per stride. The faster, but smaller,
cheetah covers at least as much distance—about ten times its shoulder height.
How do runners manage such long strides?

IMPALA
Aepyceros

GREYHOUND
Canis

JACKRABBIT
Lepus

Figure 23-3 EXAMPLES OF MAMMALIAN CURSORS showing flexion and extension of spine and legs, passage of hind feet outside of forefeet, and varied positions
of the shoulder.

Figure 23-4 THE RELATIVELY LONGER LEGS OF THE CURSOR shown by the
very fast cheetah, *Acinonyx,* standing behind the moderately fast mountain lion, *Felis.*

LENGTH AND PROPORTIONS OF LEGS The longer the leg, the longer the
stride. One might think, therefore, that speed could be increased merely by
enlarging the body uniformly in all dimensions. Many cursorial animals are large;
however, this must be for other reasons, because enlarging body size also reduces
speed in several ways. To contribute to speed, it is necessary to make the legs
relatively long in relation to the other parts of the body (Figure 23-4).

For reasons explained on page 432 and noted later in this chapter, the distal
segments of the legs usually lengthen more than the proximal segments. One
cannot offer an exact formula, but cursorial ungulates usually have a radius that

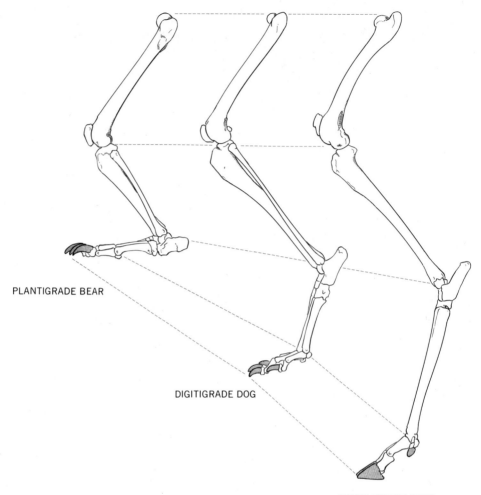

PLANTIGRADE BEAR

DIGITIGRADE DOG

UNGULIGRADE DEER

Figure 23-5   CONTRAST IN PROPORTIONS AND FOOT POSTURE in the left hind leg skeleton of a noncursor (left), moderate cursor (center), and highly specialized cursor (right).

is as long, or a little longer, than the humerus, and a tibia that is as long, or sometimes markedly longer, than the femur. The foot skeleton, which in walkers and climbers is shorter than the corresponding middle limb segment, is, in cursorial ungulates, about equal to, or even longer than, the middle limb segment. It is the metacarpals or metatarsals that lengthen most (Figure 23-5). The carpals never lengthen, and the tarsals lengthen only in several jumpers (see figure on p. 494).

(Among birds, cursorial ability is not easily determined from the length or proportions of the legs. All birds evolved from small reptiles that hopped in trees, and the distal segments of the legs are nearly always elongated. The legs are longest among wading birds, not runners. Furthermore, although cursorial lizards have relatively long hind legs, their limb proportions do not follow the general rule stated above. Seemingly, this is because their legs move in eliptical arcs lateral to the body.)

FOOT POSTURE   We have just seen that runners have relatively long leg bones. But it is the *effective* length of the leg—the part that contributes to stride length—that is important, and this can be further increased in other ways. The human foot does not contribute to the length of the leg unless one rises on "tip-toes." The heel is on the ground as one stands, and strikes first in each stride. Bears, opossums, raccoons, and most other vertebrates that walk well but seldom run, have similar feet. Such feet are called **plantigrade** (= sole + walking).

Running dinosaurs, birds, carnivores, and extinct ancestors of hoofed mammals increase effective leg length by standing on what corresponds to the ball of the human foot. These animals are **digitigrade** (= finger + walking).

Perissodactyls, artiodactyls, and the cursorial representatives of several extinct orders of mammals have, like the ballet dancer, further increased effective leg length by standing on the tips of the digits. This foot posture is **unguligrade** (= hoof + walking), which is why these animals are called ungulates.

Where foot posture and limb proportions are each modified for the cursorial habit, the enhanced length and slenderness of the leg skeleton is striking (Figure 23-5).

ROLE OF THE SHOULDER   The effective length of the forelimb of many runners is increased further by altering the structure and function of the shoulder. In amphibians and birds, and in some reptiles and mammals, the position of the shoulder joint is virtually immobilized by the clavicle and coracoid (if present) which run like struts from the sternum to the scapula. In most mammals, however, the scapula is free to move somewhat, and cursors increase this freedom by (*1*) reducing the clavicle to a vestige (carnivores) or abandoning it altogether (ungulates), and (*2*) reorienting the scapula so that it lies not flat against the back of a broad chest (as in man) but flat against the side of a deep, narrow chest, where it is free to rotate in the same plane in which the leg swings. The shoulder joint then moves in the sagittal plane, which is the equivalent of lengthening the leg by moving its pivot from the shoulder joint to a point part way up the scapula (Figures 23-6 and 23-11). Inability of the cursorial dinosaurs to use the shoulder in this way may have contributed to their "preference" for the bipedal habit.

ROLE OF THE SPINE   Cursorial lizards undulate the spine in the horizontal plane as they walk or run. (Amphibians, and to a lesser extent some mammals, do the same when walking.) The pivoting of each girdle is timed to help advance the related unweighted leg and to help swing the weighted leg to the rear.

The smaller and swifter quadrupeds among mammals—particularly carnivores—have their legs positioned under the body instead of to the side, and consequently undulate the spine in the vertical plane; the back advances like the body of a measuring worm at the same time that the legs are swinging back and forth. The body of the animal is longer when the back is extended than when it

BEAVER
Castor

DEER
Odocoileus

Figure 23-6   CONTRAST BETWEEN THE SHAPE OF THORAX, POSITION OF SCA-PULAS, AND DEVELOPMENT OF CLAVICLES IN A CURSOR (left) AND A NON-CURSOR (right). Only the first five thoracic segments are shown.

is flexed. Were the animal to extend its back while its body is suspended in the air, the hindquarters would move backward as the forequarters move forward, and the center of mass of the body would not be affected (an expression of Newton's Third Law). However, the galloping animal extends its back only when its hind feet are on the ground. Muscles of the legs pressing on the ground (and friction at the ground) prevent the hindquarters from moving backward (decelerating), so all of the increase in body length is added to stride length. Similarly, the forelimbs prevent, or at least reduce, deceleration of the forequarters as the back is flexed; therefore, the shortening of the body is also added to stride length. The cheetah is so adept at this maneuver that it could theoretically run nearly 10 km/hr without any legs at all (Figure 23-7). (Flexion and extension of the spine, and its timing relative to the thrust of the legs, also assures that the speed of the animal, that is, of its center of mass, is slightly greater than the speed of the associated girdle during the time that a pair of legs is propelling the body.)

The extra rotation of the hip and shoulder girdles that is added by flexion and extension of the spine increases the swing of the legs proper so that the limbs reach out farther, front and back, and strike and leave the ground at more acute angles than they would if the spine were held rigid. Again, this increases stride length.

UNSUPPORTED INTERVALS   There is another important way to lengthen stride. Running gaits usually include periods of suspension when all feet are off

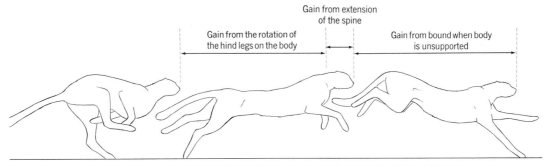

Gain from extension
of the spine

Gain from the rotation of
the hind legs on the body

Gain from bound when body
is unsupported

Figure 23-7   SOURCES OF THE LENGTH OF STRIDE OF A FAST-RUNNING CHEE-TAH shown for half of a cycle. Each factor is repeated in the other half cycle except that flexion of the body substitutes for extension.

the ground. The distance the body moves forward while it is unsupported is added to the length of the step to give the length of the stride. The bear (which runs fairly well, but is hardly cursorial) scarcely gets all its feet off the ground at the same time when it runs. Most ungulates have one unsupported period in each stride. The galloping canid has two unsupported periods, one when the body is flexed and another when it is extended. The proportion of the stride interval during which all feet are off the ground is nearly twice as great for the cheetah as for the horse. Bipedal runners are also unsupported for much of the duration of the stride. The hopping African springhare may have its feet off the ground 85% of the time!

MUSCLE MECHANICS: THE MOST FOR THE LEAST   Muscles can move the joints through wider angles when they insert close to the joints than when they insert farther away. Thus, in Figure 23-8, muscle A moves the foot only over distance a, whereas muscle B, contracting an equal distance, moves the foot over the greater distance b. This arrangement also benefits the rate of the stride (see below) and enables the muscle to function with less shortening, so it is characteristic of cursorial animals to have limb muscles that insert close to joints.

The short-legged mouse can travel faster than the long-legged frog. Speed is the product of length of stride times rate of stride, and the mouse takes many strides for each plop of the frog. Fast runners must take long strides but take them rapidly. The galloping race horse completes about $2\frac{1}{4}$ strides/sec, and the fast-running cheetah completes about $3\frac{1}{2}$ strides/sec. How are cursorial animals adapted to take rapid strides?

RATE OF MUSCLE CONTRACTION   As an animal gathers speed, it increases the rate of its stride by causing its muscles to contract faster. One might infer that cursorial animals as a group would have evolved the ability to contract their muscles faster than other animals. It is true that there is some variation in this ability among species and individuals, but for various reasons the muscles of large vertebrates tend to contract slower than equivalent muscles of small ver-

RATE OF STRIDE

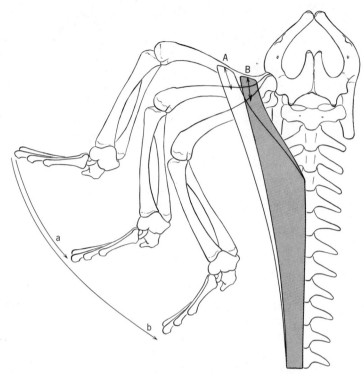

Figure 23-8   EFFECT OF DISTANCE OF MUSCLE INSERTION FROM THE JOINT TURNED ON THE RATE AND DEGREE OF ROTATION. The drawing shows in ventral view the caudofemoralis muscle, base of tail, pelvis, and right leg skeleton of the tegu lizard, *Tupinambis*. (Position B approximates the actual condition. A small insertion of the same muscle onto the knee is omitted.)

tebrates. Earlier in this chapter it was noted that the larger the animal, the longer its legs and stride, and hence the faster it might be expected to run. Now we see that longer stride and slower stride tend to cancel one another as body size increases, so that speed remains about the same. We must look elsewhere for ways to increase rate of stride.

MUSCLE MECHANICS: HIGH "GEAR RATIO"   High gear is used when a car is driven fast: The car is made to move rapidly in relation to engine speed. Cursors have evolved higher "gears" than most other animals. It was noted above that muscles can move joints through wider angles when they insert close to joints than when they insert farther away. Following principles given on page 430, $v_o = v_i l_o / l_i$, and therefore out-velocities are also increased by moving insertions close to joints (i.e., by reducing the value of l_i). Thus, it was shown in Figure 23-8, that muscle B, contracting at the same rate as muscle A, and having equal

Figure 23-9   CONTRAST BETWEEN THREE BONE-MUSCLE SYSTEMS OF MAMMALIAN CURSORS (left) AND NONCURSORS OF THE SAME ORDERS (right). The ratio of the out-lever (l_o) to the in-lever (l_i) is greater for cursors.

CHEETAH, *Acinonyx*
$\frac{l_o}{l_i} = 4.45$

Teres major

LION, *Felis*
$\frac{l_o}{l_i} = 3.36$

l_i

l_o

AGOUTI, *Dasyprocta*
$\frac{l_o}{l_i} = 5.48$

Medial head of triceps

MUSKRAT, *Ondatra*
$\frac{l_o}{l_i} = 4.30$

l_i

l_o

KANGAROO, *Macropus*
$\frac{l_o}{l_i} = 4.64$

Gastrocnemius

WOMBAT, *Lasiorhinus*
$\frac{l_o}{l_i} = 3.50$

l_o l_i

l_o l_i

$\dfrac{I_o}{I_i} = 44$

Middle gluteus

Semimembranosus

$\dfrac{I_o}{I_i} = 11$

Figure 23-10   HIGH-GEAR AND LOW-GEAR MUSCLE SYSTEMS shown by a lateral view of the left hind leg of the vicuna, *Vicugna*. The muscles are diagrammatic.

loading, moves the foot through the distance b in the same time interval required for muscle A to move the foot through the shorter distance a. Because it inserts closer to the joint, muscle B can move the foot a greater distance in the same time or an equal distance in less time. The muscle also gains an important physiological advantage by doing the job without shortening as much.

The ratio I_o/I_i for limb muscles of cursors and saltators tends to be larger than for corresponding muscles of unspecialized tetrapods, and much larger than for those of diggers and swimmers. "Larger" and "much larger" are useful terms only when speaking in generalities; with specific bones at hand for analysis, more specific values are required. Figure 23-9 shows characteristic differences between cursors and noncursors in regard to representative bone-muscle systems.

(Since different kinds of animals emphasize different mechanisms to attain similar ends, it is preferable, when assessing cursorial ability, to consider the mechanical advantages of several muscles, or to compare the same leverages in related animals, as in Figure 23-9: for the teres major muscle, the swift horse has a lower I_o/I_i value than the slow porcupine, but for the gastrocnemius muscle it has a higher value than even the kangaroo.)

As shown on page 430, there is a reciprocal relationship between the velocity and the force that a given muscle can produce through a series of skeletal levers. Like the car, which in high gear can move fast but is correspondingly handicapped in climbing grades, the cursorial animal can add further speed only by sacrificing torque. However, runners, like race cars, retain some relatively low gears. For example, Figure 23-10 shows that the semimembranosus, with its relatively long effective lever arm, is a low-gear system relative to the middle gluteus muscle system that has the same action but a much shorter effective lever arm. The animal may use the lower gear to initiate the swing of the femur, and then, like a car driver, shift to the higher gear to speed the action. Finally, cursors have reduced requirements for force by reducing the loads on their muscles. We will come to that shortly, but first, there is another way of increasing rate of stride.

SUMMATION OF VELOCITIES   Having achieved leverage that is optimum for speed, there is little further that the musculature, acting on a single joint, can do to increase the velocity of the particular action it controls. If these muscles become larger, or if additional muscles are added to help them, force is increased, but the velocity of the action remains about the same. (Several men can lift more weight together than one can lift alone, but several equally skilled sprinters cannot run faster together than one of them alone.) But if different limb muscles simultaneously move different joints in the same direction to achieve a greater total motion, the independent velocities they produce are added to derive the total velocity at the foot (Figure 23-11). (When a man walks down an escalator, his motion in relation to the steps and their motion in relation to the building are added to give his rate of advance in relation to the building.) The trick is to move

Figure 23-11 PRINCIPLE OF THE SUMMATION OF INDEPENDENT VELOCITIES shown by the travel of the leading forefoot of a running cheetah from the initiation of the backswing to the instant the foot leaves the ground. Arcs shown approximate amount of rotation around the respective pivots (indicated by spots), not relative velocities.

as many joints as possible in the same direction at the same time without interfering with the support role of the limbs. We have already seen that cursorial vertebrates add an extra pivot to the limbs by abandoning the flat-footed, plantigrade foot posture in favor of digitigrade or unguligrade posture. Furthermore, their scapulas can rotate through 20–25° on the chest. Finally, cursorial carnivores time the flexing of the spine in such a way that the chest and pelvis are always rotating in the direction of the swinging limbs. The net benefit to speed is considerable.

The relationship between body size and the requirements for strength of body framework was given on page 184, where it was shown that if body size were increased without altering body proportions, then the load on the locomotor system would increase faster than its capacity to provide support and power. This principle is significant to the analysis of structure of large cursorial vertebrates. It explains why elephants (and probably dinosaurs—in spite of speculation to the contrary) cannot gallop or jump, why some small runners, such as foxes, can travel as fast as race horses without having marked structural adaptations for speed, why no saltatorial animal is as large as typical ungulates, and why the

MASS, ENDURANCE, AND DESIGN FOR ECONOMY OF EFFORT

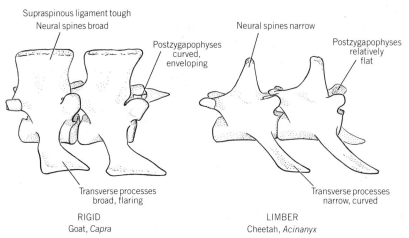

Supraspinous ligament tough
Neural spines broad

Postzygapophyses
curved,
enveloping

Neural spines narrow

Postzygapophyses
relatively
flat

Transverse processes
broad, flaring

Transverse processes
narrow, curved

RIGID
Goat, *Capra*

LIMBER
Cheetah, *Acinanyx*

Figure 23-12   LUMBAR VERTEBRAE OF LIMBER AND PASSIVELY RIGID SPINES.

larger cursors must have great adaptation for speed and endurance in order to run well. The adaptations of the larger runners must include not only many already discussed, but others that reduce the loads on locomotor structures, providing economy of effort. The evolutionary process has been so effective at fulfilling these requirements that the energy cost of locomotion is, in fact, inversely related to body weight, and large cursors have superior endurance. What are the elements of design for economy of effort?

First, large cursors reduce or eliminate many oscillating motions. The legs must swing back and forth, but the feet are not lifted so high, the back is relatively stiff (Figure 23-12), and the center of mass has less vertical displacement. Unsupported periods can follow only vertical acceleration, which is costly, so suspensions are reduced or avoided.

Next, the mass of the limbs is reduced in several ways. Adductors and abductors of the legs are reduced or are converted to move the limbs in the direction of travel. Muscles that manipulate the digits, rotate the forearm, or twist the feet inward or outward are also reduced or even eliminated. The ulna and fibula, which in other animals take part in foot-twisting and rotation mechanisms, are reduced as these functions are lost. The ulna must be retained where it completes the elbow joint, but distally may become a sliver that fuses to the radius (Figure 23-13). The fibula may become an attenuated splint (even in small mammalian cursors), and in artiodactyls is sometimes represented only by a nubbin of bone at the ankle (Figure 23-14).

Furthermore, since the forces that must be developed in oscillating systems with every reversal in the direction of their action are also in proportion to the square of their angular velocities (see again p. 434), the loads on muscles causing such motions can be reduced not only by reducing the masses of the systems but also by reducing their velocities. This is a principal reason why it is the lower limb segments that lengthen in the evolution of the long legs of the runner: When

Radius somewhat lateral
to ulna

Radial notch on ulna
shallow, smooth, permitting
rotation of radius on
ulna

Radius anterior to ulna

Contact between
bones broad, provided
with spline and
groove preventing
rotation of radius

Distal ulna strong,
free

Distal ulna reduced,
partly fused to radius

KINKAJOU
Potos

PRONGHORN
Antilocapra

Figure 23-13 CONTRAST BETWEEN THE FOREARM SKELETONS OF A NON-CURSOR (left) AND A CURSOR (right) shown by the left radius and ulna in lateral view (below) and the same bones at the elbow seen from above and behind (above).

devoid of the muscles and bones related to twisting, rotating, and digit manipulation, those segments are relatively light; fleshy parts of the limbs are kept close to the body, where they do not move so far, and hence not so fast, as the more distal segments. The advantage gained is particularly important during acceleration.

Other elements in design for economy of effort provide further saving of weight without loss of strength. The feet of unspecialized vertebrates tend to be broad and pliable. Their metapodials are rounded in cross section and are well-separated. Some runners (some dinosaurs, carnivores) provide more strength by crowding these bones together into a compact unit; each bone becomes somewhat square in cross section. Still more strength can be provided with the same weight (or the same strength with less weight) if the skeletal material is distributed among fewer bones; hence, some of the best runners and jumpers (kangaroo, jerboa, perissodactyls, artiodactyls), and particularly the large ones, tend to lose

Distal part of
fibula fused
to tibia

Fibula strong,
free

Only a rudiment
of the fibula
retained where
it articulates
with the foot

RACOON
Procyon

JACK RABBIT
Lepus

PRONGHORN
Antilocapra

Figure 23-14   CONTRAST BETWEEN THE FIBULAS OF A NONCURSOR (left) AND
TWO CURSORS (center and right) shown by the left leg in lateral view.

the lateral toes and to fuse the basal elements of the remaining toes into a single
bone of compound origin. This process has produced the cannon bone of ungulates and, in response to a hopping habit, the tarsometatarsus bone in the
ancestors of all birds. The result is a slender, light, strong foot (Figure 23-15 and
figure on p. 15).

To compensate for the bracing lost as bones and muscles of the lower limbs
are reduced or eliminated, and to guard against dislocations, cursorial vertebrates
have evolved joints modified to function as hinges, allowing motion only in the
line of travel. This has been done by introducing (or enlarging) interlocking splines and grooves in the joints (digits, ankle or hock, elbow) (Figure 23-16) or by
substituting flat or cylindrical shapes for spherical shapes at the articulations (wrist
and, to lesser extent, shoulder) (Figure 23-17).

Finally, hoofed mammals have evolved still another mechanism for relieving
the muscles. When the foot of the running animal strikes the ground, the impact
bends the joint between the phalanges and the metapodial bones (fetlock joint).
This stretches the ligaments that are called suspensory, or springing, ligaments
(see figure on p. 426). Since ligaments are elastic, the energy of deformation is

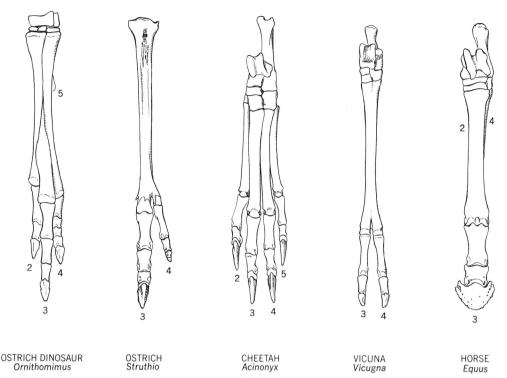

OSTRICH DINOSAUR	OSTRICH	CHEETAH	VICUNA	HORSE
Ornithomimus	*Struthio*	*Acinonyx*	*Vicugna*	*Equus*

Figure 23-15 LENGTHENING, COMPACTION, AND FUSION OF METATARSALS, LOSS OF LATERAL DIGITS, AND SQUARING OR FUSION OF TARSALS in the left hind foot of selected cursors of three classes. Digits are numbered.

recovered as the system is unloaded, thereby relieving muscles in straightening the joint and giving an initial upward impetus to the entire body (Figure 23-18). Nearly 700 kg are required to break the major springing ligament in the front leg of the horse. The evolution of these ligaments has been traced in horses from the scars they left on fossil bones. They evolved from foot muscles at the time the animals were becoming large and fast residents of the open plains.

It is probable that some cursorial quadrupeds also store energy in the tendons of certain muscles in the manner described for kangaroos in the last section of this chapter.

Large grazers and browsers that are relatively free from predation need not be particularly agile. Water buffalos, camels, giraffes, and rhinoceroses cannot start, wheel, dodge, and stop quickly. Small antelopes and certain carnivores, on the other hand, can maneuver their bodies with almost unbelievable skill. A fairly light and supple body, alertness, and rapid neuromuscular coordination are primary requisites.

A consequence of the mechanics of curvilinear motion is that when a moving object turns, there is an outward, or centrifugal force that must be opposed by an equal inward, or centripetal force if the object is to avoid skidding out of its

STABILITY AND MAN-EUVERABILITY

Figure 23-16 FETLOCK JOINT OF A PRONGHORN, *Antilocapra*, SHOWING STRENGTHENING OF A HINGE JOINT BY SPLINES AND GROOVES.

curved course (Figure 23-19). Centrifugal force is directly proportional to mass and the square of velocity, and is inversely proportional to the radius of the turn. In order to turn sharply, a running animal must lean into the turn, adjusting the angle of its body so that the outward component of the thrust of its feet onto the ground equals the centrifugal force. This is opposed by the inward component of the force of the ground on the feet. There must be sufficient friction between feet and ground to prevent slipping—a requirement easily met by hoof and claw on rough ground, though a turning mouse may slip badly on a polished floor. Some animals can turn in several body lengths, even when running fast. Their bodies may make an angle of only 25–30° with the ground as they spin around.

Any force, **F,** that acts against the side of an animal (e.g., centrifugal force, wind pressure, or the jostling of another animal) tends to upset the animal with a torque of **F**h, where h is the height of its center of gravity above the ground (Figure 23-20). In order for stability to be maintained, **F**h must be less than **M**w, where **M** is the mass of the animal and w is half the width of the animal's stance. It follows that an animal that stands or moves straight ahead (keeping feet of both sides of the body on the ground at once) is most stable if it is broad and short-legged like the hippopotamus. Cursors must be long-legged, however, and often must lift both right or both left feet at once to move fast. Furthermore, as the previous paragraph explained, they must lean into turns when running. This is difficult for a highly stable animal. It follows that stability must be sacrificed for maneuverability. Agile cursors control lateral forces actively by adjusting posture rather than passively by virtue of body proportions.

Moreover, in order to turn sharply, a galloping mammal must lead with its inside front leg. That is, the foot toward the inside of the turn must strike the ground after its opposite in each couplet of footfalls. This provides that successive footfalls are in the direction of the turn and the animal can balance over its support. The fast carnivores, unlike ungulates, can change lead immediately before the forefeet are brought down. The hapless antelope cannot evade the cheetah's dash by dodging, for the cheetah follows almost in its tracks.

GAITS   A regularly repeating sequence and manner of moving the legs in walking or running is called a gait. Gait selection relates to rate of travel, maneuverability, stability, and to the size and structure of the body. If the footfalls of a pair of legs, fore or hind, are evenly spaced in time, as in pacing, walking, and trotting, the gait is said to be symmetrical. In walking gaits, each foot is on the ground more than half the time; in running gaits, each foot is on the ground less than half the time.

In the pace, the two feet on the same side of the body swing more or less in unison, which avoids interference between fore and hind feet. The camel family and some dogs pace naturally when moving moderately fast, and some harness horses are trained to race at this gait (Figure 23-21). All pacers are long-legged; the gait would be unstable for short-legged animals.

JERBOA
Jaculus

FLYING SQUIRREL
Petaurista

JACK RABBIT
Lepus

KINKAJOU
Potos

VICUNA
Vicugna

SLOW LORIS
Nycticebus

Figure 23-17 CONTRAST BETWEEN SELECTED JOINTS OF SALTATORS AND CURSORS (left) AND CLIMBERS (right) showing that the former have a deeper patellar groove at the distal end of the femur (above), more marked trochlea and grooves at the distal end of the humerus (center), and more blocklike carpals at the wrist (below). Each drawing is an anterior view of the left side of the body.

Figure 23-18   SPRINGING ACTION OF THE PRINCIPAL SUSPENSORY LIGA-MENTS IN THE FOOT OF THE HORSE.

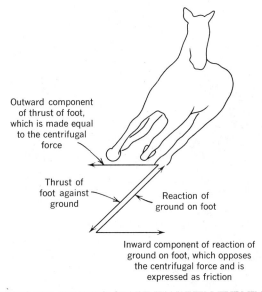

Outward component of thrust of foot, which is made equal to the centrifugal force

Thrust of foot against ground

Reaction of ground on foot

Inward component of reaction of ground on foot, which opposes the centrifugal force and is expressed as friction

Figure 23-19   FORCES RELATED TO STABILITY WHEN TURNING.

Figure 23-20 FACTORS INVOLVED IN STABILITY. (CG = center of gravity).

For stability,
Mw > **F**h

At the usual walk, the four footfalls are independent; a forefoot strikes the ground next after the hind foot on the same side of the body (lateral sequence) in most tetrapods except primates, because interference between forefeet and hind feet is then avoided. The same gait can be done at the run but is unusual except for certain show horses. Gaits having independent footfalls provide the most stability.

In the trot, which like the pace is usually performed at moderate speed by mammals, a fore and hind foot on opposite sides of the body swing about in unison. Since a line between the supporting feet passes close under the center of gravity, the gait is favorable for animals with broad bodies or, as for lizards, with legs splayed to the side of the body.

Most primates, and several other tetrapods do a diagonal sequence walk; a forefoot strikes the ground next after the hind foot on the opposite side of the body. The hind foot passes to the side of the forefoot to avoid interference. Some lemurs and small artiodactyls do the same gait at the run.

Galloping and bounding gaits are said to be asymmetrical because the footfalls of the two feet of a pair are unevenly spaced in time (Figure 23-22). The foot of a pair that strikes the ground first in each couplet of footfalls is called the trailing foot; the other is the leading foot. If the lead is the same, fore and hind, the gait is a transverse gallop (horse); if it is different, fore and hind, the gait is the more maneuverable, but less stable, rotary gallop (cheetah). Unless performed slowly, asymmetrical gaits increase length of stride by introducing periods of suspension when all feet are off the ground. The suspension may come when the legs are gathered under the animal (horse), stretched out fore and hind (deer), or both (cheetah, pronghorn). Rabbits, weasels, and many other mammals of comparable size place the forefeet on the ground alternately, but place the hind feet on the ground more or less in unison. This gait is the half bound. Many squirrels, mice, and other small mammals place each pair of feet in unison, thus doing the bound. Several artiodactyls may place all four feet in unison—a gait called the pronk. In the half bound and bound, each hind foot must be swung forward lateral to its corresponding forefoot to avoid interference; that is, the hind feet have a wider track than the forefeet.

Ricochetal mammals can accelerate faster from rest and can alter both the speed and direction of their motion faster than their quadrupedal relatives. These are important escape mechanisms that are purchased at the price of efficiency. Since the body must repeatedly be lifted against gravity, much energy is required. Saltators rarely maintain fast progression for long.

The height (*h*) to which an animal can jump is limited only by its upward velocity (**v**): $h = \mathbf{v}^2/2g$, where *g* is the pull of gravity. A 250 g galago has been accurately observed to jump vertically 2.26 m (7 ft $4\frac{3}{4}$ in.) from a crouch. Since its center of gravity was lifted more than 2 m, this performance is remarkable and must be close to a record for any animal, although a rat kangaroo, in jumping

SALTATION AND BIPEDAL RUNNING

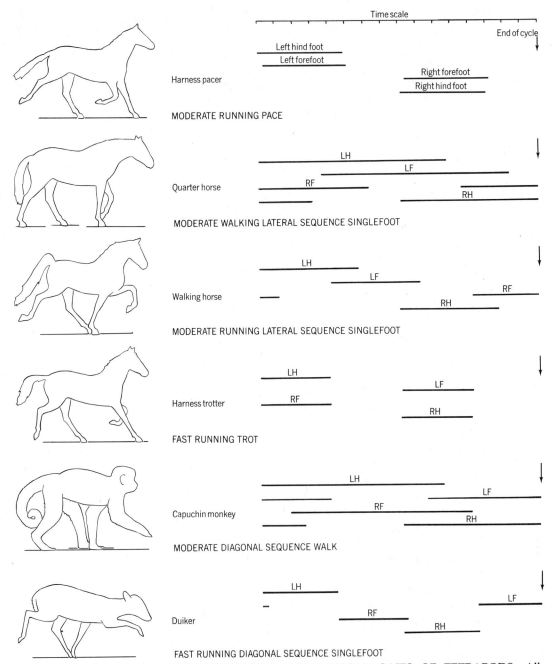

Figure 23-21 REPRESENTATIVE SYMMETRICAL GAITS OF TETRAPODS. All drawings show position of the body the instant the left hind foot strikes the round. In moving from top to bottom of the page the two forefeet rotate counterclockwise relative to the hind feet. The gait diagrams show by the length of the lines the duration of contact of the respective feet with the ground. Each diagram shows one complete cycle starting with the instant the left hind foot strikes the ground.

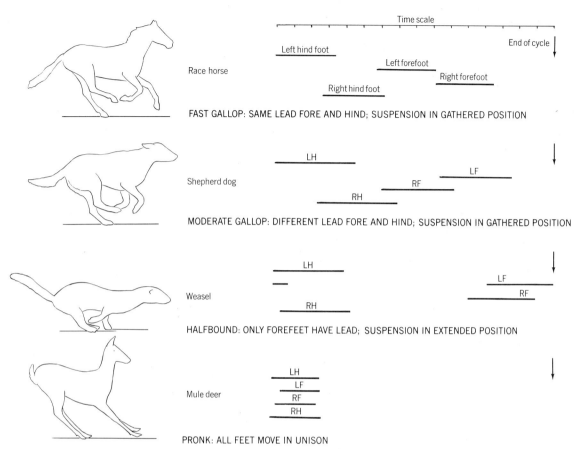

Figure 23-22 REPRESENTATIVE ASYMMETRICAL GAITS OF MAMMALS. All drawings show position of the body the instant the left (and trailing) hind foot strikes the ground. Notations for the gait diagrams are the same as for Figure 23-21.

2.4 m high, did about as well. The human high-jumper runs up to the takeoff in order to translate horizontal velocity to vertical velocity, yet lifts his center of gravity only a little more than 1 m. The takeoff velocity needed to lift the center of gravity 2 m is 625 cm/sec; that to lift it 1 m is 442 cm/sec.

The range (R) of an animal's jump depends on takeoff velocity and the angle of takeoff (Θ): $R = (\mathbf{v}^2 \sin 2\,\Theta)/g$. Theoretically, maximum range is attained when Θ is 45°, which is about the angle adopted by frogs and galagos for long jumps. The vertical jump of the galago as described above is the equivalent of a standing long-jump of about $4\frac{1}{4}$ m. The human long-jumper runs up to the takeoff, and because at maximum speed he cannot change the direction of his motion enough to take off at 45°, he takes off instead at 25–35°. Kangaroo rats and springhares usually ricochet with a takeoff angle that is also below 35°.

The acceleration required to achieve takeoff velocity is $\mathbf{v}^2/2s$, where s is the distance through which the applied force acts. This, in turn, is the difference in

Figure 23-23   RIGHT HIND LEG OF A RICOCHETAL RODENT, the jerboa *Allactaga*, showing proximal position of musculature, lengthened distal limb segments, reduced lateral digits, and some principal superficial muscles (left) and deeper muscles (right).

the functional length of the springing hind leg between its initial flexed and final extended positions. As their proportions attest, long hind legs are a great asset to saltators.

The mass (*m*) of the body does not enter directly into the formulas for height and range of jump. However, the force that must be applied to the ground to accelerate the body to takeoff velocity equals $m\mathbf{v}^2/2s$. As mass increases, the power plant increases in rough proportion, so that within limits, animals of different size but equal proportions can perform about the same. Nevertheless, the body cannot endure excessive forces, and when animals widely disparate in size are compared, complex physiological considerations must be considered. It is not possible to decide if the locust or kangaroo is the better jumper.

Relative lengthening of the hind limbs and of their distal segments is more extreme for saltators than for cursors (Figure 23-23). The tibia may be one and one-half or even two times as long as the femur (see kangaroo, Figure 23-9). Several tarsal bones may lengthen (frogs, tarsier, galago—see figure on p. 494). Lengthening of the hind limbs of ricochetal mammals is the more striking because the forelimbs are not modified for speed. They are usually used for slow progression and for handling food, so they do not become vestigial but often they are reduced in size. Some bipeds that swing the legs alternately, increase the length of the stride by oscillating the pelvis around the long axis of the vertebral column—man is a notable example.

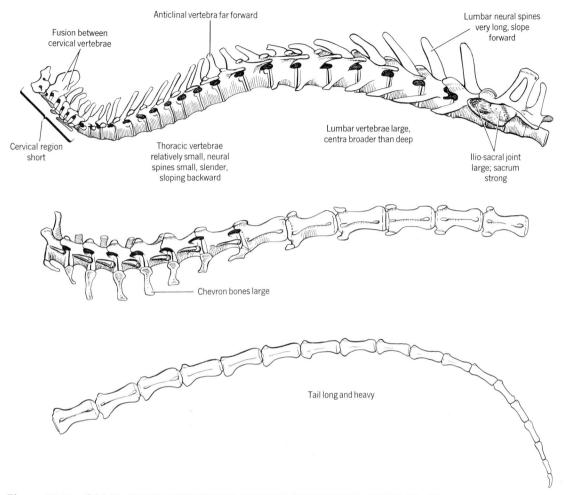

Figure 23-24   SOME CHARACTERISTICS OF THE VERTEBRAL COLUMN OF A RICOCHETAL MAMMAL shown by the springhare, *Pedetes*.

As kangaroos land from each hop, the tendons of their gastrocnemius and plantaris muscles (and to a lesser extent the muscle tissue itself) stretches, thus cushioning the impact and storing energy, springlike, for release during propulsion. The energy saving reaches an impressive 40%, and actually increases with speed. The mechanism is probably also developed, though to a lesser extent, in some other bipedal and quadrupedal hoppers and runners.

Bipedal reptiles share with ricochetal mammals the need for a long tail. It is elevated while running so that it will counterbalance the fore part of the body and thus bring the center of gravity over the hind feet. Bipedal lizards are unable to run if part of the tail is amputated; rabbits could not become ricochetal with so short a tail; absence of a tail correlates with man's upright posture. Most ricochetal mammals also use the tail as a prop, thus providing a third point of support when standing on the hind legs. Although once a jumper has left the ground its center of mass must move in a predetermined parabolic path until the animal touches the ground again, a saltator can change the orientation of its body in midjump by lashing its tail. The kangaroo rat can even reverse its field in the air, ready to bounce back along its own track on the next hop. The tail is often tufted with hair; its greater distal weight and air resistance then improve its effectiveness for controlling its jump.

Concentration of weight over the supporting hind legs is also achieved by shortening somewhat the presacral part of the spine, particularly in frogs and mammals. The lumbar region of the spine of ricochetal mammals is robust and has prominent neural spines. The thoracic region is slight and has small neural spines. Associated with these proportions is the relatively anterior position of the **anticlinal vertebra** (Figure 23-24), which is transitional between those with backward-sloping neural spines and those with forward-sloping neural spines. The cervical vertebrae of ricochetal rodents tend to fuse, apparently to reduce bobbing of the head.

Rodents that hop or bound tend to develop two ligamentous shock absorbers that limit whiplash of the spine. One is a fan of ligaments that runs down and forward from the summit of the enlarged neural spine of the second thoracic vertebra to support the cervical vertebrae. The other is an enlarged supraspinous ligament that runs from the forward-sloping neural spine of the last lumbar vertebra to the backward-sloping neural spine of the first (or second) sacral vertebra (Figure 23-24).

24

Digging; and Crawling without Appendages

Vertebrates that spend all, or most of their lives underground are said to be **fossorial.** Many other vertebrates are active above ground, yet are likewise highly adapted to dig for food or shelter, and are also included in this chapter. Some vertebrates live in burrows dug by other animals, and many tetrapods can dig somewhat, even without marked structural adaptation. Thus, the alligator lizard pushes its way into litter to hide, the thrush scratches leaves away to uncover food, the caribou paws snow from the lichens it eats, and the elephant scrapes holes with its forefeet to reach ground water where surface sources have gone dry. This chapter will stress the more highly adapted diggers.

The fossorial habit may have evolved in every class of vertebrates, though little is known of digging by ostracoderms and placoderms. Fossorial adaptations evolved independently in many orders of fishes and mammals and more than once in several orders. Legless progression on land is often related to burrowing and is described at the end of the chapter.

ADVANTAGES OF DIGGING

Fossorial vertebrates have various advantages:

1. Digging establishes microhabitats that are suitable for resting, aestivating, or hibernating. The burrow is cooler and more humid than the desert air (most desert rodents are active on the surface of the ground only at night and the remainder must retreat periodically to cool off), warmer than the winter storm (mountain chipmunks could not sleep above the snow without freezing), and relatively safe from lightning fires (many small forest tetrapods survive the flames).

2.   Many diggers, and most that are small, secure from the ground foods such as insects, insect larvae, earthworms, roots, and tubers. Several predators dig to secure smaller diggers as food.

3.   Diggers can store food underground where it is safe from other animals and the weather, and where it will be available during another season. Pikas and many kinds of rodents make large stores of dry grass or seeds; the arctic fox buries caches of birds and eggs.

4.   Nearly all fossorial vertebrates escape from predators by retreating underground. Many do not venture far from their holes and scamper back on the slightest sign of danger.

5.   Digging provides protected nests and dens in which to lay eggs or rear young. Many vertebrates, from the 10 g shrew mole to the 60 kg aardvark, raise their families underground.

SEVERAL WAYS OF DIGGING

Vertebrates dig in various dissimilar ways. A first method may be called **cover-up digging.** The animal merely covers itself with sand or soft mud. It does not make an open hole or progress through the substrate, and rarely digs deep. The animal may cover itself to escape from the environment at the surface, but usually does so either to lie in wait for prey (with protruding eyes uncovered) or to escape or hide from predators. The digger may shuffle into sand or mud (some fishes and anurans), vibrate its body for the few seconds it takes to submerge in sand (some desert lizards), run or swim rapidly and then dive into sand (several lizards, various fishes), or sway its body from side to side thus creating a trench into which it sinks (several snakes). Most of these animals have flat bodies and somewhat modified sense organs. Nevertheless, their adaptations are largely behavioral, and cover-up diggers are usually not said to be fossorial.

Soil-crawling is the term I shall give to a second general technique that is used by synbranchiform eels, caecilians, some salamanders, amphisbaenians, legless and short-limbed lizards, and burrowing snakes (Figure 24-1). The animal moves through soil that is usually soft or sandy but may be quite firm. Often the substrate closes behind the animal so a permanent burrow is not established, but the soil may instead be compacted, thus opening a burrow. A high degree of specialization is required. The body is long and slender, which reduces the amount of soil that must be displaced. Legs are reduced or absent. The head is relatively firm and usually serves as the digging instrument.

Another method of digging is **scratch-digging.** By alternately flexing and extending its limbs in the manner of a dog with a bone to bury, the animal cuts and loosens the soil with its claws and pushes or flings it to the rear. Some turtles, some birds, carnivores, ground squirrels, and a variety of other mammals dig in this way. Nests, dens, and burrows are excavated, often in hard soil.

A fourth method, **chisel-tooth digging,** is followed by gophers, mole-rats, and various other rodents. Huge gnawing incisors and powerful jaw and neck muscles are used to dislodge the soil, which is then moved with head or feet. Hard soil can be excavated, but somewhat damp or otherwise tractable soil is usually preferred.

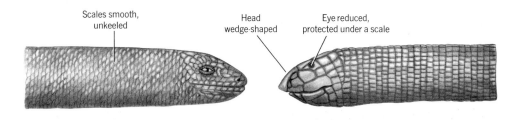

Figure 24-1 REPRESENTATIVE SOIL-CRAWLERS showing some adaptations for this mode of digging.

A fifth method is **humeral-rotation digging.** The method is best exemplified by the true moles. A mole placed on the surface can dig out of sight in the damp soil it prefers in about 6 sec. These subterranean digging machines have broad shovel-like forefeet and short powerful forelimbs. There is no pronation and supination of the forearm, and (in sharp contrast to scratch-diggers) movement at the elbow merely positions the hand without providing a forceful stroke. The elbow is positioned high above the shoulder, and the power for digging comes from the rotation of the uniquely short but broad humerus around its own long axis (Figure 24-2). This is accomplished in moles primarily by the relatively enormous teres major muscle. The echidna holds its wide humerus horizontal to the ground and rotates the bone around its long axis when it walks. It is probable that these strong animals are also humeral-rotation diggers.

Some vertebrates dig by more than one method. Many rodents are both scratchers and tooth-chiselers; several reptiles either scratch with short forelimbs or fold them away and use soil-crawling. Also, various diggers do not fit into this classification (e.g., the dipnoan that digs into the mud using body and fins, the jaw fish that moves and arranges rocks with its powerful jaws, the mole-rat that moves earth with its nose pad or head, and the young crocodile that bites chunks of clay from a muddy bank). Some vertebrates make themselves uninvited guests in burrows (vacant or occupied) dug by their more fossorial cousins, doing only house cleaning and slight alterations on their own.

Figure 24-2 THE MECHANISM OF HUMERAL-ROTATION DIGGING shown by the mole, *Scapanus,* in anterior view.

FOSSORIAL VERTEBRATES

Agnaths and Fishes.   The flattened bodies and dorsal eyes of cephalaspids and antiarchs indicate that they were bottom feeders. Perhaps some were cover-up diggers when it was best to get out of sight. Skates and rays are dorsoventrally flattened and may cover themselves up lightly, leaving eyes and spiracles free. Flounders are bilaterally flattened and lie on one side. The eye of the down side migrates to the up side during ontogeny. Light cover-up digging may supplement protective coloration in making these fishes hard to see. Among the more fossorial bony fishes are some gobies and catfishes, the jaw fish, and various eels. Cichlid and centrarchid fishes dig breeding holes in the substrate. Synbranchiform eels dig deep and extensive burrows. Dipnoans dig vertical burrows into the mud to wait out times of draught.

Amphibians.   Caecilians (see figure on p. 59) are soil-crawlers. Some short-legged urodeles are also soil-crawlers that wriggle through litter and loose soil. Other urodeles, including the ambystomids, or mole-salamanders, have stout bodies and legs with which they dig. Anurans commonly dig into mud, and various species dig holes in soil using their hind legs or (more rarely) their forelegs and heads.

Reptiles.   Many snakes are fossorial. Amphisbaenians and legless lizards (Figure 24-1) have adopted similar habits and convergent structure. Many skinks and other lizards also burrow or cover themselves up, and the tuatara retreats underground. All turtles bury their eggs using the hind feet as scoops. Tortoises have modified the forelimbs as effective digging tools.

Birds.   No bird has evident structural adaptations for digging, but shearwaters, puffins, an owl, and several other birds nest in holes or burrows that they either appropriate from mammals or laboriously construct themselves with beak and feet.

Monotremes and Marsupials.   The platypus and the echidnas are powerful diggers; the playpus exposes its claws by folding away the webs used when swimming. The marsupial mole is among the most highly adapted of subterranean mammals; it is a small scratcher (Figure 24-3). One bandicoot digs, and the wombat, which weighs about a third as much as a man, may excavate a burrow 30 m long.

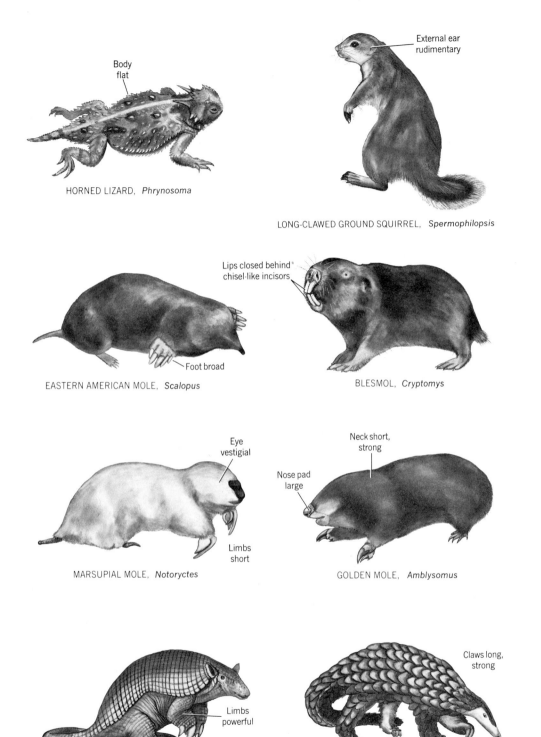

Body
flat

HORNED LIZARD, *Phrynosoma*

External ear
rudimentary

LONG-CLAWED GROUND SQUIRREL, *Spermophilopsis*

Lips closed behind
chisel-like incisors

Foot broad

EASTERN AMERICAN MOLE, *Scalopus*

BLESMOL, *Cryptomys*

Eye
vestigial

Limbs
short

MARSUPIAL MOLE, *Notoryctes*

Neck short,
strong

Nose pad
large

GOLDEN MOLE, *Amblysomus*

Limbs
powerful

GIANT ARMADILLO, *Priodontes*

Claws long,
strong

PANGOLIN, *Manis*

Figure 24-3 FURTHER EXAMPLES OF DIGGERS showing some of their adaptations.

Insectivores. About a dozen genera of true moles (family Talpidae) are extremely effective humeral-rotation diggers. Moles can progress just under the surface in the damp soil they prefer at 2 body lengths/min and 200 body lengths/day. The unrelated golden moles (family Chrysochloridae) are among the most specialized of subterranean scratch-diggers. Hedgehogs, mole shrews, and tenrecs also make burrows.

Edentates, Pangolins, and the Aardvark. Armadillos (nine genera), pangolins, and the aardvark are the most powerful of scratch-diggers. Anteaters do not burrow but rip into termite nests and the soil to secure insects.

Carnivores. The six genera of badgers and the ratel are fossorial; some of them dig out burrowing rodents for food. Various canids excavate dens or dig to cache food, yet have scant structural adaptation for digging.

Rabbits and Rodents. The plains pika and the less cursorial of the rabbits are moderately good diggers. The order Rodentia has more burrowing representatives than any other; only some of them are mentioned here. The mountain beaver digs in stream banks. Ground squirrels, marmots, and prairie dogs (all of the family Sciuridae) are avid scratch-diggers. Many of them make extensive burrow systems in hard soil. Gophers (eight genera in the family Geomyidae) are subterranean rodents of North America that make large burrow systems. Kangaroo rats, jerboas, and springhares are all ricochetal, yet manage to dig daytime retreats in sandy soil. Four genera of African mole-rats or blesmols (family Bathyergidae) are expert chisel-tooth burrowers, and the fifth genus of the family is a scratcher. The Mediterranean mole-rat (Spalacidae), bamboo and root rats (Rhizomyidae), and all but one of the mole voles (Muridae) are chisel-tooth burrowers. The Asian mole-rat, or zokor (Cricetidae), has enormous foreclaws and is a scratch-digger. The related shrew mouse, burrowing mouse, and mole mouse, and the unrelated tuco-tuco (Ctenomyidae), all of South America, are also scratchers. Their neighbor, the coruro (Octodontidae), is probably a chisel-tooth burrower.

To better appreciate the prowess of fossorial animals, imagine a sports event that would determine which human contestant could first dislodge 30 times his body weight of firm soil using only his finger nails, and then, using his hands and feet, transport all the dirt 10 m distant and pile it onto a platform that is as high as he can reach. The little tuco-tuco does this daily, not in competition but in his morning grounds. And, who knows, in the animal olympics the tuco-tuco might not make it past the gophers, mole-rats, and blesmols to the finals.

GENERAL REQUIREMENTS OF DIGGERS

Fossorial vertebrates must be effective in meeting certain requirements: (*1*) All diggers must exclude sand, dust, and earth from the mouth, eyes, ears, respiratory passages, and cloaca or anus. (*2*) All except some cover-up diggers must maneuver within the soil or inside confined spaces. Many must find their way, detect and avoid predators, and for some, locate mates and protect eggs in complete darkness. (*3*) The energy cost of digging being very high, the better diggers must conserve energy and dissipate heat. (4) All that do not dig in sand, litter, or soft mud must break or loosen the soil. (5) All that make open burrows must either transport and dispose of loose soil or compact it. How do the different diggers meet these requirements?

Large diggers (wombat, aardvark, badger) probably can keep their mouths out of the dirt, at least most of the time. Fossorial amphibians, reptiles, and insectivores have close registration of the closed jaws; the margin of one jaw often fits into a groove in the other to make a tight seal. The lower jaw of some burrowing reptiles is recessed behind the upper jaw. The furred lips of chisel-tooth diggers meet behind the protruding incisors, thus enabling the animal to exclude dirt from the mouth at the same time it is gnawing.

KEEPING DIRT OUT OF MOUTH, SENSE ORGANS, AND LUNGS

Diurnal vertebrates that burrow to nest, rear young, or hibernate, yet do their foraging above ground have eyes of normal size (tortoises, hedgehogs, ground squirrels, canids). Nocturnal rodents that burrow to escape daytime heat have large eyes (kangaroo rats, jerboas). Presumably these animals close their eyes as needed when digging. Possibly some have evolved improved mechanisms for cleansing the eyes. Snakes and certain lizards have fused but transparent eyelids. Several genera of burrowing snakes have "horns" that protrude over the eyes and are thought to provide protection. The real specialists have small to minute eyes (monotremes, armadillos, pangolins, gophers, African mole-rats, root rats, tuco-tuco) or vestigial eyes that may differentiate light from dark but form no image and often are hidden under the skin (caecilians, amphisbaenians, marsupial mole, true moles, golden moles, Mediterranean mole-rat).

Many burrowing amphibians and reptiles have no external auditory canal. If there is a tympanum, it is thick. Sound reception is often mediated by a special mechanism involving the skin, jaw, or other structures. The external auditory canal of burrowing mammals tends to be small. It is probable that some diggers can close the canal.

The external nares of diggers are also small. In digging reptiles the outer part of the nasal passage is narrow and slopes upward. Moisture around the nostrils may cause adhesion of sand and prevent the inhalation of single grains. The openings can be closed or at least constricted by muscular valves or erectile tissue and may be covered by a fold. The armadillo is able to suspend breathing for 3 or 4 min while digging vigorously in dust and sand. Some shrews have peculiar diverticula from the lungs that seem moderately effective at trapping and disposing of foreign material.

Maneuvering in the confines of a narrow burrow is facilitated in several ways. Some fossorial amphibians and reptiles and all fossorial mammals are short-legged. The spine is relatively stiff in burrowing snakes, but flexible in fossorial mammals. Some can turn very sharply by doubling into a ball and then unrolling in the new direction. Several fossorial vertebrates have such loose skins that the animal can to a degree turn within its skin and then let the skin follow. All burrowers are adept at backing up.

MANEUVERING AND WORKING UNDERGROUND

Some diggers have special reasons for having long tails: Ricochetal species use theirs for balance, the pangolin uses his as a grasping organ when climbing, and numerous species use theirs as a prop when digging. Otherwise, the evolutionary process seems often to have found a tail to be in the way underground and it tends to be short, even in burrowing snakes and some other legless diggers (though some amphibians and lizards are exceptions). Some small subterranean vertebrates have little or no tail (caecilians, golden mole, Mediterranean mole-rat, and various others).

Virtually nothing is known about how subterranean animals find their way in the absolute darkness of a deep burrow system. Fossorial reptiles seem to smell or hear their prey. Bone conduction of sound is usual. Many burrowing mammals are particularly sensitive to low frequency sound. The middle ear and tympanum tend to be large. Are the vibrissae of the mammals particularly effective? Do they use odors or a tactile memory? Is the tail, when retained, particularly sensitive, as has been claimed?

Many burrow systems are nearly saturated with warm vapor. Since digging is hard work, the animals have a problem in dissipating excess heat. All but the smallest have a low basal metabolic rate and a wide range of thermoneutrality (temperature at which oxygen consumption is minimal). Gophers radiate heat from their nearly hairless tails, and one genus of African mole-rat is entirely naked. Carbon dioxide levels tend to be high in closed burrows (to 4% and higher), and fossorial vertebrates are relatively tolerant of these concentrations. Likewise, they are adapted to the low levels of oxygen (to 15% and lower) that may prevail.

BREAKING THE SOIL

SELECTION AND MAINTENANCE OF TOOLS   When a man shovels dry sand or forest litter, he must expend energy to transport the material but virtually none to break or loosen it. Similarly, animals that use the cover-up and soil-crawling methods of digging usually confine themselves to sand, litter, or soil that is loose enough to be compacted or pushed aside without first being broken up. Accordingly, they may have no tools for breaking compact soil.

When a man shovels damp earth he must break it free before he lifts it, but it is not difficult to do so. A large shovel blade is satisfactory. Similarly, humeral-rotation diggers confine their activities to moderately soft soil. They compact, break, and cut the soil by pushing on it or scraping it with their stout claws. Again, a large "shovel blade" is effective. Moles have broad claws and very wide forefeet (Figures 24-2 and 24-9). Similarly, amphisbaenians dig by ramming the wedge-shaped head into the soil and then (according to species) lifting the head to compact the soil into the wall of the burrow.

The man who digs in dry compacted soil must expend much energy on breaking the soil prior to moving it. The shovel cannot be forced into the undisturbed material, so a pick is first used to loosen it. The pick is effective because it delivers great force to a restricted area; that is, with each blow it applies high pressure to a limited area. Chisel-tooth and scratch-diggers among vertebrates tend to avoid rocky and otherwise intractable soils, yet many of them burrow in remarkably hard earth. They gnaw with their incisors or scratch with their sharp

Figure 24-4   FOOT OF THE SPADEFOOT TOAD, *Scaphiopus*, showing tubercle for digging.

claws, thus applying great pressure to a small area before going on to another spot. The blade must be long and strong to do its job. The badger has five claws, the longest of which is half the length of its forearm between elbow and wrist joints. The mountain beaver, tuco-tuco, gophers, and ground squirrels emphasize three or four claws, the longest of which may (in some gophers) be three quarters of the length of the forearm. The anteaters, marsupial mole, and golden moles emphasize one or two claws, the longest of which may (in some golden moles) be longer than the forearm (Figure 24-7). The "shovels" of moles can exert against soft soil more force in relation to body weight than can the "blades" of ground squirrels, but the ground squirrel exerts against harder soils about twice as much pressure as does the mole.

The tools of diggers are subject to tremendous wear. Burrowing reptiles compensate by molting successive generations of the epidermis, thus exposing new surfaces to the substrate. The upper incisors of some gophers grow out at the rate of 248 mm/yr. The lower incisors, which are maneuvered more as the condyle of the mandible slips in its loose groove, may grow at the rate of 445 mm/yr. These rates are $2\frac{1}{2}$–3 times those recorded for some nonfossorial rodents of comparable size. Likewise, the center foreclaw of a gopher grew out at 90 mm/yr, and that of a tuco-tuco at 72 mm/yr.

The incisors of rodents and of the wombat have enamel only on their forward surfaces. The softer dentine wears away behind, thus providing a self-sharpening mechanism.

The spadefoot toad digs with a different kind of tool. There is a horny epidermal tubercle at the edge of the hind foot that is used as the animal progresses backward into loose soil (Figure 24-4).

DESIGN FOR LARGE OUT-FORCES   Fossorial vertebrates that dig in firm soil must be capable of applying great force against the substrate. Therefore, unlike cursors and climbers, they are constructed so that their relevant bone-muscle systems (particularly of the forelimb) produce large out-forces (F_o). In Chapter 22 it was shown that $F_o = F_i I_i / I_o$, where F_i is the in-force and I_i and I_o are, respectively, the in- and out-levers.

It is evident that one way to increase F_o is to reduce I_o. Consequently, the more expert diggers all have short legs and necks. In sharp contrast to the limbs of cursors, the limbs of diggers have relatively short distal segments. The radius is nearly always shorter than the humerus, and the manus, exclusive of the terminal phalanx with its claw, is markedly shorter than the radius. Although strong, the metacarpals may be very short (tortoises, echidna, moles) (Figure 24-5) and the proximal phalanges may be even broader than long (echidna, pangolin, anteaters, moles).

A second way to increase an out-force is to increase the related in-lever. Accordingly, muscles used in digging tend to insert far from the joints they turn. The insertion of the mammalian deltoid muscles commonly extends more than

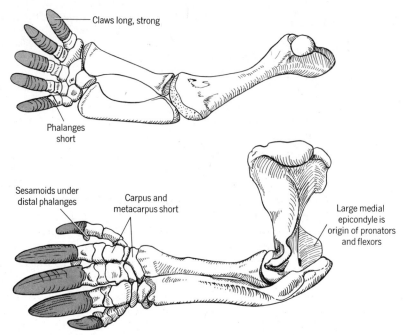

Figure 24-5   SOME ADAPTATIONS FOR DIGGING shown by dorsolateral views of the left forelimb skeletons of the tortoise, *Gopherus* (above), and echidna, *Tachyglossus* (below).

halfway down the length of the humerus (away from the shoulder joint which pivots) (Figure 24-6). Part of the latissimus dorsi of the golden mole increases its in-lever to the shoulder joint by shifting its insertion from the proximal part of the humerus (the usual position) nearly to the elbow joint (Figure 24-7), and similar leverage is gained by the unusual thoracic head of the triceps of the armadillo. The wide median epicondyle of the humerus, which is a feature of all scratch-diggers, increases the in-lever of the pronator of the forearm. A relatively proximal origin on the humerus of the long supinator muscle increases its in-lever and enables it to flex the manus as well as supinate the forearm (Figure 24-8). A relatively long pisiform bone at the carpus increases the in-lever of one of the flexors of the manus. Crucial to the special mechanism of the humeral-rotation digging of moles is a very wide flaring tubercle for the insertion on the humerus of the enormous teres major muscle. This carries the insertion away from the central long axis of the bone and thus increases the in-lever that makes possible powerful rotation of the humerus around its own long axis (Figure 24-9).

These kinds of adaptations of diggers are particularly striking when in-levers are expressed as fractions of their related out-levers. Thus, in measuring representative skeletons of 27 genera of expert diggers belonging to seven mammalian

Figure 24-6   SOME ADAPTATIONS FOR SCRATCH-DIGGING shown by lateral views of left forelimb skeletons.

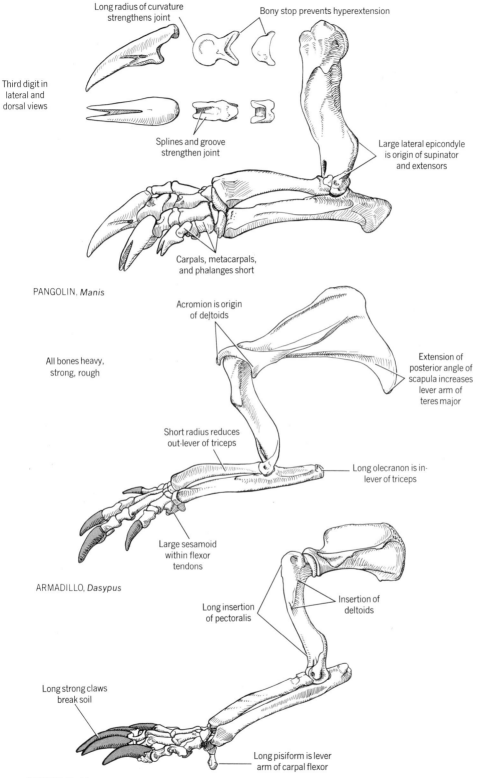

Long radius of curvature
strengthens joint

Bony stop prevents hyperextension

Third digit in
lateral and
dorsal views

Splines and groove
strengthen joint

Large lateral epicondyle
is origin of supinator
and extensors

Carpals, metacarpals,
and phalanges short

PANGOLIN, *Manis*

Acromion is origin
of deltoids

All bones heavy,
strong, rough

Extension of
posterior angle of
scapula increases
lever arm of
teres major

Short radius reduces
out-lever of triceps

Long olecranon is in-
lever of triceps

Large sesamoid
within flexor
tendons

ARMADILLO, *Dasypus*

Long insertion
of pectoralis

Insertion of
deltoids

Long strong claws
break soil

Long pisiform is lever
arm of carpal flexor

BADGER, *Taxidea*

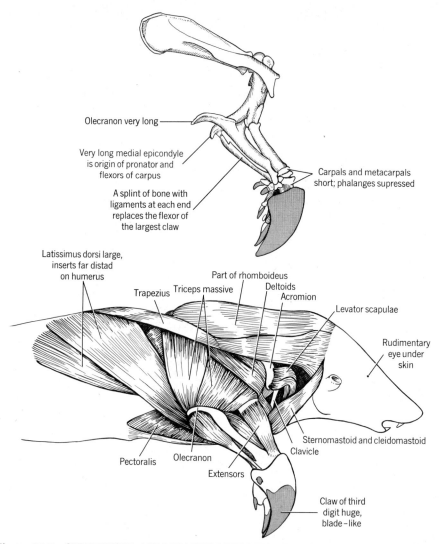

Figure 24-7 STRUCTURE ASSOCIATED WITH THE SCRATCH-DIGGING OF THE GOLDEN MOLE, *Amblysomus.*

orders I find the olecranon process (the in-lever of the triceps) to be about $\frac{1}{5}$ (a ground squirrel), $\frac{1}{3}$ (gopher, African mole-rat), $\frac{1}{2}$ (aardvark, pangolin, mole), $\frac{2}{3}$ (Mediterranean mole-rat, armadillos), or even $\frac{3}{4}$ (marsupial mole, golden moles) the length of the ulna distal to the pivot at the elbow joint (Figure 24-7). A complete listing (if the data were available) would doubtless show that diggers differ from their nonfossorial relatives in having larger values of I_i/I_o for every bone–muscle system used in digging.

A third way to increase out-force is to increase in-force. The relevant muscles of diggers are enormous. To accommodate such muscles, origins and insertions are large. This, together with their proportions, makes the forelimb bones of

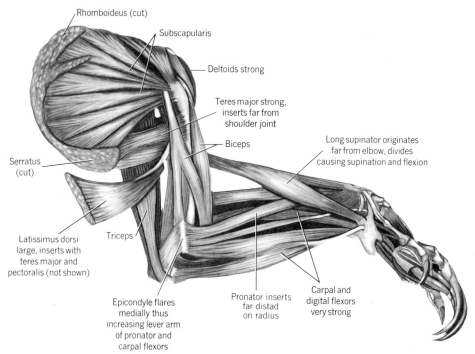

Rhomboideus (cut)

Subscapularis

Deltoids strong

Teres major strong, inserts far from shoulder joint

Biceps

Long supinator originates far from elbow, divides causing supination and flexion

Serratus (cut)

Latissimus dorsi large, inserts with teres major and pectoralis (not shown)

Triceps

Epicondyle flares medially thus increasing lever arm of pronator and carpal flexors

Pronator inserts far distad on radius

Carpal and digital flexors very strong

Figure 24-8 STRUCTURE ASSOCIATED WITH THE DIGGING OF THE GIANT ANTEATER, *Myrmecophaga,* shown by a medial view of the left forelimb. (Drawn from an air-dried dissection and hence somewhat shrunken.)

diggers rugged and rough. The medial epicondyle of the humerus (origin of flexors of digits) and deltoid crest (insertion of deltoids) are particularly prominent. The posterior angle of the scapula may be enlarged to accommodate the origins of the teres major and long head of the triceps. The anterior segment of the sternum of true moles and golden moles is long and deep to receive their great pectoral muscles (Figure 24-9). Chisel-tooth diggers have large areas of origin and insertion for their powerful jaw muscles. Diggers that push dirt with their heads (many amphisbaenians, golden moles, marsupial mole, Mediterranean mole-rat) have a large flat occipital area for the insertion of strong neck muscles (Figure 24-10).

There is another factor in design for large out-forces. When starting a burrow, some soil-crawlers loop the body over the head thus weighting it so it can be thrust into the soil. Large scratch-diggers hunch the back over the forelimbs and may prop the body with the tail, thereby applying the weight of the body to the digging tools. Subterranean diggers, however, are usually small and hence light. (They can then dissipate heat more easily, feed on the small food found in the earth, keep within one stratum of the soil, and avoid rocks and large roots.) In order to prevent motions of digging from merely pushing them away from the soil, they must force their bodies against their digging tools. Humeral-rotation diggers brace against one side of the burrow with one forepaw while digging

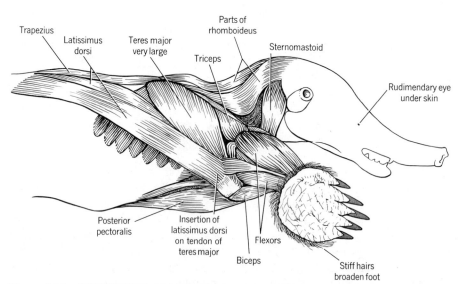

Figure 24-9 STRUCTURE ASSOCIATED WITH THE HUMERAL-ROTATION DIG-GING OF MOLES. Above, ventral view; center and below, lateral views. Above and center, the Western American mole, *Scapanus;* below, the Old World mole, *Talpa.*

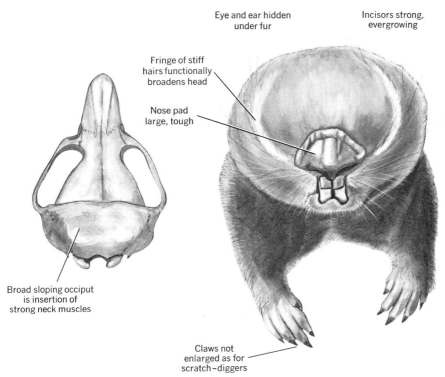

Eye and ear hidden
under fur

Incisors strong,
evergrowing

Fringe of stiff
hairs functionally
broadens head

Nose pad
large, tough

Broad sloping occiput
is insertion of
strong neck muscles

Claws not
enlarged as for
scratch–diggers

Figure 24-10 SOME ADAPTATIONS OF A CHISEL-TOOTHED DIGGER THAT
MOVES DIRT WITH ITS HEAD, the Mediterranean mole-rat, *Spalax*.

with the other. To make this possible, the forelimbs are positioned laterally,
opposite to one another. Human weight lifters lift overhead about twice their
own body weight. It has been shown that by its thrust, a mole can move as much
as 32 times its own body weight.

If an amphisbaenian presses up with its head it presses down with its "chest."
As chisel-tooth diggers press up with their lower incisors they press down with
their forefeet. Small burrowers also brace themselves with their hind legs. Xen-
opid frogs can at the same time lengthen the back at the sacroiliac joint, thus
forcing the head forward in the mud. In response to the use of the hind legs for
bracing, the innominate bones of the mammals tend to be nearly horizontal (in
line with a forward thrust). The hip joint is relatively far dorsal to be on a level
with the spine. This reduces compressive forces at the pelvic symphysis, which
is nearly always weak and sometimes absent (moles, anteaters, pangolins, and
some gophers) (Figure 24-11). The innominate bones are firmly sutured with, or
fused to, a relatively large number of vertebrae, and the sacrum is long.

RESISTANCE TO THE FORCE OF THE SOIL ON THE BODY   Soil-crawlers reduce or resist the force of the soil on the body in various ways. The skins of
burrowing fishes are commonly particularly rich in mucous glands. Scales are
absent (various fishes, amphibians) or, if present, smooth and unkeeled (reptiles)

Acetabulum faces
laterally

Pelvic symphysis
absent

POCKET GOPHER (♀)
Thomomys

Innominate parallel
with sacrum,
fused

Acetabulum nearly
on level with spine

Ischium fused
to sacrum

WESTERN AMERICAN MOLE
Scapanus

Many sacral
vertebrae

Ischium
articulates
with sacrum

ANTEATER
Tamandua

Ilium
fused to
sacrum

Pelvic
symphysis
weak

PANGOLIN
Manis

Figure 24-11   SOME CHARACTERISTICS OF THE PELVES OF FOSSORIAL MAM-
MALS. Ventral views above; dorsal views below.

to reduce the friction of soil against the body. Friction is further reduced in
uropeltid snakes by microscopic ridges that reduce wetting in damp soil. The
head is short and narrow. The skull is relatively solid; most sutures are obliterated
and others are serrate. There is no neck or shoulders. The reduction and loss of
legs is itself a major accommodation to streamlining. In order to displace as little

soil as possible the trunk becomes long and slender—in other words, snakelike. Unspecialized lizards have about 23 presacral vertebrae whereas fossorial lizards may have 60, and amphisbaenians more than 100 presacral vertebrae. Caecilians and snakes may have 250 or more such vertebrae. Cervical and lumbar ribs are usual.

Vertebrates that move along narrow burrows must avoid snagging and abrading themselves against burrow walls. Subterranean diggers have small external ears or none at all. Fur is lax, often short, and sometimes nearly upright so it can brush in any direction. At least some of the mammals clean their fur by shaking the body, dog fashion. Others groom away mud and dirt with the paws.

There must be provision against dislocation and hyperextension of joints of the forelimb and manus when a digit snags on a rock or root. Hyperextension of the phalanges of echidnas, pangolins, and anteaters (at least) is prevented by squared articulatory surfaces or bony stops that limit rotation of the joints (Figure 24-6). Dislocation of the digits is prevented in pangolins and anteaters (and others?) by large areas of bone-to-bone contact at the joints and by deep interlocking splines and grooves, recalling those of the phalanges of ungulates. The scapula of gopher tortoises is braced against the plastron more firmly than is that of turtles. Protection is also afforded by generally heavy and rugged construction, and in some instances by structural unity (i.e., common firmness) of the palm (tortoises, echidna, moles). Some ligamentous checks against dislocation have also been described.

TRANSPORTING AND DISPOSING OF SOIL

Diggers that excavate dens or burrows must transport and dispose of earth after they loosen it. Many snakes use coils of the body to sweep sand and other debris out of burrows. Lizards and turtles sweep with their feet. Some reptiles create an air space under the body so they can breathe in spite of an overburden. As mammals break the soil with forefeet or teeth it is pushed underneath or beside the body. From time to time the animal kicks this loose soil back out of the way with the hind feet. When the burrow behind becomes choked with soil it is time to take it away, and several methods are used. Some diggers back up in the burrow vigorously kicking the dirt back as they go. The hind feet are used simultaneously (tuco-tuco, African mole-rats, armadillo) or alternately (hedgehogs). Other burrowers turn around to push the dirt forward using the forefeet only (moles); forefeet, chest, and chin (gophers); or nose and the top of the head (gopher tortoise, Mediterranean mole-rat). The gopher tortoise may push dirt with the upturned front of its plastron. Leaf-nosed and hog-nosed snakes probably also use their heads as shovels, as do synbranchiform eels and amphisbaenians. Most dirt is moved to the surface, but some is used to plug abandoned side tunnels of the burrow system.

Whichever foot, fore or hind, is used for moving the soil, it is made broad in one or more ways: The toes may be webbed (toads, sea turtles and others, moles, golden moles), the pad of the foot may be widened by cartilages or bones placed lateral to the first digit (mountain beaver, moles, gophers, tuco-tuco, etc.), or the pad of the foot may be fringed with stiff hairs (nearly all subterranean mammals). The Mediterranean mole-rat similarly increases the effectiveness of its broad flat head as a dirt pusher by adding lateral fringes of stiff hairs (Figure 24-10).

Some fossorial vertebrates find it necessary to tamp the earth of a surface mound to prevent back-slip, or of the walls of a burrow to make them smooth. The Mediterranean mole-rat has a broad, tough, muscular nose pad that it uses assiduously for this purpose. Golden moles and the marsupial mole may use their nose pads in like manner. One of the African mole-rats has a peculiar way of extending its hind feet on a surface mound of earth and then vibrating them at about 22 cycle/sec to tamp the soil.

CRAWLING WITHOUT APPENDAGES

Terrestrial locomotion without limbs has evolved independently several times among amphibians and lizards, but is best exemplified by snakes and amphisbaenians. Several thousand species are involved, so the adaptation must be considered highly successful. It is probable that the ancestors of these animals first evolved long slender bodies to enable them to enter crevices or move by undulation, and then secondarily lost their limbs.

There are four principal ways in which legless vertebrates move, though more than one may be used at a time, and variants, thrashing, and even jumping have been noted. The first, and most common method, is **lateral undulation.** The body is thrown into serpentine loops, right and left. The animal locates with its coils several projections such as pebbles or plant stems. The body then presses sideways (not downward) against these objects in a direction that is obliquely backward in relation to the direction the snake is to move. The mechanical analysis for the forces at any one projecting object resembles that for the action of a fish tail: The thrust of the snake against the object (F_t in Figure 24-12) is opposed by an equal and opposite force (F_o) exerted by the object against the body. This force has a forward component (F_f) in the direction that the snake as a whole is moving, and a lateral component (F_l). (The value of F_f is somewhat reduced by friction—see below.) Motion is continuous as waves travel along the body. Each coil stays in the same place, the body following in the trail established by the head.

A snake cannot move forward by sliding a coil past a single object against which it pushes as it goes. This is because that particular coil does not move in the direction in which the snake as a whole is moving, but instead moves along its own long axis as shown in Figure 24-12. That direction being at right angles to F_o, F_o can have no component in the coil's direction of motion. If F_o (or a component thereof) were in the direction that a second coil moves past a second object, then the snake could push or pull past the second object by pushing the first coil against the first object. For continuous motion, the snake requires three or more objects that cannot all be on the same side of the body. The action is more efficient if few objects are used (3–5 unless the snake is very long and slender). The lateral components of the various thrusts add up to zero, and the sum of the forward components (less frictional resistance) is the propulsive force of the animal. It is seen that all the forces are interconnected. As the "neck" contacts new objects and the tail slides away from others, or as any object slips, the animal must instantly adjust both the magnitude and direction of the thrusts

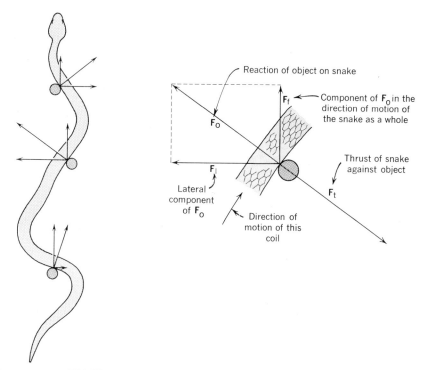

Figure 24-12 DIAGRAM OF LATERAL UNDULATION OF A SNAKE. (The value of F_f is somewhat reduced by sliding friction.)

of all active coils. One must be impressed by the complexity of the feedback mechanism and nervous control required.

Snakes are unable to progress along narrow burrows by lateral undulation because then they cannot thrust obliquely backward with their coils. Further, they cannot progress by this method on smooth surfaces because they must thrust laterally, not vertically. There is, of course, a vertical force from the animal's weight. This causes friction, which in this instance is undesirable. The coefficient of friction (see p. 491) is minimized by the smooth nature of the snake's ventral scutes. There is also sliding friction against the projections where the snake thrusts, and this also counters the propulsive force of the animal. Snakes can move very freely among rotating pegs placed on lubricated glass.

The second method of legless progression is **rectilinear movement.** It is used by various snakes and all amphisbaenians—particularly if short bodied. The skin fits loosely over the ventral part of the body and is very distensible. Muscles that slant back and down from the ribs to the scutes cause the ventral skin to bunch at several regions so that the scutes overlap (Figure 24-13). Between these regions the skin is stretched. Where scutes are bunched they rest on the ground; where stretched they are lifted clear of the ground. One by one, additional scutes are drawn into each bunched region from behind as others are stretched away in front. Thus, the scutes move along somewhat in the manner of the feet (or prolegs) of caterpillars—each starts and stops as it goes. The bunched scutes

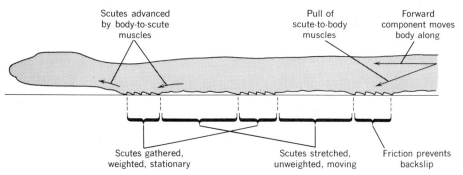

Scutes advanced by body-to-scute muscles

Pull of scute-to-body muscles

Forward component moves body along

Scutes gathered, weighted, stationary

Scutes stretched, unweighted, moving

Friction prevents backslip

Figure 24-13   DIAGRAM OF RECTILINEAR MOVEMENT OF A SNAKE.

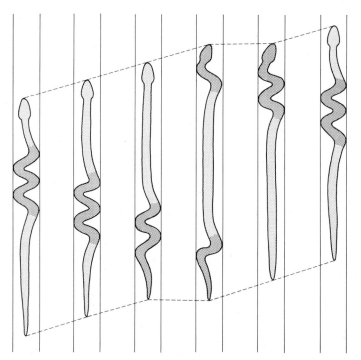

Figure 24-14   DIAGRAM SHOWING SUCCESSIVE POSITIONS IN THE CONCER-TINA MOVEMENT OF A SNAKE PROGRESSING IN A TUNNEL. Dark shaded parts of the body are stationary; light shaded parts are moving.

thrust obliquely backward against the substrate, and friction is required to prevent slipping. Muscles that slant backward and up from the scutes to the ribs haul the body along within its skin in continuous motion. The body is held in a straight line. Motion is symmetrical and directly ahead. It is slow. This kind of motion may be used for stalking prey or for moving in a narrow tunnel.

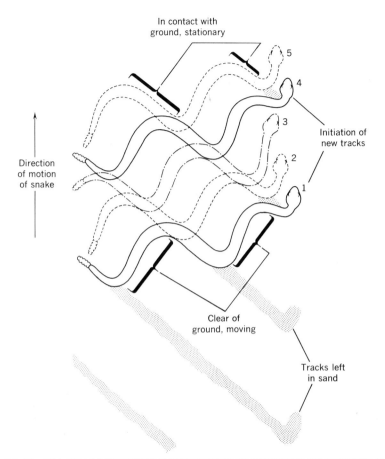

In contact with
ground, stationary

Initiation of
new tracks

Direction
of motion
of snake

Clear of
ground, moving

Tracks left
in sand

Figure 24-15 DIAGRAM SHOWING SUCCESSIVE POSITIONS IN THE SIDEWINDING OF A SNAKE.

The third way of moving without limbs is **concertina movement,** which is commonly used by caecilians, snakes, and amphisbaenians. The animal draws itself into one or more S-shaped coils. The posterior coils then press downward and backward against the substrate, relying on friction to prevent slipping (Figure 24-14). The forward component of this thrust is used to advance the head and anterior part of the body, which is held clear of the ground to avoid resistance from sliding friction and to increase the static friction of the stationary posterior coils by increasing the loading on them. Before stability is threatened, the anterior part of the body touches the substrate, builds coils, and ceases motion, so that it, in turn, can draw up the posterior part of the body before the cycle is repeated. The firmness of the base provided by the stationary coils is increased if coils are forced outward to wedge the animal within a tunnel, or between rocks or crevices in bark, or, if they are forced inward, to constrict a branch. Concertina movement is common, particularly among climbing and burrowing species, and often is combined with lateral undulation.

Some of these animals use a variant of concertina movement in which the anterior part of the spine coils or bunches within the integument. The body contour widens, but does not bend. The widened region braces the body within its burrow, providing a purchase for the head that then thrusts forward into the soil.

The last method is **sidewinding,** a kind of progression that probably evolved from the concertina method and is adapted for fast and efficient travel over loose or sandy soil. Figure 24-15 shows that the snake makes a series of tracks that are more or less straight lines, parallel to one another, angled to the direction of travel, and each is about as long as the animal. The snake contacts two or three tracks at a time. Parts of its body are within the tracks and parts are arching between tracks. As successive segments of the body are laid down to extend one track, successive segments are released from the previous track. The parts of the body that are along tracks are stationary, whereas those that span between tracks are moving and are held clear of the ground (which may be hot). Periodically, the head and neck reach forward to initiate new tracks. The entire action is very rapid. Most snakes can move in this way, but desert species are the most adept.

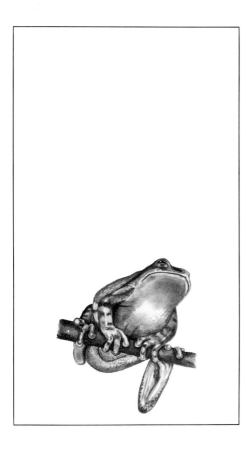

25

Climbing

Tetrapods that are adept at climbing may be called **scansorial.** Climbers are often **arboreal,** but this term means living in trees and does not directly indicate manner of locomotion; most birds are arboreal, yet few climb.

Many tetrapods are expert climbers and many more climb moderately well on occasion. The climbing habit evolved independently more times than can be traced, and several times in each of several orders. Adaptations for climbing by primates and by some of the other more strikingly modified climbers have been analyzed. However, the adaptations of many small scansorial tetrapods have scarcely been studied.

ADVANTAGES OF CLIMBING

The selective advantages of climbing include the following:

1. Climbers can secure in shrubs and trees such foods as leaves, shoots, flowers, fruits, cambium, honey, spiders, insects, and birds' eggs.

2. Many climbers avoid predation by remaining off the ground or by returning to the safety of rocks or vegetation when danger threatens. Also, climbing affords vantage points from which to look out for danger.

3. Several predators follow their prey into the trees: The fisher captures tree squirrels; the arboreal viper lies in wait for scansorial rodents. The leopard hauls his kill into a tree partly to keep it safe from jackals and hyenas when he is not in attendance.

4. By climbing, many animals find sheltered places to rest during the part of the day when they are inactive. Similarly, they find or make safe secluded nests in which to rear their young.

5. Where ground vegetation is dense climbers may be able to travel more freely and rapidly in the open upper story of the trees than they could on the ground.

6. Animals that glide must climb to reach takeoff points.

SCANSORIAL
VERTEBRATES

Fishes and Amphibians. Several kinds of air-breathing fishes move about on land and may scramble into low vegetation using strong mobile fins and perhaps fin spines as well. None, however, is really scansorial. Among amphibians, the many species of tree frogs in at least seven families are expert climbers (Figure 25-1). There are many arboreal salamanders (family Plethodontidae), some of which are so skilled that they can walk along a string.

Reptiles. Many lizards and snakes are climbers; some rarely descend from shrubs or trees. Noteworthy are the chameleons (family Chamaeleontidae), geckos, (Gekkonidae), various iguanids (Iguanidae) including the anoline lizards, and the tropical tree snakes.

Birds. Excluding birds that merely perch in trees or forage by flitting from twig to twig, some remain that are truly scansorial (Figure 25-2). These are the woodpeckers, woodhewers, creepers (of two families), nuthatches (of three families), parrots, crossbills, some of the ovenbirds, and the hoatzin. In addition, rock nuthatches and wall creepers climb on rocks. The nuthatch climbs with feet only, the woodpecker with feet and tail, the parrot with feet and beak, and the hoatzin with feet and wings.

Marsupials and Insectivores. Eleven of the twelve genera of opossums are fine climbers and all seventeen genera of phalangers climb with ease (Figure 25-3). Some marsupial mice climb, and, surprisingly, one genus of kangaroo has secondarily become a climber. Among insectivores, the five genera of tree shrews are highly scansorial.

Colugo and Bats. Like other gliders, the colugo is an arboreal climber (see figure on p. 533). Many bats roost in trees without doing much climbing, but several genera are nimble climbers.

Primates. Nearly all primates are skilled climbers, and the few that climb little or not at all (baboons, some lemurs, gorilla, man) had arboreal ancestors. Particularly noteworthy for their structural adaptations or climbing behavior are certain lemurs (family Lemuridae); indris (Indridae); lorises, pottos, and galagos (Lorisidae); tarsiers (Tarsiidae); spider and woolly monkeys (Cebidae); langurs (Cercopithecidae); and gibbons and the orangutan (Pongidae).

Edentates and Pangolins. The two smaller genera of anteaters are climbers, as are the two genera of sloths. Some pangolins are arboreal.

TREE FROG
Hyla

CHAMELEON
Chamaeleo

HARLEQUIN SNAKE
Chironius

ANOLINE LIZARD
Anolis

Figure 25-1 REPRESENTATIVE SCANSORIAL AMPHIBIANS AND REPTILES.

NUTHATCH
Sitta

PARROT
Pyrrhura

FLICKER
Colaptes

Figure 25-2   REPRESENTATIVE AVIAN CLIMBERS.

Rodents.   Scansorial rodents are in general less structurally modified than are the better climbers of other orders. Nevertheless, there are more climbers in this order than in any other. The common names of many are unfamiliar or poorly established. Climbing rodents include tree squirrels, ''flying'' squirrels, chipmunks (Sciuridae); scaly-tailed squirrels (Anomaluridae); vesper rats, harvest mice, gerbil mice, red-nosed mice, tree mice, pine mice, wood rats (Cricetidae); climbing rats and mice, forest rats and mice, tree rats, cloud rats (Muridae); dormice (Gliridae); New World porcupines (Erethizontidae); and echimyid rats (Echimyidae).

Figure 25-3   REPRESENTATIVE MAMMALIAN CLIMBERS.

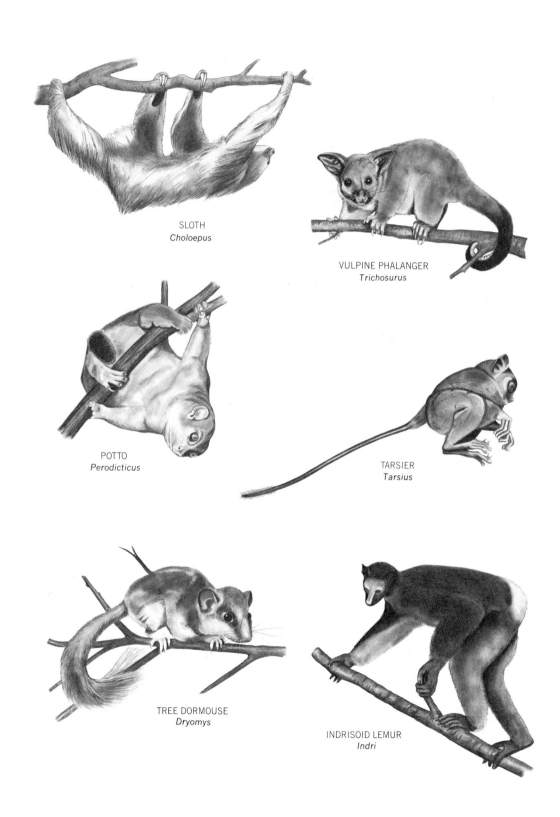

SLOTH
Choloepus

VULPINE PHALANGER
Trichosurus

POTTO
Perodicticus

TARSIER
Tarsius

TREE DORMOUSE
Dryomys

INDRISOID LEMUR
Indri

Carnivores, Hyraxes, and Ungulates.   Climbing carnivores are little modified structurally for the habit, yet they include all members of the raccoon family, some members of the weasel, bear, mongoose, and cat families, and two kinds of foxes. One genus of hyrax is remarkably skilled at climbing smooth tree trunks, and others scramble among rocks. Excepting the occasional acrobatics of some goats, no ungulate climbs trees, yet mountain goats and sheep, chamois, tahrs, and the klipspringer are master rock-scramblers.

REQUIREMENTS AND BASIC MECHANISMS OF CLIMBERS

Climbers have two basic requirements: (1) They must propel themselves on a uniquely discontinuous and three-dimensional substrate, and (2) they must avoid falling, both when moving and at rest, under particularly difficult circumstances.

Runners and diggers must also propel themselves on a somewhat uneven substrate and must also avoid falling, so it is not surprising that the structural and behavioral adaptations of climbers differ in degree, but usually not in kind, from the adaptations of many other animals. Indeed, many climbers are not markedly modified for their locomotor habit. The tree mouse looks about like the terrestrial mouse, the scansorial and terrestrial species of murine opossums are similar, and the climbing fennec and nonclimbing red fox are constructed in nearly the same way. Furthermore, of all the locomotor specializations, ability to climb combines with the most other specializations: The climbing leopard also sprints, the tree kangaroo also hops, the anteater also digs, the tree frog also swims, the colugo also glides, and the parrot also flies. Nevertheless, some climbers (tree frogs, various salamanders, geckos, anoles, several bats) do utilize unique mechanisms, and scores of others have modified more familiar mechanisms to a striking degree.

The requirement that climbers propel themselves over a discontinuous substrate leads to adaptations for leaping, springing, swinging, and reaching and pulling. The requirement that climbers avoid falling under difficult circumstances leads to adaptations for grasping, balancing, bracing, cushioning, applying suction, clinging, hooking, and adhering. These adaptations will be analyzed, but since they are numerous, and combine in many ways (springing with grasping, swinging with hooking, running with clinging, walking with adhering, running and leaping with grasping and cushioning, etc.), it will be useful to discuss first the few basic principles a climber can utilize to remain in contact with sloping rocks, branches, or twigs.

THE ROLE OF FRICTION   When a monkey stands on a horizontal log with each foot directly under its girdle, the downward force of the body is countered by an equal upward force of the support. There is compression at the interface between foot and log, but virtually no shear. Accordingly, there is no tendency for the foot to slip. The thrust of the foot against the log **(T),** and the force that is normal (i.e., perpendicular) to the surface of the Log **(N)** are the same (Figure 25-4A). This would also be true of a standing horse, and the monkey is at the moment no more climbing than the horse.

If the monkey walks along the horizontal log, then the foot pushes against the surface at an angle to the vertical. This thrust can be divided into a component that is normal to the surface and one that is shearing, or parallel to the surface

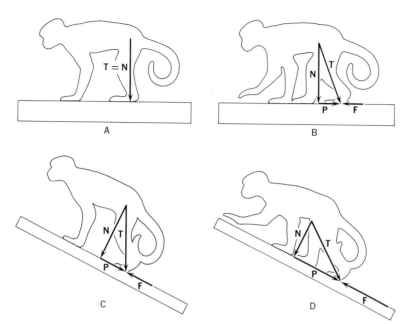

Figure 25-4 THE ROLE OF FRICTION IN PREVENTING SLIPPAGE.

(**P** in Figure 25-4B). If the monkey is not to slip, **P** must be countered by an equal frictional force (**F**), which is the mechanical resistance to motion of the foot along its support. Likewise, the horse is able to walk only because friction opposes the propulsive forces along the ground.

If the monkey stands on a branch that is inclined to the horizontal, then **T** can be divided into **N** and **P** even though there is no motion (part C), and friction is required to keep the foot from sliding down the branch. If the monkey walks up the inclined branch, **N** is decreased and **P** is increased (part D). Consequently, friction must also increase if the monkey is not to slip. The climbing monkey and terrestrial horse have now parted company. It is evident that adaptations for climbing include mechanisms for maximizing friction.

How can that be done? Friction is a complicated phenomenon that does not lend itself to exact analysis. The maximum friction that can be developed before sliding starts depends on the kinds and textures of the materials in contact, and on the force acting between them. The approximate relationship for dry and rigid surfaces is $F = \mu N$, where μ is the coefficient of friction. (Since the footpad is somewhat elastic, **F** is in fact proportional to a fractional power of **N**. Few actual values are available for the materials involved.) If sliding starts, μ decreases. As drivers may learn, it is easier to prevent a skid than to stop one.

It is obvious from the formula $F = \mu N$, that climbers could increase **F** by (1) selecting substrates that would give high values for μ, (2) evolving integumentary surfaces that would increase μ, and (3) developing mechanisms for increasing **N**. Methods 1 and 2 invite further study. Presumably, extra care is taken when the substrate is wet because lubrication greatly reduces friction (and alters the formula for its calculation). As we shall see, climbers are efficient at increasing **N** in various ways. These do not include, however, marked increase in body weight. Climbers are of medium to small size so they will not break the branches that must support them and so they can be agile.

As the formula shows, maximum friction tends to be independent of apparent area of contact. Even flat, polished surfaces actually touch at only a limited number of microscopic high points that constitute a small fraction of the visible area of apparent contact. If the visible area of contact is decreased, pressure on the remaining area is increased and more microscopic points are forced into contact, thus maintaining about the same area of actual interaction between the surfaces. This is why the klipspringer (an antelope) can perform feats of rock climbing on the tips of very tiny hoofs. Nevertheless, some climbers have large footpads. This reduces abrasion per unit area of the integument and, because large pads that are also flexible tend to touch the substrate in several planes, increases stability by preparing the foot to resist, with friction, disrupting forces coming from various directions in space. (Large pads may also increase interlocking—see below.)

THE ROLE OF INTERLOCKING   If the flat surface between an object and its support is inclined to the horizontal, then, as noted above, there is a force, **P,** parallel to the surface that must be opposed by an equal frictional force, **F,** if slipping is to be avoided. Force **P,** however, has a horizontal component (**H** in Figure 25-5A), and slipping could also be prevented by a counterforce (**H'**). If the support has a side wall, this counterforce is provided (part B). Interlocking has then substituted for friction in providing stability. When the claw of an iguana, parrot, or chipmunk lodges in a crevice of a rock or branch, this method of support is operative.

If the object instead contacts its support on a dozen or so small sloping surfaces and on as many small side walls, then the same interlocking principle applies (part C). The numerous stiff tail feathers of a woodpecker and the horny plates under the tail of a scaly-tailed squirrel exemplify this kind of support when they press into rough bark (Figures 25-2 and 25-6).

Intermeshing of fine points of contact is also one basis for frictional force,

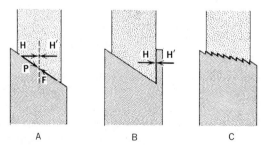

Figure 25-5   THE ROLE OF INTERLOCKING IN PREVENTING SLIPPAGE.

Figure 25-6 VENTRAL BASE OF THE TAIL AND FLIGHT MEMBRANE OF THE SCALY-TAILED SQUIRREL, *Anomalurus*.

so as interlocking surfaces become quite small (e.g., scales on the foot of a small lizard, or "fingerprint" ridges on the finger balls of a galago), the boundary between friction and interlocking as methods of support becomes fuzzy. A cushionlike footpad that contacts both microscopic and macroscopic projections on a branch utilizes both friction and interlocking to prevent slipping. The hooklike claw of a sloth might be prevented by friction from slipping along a smooth pole, but if its sharp edge were to cut slightly into the pole, interlocking would also contribute support.

THE ROLE OF ADHESION   When one smooth metal slides over another, great pressure and high temperature at the microscopic points of contact may cause molecular bonds to form. This is another basis for frictional force. Ordinarily, however, dry materials cannot be brought close enough together at enough points, even if polished and clean, for intermolecular forces to establish a significant amount of bonding between them. If a thin layer of a suitable adhesive is introduced between the materials, then they adhere because the adhesive wets (i.e., broadly contacts) each surface and is attracted to each. Thus, a glass coverslip adheres to a vertical windowpane when bonded by a thin film of water. Common experience tells us that sticky materials are better adhesives than water, and the fingerpads of some amphibians secrete a sticky material that bonds them to vertical leaves and stems.

The ability of the dry-footed geckos and anoline lizards to walk upside down on glass long defied explanation. Friction and interlocking are out (despite published claims to the contrary). There is now experimental evidence that the highly specialized structure of their toes (see below) enables them to establish intermolecular attractions (or Van der Waals forces) with the substrate; that is, adhesion without glue or sticky adhesive. The explanation makes the fact no less miraculous.

WALKING, RUNNING, LEAPING, AND SPRINGING   Animals that commonly walk or run along more or less horizontal branches (iguanas, tree mice, anteaters) have no problems of propulsion not shared by their terrestrial relatives. The feet, and sometimes the tail, may be modified to grip the substrate, but the remainder of the body is not distinctive.

Some climbers commonly, though not exclusively, propel themselves by leaping from one support to another. The jump may be somewhat upward, but usually is outward or partly downward. The animal may be moving when it leaps, and the body is more or less horizontal at takeoff. Examples are certain arboreal snakes; numerous lizards; tree squirrels; capuchin, howler, vervet, and proboscis monkeys; langurs; and arboreal mangabeys. These climbers (except the snakes) tend to have long limbs, slender bones, and muscle mechanics similar to that of cursors, though without relative shortening of proximal limb segments. The back is relatively long, strong, and flexible.

ADAPTATIONS FOR PROPULSION

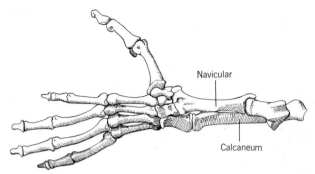

Figure 25-7   LENGTHENED TARSUS OF A SPRINGER shown by the left foot of the primate *Galago*.

Several primates propel themselves primarily by springing. When at rest they tend to hold the body in a vertical position, and they are often stationary before takeoff. The jump may be in any direction including steeply upward. These highly specialized animals are the tarsiers, galagos, indris, and two genera of lemurs. The smaller of these prodigious springers can jump 2 m straight upward, and the larger ones may jump 10 m out and down from one tree to another. All have the long hind legs, flexed knee posture, and general limb mechanics of saltators (see p. 457 to 460), although the femur is not shorter than the tibia. The foot (unlike that of jerboas and kangaroo rats) must be adapted for gripping the substrate, but in tarsiers and galagos it is also much elongated for springing by lengthening of the navicular and calcaneum bones (Figure 25-7).

REACHING AND PULLING   Many of the more adept climbers propel themselves entirely or in part by reaching from one support to another and pulling themselves along. The orangutan, pottos, lorises, and (in their upside-down way) sloths are noteworthy examples, though some frogs, chameleons, opossums, phalangers, various monkeys, the colugo, echimyid rats, and many others also use this method of propulsion. There is some evidence that orangutanlike climbing was in the ancestry of man.

These animals must meet three principal requirements. The first is a long reach. Hence, (with birds excepted) the limbs are longer than for any other locomotor adaptation except gliding, springing (hind limbs), and flying (forelimbs). Proximal and middle limb segments are nearly equal in length. The feet are large, yet must respond more to the needs of gripping than propulsion, so they do not lengthen as they do in cursors and terrestrial saltators.

Various tree snakes provide for reaching in a very different way. The morphology of the anterior vertebrae, together with a complex musculature, prevent dorsoflexion of the spine so the body will not sag as it is extended from one support toward another.

The second requirement is flexibility and agility. To gain strength with a wide range of movement, the heads of humerus and femur are not only spherical in curvature but also represent larger portions of complete spheres than is usual.

(This contrasts with the conditions in cursors, diggers, and flyers.) The girdles, even of some scansorial reptiles, are modified to allow freedom of movement. Toward this end, the scapula and clavicle of the mammals tend to become modified in ways which, being even more extreme for arm swingers, are described in the next section. To assure maximum pronation and supination of the forearm, the ulna and radius are free and about equally developed, the proximal head of the radius is round, the radial notch on which the radius rotates is evenly curved and is lateral in position (not anterior as for cursors), and a styloid process at the distal end of the ulna usually forms a pivot around which the carpus turns (see figure on p. 451). Similarly, the fibula is free and relatively large. The wrist joint is ellipsoid, not hingelike. Considerable rotation, adduction, and abduction may be possible within the tarsus. Splines and grooves are relatively little developed at limb and foot joints (see figure on p. 455).

Third, these climbers require appropriate bone-muscle mechanics. Marked strength is not needed, so muscles and muscle attachments are not prominent, and the bones are light and slender. Many of these animals (and particularly the sloths) commonly assume postures in which extensor muscles do not oppose gravity in the usual manner. Hence, these muscles are less developed, and their in-levers are shorter than is characteristic in terrestrial mammals. Thus, the in-lever of the triceps (the olecranon process) ranges from about $\frac{1}{8}$ (an opossum) to $\frac{1}{12}$ or less (lorises, sloths) of the out-lever (the length of the remainder of the ulna). Flexors, pronators, supinators, and abductors are better developed.

ARM SWINGING   Some primates propel themselves by arm swinging (or **brachiating**) under the branches. Unlike other methods of propulsion, only the forelimbs are used for support. Arm swinging is best exemplified by gibbons and the related siamang, though it is used on occasion by spider, woolly, howler, langur, colobus, and proboscis monkeys and by the chimpanzee, orangutan, and (more rarely) gorilla.

With one hand around a support that is above the body and ahead in the line of travel, the climber swings pendulum-fashion down under the support, rotates the body nearly 180° on the supporting arm (i.e., advances the unweighted shoulder), and, at the end of the upswing, stretches the other arm to another overhead support. The free arm and (in the siamang) also the legs are extended downward on the downswing. This moves the animal's center of gravity away from the pivot at the supporting hand, thus maximizing the velocity and kinetic energy gained. On the upswing the free arm and legs are flexed to shorten the pendulum, decrease its moment of inertia, and increase its angular velocity. As with a child pumping on a swing, there is a net gain in momentum. In rapid locomotion each support may be reached only after a period of free-floating travel. The gibbons and siamang do most of their climbing by arm swinging and are remarkable acrobats.

Swingers, like springers, are highly adapted for their specialty. They have the same adaptations for reaching, agility, and use of arms under tension as the reach-and-pull climbers, only the modifications are more extreme (Figure 25-8). The hind limbs are long relative to the trunk, particularly in gibbons and spider

Digits subequal
in length

Joints rounded,
without splines and
grooves

Wrist joint elipsoid

Radius and ulna
subequal, free

Joints allow
pronation and
supination of
foot

Fibula large,
free

Radial notch
lateral

Head of radius
rounded

Olecranon short

Patellar groove
shallow

Bones long,
slender, light,
not rugged

Greater trochanter small

Head of humerus
a relatively large
part of a sphere

Figure 25-8 FEATURES OF THE APPENDICULAR SKELETON OF CERTAIN CLIMBERS shown by the left leg of a sloth, *Choloepus* (left), and left arm of a spider monkey, *Ateles* (right).

monkeys, but the forelimbs are disproportionately long, becoming even twice or more as long as the trunk in the orangutan, gibbons, and siamang. The fossa on the humerus that accommodates the very short olecranon process is deep so the elbow can be completely straightened.

Supination of the supporting forearm coupled with twisting of the trunk advances the leading (nonsupporting) shoulder during a swing, adding reach and force to the action. To help accomplish this the supinator is strong, its in-lever is increased by bowing of the radius, the sternum is broad, and the chest is broad rather than deep (Figure 25-9).

The clavicle is long, reaching over the broad chest to the large acromion process. The scapula lies relatively flat on the back, rather than against the side of the thorax as for animals that do not arm swing. The glenoid cavity is oriented forward and sideward, not downward (Figure 25-10). The latissimus dorsi, pectoralis, biceps, and long head of triceps brace the shoulder against tension. Also strong are the trapezius, which pulls the acromion process in toward the neck, and the anterior serratus, which pulls the posterior angle of the scapula out toward the side of the chest. The insertion and orientation of these muscles are also modified so that together they rotate the scapula on the trunk to a unique degree. This raises the arm on the body, and, during the swing, probably rotates the body on the outstretched arm.

In contrast to cursors and quadrupedal leapers, arm swingers have short compact backs so the trunk can swing as a unit. The lumbar area contributes the least to locomotion, so it tends to be inflexible, has relatively few vertebrae, and these vertebrae have short centra (Figure 25-9). The zygapophyses of mammalian lumbar vertebrae are constructed to allow flexion and extension but to limit rotation of the spine around its long axis. The zygapophyses of anterior thoracic vertebrae do allow rotation. The transition occurs within one vertebra. This vertebra, the **diaphragmatic vertebra,** is farther posterior in swingers (which rotate the spine with each swing) than in cursors and leapers (which flex and extend but do not rotate the spine).

Some climbers are slow-moving (chameleons, lorises) whereas many runners, leapers, and arm swingers are very quick. The latter require remarkably rapid and precise neuromuscular responses. The morphological bases for such control may be assumed to include the relative prominence of cerebellum, olivary nuclei, ruber nucleus, primary motor and sensory cortex, and optic tracts and centers. The eyes are large and face forward to provide overlapping fields of vision, and hence depth perception.

GRASPING   When the fingers and palm of a chameleon, potto, or man encircle a twig or pole and grip firmly, muscular effort creates forces that are normal to the surface of the support. These forces increase frictional resistance to slipping: the tighter the grip the more resistance. An animal with strong digital flexors can

ADAPTATIONS FOR MAINTAINING CONTACT WITH THE SUBSTRATE

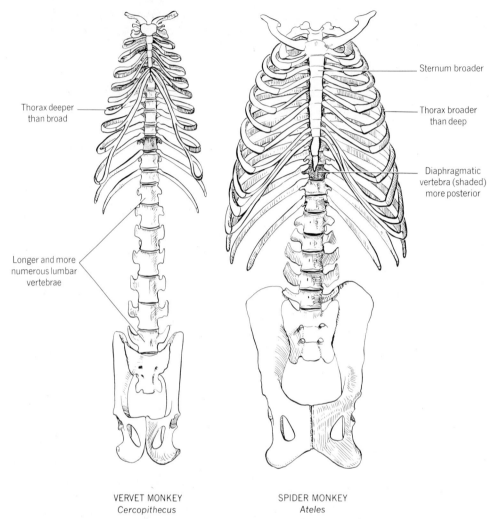

Thorax deeper than broad

Longer and more numerous lumbar vertebrae

Sternum broader

Thorax broader than deep

Diaphragmatic vertebra (shaded) more posterior

VERVET MONKEY
Cercopithecus

SPIDER MONKEY
Ateles

Figure 25-9 CONTRAST BETWEEN THE TRUNK SKELETONS OF A MONKEY THAT LEAPS (left) AND ONE THAT ARM SWINGS (right).

thus supplement the normal forces that it can develop using only its body weight. Furthermore, by grasping it can resist slipping in any direction.

Grasping is a particularly versatile and effective way of maintaining contact with the substrate, and the different climbers have independently evolved many grasping mechanisms. Various snakes grasp branches with coils of the body to provide support. The first digit is opposed to the others in one or both pairs of feet of some tree frogs, salamanders, birds, opossums, and many primates (Figures 25-11 and 25-12). The second digit of the hand tends not to be very effective in a power grip (even in man) and in the potto and lorises has become short and weak. The koala, and some other phalangers, grasp between the second and

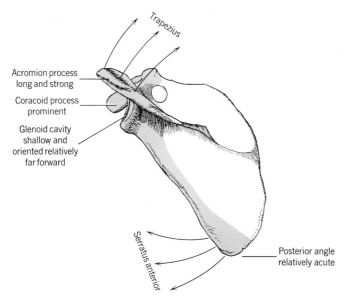

Trapezius

Acromion process
long and strong

Coracoid process
prominent

Glenoid cavity
shallow and
oriented relatively
far forward

Serratus anterior

Posterior angle
relatively acute

Figure 25-10 ADAPTATIONS OF THE SCAPULA FOR REACHING AND ARM SWINGING shown by the spider monkey, *Ateles*. Shading indicates areas of relatively heavy bone, which includes the lever of the mechanism that pivots the scapula.

third digits, as do the chameleons with the hind foot. The forefoot of these lizards grasps between the third and fourth digits, and echimyid rats do the same. Parrots and some other avian climbers oppose digits two and three with digits one and four. The palms, soles, and digits of graspers are naked and sensitive.

Terrestrial mammals often irretrievably lose one or more lateral digits. If their descendants become secondarily arboreal, those digits cannot be used for grasping, and nature has produced interesting compensations: The two-toed anteater can depress its heel into strong opposition to its remaining digits, and one Central American porcupine can fold the pad of its foot lengthwise and forcefully grasp between the lateral edges of the footpad.

The tail often evolves into a grasping organ. Such a tail is said to be **prehensile** and is characteristically long, strong, sensitive, and curled at the end. Animals with prehensile tails include some salamanders; chameleons, and several other lizards; some arboreal snakes; opossums; some phalangers; capuchin, spider, and woolly monkeys; anteaters; pangolins; the kinkajou; various rats and mice; and one porcupine. Most of these animals curl the tail ventrally, but the porcupine curls its tail dorsally. Prehensile tails tend to be flexible at the base and to have short broad vertebrae near the end. Often they have naked pads where they grasp.

BALANCING, BRACING, CUSHIONING, AND SUCKING   Climbers that move quickly must be effective balancers. It follows from principles given on page 454 that scansorial mammals that walk or run on top of branches can increase stability by lowering their center of gravity. Arboreal salamanders, snakes, and lizards

TREE FROG
Hyla

GECKONID LIZARD
Gecko

POTTO
Perodicticus

TARSIER
Tarsius

Figure 25-11 LEFT HANDS OF SOME CLIMBERS.

Arboreal salamander
Bolitoglossa

Chameleon
Chamaeleo

Parrot
Pyrrhura

Indrisoid lemur
Indri

Figure 25-12 LEFT FEET OF SOME CLIMBERS.

keep their center of gravity very low. Tree squirrels, rats and mice, marmosets, and the kinkajou either have short legs or flex the legs to hold the body low. Birds that forage on tree trunks have short legs (the tarsometatarsus shortening the most) to keep their center of gravity close to their support. The girdles of the chameleon permit the legs to come vertically under the body (unusual in reptiles) so the animal can balance over narrow stems. Climbers that swing or hang under their support using hooklike appendages (see below) are in stable equilibrium—like rocking chairs they tend to maintain position. Most climbers have long tails that contribute to the maintenance of balance; climbers without tails swing, hang, or move slowly.

Numerous climbers use their tails as braces, struts, or props. Woodpeckers, woodhewers, and creepers prop themselves with their stiff tails; the terminal

Figure 25-13   SUCTION DISK ON THE WING OF THE BAT, *Thyroptera.*

segment of the spine (pygostyle) and its musculature enlarge for the purpose. Some species of tarsier have a naked and roughened area at the base of the tail to make it a better prop. The scaly-tailed squirrel has horny ventral scales on its tail (Figure 25-6).

The appendages of most climbers have broad, soft, cushionlike pads. These are often roughened by small grooves or fingerprints that increase friction and interlocking. Footpads are particularly well-developed in primates, anteaters, and porcupines. Fingerpads are conspicuous in tree frogs, some arboreal salamanders, opossums, phalangers, indris, pottos, galagos, and tarsiers. Climbers with prehensile tails have tail pads. Feet having such cushions tend to be broad and loose. Metapodial bones are well-spaced, round in cross section, rounded on their distal ends, and devoid of splines at the joints. Phalanges are similarly rounded except that the terminal phalanx may be somewhat spatulate. Claws or nails are positioned so as not to interfere with the action of the pads.

The suction cup of human technology is usually a shallow cup of rubber, with the rim pressed against a smooth surface. The elasticity of the rubber keeps the rim tightly pressed against the surface and thus reduces pressure inside the cup so that atmospheric pressure presses the cup against the surface. Shearing force is resisted by friction. Disk-winged bats of two genera have suction cups on knuckles and ankles that function in the same way (Figure 25-13). Elastic tissue, not muscular tension, maintains suction within the disk once it is seated. The tree hyrax clings firmly to smooth vertical tree trunks without grasping or using claws. It is claimed that muscles pull inward on a cleft at the center of the soft footpad to create suction. Glands keep the pad moist to establish a good seal.

CLINGING AND HOOKING   Most climbers that do not grasp use strong, much-curved claws to cling to the substrate. The tips of the claws interlock with small cracks and crevices. On large smooth stems, claws are more secure than grasping digits. Usually the weight of the animal is insufficient to maintain firm interlocking, so one set of claws must be pulled against another. The two feet of a pair may sprawl wide on opposite sides of the body (rock lizard), or the hind feet may be turned toes-backward so foreclaws can pull against hind claws (tree squirrel). In addition, scansorial birds pull with one or two claws of the foot against the other claws of the same foot. The digital flexors are strong and the terminal phalanx is designed to provide a good in-lever (Figure 25-14).

Some climbers modify the appendages as hooks and swing or hang under their support. Sloths and pangolins use 1–3 very long, strong, curved claws as their hooks. Bats and the colugo use 5 nearly equal claws. Primates that swing use the four fingers of the hand together as a hook. The hand is very long, and the phalanges are curved to conform to the round cross section of the branches. Prehensile tails can be used as hooks as well as grasping organs. In all these instances, the flexor tendons of the hook are short enough to passively prevent the hook from opening up: a dead gibbon can be hung by an upstretched arm; a dead spider monkey can be hung by the end of its tail.

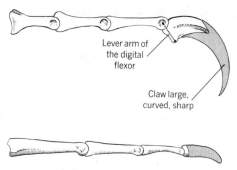

Lever arm of
the digital
flexor

Claw large,
curved, sharp

Figure 25-14 ADAPTATIONS OF THE AVIAN FOOT FOR CLIMBING shown by third toes of a flicker, *Colaptes* (above), and for contrast, of the nonclimbing, nonperching merganser, *Mergus* (below).

ADHERING   The expanded finger balls of tree frogs enclose glands having a sticky secretion that these little frogs use to glue themselves to rocks, leaves, and stems. They usually press their moist tummies against the substrate so these will also adhere.

Many tropical salamanders have webs between their well-spread digits. The result is a large common adhesive pad. To break contact, the pad is curled up from its margin or lifted from the back. Frogs may have an "extra" segment of cartilage or bone just proximal to the terminal phalanx. This seems to assist the animal in feeling about with the tips of its toes for the best spot to make contact with the substrate.

The gecko has sharp claws with which to cling to rough substrates. If a smooth support is moderately inclined, the animal can maintain its position by friction. On steep smooth surfaces and overhangs, however, this creature brings into play one of nature's most remarkable adaptive mechanisms. Under each toe are 16–21 broad imbricated scales or lamellae (Figure 25-15). On the exposed surfaces of the lamellae of each toe are up to 150,000 hair-like setae ranging from 30 to 130 μm long. Each seta branches into about 2000 bristles, and each of these has a saucerlike endplate measuring about 0.2 μm in diameter. There are in all some 100 million of these endplates that touch the substrate at points on their rims. (All these numbers vary by species.) A slight shearing force (from body weight or muscular pull) is required to give the setae the S curve that positions the endplates on the substrate. Blood sinuses under the lamellae cushion the toes so a maximum number of endplates can reach into any irregularities of the surface. Collectively there are so many close contacts that the animal adheres by surface tension.

The contact is firm: When one investigator tried to pull a large gecko from a vertical pane of glass, the glass broke. Adhesion to vertical glass continues even when the animal is dead and the body is in a vacuum. However, the lizards adhere with difficulty to materials having low surface tension (e.g., Teflon), and if the setae become dirty or mussed, climbing ability is impaired until after the

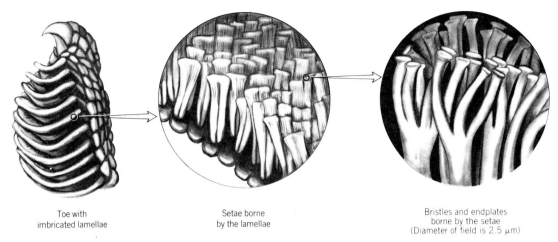

Toe with
imbricated lamellae

Setae borne
by the lamellae

Bristles and endplates
borne by the setae
(Diameter of field is 2.5 μm)

Figure 25-15   ADAPTATIONS FOR CLIMBING BY DRY ADHESION SHOWN BY A TOE OF THE LIZARD, *Gecko*.

skin is shed and replaced by new setae. To break the contacts, the lizard peels its extra flexible toes off the substrate by rolling them up "backwards," starting from the tips.

The unrelated anoline lizards have shorter unbranched setae, but they function the same way. This is an amazing example of convergent evolution.

26

Swimming and Diving

Vertebrates that live in water are said to be **aquatic.** Unfortunately, there is no term for the experts among them that swim with particular skill, speed, or endurance. All fishes are **primary swimmers**—their ancestors also swam. Other swimming vertebrates are **secondary swimmers**—their ancestors passed through a terrestrial state, and consequently they have structural and physiological handicaps that have prevented most of them from becoming entirely aquatic again.

Nearly all vertebrates can swim somewhat, and there are many expert swimmers in every class; there can be no sharp division between swimmers and nonswimmers. However, most that will receive attention here seek food and refuge in the water and can swim below the surface.

Relatively much has been published about swimming, yet detailed analysis is difficult, and much remains to be learned.

ADVANTAGES OF SWIMMING AND DIVING

It is academic to ask about the survival value of an aquatic life to primary swimmers: It is a successful way of life, and nature has provided no alternative to most of them. As we saw in Part II of this book, the change to terrestrial life was so profound that it took about 100 million years to complete. The reverse trend back to water has been "easier" and has occurred many times. Secondary swimmers and divers may have any of the following advantages over nonswimmers:

1. They gain access to a wide variety of aquatic foods including fishes, plankton, and larger invertebrates and plants.

2.   They escape terrestrial predators. They may also subject themselves to aquatic predators, of course, but the process of evolution has in specific instances moved in the direction of greater safety.

3.   The oceans and major inland waterways are favorable avenues for dispersal and migration.

THE SKILLS OF SWIMMERS AND DIVERS

SPEED   A first skill shared by many aquatic vertebrates is speed. The mako shark is probably the fastest cartilaginous fish, and the families Scombridae (tunas, mackerels), Istiophoridae (marlins), and Xiphiidae (swordfish) hold the honors for

Aspect ratio $= \dfrac{\text{span}}{\text{chord}} =$ high

Chord

Span

BLACK MARLIN, *Makaira*

Dorsal and ventral finlets reduce cross–flow

Shape of peduncle and its keel equalize pressure over fin

MEXICAN BONITO, *Sarda*

Body spindle-shaped

Peduncle narrow, stiff

Tail stiff

BLUEFIN TUNA, *Thunnus*

Figure 26-1   EXAMPLES OF FISHES SHOWING ADAPTATIONS FOR FAST SWIMMING.

bony fishes: The tuna can reach 75 km/hr (Figure 26-1), and the wahoo 77 km/hr (48 mile/hr or 41.6 knots). These are rates of 18–21 body lengths/sec, and an order of magnitude faster than humans can swim. Much greater speeds have been claimed for marlin, sailfish, and swordfish, but the records need verification.

Many living amphibians and reptiles are aquatic, but none is remarkably speedy. The extinct sea lizards called mosasaurs and the dolphinlike ichthyosaurs (see figure on p. 65) may have been much faster. Some penguins attain 36 km/hr (Figure 26-2).

Among aquatic mammals, the platypus; water opossum; water shrews, tenrec, and desmans; beaver, capybara, and nutria (Figure 26-3); manatee; and hippopotamus are among the best swimmers of their respective orders, yet are not speedsters. Sea lions and the related fur seal can sprint at about 22 km/hr. The spotted dolphin can approach 40 km/hr, and the finback whale and rorqual are thought to be able to sprint to 65 km/hr (40 mile/hr).

DIVING Various primary swimmers dive to the depths of the ocean. In spite of their dependence on breathing air at the surface, many secondary swimmers dive deeply and remain submerged for long periods. The green sea turtle dives to 290 m, and sea snakes remain under water for up to 2 hrs. Pelicans, diving petrels, tropic birds, boobies, and terns dive from the wing. Penguins use their paddlelike flightless wings underwater, and at least one species descends to 265 m on dives up to 18 min. Auks, murres, puffins, and diving petrels both swim and fly with their narrow wings. Cormorants, loons, grebes and some ducks swim underwater using their feet. Both feet and wings are used in swimming by some of these birds. Loons may dive to 55 m.

Various rodents can remain submerged for 6–10 min., and the manatee for 16 min. The bush dog, otter-civet, water mongoose, river otter, and sea otter are among the diving carnivores. The sea lion can dive to at least 146 m, the harp seal to 273 m, and the Weddell seal to 600 m. The latter, which forages under the ice as far as several kilometers from its blow hole, can remain submerged for 70 min. Dolphins can remain submerged for 10–20 min. Other records are 50 min for the blue whale, 90 min for the sperm whale, and 120 min for the bottlenose whale. A dolphin was trained to dive repeatedly to 300 m. Fin whales dive to 500 m. Sperm whales have gone as deep as 1100 m, where the pressure is 110 atm (1617 lb/in.^{2}).

ENDURANCE Salmon swam 1000 km up one river with an expenditure of energy equivalent to an average speed in quiet water of 4.2 km/hr. Most scombroid fishes and the mako shark swim continuously. Fur seals commonly migrate 12,000 km/yr. Gray whales cruise at about $5\frac{1}{2}$ km/hr on their annual round trip migration of about 19,000 km. The blue whale is reported to be able to swim at 27 km/hr for 2 hr (with a harpoon wound).

ACCELERATION AND MANEUVERABILITY Other skills of aquatic vertebrates are less well documented. Trout can accelerate at 50 m/sec^2, achieving maximum speed of 9 body lengths/sec in 0.15 sec. The spotted dolphin can reach 40 km/hr in only 2 sec. The agility of schooling fishes, reef fishes, seasnakes, penguins, otters, and many other swimmers is both beautiful and astounding.

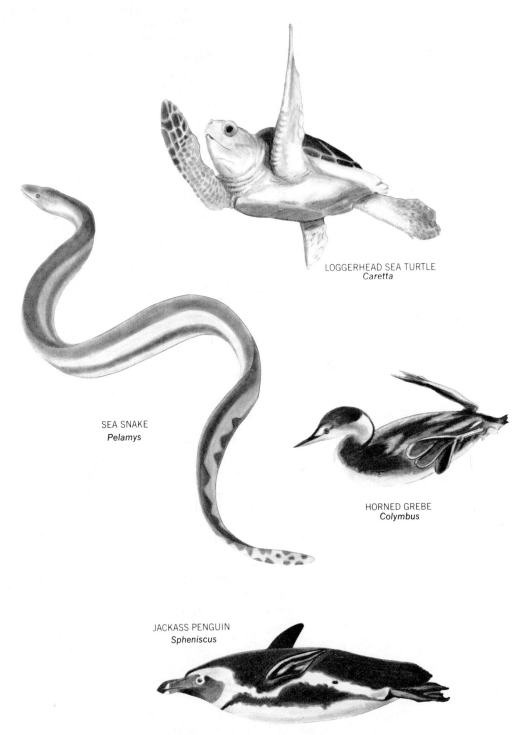

LOGGERHEAD SEA TURTLE
Caretta

SEA SNAKE
Pelamys

HORNED GREBE
Colymbus

JACKASS PENGUIN
Spheniscus

Figure 26-2   EXAMPLES OF UNDERWATER SWIMMERS AMONG REPTILES AND BIRDS.

NUTRIA
Myocastor

OTTER
Lutra

MANATEE
Trichechus

Figure 26-3 EXAMPLES OF AQUATIC MAMMALS of three orders: Rodentia (above), Carnivora (center), and Sirenia (below).

GENERAL REQUIREMENTS OF SWIMMERS AND DIVERS

All proficient swimmers and divers must (*1*) reduce the resistance that water offers to motions of the moving body, (*2*) propel themselves in a relatively dense medium, (*3*) control vertical position in the water, and (*4*) maintain orientation and steer the body. Secondary swimmers must also (*5*) exclude water from their respiratory passages and ears, (*6*) avoid harm from crushing of gas-filled spaces, (*7*) alter ears and eyes to function (again) underwater, and (*8*) modify their respiratory and circulatory physiology to permit suspension of breathing and avoidance of the bubbling of gas in the blood (bends) on returning to the surface after a dive. Some aquatic birds and mammals must also (*9*) control body temperature in a medium having high thermal conductivity, and (*10*) adapt their reproductive biology to life in the water.

How have the various swimmers and divers met these many requirements?

DRAG

ORIGINS AND NATURE OF DRAG   The resistance that a medium (here water) offers to the motion of an object is called drag. There are several sources, or kinds, of drag. They are interdependent, but can best be presented one at a time. First is **frictional drag.** Imagine a smooth, spindle-shaped, rigid object moving underwater. A film of water wets its surface and moves with it, yet a short distance away, the water does not move with the object at all. Between the object and the still water is the thin **boundary layer** where successive layers of water slide past one another; those nearest the object move nearly as fast as it does, and those more and more distant move slower and slower. The shearing forces thus produced tend to slow the moving object and are a source of drag. The boundary layer gets thicker toward the posterior end of the moving object (Figure 26-4).

If a smooth spindle-shaped object moves slowly through the water, successive layers or lamina of the boundary layer slip past one another without any eddies. Flow is said to be **laminar.** However, if the object moves fast, then where

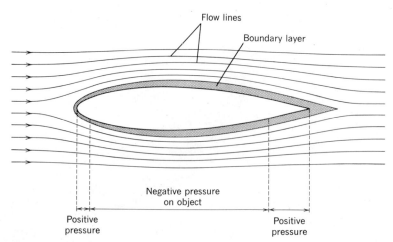

Figure 26-4   FLOW LINES, BOUNDARY LAYER, AND PRESSURE DISTRIBUTION WHEN FLOW AROUND A STREAMLINED OBJECT IS LAMINAR.

Figure 26-5 FLOW LINES SHOWING THE TURBULENCE CREATED BY A SWIM-MING FISH as seen from above. Various other patterns might also be created, depending on the variables.

the boundary layer reaches a certain thickness, or where there are even slight roughnesses on the surface of the object, the water curls into complex eddies (Figure 26-5). The energy that moves the water in these eddies comes from the moving object and greatly increases drag. Such flow is said to be **turbulent.** Turbulent flow produces a thicker boundary layer and much more drag than does laminar flow. Many fishes have nearly laminar flow when moving slowly, but when moving fast flow is turbulent over the posterior part of the body—and we shall see that this is advantageous.

To be realistic, let us examine the variables in some detail before seeking a simplification. Frictional drag (which is expressed in dynes) is the product of half the density of the fluid medium, times the area of the object, times a value called the drag coefficient. The drag coefficient must be calculated in each instance and varies with the shape and surface texture of the object (hence with the laminar or turbulent nature of flow) and with a value called the **Reynolds number.** This number, in turn, is the ratio of the inertia of the medium to its viscosity, and is calculated using the speed of the object, its length, and the density and viscosity of the medium. It is a dimensionless number having high magnitude (about $10^{5.5} - 10^{7.5}$) for large fast-swimming vertebrates.

This seems very involved, but let us unscramble it. Speed is squared in the basic formula and occurs again in the calculation of the Reynolds number. It follows that slow-swimming vertebrates have negligible drag no matter how the other variables change: Witness the unstreamlined bodies of the sluggish sea horse and trunkfish. Conversely, drag on rapid swimmers increases very fast with each increment of speed. Metabolic rate must be about doubled every time speed is increased by 1 body length/sec. It appears that the fastest swimmers closely approach the biological limits. Furthermore, in order to swim fast, the experts must reduce as much as possible all factors other than speed that increase drag.

With few exceptions, swimmers can do little or nothing to alter the density or viscosity of the water, so these elements of the formula usually can be disregarded. Drag increases with body size, but so does the output of the animal's power plant, and these factors nearly cancel one another. The consequence of some rather complicated physiological considerations seems to be that moderately large swimmers have some advantage. The fastest swimmers are large fishes and small whales. There remain the important variables of body shape and the nature of the surface of the body. Before we see how these are adapted for speed and efficiency, there are other sources of drag to consider.

As a spindle-shaped object moves through the water it displaces a volume of water equal to its own volume plus a volume equivalent to about $\frac{1}{3}$ that of the boundary layer. Also, there is a backflow of water behind the object as it moves along. Under some conditions of speed, size, and shape of object, the backfill is incomplete and separation of the boundary layer occurs, which creates suction behind the object and causes water to follow in its wake. (An observer at the stern of a ship easily sees that water in the wake falls back less rapidly than water to the sides of the wake.) Also, there is positive pressure against the anterior and posterior parts of the object and negative pressure at intermediate levels. The energy needed to cause these motions of water and pressure changes is taken from the object. The resultant drag is **pressure drag.** Pressure drag is negligible when the Reynolds number is low, but important when it is high. This kind of drag is complicated to calculate.

Finally, if the object moves on the surface of the water, like a ship or duck, or close enough to the surface to cause surface waves, then energy is extracted from the object to create the waves and **wave drag** occurs. Resistance is greatest when the object moves just below the surface. The variables associated with wave drag are not well-known.

These considerations of the drag on a rigid object make it evident that fast-swimming vertebrates require adaptations of body form and of the nature of the surface of the body. Since swimmers are not rigid, drag can also be reduced by certain behavioral adaptations.

REDUCTION OF DRAG BY ADAPTATIONS OF BODY FORM   Pressure drag is low when the body is long and slender, like that of a snake or eel, because there is then little displacement and backfill. Frictional drag is minimal, however, when the body is short and plump, because surface area is then minimal. The best compromise is a spindle that is circular in cross section and thickest near the center of its length where its diameter is $\frac{1}{4}$–$\frac{1}{5}$ of its length. The bodies of tunas, swordfishes, and dolphins closely approach this shape. Absence of a functional neck (primary swimmers, cetaceans, sirenians), symmetry of the head, molding of thorax and body musculature, and the distribution of fat and blubber all may contribute to streamlining.

Projections from the basic spindle usually cause turbulence and eddies, and increase drag. Accordingly, expert swimmers reduce or eliminate all major projections not needed for propulsion and steering: Swimmers other than mammals have no external ears or external genitalia in their ancestry. Aquatic mammals secondarily lose their external ears and move the testes back into the abdomen. Nipples or teats and the penis may be withdrawn within the body contour when not functioning. Fast primary swimmers have no limb segments between their fins and bodies. Fast secondary swimmers have very short proximal limb segments to again bring the feet or flippers close to the body. The humerus of cetaceans may be only about as long as it is wide; the femur of pinnipeds may be less than twice as long as wide; the femur of diving birds is short and most of the leg musculature is contained within the contour of the body. Cetaceans and sirenians have reduced the pelvic appendages to internal vestiges, and some other swimmers position the hind limbs in such a way that they do not protrude but instead extend the contour of the spindle-shaped body. The knee joints of pinnipeds, beavers, and many diving birds are constructed to allow the necessary reorientation of the limb.

Salamanders, crocodilians, and aquatic lizards hold their limbs against the body as they swim with tail and trunk. Lateral fins and flippers that propel the body, on the other hand, must protrude and present a flat surface to the water on the power stroke. Sea turtles and penguins commonly achieve a power stroke on both the downswing and upswing of the flipper by rotating it around its own long axis as the direction of the stroke changes. However, most swimmers with lateral paddles have a power stroke and a recovery stroke. Drag is reduced in various ways during the latter. Rotation of the entire appendage at its base may cause the appendage to cut the water edge-on (flipper of sea lion). The median lobes on the toes of grebes are similarly rotated on the recovery stroke, and the lateral lobes passively fold. Such flippers and lobes, and also median fins and paired appendages that are used primarily for steering, are streamlined in cross section so that the flow of water over them is nearly laminar when they are presented to the water edge-on. Pectoral flippers of pinnipeds, wings of most diving birds, and paired fins of fishes are pressed against the body when the animal glides. Bony fishes also can reduce the area of their fins by folding. Dorsal and anal fins (including the "sail" of the sailfish) may be retracted into grooves on the body surface during fast swimming. Web-footed tetrapods flex their limbs and adduct and curl their toes on the recovery stroke.

REDUCTION OF DRAG BY ADAPTATIONS OF BODY SURFACE AND BEHAVIOR It is advantageous for most swimmers to achieve laminar flow over as much of the body as possible. To maximize laminar flow, fishes evolve small smooth scales, or none at all, and become covered with slime. The small scalelike feathers of penguins and the hair of seals and otters form remarkably smooth coverings. Cetaceans and sirenians (and probably ichthyosaurs) are (or were) secondarily naked and slick-skinned.

The larger and faster swimmers are unable to prevent turbulent flow over most of the body and, in fact, benefit from it because turbulence greatly reduces separation of the boundary layer and backfill in the wake of the body. The "strategy" of the swimmer, therefore, is to cause, but control turbulence. In sharp contrast to the large eddies associated with pressure drag, eddies of the desired turbulence are very small and close to the body. As with the dimples on a golf ball, the increase in frictional drag that is caused by such turbulence is more than offset by the reduction in pressure drag. Associated adaptations appear to be numerous, but experimental verification of theory is as yet scanty.

Many fishes have projections on their scales that are calculated to be large enough to cause turbulence. These are found even on the sword of the swordfish. Commonly other scales have microrelief that forms longitudinal "runoff grooves" that are thought to control flow in the boundary layer. The skin of cetaceans and the basking shark has a spongy layer that is capable of elastic deformation and probably dampens pulsations of turbulence. The finlets on the caudal peduncle of many fast-swimming fishes (Figure 26-1), and the lateral keels on their tails, are described as damping screens and deflectors that direct the flow of water past the caudal fin.

It is not only the smoothness of fish slime that reduces drag. Slime is soluble in water but only when stirred and, hence, is not easily washed away. In concentrations as low as 1% it reduces flow friction by as much as 60%. That is, it reduces the viscosity of water in the boundary layer. Fishes that accelerate fast, such as trout, have the most slime.

Importantly, fast swimmers outperform submarines and torpedos in ways that involve behavior. The drag on a swimming salmon is about half the drag on a dead salmon that is towed. The factors are complex and are understood only in general terms. It was noted that separation of the boundary layer creates suction that causes water to follow after a swimmer. However, in propelling the body, the tail fin pushes water back, thus tending to cancel this source of drag. The opercula of the fastest bony fishes open alternately in synchrony with the undulations of the body. The result is that water may be ejected from the gills fast enough to reduce drag, or slow enough to initiate (desirable?) turbulence.

When animals swim just below the surface, wave drag increases total resistance by as much as five times the minimum. It is unlikely that many vertebrates attempt to swim rapidly in that position. The "playful" leaps of dolphins may be made in part to avoid swimming at the surface while breathing.

PROPULSION   **NATURE AND PRINCIPLES OF PROPULSION IN WATER**   Vertebrate swimmers propel themselves only with undulating or oscillating mechanisms; they have no analogs of sails, screw propellers, or jet engines (except for small forces acting at the gills). Two kinds of mechanisms are used. In one, an effective power stroke is followed by a dissimilar and less effective or ineffective recovery stroke. Examples are paired appendages such as the flippers of sea lions, the webbed feet of ducks and the platypus, and the wing-paddles of puffins. In the other kind of mechanism, each stroke is effective so the oscillations are symmetrical (or nearly so) and there is no recovery stroke. Examples are the undulating bodies of eels and water snakes, the undulating longitudinal fins of ribbon fishes, and the sweeping tail fins of trout and dolphins.

The more powerful swimmers all propel themselves by oscillations of the tail, and this mechanism will serve to illustrate the general principles of propulsion in water. Either a vertical tail fin sweeps back and forth from side to side (fishes, ichthyosaurs) or horizontal tail flukes sweep up and down (cetaceans, sirenians). Seals trail the hind feet behind the body and use them much as though they were a single vertical fin. The mechanism is the same regardless of direction of motion, so to simplify terminology in relation to space, I shall describe the action of the vertical tail of a fish.

The following factors are basic to the function of the caudal fin:(*1*) It is broad and flat so water cannot easily flow around it but instead will resist its lateral motion. In other words, it has high drag when presented broadside to the water. (*2*) Pivoting at its base, the fin membrane angles back and forth from right to left (Figure 26-6A). (*3*) The base of the tail, or supporting peduncle, moves from side to side carrying the caudal membrane with it (part B). (*4*) These two motions are

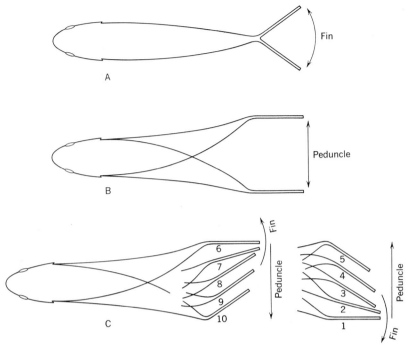

Figure 26-6 MOTION OF THE CAUDAL FIN OF A SWIMMING FISH.

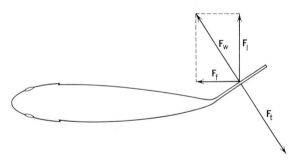

Figure 26-7 FORCES ACTING AT THE CAUDAL FIN OF A SWIMMING FISH. (The orientation of F_w would be somewhat altered by drag.)

timed so that the fin angles back to the left as it sweeps from left to right, and back to the right as it sweeps from right to left (part C).

As a first step in the analysis of the function of the tail we note that a consequence of the above is that except at the limits of its travel the caudal fin is constantly thrusting obliquely against the water (force $\mathbf{F_t}$ in Figure 26-7). The inertia of the water causes it to push with equal force in the opposite direction ($\mathbf{F_w}$ in the figure). This force can be broken down into a forward component ($\mathbf{F_f}$) and a lateral component ($\mathbf{F_l}$). Because of the streamlined shape of the fish, the water offers little resistance to $\mathbf{F_f}$, so the body glides forward. Force $\mathbf{F_l}$ tends to cause the fish to pivot around its center of mass; the posterior part of the fish

Figure 26-8   PATH OF A SWIMMING TROUT as seen from above.

moves right (in the illustration) and the anterior part moves left (the in-torque equaling the out-torque). However, the water offers considerable resistance to sideways motion of the body. The entire body does undulate from side to side, but lateral motion of the heavier, stiffer, flatter, anterior part of the body is much less than the lateral motion of the peduncle of the tail.

In fact, the function of the tail is somewhat more complicated. Frictional forces, pressure fields, and "lift" (as explained in Chapter 27) are created around the fin. These are difficult to quantify. When the forward and undulating motions of the body are combined with the motions described above for the caudal fin and its peduncle, the result is the graceful glide shown in Figure 26-8. It is seen that the tail does not thrash back and forth creating great turbulence, but instead largely follows the path of the body, avoiding marked cross-flow of water and separation of the boundary layer.

Various aquatic vertebrates swim by undulating the body. The body is then usually long, with gradually increasing flexibility toward the tail. Examples are lampreys, eels, and water snakes. A long robust tail may functionally extend the body and continue its undulations as for salamanders, aquatic lizards, and crocodilians. River and sea otters, desmans, and various rodents swim in this way but also propel themselves with their feet. Long dorsal and ventral fins commonly serve to increase the effectiveness of the body as a waterfoil (lampreys, eels). Similarly, the tail may be bilaterally compressed (salamanders, crocodilians, sea snakes, desmans, giant water shrew, muskrat). Some relatively slowswimming fishes (skates, some ribbon fishes, pipe fish) propel themselves by keeping the body straight and undulating only their fins.

Propulsion by undulation follows basically the same principles as propulsion by tail fin. Indeed, the two "methods" are not sharply demarcated. Any given segment of an undulating body or fin sweeps back and forth thrusting obliquely against the water much like an oscillating caudal fin. Compare Figure 26-9 with Figure 26-6C. Undulations move along the body as traveling waves from anterior to posterior, with the amplitude of the waves increasing to the rear. Many of these swimmers can reverse the direction of the waves and swim backward.

The action of paired flippers and wing-paddles differs from that of caudal fins in that their broadsides are presented more nearly flat-on to the water. (They cannot thrust in quite the same direction throughout the power stroke because they pivot at the shoulder or knee and hence describe arcs.) Accordingly, these

Figure 26-9   MOTION OF A SWIMMING EEL DURING HALF OF A CYCLE.

paddles thrust in a direction more nearly opposite to the forward motion of the body. However, since the thrust is applied beside, not behind, the body, a torque is established that tends to turn the head toward the opposite side. This torque must be countered by the stroke of a paddle on the other side of the body. The better swimmers among these animals (sea turtles, diving birds that swim with the wings, sea lion) have the paddles of the two sides of the body placed directly opposite to one another and they usually use them simultaneously. Those that swim with the hind feet may kick them together (frogs) or alternately (ducks, muskrats). If all four limbs are used in swimming (pond turtles, capybara, polar bear), diagonally opposite feet may be kicked in synchrony, though other sequences also serve. Actions of these various paddles to reduce resistance on the recovery stroke were noted above under the heading of drag.

Another kind of propulsion is used on occasion by some swimmers. If the animal can find water that has either motion (a velocity field) or a suitable pressure gradient (pressure field), it can then freeload, at least in part. One fish or whale may station itself beside and a little behind another, often larger, fish or whale, and thus benefit from the pressure drag created by the lead animal. It is probable that waves caused by the wind are sometimes briefly used in similar fashion.

The most spectacular example of aquatic freeloading is the wave-riding of dolphins. Groups of dolphins may move along for many kilometers in the bow wave of a ship, seemingly without exertion. The pressure field in the front slope of the wave is parallel to the surface of the water, not to the horizontal, and hence has a forward component. Upwelling water thrusting against the obliquely oriented tail flukes also provides a forward impetus. The tendency to pitch forward that is created by this pressure on the tail may be compensated by the pectoral flippers. It is clear that the animal is remarkably sensitive to the pressure and velocity fields of its immediate environment and instantly compensates for every change.

STRUCTURE AND FUNCTION OF THE LOCOMOTOR SYSTEM OF SWIMMERS   The span of the caudal fin from tip to tip divided by its chord, or average width in the direction of forward motion, is called the **aspect ratio** (Figure 26-1). (If it is easier to calculate, the equivalent equation, span^2/area, can be used.)

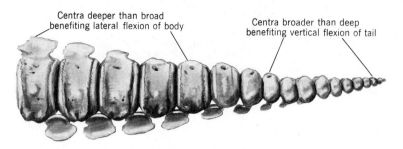

Horizontal tail flukes supported
by connective tissue, not bone

Centra deeper than broad
benefiting lateral flexion of body

Centra broader than deep
benefiting vertical flexion of tail

DOLPHIN, *Delphinus*

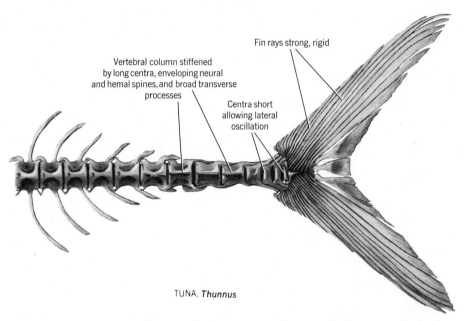

Fin rays strong, rigid

Vertebral column stiffened
by long centra, enveloping neural
and hemal spines, and broad transverse
processes

Centra short
allowing lateral
oscillation

TUNA, *Thunnus*

Figure 26-10 CAUDAL SKELETONS OF TWO VERY DIFFERENT FAST SWIM-MERS.

The fastest swimmers have an aspect ratio of 4 to 6, whereas slow swimmers have values of 1–2 (see bowfin, figure on p. 161). By having a large span, much of the tail fin gains effectiveness by extending above and below the turbulence that follows the swimmer. The caudal fin of fast swimmers is streamlined in cross section, has low mass relative to the body, and is stiff—particularly on its leading edges. Such tails oscillate with low amplitude. Frequency, but not amplitude, increases with speed; during a burst of speed the tail virtually vibrates (12 Hz being common for a 2 m tuna).

Rapid swimmers that propel themselves with the tail tend to have stiff spinal columns: Centra may be long to reduce the number of intervertebral joints (sailfish), the centra may be large and platyan to reduce flexibility at the joints (some cetaceans), among fishes the zygapophyses or spines may be unusually broad and strong to brace the joints (marlin, tuna, Figure 26-10), or a longitudinal vertebral ligament may add rigidity (cetaceans).

There must be adequate flexibility, however, at the base of the tail and peduncle. Bony fishes usually have diarthroses where the tail fin joins the spine, and the spine of cetaceans is dorsoventrally compressed in the anal area to allow vertical flexibility at that place.

The axial musculature, which is the power plant of the tail swimmer, may weigh half as much as the entire animal. Fishes that swim constantly (tunas) or migrate long distances (salmon) have on their flanks red muscle that contracts relatively slowly, contains myoglobin, is aerobic, and has an unusually rich blood supply. The temperature of such muscle is commonly 10°C above water temperature and is not dissipated because of a countercurrent exchange mechanism (see p. 232). The white muscle, which accounts for most of the bulk of the muscular system, is capable of faster, anaerobic contraction.

The tough fibrous skin of sharks is stiffened by the hydrostatic pressure of underlying muscles when the animal swims fast. Thus it becomes a whole-body exotendon and stretches on the convex side of the undulating body to store energy for release as that side flexes. It is probable that the muscles, tendons, and ligaments of certain bony fishes also store energy.

The myomeres of fishes fold to form zigzags as seen on the surface of the body, but series of interrelated cones as seen in three dimensions (Figure 26-11). The mechanics of these myomeres is complex and is still debated. The cones may extend the force of contraction of one muscle segment over several segments of the skeleton. They also ensure that muscle fibers will insert on the myosepta at oblique angles, thus providing some of the advantages of pinnate muscles. Small bones, which are a great nuisance at the dinner table, may run in the myosepta from the apex of one cone to the apex of another. The apexes of cones of the more posterior myomeres of the fastest fishes are extended into the tail as longitudinal tendons. The peduncle is therefore slender and tendinous, rather than broad and fleshy as in other fishes (Contrast Figure 26-11 with the figure on p. 189).

Figure 26-11 AXIAL MUSCULATURE OF THE FAST-SWIMMING MACKEREL, *Scomber*, with myomeres removed at successive levels.

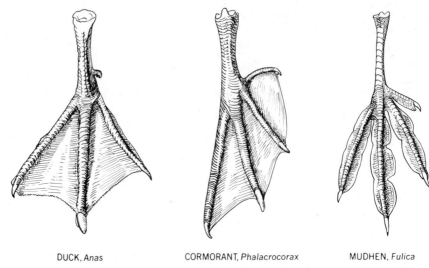

DUCK, *Anas*   CORMORANT, *Phalacrocorax*   MUDHEN, *Fulica*

Figure 26-12   FEET OF SOME AQUATIC BIRDS.

Secondary swimmers that propel or steer themselves with paired appendages need large broad paddles. If the paddle is in some way folded on the recovery stroke to reduce drag, then it must be enlarged on the power stroke by a flexible membrane. Hence the webs between three toes of ducks, flamingos, gulls, auks, loons, and penguins; between four toes of cormorants, boobies, and pelicans; and between all five toes of frogs, pond turtles, platypus, beaver, and sea otter. Grebes, mudhens, and finfoots have lobes on the toes instead (Figure 26-12). Small aquatic mammals usually have fringes of long stiff hairs that functionally broaden the foot.

If the paddle is canted instead of folded on the recovery stroke, then there is wide range of motion at the shoulder or knee (those joints being modified accordingly), but the waterfoil itself is enlarged by rigid structures so it will stand against water pressure without muscular effort. Sea turtles and pinnipeds flatten and greatly lengthen the usual complement of metapodials and phalanges, particularly at the leading edge of the flipper (Figure 26-13). The wing skeleton of penguins, and to a lesser extent of auks, is also flattened. Ichthyosaurs evolved extra digits to broaden the paddle, one species having nine digits in all. Ichthyosaurs, plesiosaurs, and cetaceans add phalanges to one or more digits, bringing each series to from 4 to as many as 26 units. Pinnipeds extend some of the bony digits with cartilages. The integumentary membrane extends beyond the skeleton in flippers of some pinnipeds (Figure 26-14). The paddle may also be broadened by sesamoid bones, as at the elbow of penguins, or by the flattening and lateral extension of existing bones such as the pisiform of sea turtles and ulnare of penguins. Diving birds and cetaceans incorporate the forearm into the paddle; the radius and ulna become short, flat, and positioned in the same plane.

Figure 26-13   ARM SKELETONS OF SOME AQUATIC VERTEBRATES that use the pectoral appendage as a paddle, Dorsal (lateral) views of right appendage.

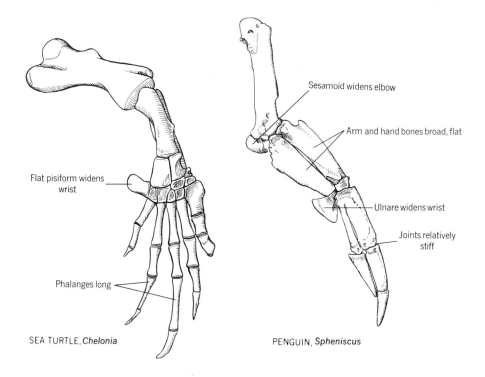

Flat pisiform widens
wrist

Phalanges long

SEA TURTLE, *Chelonia*

Sesamoid widens elbow

Arm and hand bones broad, flat

Ulnare widens wrist

Joints relatively
stiff

PENGUIN, *Spheniscus*

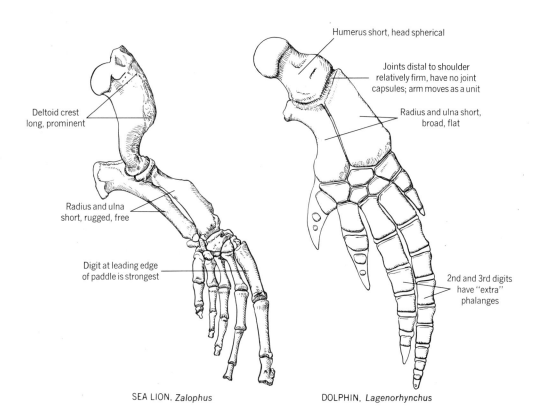

Deltoid crest
long, prominent

Radius and ulna
short, rugged, free

Digit at leading edge
of paddle is strongest

SEA LION, *Zalophus*

Humerus short, head spherical

Joints distal to shoulder
relatively firm, have no joint
capsules; arm moves as a unit

Radius and ulna short,
broad, flat

2nd and 3rd digits
have "extra"
phalanges

DOLPHIN, *Lagenorhynchus*

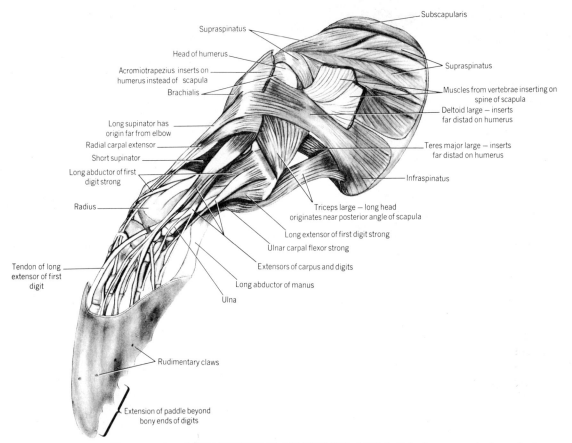

Figure 26-14 LEFT FORELIMB OF THE SEA LION, *Zalophus,* seen in lateral view. (Drawn from an air-dried dissection and hence somewhat shrunken.)

These paddles are made rigid in various ways, though some resilience remains. In flippers of cetaceans, amphiarthroses replace diarthroses, and bones are flat-ended. Spaces between the bony digits are sufficiently filled with firm tissue to brace the digits and make the surface contour of the paddle smooth.

CONTROL OF VERTICAL POSITION
Vertebrates that swim only on the surface are light so they will float high in the water like swans and gulls. Nondiving ducks have a specific gravity of only about 0.6. A light skeleton, fat deposits, gas in air sacs or lungs, and air trapped in feathers or fur contribute to buoyancy. Vertebrates that rest on the bottom, by contrast, need to be more dense than water to maintain their position. Flat fishes and skates, which have no gas bladders, have a specific gravity of about 1.09.

Swimmers that vary their vertical position in the water maintain place in any of several ways. The hydrostatic function of the gas bladder of bony fishes was noted on page 238. Some sharks may exercise some control by selectively producing in the liver either of two metabolites which have different densities. A similar selective production of lipids occurs in a number of teleosts, their storage being in muscle, skin, or skull, and in the surviving coelacanth, which has a fat-filled gas bladder. Various sharks and bony fishes that have no gas bladders and are slightly more dense than water (e.g., leopard shark, mackerel) maintain position by swimming slowly all the time, just as tetrapods breathe all the time. The source of their lift is discussed below.

Avian and mammalian divers must find ways to make their bodies more dense than those of their nondiving relatives. The bones of diving birds are less pneumatic. Their air sacs are reduced (loons, penguins). They press their feathers against the body to exclude air: Auks bubble constantly when underwater, and the feathers of some diving birds become wet. Penguins achieve a density of 0.98.

Mammals that dive deep may hyperventilate before submerging, but they do not fill their lungs. Indeed, they may exhale during the dive. Deep-diving whales have relatively small lungs. Sirenia, which may feed while resting on the bottom or standing on their tails, have unusually heavy skeletons; their ribs are swollen and solid.

Rotation of a swimmer (or ship) around its long axis is called **roll,** rotation around its transverse axis is **pitch,** and rotation around its vertical axis is **yaw.** The body is stable if it passively tends to correct for displacements from a given position; it is unstable if a small displacement tends to increase to become a larger displacement. As for cursors (see p. 453), increased stability of swimmers reduces muscular effort but also reduces maneuverability.

Of several factors influencing stability, one is independent of forward motion. Two forces act on any submerged object: Gravity tends to make it sink, and buoyancy tends to make it rise. If the object has the same density as water, then the two forces are equal and the object neither sinks nor rises. Gravity acts on an object as though all its mass were at its **center of gravity** (CG). Buoyancy acts as though all lift were applied at its **center of buoyancy** (CB). The CB is located where the CG would lie if the object were uniformly dense throughout (like the displaced water). Vertebrates, however, are not uniformly dense: Bone, cartilage, and muscle are more dense than water, whereas fat, oil, and gas in lungs or gas bladder are less dense. The CG and CB are, therefore, usually in different places, and gravity and buoyancy act to turn the object in the water (Figure 26-15).

The diaphragm of cetaceans is oriented diagonally under long lungs placed high in the body, rather than behind short lungs placed forward in the body. This raises the CB relative to the CG and places it near, and usually a little above, the CG, thus giving the animal slight positive stability. The CG and CB of sharks are at about the same vertical level, so there is scant tendency to roll, but the CB may be a little more anterior, thus tending to lift the head. The gas bladders of bony fishes are located high in the coelomic cavity, but the heavy spine and epaxial muscles are still more dorsal, so the CG tends to lie above the CB, and the fish is unstable in regard to roll. Dead fishes float belly up.

Any fins or flippers may be extended to function as brakes, but the pectoral

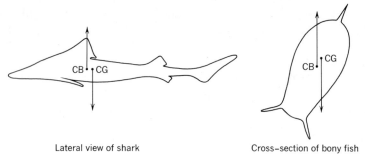

Lateral view of shark   Cross—section of bony fish

Figure 26-15   INSTABILITY RESULTING FROM DIFFERENT LOCATIONS OF CEN-
TER OF GRAVITY (CG) AND CENTER OF BUOYANCY (CB).

fins are most commonly used for this purpose. If they are lower than the CG,
then the head tends to pitch downward. This tendency may be countered by
canting the pectoral fins to give lift (the entire body then rising as it slows), by
extending the dorsal fin, or by actions of other fins. The caudal fin may be curled
into a hook as an effective supplemental brake.

Caudal fins of various shapes were named according to evolutionary relationships in the figure on page 161. Following much experimentation with am-
putation of fins and with the forces exerted on models in water and in wind
tunnels, tails of different shapes have also been classified according to function.
Assuming the notochordal or spinal axis to be adequately rigid and the fin membranes to be passive and uniformly flexible, then symmetrical caudal fins do not
cause pitch (parts A–C of Figure 26-16). Caudal fins with a down-tilted spinal
axis, or with more membrane below than above the longitudinal axis of the body,
or with the dorsal lobe stiffer than the ventral lobe, cause the tail to rise and the
head to pitch downward (parts E–G). Tails with the reverse structure have the
reverse function (parts H–J).

The traditional view has been that tails of sharks resemble part G; constant
lift by the pectoral fins is then needed to counter a tendency for the head to pitch
downward. However, the similar tail shape of part D causes no pitch (downward
thrust caused by the uptilted axis being cancelled by upward thrust caused by
the more flexible ventral lobe). If an edge of the fin membrane does not necessarily trail passively in the water but instead can be made temporarily more rigid,
or can be actively advanced with, or ahead of, the axis of the tail (parts K and
L and Figure 26-17) then the tail might cause no pitch, or pitch either up or down
as the fish chooses. This is a usual circumstance unless the membrane is stiff, as
in scombroids. Shapes D, G, and L get the tail up and out of the way as the fish
swims over the bottom. Most sharks are a little heavier than water, so some lift
is required much of the time.

The density of bony fishes having hydrostatic gas bladders closely approximates that of water. Hence, the tail need not be constructed to provide constant
lift. Likewise, the pectoral fins need not provide constant lift, and fold against
the body to reduce drag. Even though the density of these fishes is about the

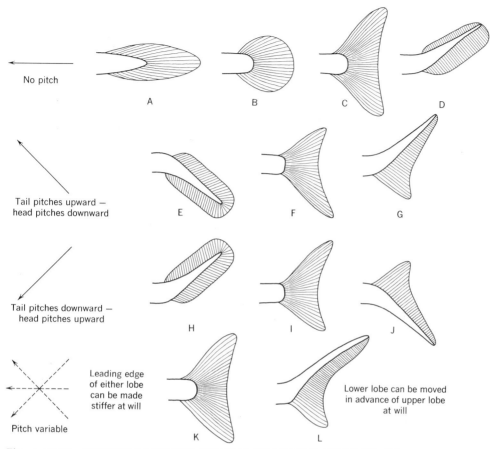

No pitch

A   B   C   D

Tail pitches upward —
head pitches downward

E   F   G

Tail pitches downward —
head pitches upward

H   I   J

Leading edge
of either lobe
can be made
stiffer at will

Pitch variable

Lower lobe can be moved
in advance of upper lobe
at will

K   L

Figure 26-16 FUNCTION OF THE CAUDAL FIN IN RELATION TO SHAPE AND OTHER VARIABLES. The membrane is assumed to be passive in A through J.

same as that of water, their equilibrium tends to be unstable, so slight corrections are constantly needed. These fishes can alter the area, extension, curvature, shape, period, and amplitude of all their fins in subtle ways; their sense organs and nervous systems instantly make computations which man would find difficult to approximate.

Flexibility of the entire body is advantageous for maneuvering. Pinnipeds have supple necks and backs; penguins have secondarily regained much of the flexibility of the trunk that was lost in the ancestry of birds. The most maneuverable fishes live among plants and in coral reefs. Their bodies are short so they can turn (yaw) quickly, and are bilaterally compressed to allow flexibility. Fins that are anterior to the CG (including the forward part of the dorsal fin) are also used for fast turns. They trip the body around just as a dart turns in the air if thrown feather-end first.

Figure 26-17 DORSAL VIEW OF HETEROCERCAL TAIL WITH AN ACTIVE MEMBRANE.

OTHER ADAPTATIONS OF SECONDARY SWIMMERS

PROTECTION OF SKIN, EARS, AND RESPIRATORY SYSTEM   Many aquatic mammals and birds trap enough air in the fur or plumage to shield the skin from wetting. The sebaceous glands of pinnipeds secrete quantities of waterproof sebum, and aquatic birds have large oil glands. The skin of cetaceans is resistant to water but not to drying. It is said that even the pouch of the water opossum is waterproof.

Cormorants and pelicans have lost the external nares. The external nares of other aquatic tetrapods, from frogs and alligators to beavers, hippopotamuses, and dolphins, are always dorsal in position, and the owner seems always to know when they are barely out of water. A ridge deflects water from the blowhole of many whales. When under water, the nares are automatically tightly closed. This is usually accomplished by sphincter muscles, but baleen whales use a large valvular plug, and toothed whales add an intricate system of pneumatic sacs so that great pressure can be resisted in each direction. Respiratory exchange may be surprisingly rapid: Whales of moderate size require only 1–2 sec to exhale and inhale. Tidal air is relatively great and residual air is relatively small.

Various aquatic tetrapods, from crocodilians to rodents, have modified the palate and glottis to permit chewing and swallowing under water without interference to the airway: The epiglottis is up within the nasal chamber in beavers as well as dolphins. The pelican, which plunges into water from the wing, has modified the larynx to exclude water under pressure.

The external auditory canal can be plugged or furled by pinnipeds. Cetaceans admit water to the outer part of the canal; the inner part grows shut in baleen whales. Protection of the middle ear from collapse is mentioned in the next section.

Sirenia, and Cetacea and Pinnipedia that dive to moderate depths, have a succession of 8–40 valves of smooth muscle in each bronchiole to hold air in the alveoli against pressure. Lung tissue of whales and manatees is relatively rigid due to the presence of cartilage and muscle. Their tracheas and bronchi are thick and strong. The tracheas of penguins, some petrels, sea lions, and the dugong are braced by a longitudinal partition.

Deep-diving mammals do not carry much air down. The lungs are not an oxygen store. If oxygen were exchanged, nitrogen would dissolve in the blood, only to bubble and cause great distress on the return to the surface. Alveolar collapse is probably complete at 70 m in the bottlenose dolphin, forcing the air into larger, stronger, and nonabsorbing airways. Lungs of the Weddell seal also shrink markedly. The trachea is bow-shaped, not round, in cross section to allow closure without injury. Whales have a short sternum and few fixed ribs. The remaining ribs have a single head. Thus, the thorax can also "collapse" without damage.

ADAPTATIONS OF SENSE ORGANS   The sense of taste appears to be "normal" in most aquatic vertebrates, but is rudimentary in Cetacea. Olfaction is considered to be poor in Pinnipedia and very poor in Cetacea. The structure of the nervous system clearly indicates the regression of these senses. Deep-diving mammals have small cranial sinuses, or none.

Hearing is acute in the more highly adapted aquatic tetrapods. Many whales have an amazing repertoire of sounds ranging from low resonant honks to high flutelike tones. In favorable circumstances some whales probably can communicate over distances of 160 km or more, though little is known of their "language." At least most cetaceans and pinnipeds are capable of remarkably discriminating echo-ranging. They make sounds up to 200,000 Hz emitted in pulses of from 16–400/sec. There is indirect evidence that penguins use the cavitation clicks produced by the turbulence of their own swimming as the sound source for echo-ranging (they can quickly locate fish in absolute darkness).

Underwater hearing is not dependent on a superficial tympanum or external auditory canal, provided that a gas-filled space functions as a resonator (see Chapter 19). The tough, fibrous auditory canal of cetaceans may be partly occluded. The oil-filled cavities of the heads of certain cetaceans may beam sounds and presumably return sound to the ear by specific pathways.

The tympanum and ossicles of cetaceans function in the usual way, yet are of distinctive structure: The ligamentous tympanum is mounted by a horny cone, and the ossicles are large, heavy, and firm. It is essential that the tympanum vibrate against an air space in the middle ear and, of course, that this space not collapse during dives. The middle ear of pinnipeds (and parts of the external canal?), and air sinuses communicating with the middle ear of cetaceans, are lined by highly vascular tissue that engorges during dives. The volume lost by compression of air is thus replaced by blood. Furthermore, in cetaceans the sinuses and parts of the middle ear not adjacent to the drum are filled with a foam consisting of small air bubbles in an oil-mucus emulsion. Experiments show that these bubbles do not collapse, even under a pressure of 100 atm. Finally, the tympanic bulla of whales is strengthened by some of the thickest, most dense bone known. The bullas are loosely attached to the remainder of the skull and are cushioned in foam and blood sinuses. This permits them to function independently of each other and of the body. Directional hearing of aquatic mammals is excellent and probably is based largely on intensity discrimination.

Sirenians have poor vision, as would be expected from their sluggish habits, stationary (plant) food, and often murky environment. Baleen whales have moderately good eyesight, but have a limited field of vision and cannot see out of water. Their food is passive, and some of them dive below the level of light penetration. The food of toothed whales and pinnipeds is active, and they have excellent vision, both below and above water.

Eyes of aquatic tetrapods with good vision have secondarily acquired characteristics of the eyes of their remote ancestors among primary swimmers: The eye is large, the eyeball short along its optical axis, the lens large and spherical, and the cornea flattish or elliptical for streamlining. Lacrimal glands are reduced (pinnipeds) or absent (sirenians, cetaceans). To protect the eye from saltwater, the cornea is cornified and is bathed by the secretion of large glands in the lids. In order to withstand wave pressure, the sclera of cetaceans is very thick and tough. Pinnipeds can change the shape of the lens more than is usual to permit vision in and out of water.

THERMOREGULATION; AND RESPONSE OF THE CIRCULATORY SYSTEM
Since the thermal conductivity of water is about 20 times greater than that of air, endothermic aquatic vertebrates must protect themselves from heat loss, particularly when inactive and when in cold seas. Air trapped in plumage or dry underfur (as of the sea otter, beaver, or fur seal) is an effective insulator. Large mammals have relatively little heat loss because of their low surface-to-volume ratio. Blubber is an effective insulator for them, coming, in extreme instances, to $\frac{1}{4}$ of the body weight. Flippers of cetaceans have slow circulation and countercurrent exchange, so warm outgoing blood gives its heat to the cold incoming blood (see p. 232). It is probable that some whales require moderate activity to maintain body temperature in arctic waters.

Conversely, swimmers must be able to dissipate heat during periods of activity when heat production may rise tenfold. The countercurrent exchange mechanism can be bypassed, and the large flat flippers (which are devoid of blubber) then serve as radiators. Also, vascular papillary ridges in the epidermis of whales dissipate heat when needed.

The circulatory physiology of air-breathing vertebrates during dives adjusts so as to supply oxygen to the brain and heart, and otherwise to avoid stress from lack of oxygen or buildup of carbon dioxide and lactic acid. Bradycardia, or slowing of the heart, is universal and occurs on submergence. The rate commonly is reduced to $\frac{1}{10}$ or $\frac{1}{15}$ of normal, and the slowing is in part preventive; its onset is faster when a deep dive is anticipated. The aorta dilates near the heart to help maintain blood pressure during bradycardia, but all arterioles constrict except those of the brain and heart. Excretion stops. The veins of pinnipeds and cetaceans (like those of fishes) have no valves. Blood volume of these swimmers, and of some diving turtles and birds, reaches two times that of comparable terrestrial vertebrates. The hepatic portal system is large, and venous sinuses may be present in the thorax and abdomen. The result is that quantities of blood stagnate in the body cavities.

Deep divers are usually not very active. Their hearts tend to be relatively small (though that of a blue whale may still weigh 600 kg!). The metabolic rate falls off a little (pinnipeds, cetaceans, alligators, ducks). The blood of fast-swimming and deep-diving dolphins is able to carry up to three times as much oxygen as that of their opposites. The myoglobin content of divers' muscles is high. They tolerate twice as much carbon dioxide in the blood as does man. Lactic acid is stored in the muscles until breathing resumes.

The circulatory systems of swimmers show convergence in other ways that remain enigmatic. Why is the postcava doubled (a turtle, pinnipeds, cetaceans, and sirenians, but also the nonaquatic edentates and slow loris)? Why do pinnipeds have a sphincter in the postcava at the level of the diaphragm? Why are the intervertebral vessels enlarged and the jugular veins reduced (pinnipeds, cetaceans)? Why is there a venous plexus in the drainage of the kidney, or a rete to damp the flow of blood to the brain (cetaceans)?

REPRODUCTIVE BIOLOGY   Most sea snakes are viviparous and give birth at sea, as did the ichthyosaurs. Other reptiles and all birds lay their eggs or give birth on land. Among aquatic mammals, the walrus and hippopotamus sometimes give birth in the water, and Cetacea and Sirenia always do so. A single, large, precocious young is born at a time. Cetacea deliver rapidly; the calf emerges tail first. The mother whirls in the water, thus snapping the relatively short umbilical cord at a predetermined point of weakness. The newborn swims to the surface to breathe, sometimes with maternal assistance. The tail flukes are soft and curled at birth, but harden in about two days, by which time the calf can keep up with the herd.

Whale milk is thick and rich in fat. It collects in sinuses and is forced out in mouthfuls during underwater nursing. Growth of young pinnipeds and cetaceans is rapid. The calf of the blue whale, which is 7 m long at birth, gains about 90 kg/day on its mother's milk!

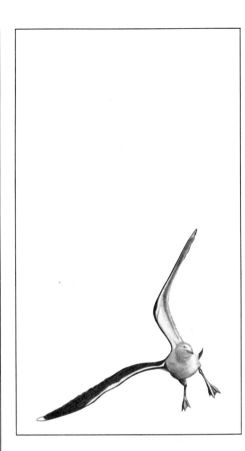

27

Flying and Gliding

Some lightweight climbers can retard a fall and travel horizontally as they move downward. If control of the path followed is minimal, and the line from takeoff to landing is steeper than about 45° to the horizontal, then the animal is said to **parachute.** If some maneuvering in the air is possible and the line from takeoff to landing is less than 45° to the horizontal, then the animal is said to **glide.** If the animal is capable of sustaining itself in the air, we say it can **fly** and is **volant.**

The principles of animal flight are understood in general, but analysis is complex and much remains to be learned. The field of study is active and conflicting theories may soon be resolved. Even when vertebrates fly under relatively uniform conditions, they must make frequent slight adjustments to compensate for alterations of external variables. When a gull maneuvers in a changeable wind, major adjustments, many of kinds that are not possible for man-made aircraft, must be constant and nearly instantaneous. Finally, it is technically very difficult to study living, fast-flying animals, and models and stuffed animals in wind tunnels often give spurious results.

ORIGIN AND ADVANTAGES OF FLYING AND GLIDING

Although much more primitive than modern birds, the first known birds (suborder Archaeornithes) were apparently already moderately good gliders (see figure on p. 67). It is possible that birds evolved from small, bipedal, arboreal archosaurs that hopped from branch to branch, steadying themselves with outstretched forelimbs. As feathers enlarged on the margins of the forelimbs, the hops were likely extended. The animals may have utilized gravity to increase air speed as they planed to a lower perch. Thus, the power of flight could have been established gradually.

An alternative theory of the origin of flight is advocated by some biologists. Citing similarities between Archaeornithes and certain small bipedal dinosaurs, they postulate that the ancestor of birds was a bipedal cursor. At dispute are interpretations of the structure of the feet, claws, and feathers, the relative size of the forelimbs, and the advantage to a ground-living vertebrate of evolving feathers.

The first known bat lived 50 million years ago, but was modern in appearance and flying ability. Presumably bats evolved from small, agile mammals that scrambled about in trees seeking insects. The flying reptiles, or pterosaurs (an order of the subclass Archosauria), survived from about 180 until 65 million years ago. The earliest known representatives were already competent flyers, so nothing is certainly known of the origin of their ability to fly.

As we shall see, birds, bats, and pterosaurs differ considerably in structure. Nevertheless, in order to fly at all, certain conditions must be met, and these three groups display much convergent evolution. Flyers are among the most specialized of vertebrates.

Parachuters may drop from tree to ground to escape from less versatile predators or to move quickly to another location. Also, an inadvertent fall is not damaging. Gliders have the same advantages to a greater degree, and by gliding can forage more quickly and over a wider area than would otherwise be possible. They have the advantage over flyers that the forelimbs, being little specialized, remain useful for climbing and the manipulation of food.

The various flyers benefit from flight in different ways. Potential advantages include the following:

1.   Flyers gain access to food that is in the air (flying insects), that must be reached from the air (terminal flowers), or that can be located from the air (rodents, fishes).

2.   Great mobility and maneuverability enable flyers to search rapidly and efficiently for food and shelter.

3.   Escape is provided from nonvolant predators.

4.   By migrating, flyers can travel, according to season, to regions where climate, food supply, and nesting sites are favorable.

5.   Dispersal is possible over distances and geographic barriers that would otherwise be insurmountable.

VERTEBRATES THAT PARACHUTE AND GLIDE

PARACHUTERS AND GLIDERS   All parachuters are arboreal. Various tree frogs are included. These launch themselves with a jump, hold the limbs out to the side, and control their orientation so that their flat ventral surface is presented to the airstream. One species (family Hylidae) has no other adaptations for slowing

Figure 27-1   EXAMPLES OF PARACHUTERS AND GLIDERS.

"FLYING" FISH
Cypelurus

PARACHUTING FROG
Rhacophorus

GLIDING LIZARD
Draco

PARACHUTING GECKO
Ptychozoon

COLUGO
Cynocephalus

"FLYING" SQUIRREL
Glaucomys

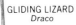

its fall, yet achieves an angle of descent of about 60°. Several other tree frogs (family Rhacophoridae) are more expert, approaching an angle of descent of 45°. Their huge feet are fully webbed, and small membranes fringe the arms and span the angle between thighs and body wall (Figure 27-1).

A genus of tree snakes is capable of controlled parachuting at fairly flat angles. The body is held horizontal, the ribs are spread to the sides, and the belly is drawn in to present a concave surface to the airstream.

Some lizards can descend at about 70°, relying mainly on behavioral adaptations. Others, having fringed tails, do a little better. Several genera of geckos (family Geckonidae) have broadly webbed toes and fringes on the head and body. They can descend at nearly 45° to the horizontal with considerable maneuvering. Various small tree squirrels also parachute to break a fall by spreading legs and tail and coming down flat to the airstream.

Several fishes, at least three genera of lizards, and representatives of three orders of mammals are gliders. All have evolved broad membranes to catch an airstream, but these can function only when adequate air speed is attained. The lizards and mammals climb a tree to gain height, and then they jump. They fall steeply until sufficient speed is attained by gravity, then they sail off following a parabolic path. Doubling the height of the takeoff more than doubles the horizontal travel.

The fishes attain adequate air speed in a very different way. A "flying" fish (family Exocoetidae) first swims rapidly just under the surface and then emerges until only the large lower lobe of its hypocercal tail is in the water. The pectoral fins, which in one genus are as long as the body, are spread to the sides as the tail thrusts rapidly back and forth in the water. The fish skims along in this manner for from 1 to 6 m until its air speed increases to an estimated 40–70 km/hr. It then spreads the somewhat smaller pelvic fins and glides free of the water, usually for 2–4 sec, but occasionally for 10 sec or more, sometimes traveling 100 m. On landing, the fish either submerges or immediately skims again with only the tail in the water preparatory to another glide. From two to twelve glides may be made in succession. The fishes are unable to glide if there is no wind, but they can sail several meters high over the water if there is a good breeze. The fins are fixed during most of the glide, but they may be seen and heard to vibrate at takeoff and landing. The purpose of this action is not known; fin musculature is not enlarged.

Some fishes of another family (Characinidae) leap high in the air and then spread their large pectoral fins stiffly as they fall back to the water. They sometimes jump into boats. Other members of the same family have larger pectoral fins which they flutter, using enormous ventral muscles as they make long arcing excursions from the water. The aerodynamics of this near-flight is unknown.

Gliding lizards of the genus *Draco* have extensive membranes reaching on each side from the thorax to the base of the hind leg. The membranes are supported by six pairs of much-lengthened ribs. When they are not in use for gliding or display they are folded against the body. These lizards commonly glide at

20–30° to the horizontal, but can even gain elevation if they move into an updraft. Glides of 24 m have been observed.

The colugo (mammalian order Dermoptera) is a cat-sized Asiatic glider with the largest "flight" membrane of all: It extends from the throat to the wrists to the ankles to the tip of the tail. Even the toes are webbed. One animal sailed 136 m at an angle of only 5° to the horizontal.

Five species of phalangers (order Marsupialia) in three genera are gliders. They range in weight from 14 to about 1360 g. All are vivacious forest dwellers with soft fur, long bushy tails, and membranes extending from elbow to knee. They launch themselves with a leap, and commonly glide to 100 m.

Fifteen genera of rodents are expert gliders. They range from chipmunk-size to cat-size. The very large membranes of scaly-tailed squirrels (family Anomaluridae) are supported in part by long cartilaginous struts (or calcars) from the elbows. Some "flying" squirrels (family Sciuridae) have shorter struts from the wrists. Usual glides are 6–10 m in length at angles of 30–50° to the ground. However, an American species was seen to glide 50 m with a drop of only 18 m. Such figures really mean little—in usual circumstances all gliders seem able to glide as far as they "want" to.

Most birds frequently intersperse gliding with flying, sometimes with wings open and sometimes with wings folded. The saving of energy may be considerable.

FLYERS AND THEIR SKILLS

PTEROSAURS had short bodies, long necks, large birdlike heads, batlike hind limbs, and long narrow wings supported by the arms and elongated fourth fingers (Figures 27-2 and 27-3). Eyes and brains were also birdlike. There were no scales and there is evidence that some had hair. It is possible that they were able to elevate the body temperature, at least when flying. One group had long tails and many homodont teeth; the other group had short tails and few or no teeth. The more than 25 known genera ranged from the size of a starling to the largest of all flyers, which had a wing span variously estimated to be 11 to 15 m. Most

Figure 27-2   RESTORATION OF THE GIANT PTEROSAUR, *Pteranodon.*

Figure 27-3 COMPARISON OF THE RIGHT WING SKELETONS OF A BIRD, A PTEROSAUR, AND A BAT. The digits are numbered.

pterosaurs lived along seacoasts and ate fish. It is probable that some could alight and take off into the wind from water. Perhaps some could even swim a little with their wings. They likely roosted in trees or on cliffs, scrambling about like bats when not in the air. The wing musculature of the larger species may have been too weak for constant flapping; like hang gliders they probably flew largely by soaring, depending at least in part on wind.

BATS are very successful flyers. There are about 175 living genera and more species than in any other mammalian order except Rodentia. The smallest bats weigh only 4 g. The largest weigh 900 g and have a wing span of 1.7 m. Bats are variously adapted for eating insects, fruits, flowers, nectar and pollen, blood, and fish. Flight membranes of skin are supported by the arms, greatly elongated second through fifth fingers, hind limbs, and usually all or part of the tail. Some bats have fast and direct flight and others have slow and erratic flight.

BIRDS, the finest of all flyers, are relatively uniform in structure compared to other vertebrate classes, yet are diverse in habits and habitats. The 2 g bee hummer is the smallest. Excluding flightless birds, the extinct vulture *Teratornis* was the largest. It probably weighed about 23 kg (50 lb) and had a wing span between 3 and 4 m. Flight feathers are supported by the long arms and one robust digit (probably the second); those borne by the hand are **primary feathers** and those on the forearm are **secondary feathers.** A feathered membrane called the **patagium** spans the angle in front of the elbow.

SPEED   Many flyers are capable of high speed, yet in spite of efforts to correct for wind and angle of flight, their true capabilities remain in doubt. Songbirds fly 16–40 km/hr, ducks cruise at 50–65 km/hr and several species probably can fly 100 km/hr when pressed. A flock of sandpipers that flew diagonally in front of a plane was estimated to be flying 180 km/hr. Even greater speeds have been claimed for several birds.

ENDURANCE   In contrast to cursors and swimmers, it is small to medium-sized flyers that have the greatest endurance. The mastiff bat remains on the wing continuously for 6 hr and more. The sooty shearwater and sanderling migrate the 11,000 to 13,000 km one way between Arctic America and Patagonia. The golden plover flies 3800 km nonstop from Labrador to South America. Various species may remain in the air for 90 hr when migrating. It was calculated that if the fat stores of the Blackpoll Warbler are equated to gasoline on a weight basis, the little bird gets 720,000 mi/gal.

HIGH FLYING AND LIFT   Certain bats fly at least 3000 m high. Most birds fly below 1500 m, but migrants occasionally travel as high as 6400 m (21,000 ft). Birds have been observed above 8000 m in the Himalayas. Even when at rest, mammals become unconscious at lesser altitudes: To compensate for low oxygen they hyperventilate. This flushes carbon dioxide from the body, which makes the blood alkaline and that, in turn, causes blood vessels, to the brain and elsewhere, to constrict. An unknown mechanism enables birds to hyperventilate without vasoconstriction.

Numerous bats can fly carrying young weighing 50% of their own weight.

One species can lift 73% of its weight. Various birds can carry loads of prey or fat stores that about equal their unladen weight.

ACCELERATION AND MANEUVERABILITY   Although scarcely documented, these skills of flyers are impressive. Certain bats and birds start, stop, and change speed and direction remarkably fast. Flocking birds do these things in unison. Many bats and birds can hover in place, and hummingbirds can fly backwards.

GENERAL REQUIREMENTS OF FLYERS

Flyers, like swimmers, move within a fluid medium. Aerodynamics and hydrodynamics are closely related fields, so the requirements of flyers parallel those of swimmers. However, air being much less dense than water, the relative importance of the variables is altered and flyers are deprived of support by flotation. All flyers must (1) derive sufficient upward force from their muscles or from the environment to counter the pull of gravity, (2) reduce drag, particularly if flights are long or fast, (3) propel themselves at various speeds and sometimes in restricted spaces, and (4) retain stability, maneuver, brake, and land according to habit. These primary requirements establish some rigid secondary requirements that focus on needs for (5) strength with light weight, (6) firmness of the trunk, and (7) the efficient production and utilization of power.

How do flyers meet their stringent requirements?

MOSTLY ABOUT LIFT

Since flyers are more dense than air, an upward force must act on them in order for flight to be sustained. During level flight, this force must just counter the pull of gravity, which is to say it must equal the flyer's weight. During ascending flight, flight in a downdraft, and flight while carrying young or prey, the upward force must exceed the weight of the flyer. Where does the upward force come from?

We shall first consider only level flight in still air with no flapping of the wings. These conditions are set so we can clearly distinguish upward force from backward force (drag) and forward force (propulsion).

As a first approximation, imagine a crude model of a flying bird with wings cut from thin slats of wood. If moved in an airstream in such a way that the wings meet the wind exactly edge-on, then air flows equally over and under the wings and there is no upward force (Figure 27-4A).

Now suppose that the model is improved by tilting the leading edge of the wing upward. The angle the wing makes with the airstream is called the **angle of attack,** or α. When α is small, air flows over the wing as shown in part B. Why the airstream divides as it does is poorly understood, but it is seen that the air passing over the wing has farther to travel than the air passing under the wing. Accordingly, it must move faster and its pressure decrease. Consequently (following Bernoulli's theorem), the pressure is lower above the wing than below. The disturbed part of the airstream is thinnest near the leading edge of each wing. Air moves fastest there and creates the lowest pressure. Forces on all other parts of the wing are also in proportion to the adjacent air pressures (part C). All the forces acting on the wing that are derived from its motion can be divided into a component called **drag (D)** which by definition is in line with the airstream and opposite to the direction of flight, and a component called **lift (L)** which is at

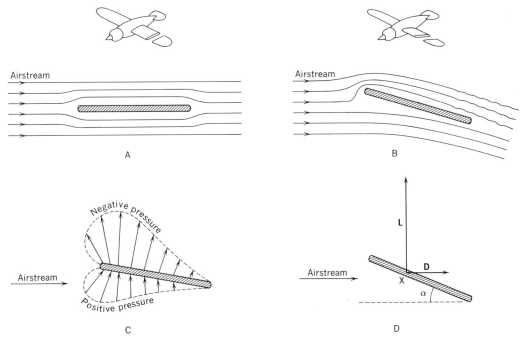

Figure 27-4 SOME FACTORS RELATED TO LIFT.

right angles to **D** (part D). These forces act at the center of pressure, **X,** which is usually $\frac{1}{4}$–$\frac{1}{2}$ of the way back from the leading edge of the wing. In other words, all the actual forces together have the same effect on the wing as **D** and **L** acting at **X.**

In level flight in still air, **L** is directly upward and is the force needed to counter the pull of gravity. In ascending or descending flight on fixed wings, "lift," in spite of the term, is not vertical but at right angles to the airstream. It does always have a vertical component (unless the flyer flies upside down or depresses the leading edge of the wing, making α negative), but it also has a horizontal component that may be forward (descending flight) or backward (ascending flight).

As α increases from 0°, **L** also increases, and the center of pressure moves forward to about $\frac{1}{4}$ of the way back from the leading edge of the airfoil. However, as α exceeds about 15°, the airstream above the wing suddenly ceases to flow smoothly and instead separates from the wing in strong eddies. Lift is then lost, and the flyer is said to **stall.**

There are also other variables. If an airfoil is convex on its upper surface as seen in cross section (Figure 27-5), it is said to have **camber.** The wings of vertebrate flyers are cambered, at least proximally. The line joining front and back edges of a cambered airfoil is called the **chord.** Moreover, the placement of bone and flesh in a vertebrate wing is such that it may be thicker along the leading edge than elsewhere, thus making the wing "streamlined" in cross section. This is particularly true of bird wings. Finally, if a small second airfoil is

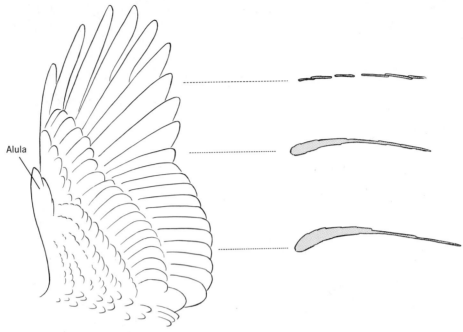

Alula

Figure 27-5 CAMBER AND STREAMLINING OF A BIRD WING AT SELECTED CROSS SECTIONS.

positioned just above the leading edge of the first, a **wing slot** is created. This slot deflects air down onto the upper surface of the main wing thus energizing the region near the boundary and preventing separation. As a result, **L** is increased and α can be greater before the wing stalls. Pterosaurs and bats are unable to use slots. The first digit in the wing of a bird, together with its feathers, is called the **alula.** It ranges from $\frac{1}{10}$ to $\frac{3}{10}$ as long as the wing, and when lifted creates a wing slot. (Some birds also have slots between the tips of the primary feathers at the end of the wing. Their function is mentioned later on.)

We can now relate these variables more precisely. Lift $= \frac{1}{2}\rho V^2 SC_L$, where ρ = the density of air, V = air speed, S = projected area of the wing (i.e., the area of its shadow), and C_L = the coefficient of lift. The two wings and the part of the body that joins them function as a unit in creating lift so they are best taken together in calculating S. When the wings are not flapped, lift is greater toward the center of the unit than toward the extremities, or wing tips. The coefficient, C_L, is a dimensionless number that usually equals about 1.5 for birds, but can range above 2 if the wings are slotted. Its value depends on the angle of attack, the camber and streamlining of the wing, the presence and nature of wing slots, and the Reynolds number. The Reynolds number, in turn, equals the density of air, times the air speed, times the length of the average chord of the wing, divided by the viscosity of air. For vertebrate flyers, usual values are 40,000–130,000 (which is lower than corresponding numbers for swimmers).

For our purposes, this complex of variables can be simplified. Since the density and viscosity of the air are small in value and scarcely subject to control by the flyer, we can discount them when considering the generation of lift. The camber and shape of the wing influence lift largely through their relation to the angle of attack. This leaves angle of attack, wing area, and air speed as of particular importance to lift. Air speed is squared in the basic formula for lift, and occurs also in the calculation of the Reynolds number. It follows that if flyers are fast, they have enough lift even when their wings are narrow, small in area, have little camber, no slots, and function with a small angle of attack. Conversely, slow flyers need large wings with camber and maximum angle of attack in order to ascend quickly.

(There is, of course, an upward component to ascending flapping flight. This will be considered below. Also, a flyer may derive upward force from updrafts. This is discussed under soaring, gliding, and formation flying.)

DRAG

Drag (the resistance the air offers to the motion of the flyer) acts horizontally backward when flight is level and in still air. It is convenient to divide the total drag **(D)** into two categories, profile drag and induced drag.

Profile drag (D$_p$) (or parasite drag) is all the drag acting on a hypothetical airfoil of infinite length. It is also all the drag acting on the wings and intermediate body (again taken as a unit) of a real flyer exclusive of the drag induced by motions of air around the tips of the wings. Profile drag is produced by energy lost to the environment through the friction of the air against the body, the displacement of air, the formation of pressure gradients in the air, and the creation of eddies. The value of $\mathbf{D}_p = \frac{1}{2}\rho V^2 S C_{dp}$, where C_{dp} is the coefficient of profile drag and the other variables are the same as those in the formula for **L**. The value of C_{dp} varies with the camber, streamlining, slotting, and outline of the wing-body unit, and with the Reynolds number. (These are the same variables that influence the value of C_L, but the relationships are different here.) On the upstroke of the wings, C_{dp} may be much reduced by birds by allowing air to pass down through the feathers. Flyers, like swimmers, can sense and control the flow of the medium over the body. In wind tunnels the energy loss by models of birds is about twice as great as that by live birds. Analysis has shown that a vulture can achieve laminar flow over much of its body.

Since air pressure below a wing that is lifting is positive relative to atmospheric pressure, and the air pressure above the wing is negative, there is a flow of air outward under the wing, around the wing tip, and inward over the wing. This flow induces a large eddy, or vortex at each wing tip and affects air flow elsewhere in a way equivalent to a slight increase in α (Figure 27-6). This takes energy from the flyer, and causes the drag called **induced drag (D$_i$).** The value of $\mathbf{D}_i = \frac{1}{2}\rho V^2 S C_{di}$, where C_{di} is the coefficient of induced drag and the other variables are as before. Since **D**$_i$ results from airflow around the tips of the wings, it increases as the pressure gradient around the wings increases and is smaller for narrow wings than for broad wings. Narrowness of wing is expressed by the **aspect ratio,** *A*, which is the span of the wings (tip of one wing to tip of the other) divided by the average width or **chord.** (Since the average chord cannot

Figure 27-6   THE NATURE OF THE ASPECT RATIO AND WING TIP VORTEX for a gliding flyer.

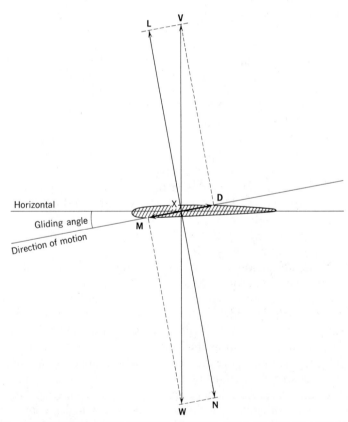

Figure 27-7   FORCES RELATED TO THE AIRFOIL OF A GLIDING ANIMAL moving with constant speed and direction when **L/D** = 5; X is the center of pressure.

be directly measured, the equivalent formula, $A = span^2/area$, may be used instead.) The value of $C_{di} = kC_L^2/A$, where k is an empirical constant equal to about 2.

There can be neither lift nor drag unless there is forward velocity. So far we have assumed forward velocity without establishing a source for it. One way for an animal to derive or maintain forward velocity is to glide downward using gravity to convert potential energy to kinetic energy. Assume that the animal (whether bird, reptile, or mammal) glides with constant speed and direction in still air. The angle that its path makes with the horizontal is called the **gliding angle,** and the rate of vertical descent is called the **sinking speed.** It is as though the glider were sliding down an incline. As for a wagon that coasts down a hill, the glider's weight, **W** (see Figure 27-7), has a component, **M,** which is in the direction of motion, and a component, **N,** which is normal to the slope of the incline.

The glider's airfoils create lift and drag. Lift, drag, and gravity are the external forces that act on the glider. In order for speed and direction to remain constant, these forces must be in equilibrium. The glider therefore adjusts the angle of attack and area of its airfoils so that **L = N** and **D = M.** Similarly, **V,** the resultant of **L** and **D,** is vertical and equal to **W.** It is evident from Figure 27-7 that the greater the value of **L/D,** the flatter the angle of glide will be. A far-ranging fast-flying albatross may sink 1 unit of distance vertically while moving 12 to 18 units horizontally, and a vulture may do as well. It is usually desirable for small birds to achieve minimum sinking speed. This is accomplished with a steeper angle of glide (about 1:8) but flight speed only a little faster than the stalling speed.

Sustained flight without flapping of the wings is soaring. There are two kinds of soaring, depending on circumstance. The first, **static soaring,** is dependent on updrafts. A flyer can stay aloft if its sinking speed relative to the surrounding air is equal to, or less than, the rate of ascent of the surrounding air relative to the ground. Like a person who walks continuously down a rising escalator, the flyer glides continuously down through rising air. Air rises when wind is deflected upward by a hill, coastline, wave, or ship. Gulls, pelicans, and some other sea birds often soar in such updrafts. Similarly, columns of air sometimes rise by convection over warm water or land. If a breeze is blowing, the columns tilt over, becoming inclined to the ground.

Vultures, hawks, and most other soaring land birds instead do their static soaring in **thermals.** The morning sun warms the ground, which warms a layer of air next to the ground. This air flows into bubbles that arch up and break away from the ground layer to rise like balloons through the higher cooler air, expanding as they go and drifting horizontally if there is any breeze. Unlike the gas in a balloon, however, the air in a thermal is not stagnant but circulates in a vortex shaped like a doughnut (Figure 27-8). Air constantly rises in the center of the thermal (the entire center, not just on the surface of the doughnut), radiates outward at the top of the floating bubble, descends in the margins, and turns in to rise again. The vulture soars only in the center of a thermal, wheeling con-

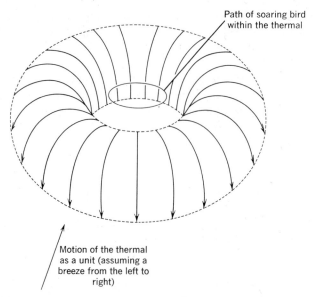

Path of soaring bird within the thermal

Motion of the thermal as a unit (assuming a breeze from the left to right)

Figure 27-8   SIMPLIFIED DIAGRAM OF THE CIRCULATION OF AIR IN A THERMAL as seen from above and to one side. Static soaring is done in the air rising throughout the center of the "doughnut."

stantly to remain within the hole of the doughnut. After rising with the thermal, sometimes for several thousand meters, the bird glides slowly away to the ground or into another thermal. In this way it remains on the wing for long periods and covers much distance with little expenditure of energy.

Thermals rarely rise over water. Albatrosses and some related oceanic birds are masters of **dynamic soaring,** which is dependent on the wind. The wind (which is brisk and steady over vast areas of the oceans) has ever-decreasing velocity from about 15 m over the water down to water level. This is because of the friction between air and water. The difference in wind velocity between the top and bottom of this shear layer commonly amounts to about 60 km/hr, and the bird soars only in this zone.

Let us follow a basic flight cycle of an albatross starting with the bird flying into the wind near the top of the shear layer. Its air speed might be 70 km/hr, but its absolute speed over the water might be only 10 km/hr because of a wind speed of 60 km/hr. Now the bird turns downwind. At the completion of the turn the velocity of the wind is added to, rather than subtracted from the bird's air speed to derive its absolute speed downwind, which might be 130 km/hr. The albatross next partially flexes its wings to increase wing loading and further increases speed as it glides steeply down to the water, converting potential energy to kinetic energy as it drops. The bird now banks sharply to execute a windward turn. It still has high absolute speed when it completes this turn, because the wind has less velocity at water level. Fully extending its wings and increasing their angle of attack, the bird now completes the basic flight cycle by rising to the top of the shear layer, converting kinetic energy to potential energy again as it climbs. Absolute speed falls off, but ever-increasing wind speed maintains adequate air speed to prevent stall.

The energy seemingly extracted from the wind is sufficiently greater than the energy lost during the basic flight cycle to provide an excess that can be used for gradually maneuvering, for instance by gliding cross-wind when halfway through a turn, or by lengthening each climb to progress directly into the wind. These birds are also expert at static soaring in the updrafts over wave fronts. In these ways albatrosses travel thousands of kilometers with only infrequent flapping of the wings.

Vultures sometimes interrupt long glides downwind to loop into the wind, thus using the principal of dynamic soaring to gain elevation for further travel downwind.

As noted in a previous section, there is an updraft of air at the wing tips. By flying in formation, ducks, geese, and some other birds utilize the upwash created by their neighbors to reduce the lift they must generate with their muscles. If the birds flew line-abreast the center birds would benefit the most. By flying in a vee formation with slightly swept back arms, all birds (including the lead bird) have about equal benefit. Nearly maximum advantage is attained when there are ten or more birds and they fly with less than half a wing span separating them horizontally. The energy saved is considerable. Birds that flock in a cluster may achieve a decrease in total drag, depending on the shape of the flock.

FLAPPING FLIGHT

Most fliers propel themselves by flapping their wings. Even expert soarers flap when landing and taking off, and when conditions needed for soaring are temporarily lost. Flapping flight is exceedingly complex. Quantitative analysis is not yet possible, though the various kinds of flapping flight are understood in general terms. The types of flight described below are representative, but they grade into one another.

HOVERING Hovering, or stationary flight in still air, represents a specialization in the evolutionary sense. The small flyers that can hover make the body motionless in midair, usually to feed on the nectar of flowers that are not accessible in another way. Hummingbirds are prominent among such flyers, though some other birds and many kinds of bats can also hover.

The body is usually held nearly vertical. The wings, therefore, beat backward and forward instead of up and down, propelling a vortex of air downward with each stroke. The wing tips describe a distorted figure eight (Figure 27-9). On the forward stroke, the wing, having a positive angle of attack, creates much lift, which is at right angles to the airstream but has a large vertical component. At the forward limit of the long wing stroke the wing rotates at the shoulder (the wrist and elbow are quite stiff in these birds), thus turning over and making its anatomical underside uppermost. With a positive angle of attack in the new direction the wing then sweeps backward, again creating lift with a large vertical component. The horizontal components of the two strokes tend to cancel, yet by slightly altering the power of one stroke in relation to the other, the flyer can

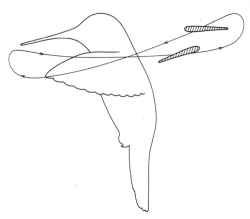

Figure 27-9   HOVERING FLIGHT OF A HUMMINGBIRD showing approximate path of the wing tip and angle of attack of wing on forward and backward strokes.

move horizontally forward or backward as needed when feeding. When the flyer is not hovering, it flies with the body horizontal.

Since the body of a hovering hummer is stationary relative to the air, the bird must move its wings very fast to establish enough air speed at the flight feathers to create lift. The frequency of the beat is 35–50 Hz. The wings (and therefore the bird) are small so that their oscillations will not generate excessive inertia. The hand part of the wing is relatively long and the slower-moving arm part is short (Figure 27-14). Since the backstroke of the wings is active, the muscles that elevate the wings are relatively larger than for other birds. The entire power plant is large: The breast muscles account for 25% of the weight of the entire bird, and the little flyer may eat about twice its body weight in food per day.

Bats that hover do not follow hummers in turning the wing over at the shoulder. They flick the hand wing in the manner described below for pigeons and gulls ascending steeply.

SLOW ASCENDING AND DESCENDING FLIGHT   Most small birds of bush and forest make frequent short flights that often include steep ascents and descents. They fly slowly but flap their wings rapidly (commonly 15–25 Hz) to attain the necessary air speed of the flight feathers. The body axis is held between 45° and vertical while ascending and descending. At such times the wings, therefore, move nearly backward and forward, with their tips describing flat ovals or loops. Small bats with erratic flight presumably fly much as small birds do. The following account, however, is based on birds.

On the forward stroke the wing is fully extended and has a positive angle of attack. Marked lift (to raise the flyer or retard its descent) and propulsion result. The backstroke, by contrast, is merely a recovery stroke having little reaction with the airstream. In order to reduce drag on the backstroke, the wing is partially flexed, thus reducing its area. The muscles that elevate the wing tend to be much smaller than those controlling the downstroke.

Wing loading is the weight of the flyer divided by the area of its wings. As was explained on page 22, small animals have more surface area in relation to volume than do larger animals of identical proportions. Small flyers, therefore, have relatively low wing loading without having particularly large wings. Representative values for small bats are $0.07–0.17$ g/cm^2 and for small birds are $0.11–0.23$ g/cm^2. (There are various methods of measuring wing area, and the effect of age, sex, and season on body weight should be considered in making comparisons.)

Pigeons, gulls, ducks, hawks, owls, pheasants, and various other strong flyers of medium size can also ascend and descend at steep angles while flying slowly. Their larger size means that their wings must beat more slowly. Frequencies of 3–10 Hz are usual. Furthermore, because of the relationship between surface–volume ratios and body size, and because of the nature of the fast flight of these birds, they have moderate to high wing loadings (i.e., $0.40–1.30$ g/cm^2).

When the outstretched wing sweeps downward and forward, it produces much lift and a little propulsion. Since the wing loading is large and the frequency of beat is only moderate, these birds cannot afford a passive backstroke (as small birds can). Their backstroke is complicated, but three motions are particularly characteristic: (*1*) The wing is flexed so the tip travels close to the body, (*2*) the outer half of the wing (only) is turned over with what is called a flick, and (*3*) the primary feathers rotate individually like the slats of a venetian blind so that each feather acts as a strong airfoil with positive angle of attack, yet air can pass between the feathers (Figure 27-10). The opening of this venetian blind is the automatic consequence of the facts that (*1*) the shafts of the feathers are not central in their respective vanes so air pressure tends to make the vanes rotate, and (*2*) the flick reverses the direction of air pressure on the feathers, thus enabling them to rotate in the direction not prevented by their overlap. This backstroke produces even more lift than the forward stroke and also some propulsion. The slower-moving inner wing is nearly useless. Pigeons can take off and land without the inner, or secondary, feathers. The wing tip describes a figure eight in each full cycle.

Some birds of comparable size, and many larger birds, having even slower wing beats and higher wing loadings (swan, 1.7 g/cm^2; heron, 2.5 g/cm^2), are unable to ascend and descend steeply while flying slowly. Such birds often run to gain speed for a fast flat takeoff and land fast on water, which cushions the impact.

FAST LEVEL FLIGHT   Fast level flight is characteristic of the large mastiff bat and of many birds of moderate or large size such as ducks, geese, shorebirds, falcons, toucans, and gulls. The body is carried horizontally, and the wings beat up and down. High speed of the body provides high air speed over the wings. Consequently, the slower-moving inner wing produces much lift on both the upstroke and downstroke. The wing beat is relatively slow, and the amplitude is small. Wing loading is usually high. Since a propulsive force is needed only to oppose drag and not to power acceleration or ascent, or to retard descent, the

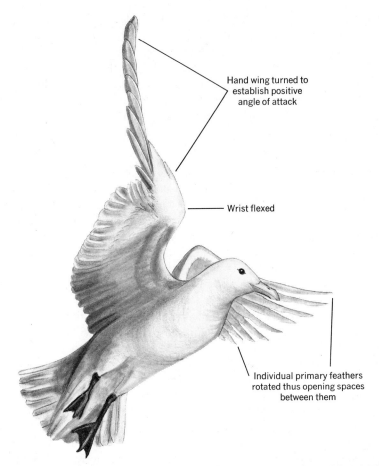

Hand wing turned to
establish positive
angle of attack

Wrist flexed

Individual primary feathers
rotated thus opening spaces
between them

Figure 27-10   BACKSTROKE OF WINGS IN ASCENDING FLIGHT OF A MEDIUM-
SIZED BIRD shown by a gull, *Larus.*

demand for power is relatively low. Virtually all propulsion is provided by the
outer half of the wing on the downstroke. The upstroke is passive; negative air
pressure over the wing may lift it without muscular effort. The outer wing does
not flick over on the upstroke as for slow flight of some of the same kinds of
birds. The position of the inner wing changes little on the upstroke; the outer
wing is flexed somewhat at the wrist. The wing tip never moves backward in
relation to the ground.

**FLIGHT
CONTROL**

STABILITY AND MANEUVERABILITY   As noted in previous chapters, animals
must choose between stability and maneuverability. The elephant opts for stability; many flyers (more than designers of aircraft) opt for maneuverability. If the
body is moved slightly out of position, aerodynamic forces usually do not restore
the original orientation but instead tend to quickly force the flyer farther out of
position. This makes it mandatory that the sensory–nervous–motor control system
correct constantly and almost instantaneously for minor displacements. The same

circumstances, however, permit rapid maneuvering. Some birds occasionally do loops or rolls, and such displays are surpassed by the aerobatics of many bats and birds as they pursue flying insects or engage in courtship flight.

A flyer can correct for roll by increasing the angle of attack of the lower wing, by increasing its surface area, or by flapping it harder than its opposite, any of which would increase its lift. (Since the same actions influence drag, compensations are needed to prevent yaw.) Also, the tail, if large, might be formed into a screw shape to correct for roll. The same kinds of behavior can, of course, be used to cause roll.

Unlike the swimmer and man-made aircraft, the flyer has no vertical fins or rudders to control yaw. Instead, it increases the drag on the wing that is tending to advance faster than its opposite. This can be done by increasing its angle of attack. Greater angle of attack, however, and also greater air speed (since it is moving ahead of the other wing) increase its lift and hence tend to cause roll. To compensate, the advancing wing is flexed to reduce its area. Dropping one foot out of the streamlined contour of the body also increases drag on that side. Furthermore, if a bird with a large stiff tail is banking to turn, it can open and twist the tail so as to form a vertical rudder.

To correct for downward pitch of the body (or to initiate upward pitch) a flyer moves its wings forward, thus placing its support farther forward in relation to its center of gravity. Also, if the tail is suitably constructed, it is bent upward. The converse behaviors correct for upward pitch.

BRAKING AND TURNING   Gliders, and flyers approaching an elevated perch, may brake by swooping upward into a stall, thus using gravity for deceleration. As noted, flyers can also flap their wings to create upward force that slows descent preparatory to landing. In sustained level flight it is desirable for the ratio of **L:D** to be maximum consistent with moderate to high speed. When braking, by contrast, **D** should be high and speed should be low. This is achieved by making the angle of attack, the wing area, and the camber maximum. Birds also depress their fanned tails and bats depress the tail membrane. This is the equivalent of lowering wing flaps of man-made aircraft. The feet are thrust forward (toes spread if they are webbed) not only to be in position for landing, but also to provide drag (Figure 27-11). Moreover, when birds brake, they lift the alula to create a wing slot. This increases the angle of attack and decreases the velocity that can be attained before stalling occurs.

Adaptations of the legs of flyers for absorbing the shock of landing have been little studied. The impact does not shear the ball-and-socket joint at the hip, but instead is transmitted to the ilium as compression by the neck and trochanter of the femoral head. Many fast-flying water birds land into the wind and plane briefly on their webbed feet as they strike the water.

Since moving bodies tend to continue to move in straight lines, when animals turn they tend to slip sideways toward the outside of the turn. The force out of the turn is **centrifugal force,** which varies directly with the square of the velocity and the mass of the body and inversely with the radius of the turn. This force must be opposed by an equal and opposite **centripetal force** into the turn. Sharply turning cursors resist centrifugal force by friction at the ground (see p.

Wings banked, have
different angle of attack

MANEUVERING
Gull, *Larus*

Wings, tail, feet
create maximum drag

BRAKING PRIOR TO LANDING
Razorbill, *Alca*

FLIPPING OVER ON LANDING
TO HANG BY FEET
Leafnose bat, *Macrotus*

Tail fanned,
elevated

Wings unequally
spread

TAKING OFF
Chickadee, *Penthestes*

Suction above stalling
wing lifts feathers

Webbed feet cushion
impact

HIGH-SPEED LANDING ON WATER
Swan, *Cygnus*

Figure 27-11 EXAMPLES OF FLIGHT CONTROL.

456), and swimmers rely on drag produced by vertical fins and a deep body. Flyers must use another method. They bank into the turn causing **L** to tilt inward (Figure 27-12). The horizontal component of **L**, or **I**, must equal the outward centrifugal force, **O**. If the flyer is not to lose altitude, the vertical component of **L**, or **V**, must equal the full value of the lift that existed before the turn started. This means that lift, and also wing loading, must increase during turns: If a 60° bank is executed, lift must be doubled. It is clear from the relation of the magnitude of **O** to mass and velocity that flyers that dodge quickly must be small, and that large flyers cannot turn sharply at high speed.

WING STRUCTURE

The various animal flyers have strikingly different flight habits. Since form must correspond to function, wing structure is also varied. Savile has described four general types of wings, it being understood that they intergrade and that within types there is great variation of detail. The aerodynamics of flight is not yet well enough understood to fully interpret the morphology of wings.

First is the **elliptical wing** (Figure 27-13). It is characteristic of most bats and of most small and medium-sized birds of shrub and forest (e.g., sparrows, robins, crows, quails, pigeons). The specialization is for high maneuverability and precise control, often in confined spaces, and for minimum induced drag. The flyer is able to fly slowly and to ascend and descend rapidly. The wing has an elliptical outline (except where it meets the body). It is short and broad, making the aspect ratio low (commonly 3–6). Camber is moderate to marked (particularly in bats). Flapping flight is usual, though there may also be some gliding or swooping. The wing beat is moderately fast, and the amplitude is relatively great. The birds have a large alula, and in the larger and more active birds the primary feathers separate

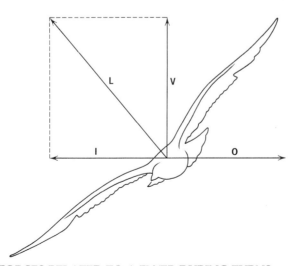

Figure 27-12 FORCES RELATED TO A FLYER DURING TURNS.

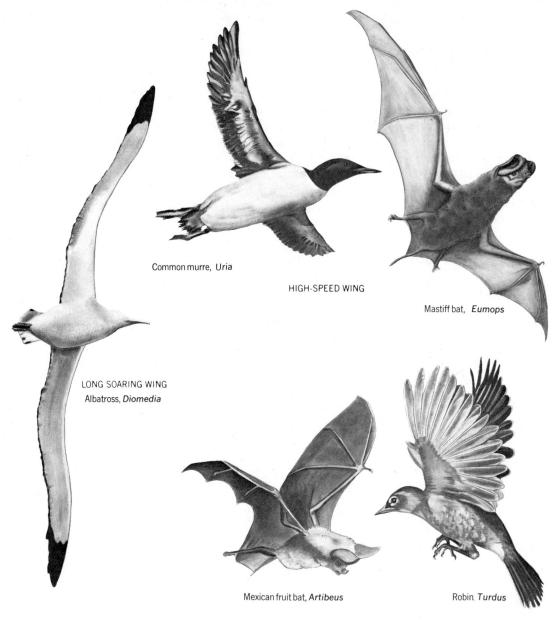

Common murre, *Uria*

HIGH-SPEED WING

Mastiff bat, *Eumops*

LONG SOARING WING
Albatross, *Diomedia*

Mexican fruit bat, *Artibeus*

Robin, *Turdus*

ELLIPTICAL WING

BROAD SOARING WING
California condor, *Gymnogyps*

Figure 27-13 THE FOUR PRINCIPAL TYPES OF WINGS.

to form additional wing slots. In all, the slots may extend along the leading edge of 15–30% of the length of the wing. Each slot speeds up the air passing over the next posterior feather, thus preventing air from breaking away in eddies. The slots serve as an antistalling device at low speeds. They can be closed in faster, level flight. The primary feathers may rotate individually when the slots open. This is the equivalent of having the entire wing twist like a propeller as it stretches out from the body, a result that is otherwise difficult to achieve with a short wing. Most of the velocity of the base of the wing is forward, whereas much of the greater velocity of the wing tip is up or down. Consequently, the wing must twist to provide the same angle of attack at base and tip. Slots increase lift by as much as 60%.

The **high-speed wing** is characteristic of mastiff and free-tailed bats and of swifts, swallows, falcons, shore birds, hummers, and ducks. The specialization is for high flight speed with low drag and low expenditure of energy. The wing is relatively small, so wing loading is large. The wing tapers to a slender tip, may be somewhat swept back along the leading edge, and is faired into the body behind without a sharp angle. The aspect ratio is moderately high (commonly 5–9). The hand skeleton is relatively long, and strikingly so in hummers (Figure 27-14). In cross section the wing is thin and has little camber. Flapping is constant except perhaps for short glides when descending. The beat is rapid in relation to the size of the flyer, and the amplitude is small. There are no wing slots.

The **long soaring wing** is characteristic of albatrosses, frigate birds, gannets, terns, and gulls. Some large pterosaurs probably had similar wings. Most such birds fly over water where long wings are not a handicap. The specialization is for a high ratio of lift to drag permitting soaring at high speed with low expenditure of energy and a low gliding angle. The wing is long, slender, and pointed; the aspect ratio ranges from 9 to 18; and the span of the wings reaches more than five body lengths. The hand skeleton is relatively short in these birds (Figure 27-14), though it is not in pterosaurs. Wing loading is high. Camber is low. There are no wing slots. Landing and takeoff speed is high. Most of the birds land on water and take off into the wind—often after running to gain speed.

The **broad soaring wing** is seen in vultures, eagles, and buteo hawks. The wings of the raven and pelican are similar. The specialization is for soaring at

Figure 27-14 CONTRAST BETWEEN THE RIGHT WING SKELETONS OF A DYNAMIC SOARING BIRD AND A HOVERING BIRD shown by an albatross, *Diomedea* (above), and a hummer, *Lampornis* (below). Hand skeletons drawn to the same length.

low speed, takeoffs and landings in confined areas, high lift, and low sinking speed. The wing is moderately long and broad, thus giving it large area, has only moderate wing loading for such large birds, and has moderate aspect ratios (e.g., 6 or 7). Camber is marked. The alula is prominent, and terminal wing slots are conspicuous. The slots seem to correlate with broad wing tips and correspondingly large tendency to induced drag. The slots usually have U- or ⊔-shaped bases rather than the less highly evolved V-shape of some elliptical wings. Each spaced primary feather functions independently as a winglet and each is streamlined in cross section. Muscles that elevate the wing are less developed than in birds that emphasize flapping flight: Contrast the supracoracoid muscle in the figures on pages 197 and 198.

FURTHER STRUCTURAL ADAPTATIONS

We have now considered the primary requirements for flight. These, however, are not enough: Even if they were provided with wings, lizards and rats (and angels) could not fly. One secondary requirement is for light weight. Flyers have light skeletons. (The skeleton of an eagle accounted for less than 7% of the total body weight—about half as much as for man.) The bones of birds and pterosaurs are hollow and air-filled. This permits the bones to have a maximum diameter and hence maximum resistance to bending forces with minimum weight. Many bones have thin inner and outer lamellae joined by spicules that tend to follow stress lines. Furthermore, sutures tend to become obliterated. These constructions provide maximum strength with minimum hard tissue. The bony axis of the avian tail is short. Reduction or loss of digits also saves weight for birds and pterosaurs. Modern birds and some pterosaurs lack teeth, which are heavy. (The feeding habits of most bats require that they have large ears and teeth. The fast-flying mastiff bat has enormous horizontally oriented and cambered ears. It is probable that these ears generate enough lift to help support the heavy head!)

The gonads of flyers regress when they are not active. Good flyers eat foods which, being nutritious, are lightweight in relation to the energy provided. Also, digestion is very rapid (often within an hour). Some flyers (hummers, swifts, many bats) use the legs only for perching or roosting. The legs are then very small and light. Feathers, being stiff, yet elastic, are remarkable for the loads they can bear in spite of light weight. They provide a smooth streamlined contour and are resistant to damage (see p. 98 and 99).

A second requirement is for compactness and firmness of the body so that the thrust of the wings will be transmitted to the body near its center of mass and will propel the body as a unit without deforming it or causing it to flap. Flyers have a short trunk that is stiff (Figure 27-15). Fusions between adjacent trunk vertebrae are common in bats and usual in other flyers. Pterosaurs had many sacral vertebrae and usually had fusions in the thoracic region. Birds have 12–20 vertebrae fused together in the synsacrum (see figure on p. 155) and may have additional fusions in the thoracic region (pheasants, falcons). Even in the absence of fusions, the shape of the joints between trunk vertebrae are such as to limit or prevent motion. Ribs tend to be broad and flat so they nearly touch edge to edge. They may fuse together proximally (some bats, many pterosaurs) or to the spine (some pterosaurs). The uncinate processes of avian ribs probably provide bracing.

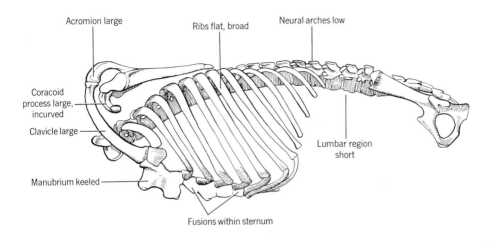

Acromion large

Ribs flat, broad

Neural arches low

Coracoid process large, incurved

Clavicle large

Manubrium keeled

Lumbar region short

Fusions within sternum

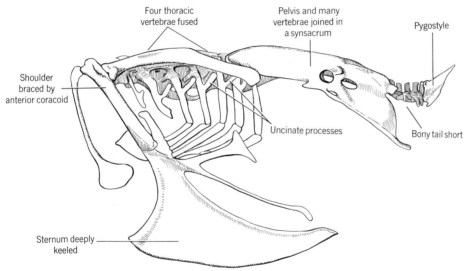

Four thoracic vertebrae fused

Pelvis and many vertebrae joined in a synsacrum

Pygostyle

Shoulder braced by anterior coracoid

Uncinate processes

Bony tail short

Sternum deeply keeled

Figure 27-15 SOME CHARACTERISTICS OF THE BODY SKELETON OF FLYERS shown by a fruit bat, *Pteropus* (above), and a pheasant, *Phasianus* (below).

The sternal ribs tend to ossify. The number of units in the sternum is reduced to three (bats) or one (birds, pterosaurs).

The body is made compact by the shortness of the bony tail (birds, some pterosaurs, and some bats) and by sharply flexing the neck during flight. The pectoral girdle is large, strong, and has unusually firm attachment to the body by the anterior coracoid (birds, pterosaurs), clavicle (bats), or even the scapula (some pterosaurs). The entire girdle is virtually immovable in birds and pterosaurs, though not in bats. The concentration of flight muscles (including elevators of the wings) on the breast of birds positions their great weight near, but a little below, the point of support of the body by the wings. The same is true to a lesser extent of bats.

Figure 27-16   USE OF A PARALLELOGRAM TO EXTEND AUTOMATICALLY THE WRIST WITH THE ELBOW, shown by a pheasant, *Phasianus.*

A third requirement is for automatic mechanisms that save muscular effort and, therefore, both weight and energy. The joints of the wings of both birds and bats are constructed so as to limit motion to one plane. There is no supination or pronation of the forearm. The ulna (which would be required for rotation of the forearm) adheres to the radius in pterosaurs, and its distal portion is nearly lost in bats. It is retained in birds because it has a new function: The radius and ulna join the humerus and carpus in such a way as to form a parallelogram. When the elbow opens out, the wrist opens automatically (Figure 27-16).

Various bats have locking devices to assist in holding the shoulder, elbow, and wrist joints at the correct angles. It is said that a bone at the elbow in the albatross can be placed in the joint to hold it open. Only a small part of the wing membrane of bats is attached to the small second digit, yet many bats use this digit and its ligamentous extension as a bow to keep the leading edge of the wing stiff.

Another requirement of flyers is for an efficient power plant. Birds have a relatively high body temperature, and it is likely that even pterosaurs could elevate the body temperature, at least when flying. Birds and bats have a high metabolic rate and efficient respiratory systems. Ventilation of the lungs is apparently synchronized with the wing beats. The heart is large, blood pressure high, and circulation rapid. Even making allowance for great efficiency, it seems certain that the energy release in avian muscles is higher than in nonflying vertebrates. Fat is the main source of energy.

For reasons explained earlier, flyers have a large cerebellum and (bats excepted) large eyes. Bats and pterosaurs roost upside down and hence have specializations of the feet to form hooks and of the pelvis and knee to allow suitable orientation of the legs.

Feeding

Adaptations for feeding are particularly diverse. This is, therefore, a good place to emphasize that each animal adapts not to a random combination of independent activities, but instead to a coordinated life style. Manner of locomotion is nearly always related to feeding habits, and reproductive, defensive, and other behavior are usually correlated with manner of feeding and locomotion. Parts of preceding chapters on the teeth, digestive system, sense organs, and locomotor adaptations relate directly to feeding; other parts relate indirectly.

Before studying the varied ways that vertebrates locate, secure, and process their food it is desirable to learn about some of the principal kinds of jaw mechanics.

JAW MECHANICS   Analysis of the action of the jaws and teeth has become a complex, yet challenging field. Most studies completed in the first half of the twentieth century treated the jaw as a simple lever pivoting at the mandibular condyle. Dissection, manipulation, and linear measurements were used to make correlations between form and function. Current research supplements those still-important techniques with behavioral observation, stress analysis using strain gauges affixed to functioning jaws, motion analysis using accelerometers, monitoring of muscle action by electromyography, and study of trajectories, velocities, and accelerations using high speed cinephotography (standard and x-ray, and often simultaneously from different angles). Motion may be analyzed in relation to the condyle, symphysis, or other joint (or joints), at the occluding teeth, at some other point

PRIMITIVE CHONDROSTEAN, *Pteroniscus*
A

LESS SPECIALIZED TELEOST, *Salmo*
B

MORE SPECIALIZED TELEOST, *Sebastodes*
C

Figure 28-1   STAGES IN THE EVOLUTION OF THE JAWS OF RAY-FINED FISHES.

determined by the forces generated, or instead, the entire system may be treated as a free body (see p. 432 and 433).

JAWS OF FISHES   The upper jaw of fishes is said to be protrusible if it not only can be opened and closed, but also moved forward and backward in relation to the remainder of the head.

Acanthodians and the most primitive of actinopterygian fishes had large mouths. The hinge of the jaws was under the back of the braincase. Premaxilla and maxilla were not movable, and the upper jaw was not protrusible (Figure 28-1A). These fishes apparently opened the mouth wide and snapped at prey; the rapidly moving lower jaw completed closure partly by inertia. The bite was not strong.

A subsequent evolutionary stage is exemplified by the bowfin and salmon. The hinge of the jaw is nearly as far back as before, so the mouth remains large. However, the adductor of the jaw is stronger and inserts farther forward on the mandible, giving the fish a strong bite. The premaxilla is still fixed, but the maxilla is now free. Its posterior end is joined to the mandible by a ligament, and its anterior end pivots on the rostrum (part B). Consequently, when the mandible opens, the lower end of the maxilla rocks downward and forward, thus preventing food from escaping at the corners of the mouth. The upper jaw still protrudes very little or not at all.

At the next stage in the specialization of fish jaws, the hinge of the mandible moves forward under the orbit, so the mouth is smaller. The premaxilla, like the maxilla, pivots forward as the mandible is opened (part C). Furthermore, the premaxilla is L-shaped; one arm forms part of the margin of the mouth, and the other, shorter arm forms an ascending process running backward toward the ethmoid region. The premaxilla can slide forward more or less along the axis of this ascending process, thus making the upper jaw truly protrusible. The lengthening of the head amounts to 10–20%. A complicated system of ligaments controls the motions of the bones, but the mechanism is not the same in all fishes. The position of the maxilla is commonly altered by motion of the mandible or by action of the adductor mandibulae, which may have one or more heads inserting on the upper jaw. Motion of the maxilla, in turn, moves the premaxilla. The ascending process usually is guided by a median ridge of the vomer or ethmoid as the premaxilla slides down and forward. A small camlike bone in the ethmoid region contributes to the control of the action in some fishes. The upper jaw usually can be closed on the lower jaw both when it is protruded and when it is retracted.

Protrusible jaws have several advantages, according to species. Importantly, they permit the mouth to be closed during or immediately after sucking food into the mouth, without blowing the food back out again (which might occur if a long lower jaw were raised). Sometimes the protruded mouth points more downward than the unprotruded mouth—an advantage when taking food, including algae, from the substrate. The jaws probably can be closed a little faster when protruded because the mandible does not have as far to travel. Finally, protrusion places the mouth a little closer to food, though swimming would do the same. Perhaps half of all teleosts have protrusible jaws. They can be recognized by a transverse fold of skin just behind the margin of the upper jaw. This fold is pulled tight as the premaxillas move forward.

Thorough analysis of the feeding of fishes also requires study of the complex mechanisms of the pharynx, floor of the mouth, and, in some instances, even of the pectoral girdle.

CRANIAL KINESIS   Relative motion of several parts of the roof of the skull is called cranial kinesis. Kinetic skulls are divided into several movable units that are linked together and usually function together to elevate and depress the upper jaw. Crossopterygians, some labyrinthodonts, caecilians, some lizards, snakes, some extinct reptiles, and birds have, or had, kinetic skulls. The mechanisms are often complex. They have been analyzed for various species, but largely on the basis of two-dimensional models.

The hinged braincase of crossopterygians (see figure on p. 130) is unique to that group of fishes and is associated with a distinctive musculature. There was some related lateral motion of dermal bones of the skull; the hyomandibula served to integrate various of the complex movements. The mechanism increased the gape of the mouth and possibly also the power of the bite.

Most lizards have kinetic skulls, and for them the mechanism is relatively simple. Four principal units (some paired) are joined on each side of the head by four principal joints (Figure 28-2A). (Some struts, not shown in the figure, articulate with the mechanism.) Motion of any one unit requires motion of two others and of all four joints. Movement on each side of the head is nearly restricted to one plane, and the two sides usually must function together. Kinesis allows the

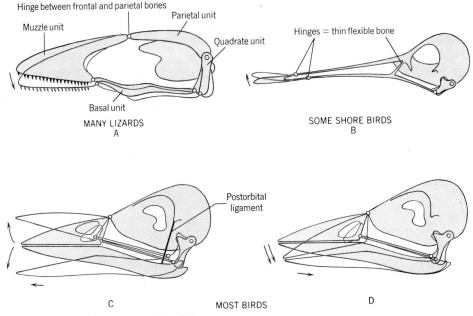

Figure 28-2   CRANIAL KINESIS.

upper jaw to bite down on food and to manipulate food by moving relative to the lower jaw.

The kinetic mechanism of snakes includes many units (eight in the boa) that are loosely joined together and independent on the two sides of the head. In consequence, an infinite variety of relative motions is possible—a characteristic of the feeding of snakes.

The cranial kinesis of birds has four principal units and four principal joints on each side of the head as for lizards, yet is quite different and more complicated. Some birds can move the quadrate units on the two sides of the head somewhat independently, and some can slide as well as pivot the mandible on the quadrate. Motion of upper and lower bills is not necessarily confined to one plane. Furthermore, the mandible is joined to the braincase by the stout postorbital ligament. This appears to link the motions of upper and lower bills in at least some birds, though a degree of independent voluntary control seems to be present. (The ligament may also keep the jaw closed passively.) The single dorsal hinge of the mechanism is always farther forward than any joint of lizards, but may be at the base of the upper bill or near the tip (compare parts B and C of Figure 28-2). Kinesis in birds allows the bill to open wide, and in so doing to maintain its longitudinal axis (part C). It provides for skilled manipulation of food by sliding one bill lengthwise in relation to the other and by creating a wedge between the bills that opens backward into the mouth, thus preventing food from escaping (part D). Kinesis permits upper and lower bills to close faster than otherwise because each is active. It probably also cushions the shock of pecking for some birds.

JAWS OF MAMMALS   The jaws of ancient fishes, most amphibians, many reptiles, and mammals are neither protrusible nor kinetic. The jaws of carnivorous and herbivorous mammals are further characterized by adaptations for a strong bite that shears, crushes, or grinds, rather than for a weaker but faster snap that catches prey which is swallowed whole.

Elephants apply force more or less evenly all along each tooth row. Most ungulates do likewise for the cheek teeth, but also can apply a lesser force at an anterior cropping mechanism (see figure on p. 16). Rabbits and rodents apply force either at their anterior gnawing incisors or at their posterior grinding cheek teeth (Figure 28-19). Likewise, Carnivora apply greatest force with their canines at the front of the mouth or with their **carnassials** (shearing teeth) at the back of the mouth. The two mechanisms do not function simultaneously. When a rabbit

| Tooth rows diverge anteriorly | Tooth rows converge anteriorly | Tooth rows long, bowed outward; condyle transverse | Tooth rows short, nearly straight; condyle longitudinal | Tooth rows medium, bowed inward; condyle ovoid |

MOUNTAIN BEAVER *Aplodontia*   GUINEA PIG *Cavia*   BUSH HYRAX *Dendrohyrax*   DOMESTIC RABBIT *Oryctolagus*   PACA *Cuniculus*

Upper tooth rows closer together, angle outward
NUTRIA *Myocastor*

Upper tooth rows farther apart, angle inward
JACK RABBIT *Lepus*

Above: tooth row bowed downward
PORCUPINE. *Erethizon*

Below: tooth row straight
CAPYBARA, *Hydrochoerus*

Figure 28-3   SOME VARIATIONS IN TOOTH ROWS AND MANDIBULAR CONDYLES AMONG THREE ORDERS OF MAMMALS. Above left, skulls in ventral view; above right, mandibles in dorsal view; below left, cross sections of skulls and mandibles at level of orbits; below right, medial views of left mandibles.

chews, its incisors do not touch; when a beaver gnaws, the mandible is moved forward and the cheek teeth do not occlude.

Carnivora bite with a chopping motion. Upper and lower tooth rows come together without sliding past one another either front to back or side to side. The joint between mandible and skull consists of a transversely oriented cylindrical condyle on the mandible that rotates hingelike in a fossa on the temporal bone. However, when the carnassial shearing mechanism on one side of the mouth is in use, the carnassials on the other side are not aligned and cannot function (see a following section on shearing).

Most mammalian herbivores (plant eaters) chew by sliding the lower tooth rows over the upper tooth rows. Motion may be back to front or side to side. Grinding occurs during one phase of the cycle—usually either as the mandible moves forward or from one side toward the center line. During the other, slower phase of the cycle, the jaw opens a little as it is returned to the starting position; the tongue meanwhile places more food between the tooth rows. If motion is back to front, upper and lower tooth rows are usually about the same distance apart and all cheek teeth function together. If motion is side to side, the lower tooth rows are commonly closer together than the upper rows, though the reverse is sometimes true. Either way, the animal must chew on one side of the mouth at a time. Whatever the relative motion of upper and lower tooth rows, it is nearly always crosswise or oblique (not parallel) to the orientation of the cutting ridges on the teeth.

Seen in side view, a tooth row may be straight or arched; seen in cross section, a row may be horizontal or may slant inward or outward (Figure 28-3). The rows of one jaw (upper or lower) may be either parallel or diverging toward the back of the mouth (though usually less so than for carnivores). These factors may guide and restrict the relative motions of upper and lower tooth rows. Differential hardness of the different materials of the cheek tooth of an herbivore causes cutting ridges to form (of which more later in this chapter). For the same reason, wear also produces larger **grinding ridges** in rabbits and many artiodactyls (Figure 28-4). These serve to position and guide the tooth rows as they slide over one another. It is clear that the chewing motions of herbivores are exceedingly diverse.

The structure of the articulation between mandible and skull of herbivores

Grinding ridges

Figure 28-4 CHEEK TEETH OF A CARIBOU, *Rangifer*, SHOWING GRINDING RIDGES THAT MAINTAIN ALIGNMENT DURING SIDE-TO-SIDE MOTION OF MANDIBLE. Lateral view of left teeth.

is correspondingly varied. The condyle may be transverse as for carnivores. When it moves instead in a less restricting fossa (wombat, hyrax, horse), it may be longitudinally oriented; when it rocks and slips in a longitudinal groove (rabbit), it may be bluntly rounded (beaver, paca), or relatively flat (many artiodactyls). The reader might find it useful to infer from the structure and functions of the tooth rows (preceding two paragraphs) how many different kinds of motions may be required at this joint.

The symphysis between the mandibles is firmly ossified in many herbivores (sloth, hyrax, perissodactyls, camels), yet is unfused and movable in rodents and most artiodactyls. The significance of this joint in herbivores invites study.

There are three adductors of the mandible. The **masseter** takes origin from the zygomatic arch and often also from the orbit or maxillary region. It inserts on the outside of the ramus and angle of the mandible (Figure 28-5). The **temporal** muscle originates on the braincase and sagittal crest, if present, and inserts on both the outside and inside of the coronoid process of the mandible. The **pterygoid** muscle originates at the base of the skull beside the palate and behind the orbit, and inserts on the inside of the ramus and angle of the mandible. Each muscle is commonly divided into several parts. These may be distinct, but they often merge with each other and (particularly for the masseter) with a complex of internal tendinous sheets. Muscle fibers are rarely parallel even in one division of one muscle. Unfortunately, the force diagrams one sees of this three-dimensional complex of muscles are necessarily two-dimensional approximations.

The temporal muscle comprises more than half the total adductor mass of carnivores. The coronoid process provides the lever arm of this muscle, and in these animals is large. The masseter is somewhat smaller and serves in part to stabilize the articulation of the jaw. The pterygoid muscle positions the carnassials, but adds little to the force of the bite and is small.

The masseter commonly comprises about two-thirds of the adductor mass of herbivores. The pterygoid muscle is next in size and joins the masseter in producing lateral grinding motions of the mandible. The temporal muscle is rel-

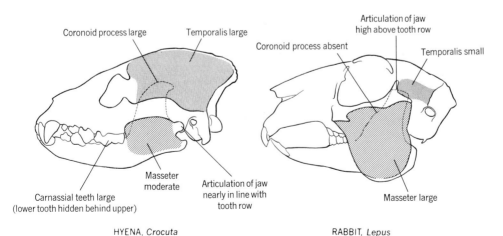

Figure 28-5 CONTRAST BETWEEN THE JAW MECHANICS OF A CARNIVORE (left) AND AN HERBIVORE (right).

atively small, and the coronoid process, though variable, tends to be small and may be absent (rabbit, capybara).

The articulation of the jaw is about in line with the tooth rows in carnivores. Its position is various in rodents but usually is somewhat above the tooth rows. It is much above the tooth rows in most other herbivores. The high position increases the lever arm of the masseter muscle. It also causes the lower teeth to approach the upper teeth obliquely rather than perpendicularly. This might tend to roll or slice plant food as it was crushed.

LOCATING AND SECURING FOOD

FINDING FOOD AND MAKING IT ACCESSIBLE　The food of vertebrates ranges from microscopic (diatoms, algae) to very large, from passive (plants) or nearly so (mollusks) to very active, from nutritious (insects, worms) to low in food value (stems), from defenseless to protected and from extensively available to highly restricted (e.g., the nectar of only certain flowers). Food may be seasonal or constantly accessible, abundant or rare, and available with or without competition. Small wonder, therefore, that the ways that vertebrates locate and process their food are exceedingly varied. Only a summary can be offered here.

The hungry animal usually searches actively for a meal, but instead may lie in wait for prey to come within range of a sudden dash or strike (groupers, vipers). Ordinarily, the feeder perceives the food object before contact is made. This is probably true in a general sense even of the vertebrates that strain minute food particles from quantities of water. However, some larger prey is not identified before capture: Nighthawks sweep the air for small insects. Apparently responding to discontinuities on the water surface, certain bats fly low over water with the large curved claws of their feet trailing in the water ready to gaff any small fish encountered. Similarly, the bird called the skimmer flies over water with its huge mandible cutting the surface ready to scoop up any small fish in its path.

Regardless of the method of search—active or passive, directed or random—the food must be recognized when it becomes available. Some vertebrates in each class are largely dependent on vision, and many rely on hearing, including echo-ranging and form-discrimination using vocal signals (many volant, aquatic, and some other mammals) or electric signals (some fishes). The chemical senses are rarely used by birds, but otherwise are commonly employed. The receptors may be outside the nose and mouth on barbles (catfish), rostrum (sturgeon), surfaces of the skin (amphisbaenians), or tip of the beak (some shore birds). Touch may also be important, the sensitive organ being the lips (manatee), snout or proboscis (elephant, pig, mole), or tongue (anteater). The use of the pit organ by certain snakes was mentioned on page 356.

Many animals move some object to uncover food. For example, the turnstone (a bird) rolls stones with its beak in search of invertebrate food, the walrus digs clams from mud with its tusks, the towhee kicks away litter to reveal insects, the bear rips logs to find animal food, the warthog uproots tubers with its tusks, and the aye-aye (a primate) cuts bark with its strong curved incisors to get at insect larvae. Some woodpeckers drill into solid wood to secure insects: The horny beak is extra long, strong, and rapid-growing, and the bony beak is reinforced. Force is transmitted by bone, ligament, and muscle to the thick and well-ossified interorbital septum of the skull.

Figure 28-6   A BROWSING GERENUK, *Litocranius.*

Numerous vertebrates make food accessible by using a long reach. The arms of apes and monkeys, and the trunk of the elephant serve this purpose. The browsing giraffe reaches with a neck so long that a plexus of arteries has evolved at the base of the brain to even the blood pressure as the head swings up and down. The strikingly long neck and legs of the gerenuk (an antelope) similarly increase its reach (Figure 28-6). Several vertebrates have structures for reaching into restricted spaces. Examples are the bills of hummingbirds and creepers, and the tongues of anteaters. The third finger of the aye-aye (Figure 28-7) and the fourth finger of two kinds of phalanger show remarkable convergence in that each is slender and longer than the other fingers to serve as a probe for removing insects from crevices.

Bill modified as worm–and–insect forceps
LONG-BILLED CURLEW, *Numenius*

Canines and upper incisor modified as scalpels
VAMPIRE BAT, *Desmodus*

Third finger modified as insect probe
AYE-AYE, *Daubentonia*

First fin spine modified as lure
ANGLER FISH, *Lophius*

Figure 28-7   VARIED DEVICES FOR MAKING FOOD ACCESSIBLE.

Tools other than parts of the body are used by several vertebrates: The chimpanzee may use a stick as a probe and leaves as a sponge, a Galapagos finch uses cactus spines to poke into holes, an Egyptian vulture and the banded mongoose drop or throw rocks to break large eggs, and as it swims on its back, the sea otter pounds shellfish on a rock anvil placed on its chest for that purpose.

Jays hold acorns against chopping blocks as they hammer at the hulls, gulls drop crabs and mollusks onto rocks from the air to crack them open, and the beaver cuts down trees to get at the bark. Various birds follow ungulates to secure the flies and ticks attracted by those beasts. Some carnivores watch vultures to locate carrion. Alligator snapping turtles have a lure at the tip of the tongue that attracts fishes. Anglerfishes attract prey by dangling in front of their jaws a lure derived from a dorsal fin spine. Those living deep in the ocean even have luminescent lures. The vampire bat makes a shallow wound on its prey using a pair of sharp, curved, upper incisors, and prevents the blood meal from clotting by the action of its saliva. The squirrel gnaws away the bracts of pine cones to get at the seeds, whereas the crossbill pries the bracts apart with its special beak. The archer fish looks up from under the water, locates insects as much as a meter away on overhanging vegetation, and then, with unerring aim, strikes them down with a forceful jet of water squirted from the mouth. The gill chamber is the force pump, and a groove on the roof of the mouth is the barrel of this water pistol.

Nature has been so inventive that many, many other devices for securing food could be listed.

CAPTURING AND KILLING ACTIVE PREY   Some fishes and snakes, and many mammals, stalk their prey, using stealth, concealment, surprise, camouflage, and appropriate postures and motions. Some spring onto the prospective meal. Most predators make a final short dash or longer chase, whether it be in water, on land, or in trees. Raccoons, some primates, and several rodents use the hands for capturing, or at least for controlling prey. Most hawks and owls dive on rodents from the air, securing the booty with their talons. Similarly, the osprey and fish owl plunge feet first into water to capture fish. The undersurface of the toes of these birds is rough, to prevent fish from slipping, and the scutes of the legs overlap so as to enter the water the "smooth way."

Several falcons dive, or "stoop" at flying birds, delivering a glancing blow with the feet and raking the victim with their strong hind talons. Various birds (pelicans, boobies, gannets, tropicbirds, kingfishers, etc.) dive head first into the water to catch fish. It is thought that the air sacs of the pelican cushion the impact. The eye is probably covered by the lower lid, and the nostrils and throat are modified to exclude water. The mandibles bow outward on impact so that the pouch temporarily holds about 10 liters of water and fish.

Various long-necked predators (snapping turtle, snakes) use the head as a catapult for capturing prey. Some snakes can even correct slightly the direction of the strike while it is in progress. The snake may hold the prey between the jaws or may pierce it with large sharp fangs and then withdraw temporarily. The long-necked, aquatic plesiosaurs may have lashed out with their small heads to strike fish, with the body wheeling about but not progressing rapidly. Herons and egrets, which wade, and snapping turtles, cormorants, and loons, which swim under water, strike at fish by suddenly straightening the long curved neck. Dorsal and ventral neck muscles are strong and somewhat consolidated. Several cervical vertebrae are elongated and modified to provide a pulley for the dorsal extensors. Most of these birds use their long straight beaks as forceps to grasp the prey, but the tropical anhinga spears fish; fine barbs at the edge of the bill prevent loss of the food.

Many devices have evolved for capturing insects. The use of the tongue for this purpose is mentioned under a subsequent subheading. Birds that take insects on the wing have large broad mouths fringed by stiff bristles. The night-feeding frogmouths, poorwills, and nighthawks have small horny bills but a prodigious gape (Figure 28-8). Another group of night feeders, the insectivorous bats, locate individual insects by echo-ranging so they do not need such large mouths. Several species use the flight membranes in catching insects. One small bat captured 175 mosquitos in 15 min. Such birds as warblers, wrens, vireos, creepers, and thrashers that take insects from vegetation or the ground, have narrow bills that are long or short, straight or curved, according to the bird's specific habits.

Some active prey is swallowed whole and killed by smothering and the action of digestive juices—a common fate of small fishes. Somewhat larger prey must be immobilized before it can be swallowed: The secretary bird stamps reptiles to death with its long heavy legs, some fishes paralyze prey with electric

Partly erected poison fangs
PUFF ADDER, *Bitis*

Flying insect trap
NIGHTHAWK, *Chordeiles*

Self-adjusting rodent trap
GOLDEN EAGLE, *Aquila*

Figure 28-8 VARIED DEVICES FOR CAPTURING ACTIVE PREY.

shocks (see p. 202), some snakes kill with a poison bite and others by constriction, and birds that eat flesh or fish may kill with their claws or beaks. Some flesh eaters bleed or strangle their hapless victims, whereas others (African hunting dogs, several sharks, piranhas) cripple or surround the prey and then eat it alive— death comes from shock and bleeding.

SEVERING AND INGESTING PLANT FOOD   Various fishes and many larval anurans eat algae that may be scraped from the substrate with filelike teeth or (for the tadpoles) horny tooth analogs. Many vertebrates are said to be **herbivorous** because they eat such plant foods as leaves or stems. Relatively few adult fishes and amphibians are primarily herbivorous, but among reptiles, the giant dinosaurs, duck-billed dinosaurs, horned dinosaurs, and many turtles are, or were herbivorous. The ostrich, hoatzin, geese, and some parrots, grouse, and finches are plant eaters. Kangaroos, wallabies, wombats, langurs, sloths, rabbits, many rodents, elephants, hyraxes, sea cows, and ungulates are examples of herbivorous mammals.

The animal may bite the plants directly (duck-billed dinosaurs, iguanas, tortoises, geese). Plants may be gathered into the mouth by sensitive, mobile lips (sirenians, perissodactyls), or a protrusible tongue (giraffe, cattle), or may be fed into the mouth with hands (langurs, some rodents) or trunk (elephant). The food may be broken from the bush or tree before it reaches the mouth, but usually the leaf or stem is severed from the plant by the mouth, and this requires a cutting or cropping mechanism.

The horny margins of the mouth function as shears for tortoises and parrots (Figure 28-9). Serrate lateral teeth serve as cutting edges for the iguana. Geese have numerous lamellae at the margins of the bill which act as cutters. A specialized pair of upper and lower front teeth form a strong shearing apparatus in wombats, some sloths, hyraxes, rabbits, and rodents. These teeth have open roots and are ever-growing. Other herbivores use a cropping mechanism: Horses pinch grass between the upper and lower rows of contiguous incisors and then break it off with a sideways jerk of the head. Most artiodactyls instead break off the grass after pinching it between their lower incisors and a horny plate on the upper jaw. The analogous mechanism of kangaroos consists of a row of upper incisors that crop down against a pair of broad, procumbent lower incisors.

FILTERING AND STRAINING   Any aquatic vertebrate that feeds on floating or suspended food particles that are much smaller than itself (i.e., on plankton) must filter or strain the particles from the water. Ammocoetes, the larva of the lamprey, is the only primitive filter-feeder among vertebrates. It traps minute food particles in the sticky mucus on its complex pharynx and passes the mucus gradually back into a short intestine. All other vertebrate strainers have secondarily become specialized in various ways for their mode of feeding.

Strainers among fishes include the paddle fish (Figure 28-10), shad, and herring, and also the huge basking and whale sharks. The menhaden can strain 26 liter of water per minute. The gill rakers of these fishes are modified to form filters. They are long, slender, closely set along the gill bars, and have secondary or even tertiary branches. The larvae of various frogs are remarkably effective

GALAPAGOS TORTOISE, *Testudo*

COCKATOO, *Cacatua*

SLOTH, *Choloepus*

HORSE, *Equus*

Figure 28-9 EXAMPLES OF SHEARING AND CROPPING MECHANISMS OF PLANT EATERS.

filter feeders. The tadpole of *Xenopus* can filter to 0.1 μm, which challenges human technology.

Among ducks, the spoonbills and other dabblers strain plant or animal food from water. The bill is swung back and forth sideways, while the mandible pumps very fast. One hundred to 300 lamellae at the edge of the bill form the strainer. The flamingos are much more highly specialized and can remove diatoms and algae measuring only 0.1–0.2 mm in their longest dimension. During feeding, the large curved bill is positioned under the water "upside down," pointing toward the bird's feet. Upper and lower bills have up to twenty parallel rows of horny platelets that intermesh to form a fine screen. The platelets are less than 1 mm long, 0.1–0.2 mm apart, and frayed into dozens of hair-like projections on one edge (but with variations according to species). The muscular tongue is the pump. Horny hooks on the tongue comb the food from the platelets.

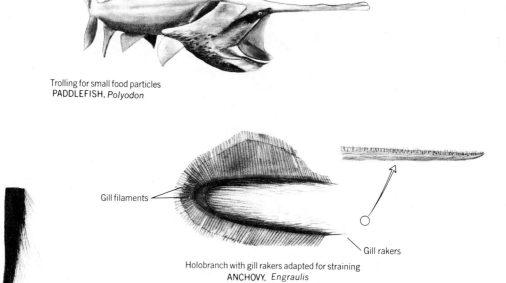

Trolling for small food particles
PADDLEFISH, *Polyodon*

Gill filaments

Gill rakers

Holobranch with gill rakers adapted for straining
ANCHOVY, *Engraulis*

Baleen
Tongue

Rows of baleen plates adapted for straining
RIGHT WHALE, *Eubalaena*

Figure 28-10 SOME ADAPTATIONS FOR STRAINING SMALL FOOD PARTICLES FROM WATER.

The baleen whales feed mostly on krill, which consists of swarming crustaceans measuring 75 mm and less in length. The whale has 230–400 plates of baleen along each side of the upper jaw. Each plate is ribbonlike, being flat and from $\frac{1}{4}$ to 3 m long, according to species. A plate is made up like a sandwich consisting of a horny lamella on each surface and a core of horny hollow tubes running lengthwise within. This makes it light and flexible but very strong. The lamellae gradually wear off of the tip and inside margin of each plate releasing the hundreds of tubes that tangle like stiff hairs to make the filter. The plates are ever-growing at their bases to compensate for wear. Some whales swim slowly along with the mouth open and water spilling out at the corners through the baleen. Others gulp in water and then elevate the enormous tongue (heavier than an elephant in large blue whales) to send water gushing out through the filter.

GAPE AND SUCK FEEDING   Many aquatic vertebrates suck food into the mouth in a stream of water. Cephalaspids and pteraspids may have fed this way. Gape and suck feeding is very common among bony fishes; examples are sturgeons, suckers, carp, and perch. Some sharks, salamander larvae, some frogs, and several turtles do the same.

The strong suction pump is provided, according to species, by expansion of the pharynx through action of opercula and gill chambers, or by expansion of the mouth and pharynx through lowering of the base of the hyoid arch, hyoid apparatus, and related structures. The volume of water taken in at one time commonly is 40% of the volume of the head, or about six times the resting volume of mouth and pharynx. Anglerfishes, however, can increase the volume by as much as twelve times, and whereas the duration of the gulp is 40 msec or more for many fishes, anglerfishes complete the act in about 6 msec. Fishes may supplement the sucking action by swimming closer to the food, but even if they do not do so, food usually can be sucked in from a distance equal to about $\frac{1}{4}$ the length of the head. The smaller the mouth, the greater the velocity of the incoming water: Sea horses and pipe fishes feed on active small crustaceans by whisking them into a tiny mouth on the end of a long pipettelike rostrum. Fishes that use gape and suck feeding commonly have mouths that are round or ovoid and usually are protrusible. Small teeth may be present in the mouth (they may shred food that rushes past them), but often the only teeth are in the pharynx.

(Another animal draws food in with air instead of water, and although it sucks it does not gape: The sloth bear protrudes its mobile lips into a cylinder and sucks up swarming insects and grubs with loud smacking noises. The inner upper incisors are absent and the palate is arched to provide an airway.)

FEEDING WITH THE TONGUE   The tongue of fishes may move up and down or forward and backward with the hyoid arch, to which it is firmly attached. In bony fishes the tongue may be provided with teeth, in which case it helps to hold and swallow food. The tongues of tetrapods tend to be larger, more fleshy (most birds excepted), and more movable. In the different taxa it contributes to feeding in a variety of ways. Some examples follow.

The bullfrog jumps at its prey, which may be a flying insect, opens its mouth wide and (with eyes shut) captures the creature with a flip of a very sticky tongue. The tongue is anchored at the front of the mouth (see figure on p. 5) over a pair of special little bones. Muscles draw the mandibles together, thus depressing the bones. This, a sudden forward shift of the hyoid, and the hardening of another muscle that then acts as a cam, flip the tongue out in 0.05 sec. The mechanism probably varies among frogs.

Lungless salamanders project a sticky tongue by suddenly folding the specialized hyoid apparatus toward the midline and shifting it forward. The tongue of various plant-eating turtles is large and sticky so it can draw leaves into the mouth.

The chameleon is a bizarre lizard that catches insects on the end of a sticky tongue, which it can protrude a distance equal to its own body length in about 0.04 sec. The hyoid apparatus has a long horn that tapers as it extends forward in the midline. The horn fits within a lubricated sleeve of connective tissue that

Figure 28-11 TONGUE OF A NECTAR-FEEDING AND POLLEN-FEEDING BAT, *Leptonycteris*, under 19× magnification. Dorsal surface with tip of tongue out of picture to left. Photograph taken with a scanning electron microscope by John Mais.

is surrounded by a specialized accelerator muscle of the tongue. The lizard opens its mouth slowly, advances the hyoid into position, aims by orienting its head just right, and then suddenly contracts the muscle that squeezes itself off of the tapering hyoid horn like a tiny man pinching himself away from a giant watermelon seed.

Having exposed an insect with its drill-like bill, the woodpecker captures it with a tongue that is sharp, cylindrical, barbed, and sticky. A highly modified hyoid apparatus enables the woodpecker to extrude its tongue as much as five times the length of its long bill.

Nectar and pollen form all or most of the diets of hummingbirds, honeyeaters, sunbirds, honeycreepers, and some other birds, and also of six genera of bats, and the little honey possum. The birds all have long slender bills, variously curved according to the structure of the preferred flowers. The mammals have much longer rostrums than related species of other habits. Teeth are small and weak (bats) or reduced in number (honey possum), and jaw musculature has regressed. The tongue is invariably quite long, slender, and protrusible. It usually terminates in a brush consisting of rows or tufts of hairlike projections which slope toward the throat. This device effectively transports nectar to the mouth (Figure 28-11). The tongues of several nectar feeders have in addition, or instead, one or two narrow tubes through which nectar can be sucked into the mouth; muscles of the throat act as a suction pump. Pollen-feeding bats have unique hair adapted to collect pollen, which is then groomed from the pelt.

Vampire bats feed exclusively on blood. Teeth not used for inflicting the pumpkin seed-sized wound on the host are reduced or absent. When blood oozes from the wound it is lapped up; when it flows freely, the edges of the tongue are

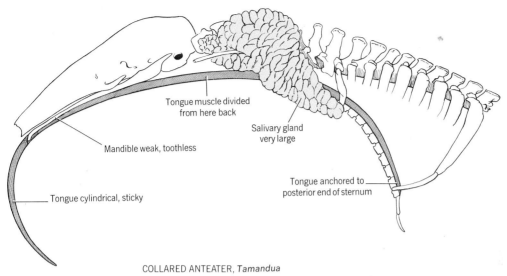

Tongue muscle divided
from here back

Salivary gland
very large

Mandible weak, toothless

Tongue anchored to
posterior end of sternum

Tongue cylindrical, sticky

COLLARED ANTEATER, *Tamandua*

Figure 28-12 SOME ADAPTATIONS OF A MAMMALIAN ANTEATER.

curled up to make a tube, and the blood is sucked into the mouth. The tongue is pumped in and out to establish the suction, but the details of the mechanism have not yet been learned.

The larger mammals that feed on swarming insects are highly specialized. These include the spiny echidna, rat-sized marsupial anteater, aardvark, pangolins, and the giant and smaller "true" anteaters. The rostrum is usually long. The tongue is cylindrical, sticky, mobile, and exceedingly long (reaching three times the length of the head). In several anteaters of two orders it is anchored at the posterior end of the breastbone rather than in the throat (Figure 28-12). Vascular spaces in the tongue of echidnas and pangolins function as erectile tissue. These animals are protected in several ways from ant bites: Pangolins (and others?) can close the nostrils. Most anteaters have a very thick, tough skin, and the eyelids and even the cornea may be thickened.

SWALLOWING FOOD WHOLE: SMALL, MEDIUM, AND LARGE Many mammals and most nonmammals swallow much or all of their food whole with little or no prior slicing, grinding, or other processing in the mouth. The nature and degree of adaptation required relates to the size of the food objects relative to the size of the feeder.

Liquid and soft foods, and food objects that are of small size relative to the size of the mouth, can be swallowed with few special provisions. This includes most of the seeds, berries, insects, and fishes eaten. Indeed, it includes most of the food taken by filterers, strainers, gape and suck feeders, and tongue feeders, and most food of all kinds eaten by nonpredatory birds. About 30% of all fishes feed on larvae or adult insects.

If all the food eaten by a kind of animal is liquid, soft, or small (e.g., blood,

PROCESSING FOOD IN MOUTH AND PHARYNX

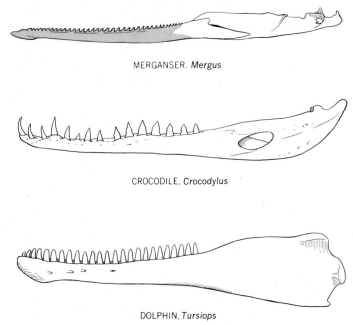

MERGANSER, *Mergus*

CROCODILE, *Crocodylus*

DOLPHIN, *Tursiops*

Figure 28-13   LEFT MANDIBLES OF VERTEBRATES THAT SWALLOW ANIMAL FOOD OF MODERATE SIZE REQUIRING HOLDING BUT LITTLE SHEARING.

honey, diatoms, ants, termites), and teeth are not needed for purposes other than feeding (defense, grooming), then the teeth tend to become small and simple (blood- and nectar-feeding bats, marsupial anteater), then peglike, wanting in enamel, single-rooted or open-rooted (aardwolf, aardvark, armadillo), and finally lost (minnows, echidna, pangolin, anteaters). Many such animals have protrusible jaws or long rostrum, small mouths, and weak mandibles.

If part or most of the food eaten is of larger size relative to the size of the feeder, then, of course, mouths tend to be larger and jaws stronger. If the food eaten is active or slippery, then teeth tend to be small and of simple structure, yet sharp and numerous to prevent the escape of prey before it can be swallowed. Examples are the teeth of many salamanders, lizards, crocodilians, ichthyosaurs, plesiosaurs, seals, sea lions, and toothed whales (Figure 28-13).

Some fishes (moray, toadfish) have teeth that are hinged so that they tilt back toward the throat as prey enters the mouth, but stand erect when pressure is in the other direction. Mergansers (aquatic birds) have horny tooth analogs at the edge of the bill to prevent the escape of fishes. Sea turtles have numerous horny, backward-pointing spines in the gullet (Figure 28-14). If the food is plant material which is to be swallowed in large pieces, then the animal is commonly toothless (various dinosaurs, turtles).

Coming to the other end of the scale, some flesh, fish, and egg eaters swallow exceedingly large food objects whole (Figure 28-15). One fish with a distensible belly gulps down prey as large as itself. Penguins, loons, pelicans, cormorants,

Figure 28-14 TONGUE AND OPENED GULLET OF A SEA TURTLE showing horny spines.

boobies, auks, roadrunners, and secretary birds swallow very large prey. Most snakes swallow large food, and several kinds that are adapted for eating eggs can manage eggs having a diameter three and more times the diameter of the head.

First, the object must be oriented end-on to the throat, with any large scales pointing the "smooth way." Pelicans (and, to a lesser extent, some other birds) have a distensible gular, or throat pouch to hold large fishes until they can be positioned for swallowing. Birds may toss a fish in the air and catch it with the desired orientation; snakes gradually work the jaws around to the correct position. Birds usually hold the bill up so gravity will assist in swallowing. They may also suddenly extend the neck, thus thrusting against the inertia of the food to aid in working the food back. Frogs draw the eyes down toward the mouth to help push on a large mouthful. They may use the hands to stuff in a particularly large insect.

Some fishes can protract and retract the upper jaw mechanism, with the small backward-slanting marginal teeth of right and left sides alternately pulling on the prey, releasing and reaching forward, and pulling again. When the prey reaches the pharynx, pharyngeal teeth function in a similar way. The remarkably adapted head of the snake is even more specialized. The teeth not only curve toward the throat, but the bones bearing them can be rotated so the teeth angle inward or outward at will. (Egg-eating snakes, however, tend to reduce or lose the teeth.) The lower jaw has no symphysis, so the mandibles can separate widely. The angles of the jaw pivot laterally from the skull, and the upper and lower jaws of many species can work forward and backward one side at a time.

Swallowers of large prey have extra large and distensible throats and huge stomachs. Snakes have no sternum so that large food items can distend the body

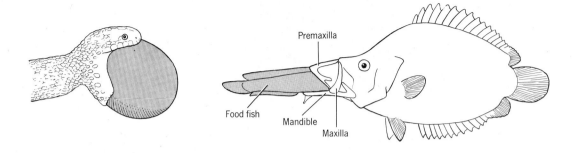

EGG-EATING SNAKE, *Dasypeltis*   NANDID FISH, *Monocirrhus*

Figure 28-15   EXAMPLES OF VERTEBRATES THAT SWALLOW LARGE FOOD OB-
JECTS WHOLE.

by spreading the ribs. The airway to the lungs is kept open during the slow
swallowing process by protruding the epiglottis out of the mouth like a snorkel
below the prey being ingested. Several egg-eating snakes have evolved a remarkable method of breaking an egg after it is swallowed. Sharp, midventral
projections from about a dozen vertebrae actually penetrate the esophagus. When
the egg is in position, powerful muscles squeeze the egg against the projections,
thus breaking the shell. Some species pass the crumpled shell back through the
gut, whereas others hold the egg contents in place with special valves and regurgitate the shell.

Road-runners, secretary birds, and some large frogs may swallow half of a
large snake or lizard and wait for it to digest so there will be room to swallow
the remainder.

CUTTING, SHEARING, AND TEARING OF ANIMAL FOOD   The subdividing
of animal food by shearing is in some instances functionally similar to the inges-
tion of plant food by severing. Predators that eat mostly flesh are said to be
carnivorous. Examples are crocodilians, various extinct reptiles, raptorial birds
(hawks and owls), the Carnivora, and one family of Marsupialia. Predators that
eat mostly fishes are termed **piscivorous.** About 30% of all fishes are piscivorous,
but most of them swallow their prey whole without first subdividing the hapless
victim. Many sharks do bite and shear their prey, however, and to a lesser extent
the long, sharp teeth of the barracuda, pike, gar, and pickerel may prepare the
prey for swallowing. Animals that eat carrion are called **scavengers.** Albatrosses,
fulmars, and their relatives are scavengers at sea; vultures, caracaras, adjutant
storks, jackals, and hyenas are scavengers on land.

Animal food requires little or no chewing to make it digestible, but many
carnivores (including some piscivores) and most scavengers must cut, shear, or
tear at least part of their food to prepare pieces that are small enough to swallow.
Animals that eat insects (and other small arthropods and worms) are said to be
insectivorous. If the feeder is small relative to the insect, then it too must sub-
divide the prey before swallowing it. What are the adaptations of such feeders?

Figure 28-16   MANDIBLE AND SHEARING TEETH OF THE PIRANHA FISH.

In general the teeth tend to be conical or bladelike, sharp, strong, thecodont (except in fishes), and relatively few in number (again with exceptions among fishes). Many sharks, piranhas, and various extinct carnivorous reptiles have (or had) teeth shaped like pointed blades with straight or serrate margins (see Figure 28-16 for piranha and figure on p. 112 for Selachii). In some instances upper and lower teeth shear past one another. Synapsid reptiles were moderately heterodont, and mammalian carnivores are markedly so. (The very long canines have a greater role in fighting and in capturing prey than in eating it.) Incisors and anterior premolars are most commonly adapted for cutting. Carnivora have one pair of teeth on each side of the mouth (upper fourth premolar and lower first molar in living forms), called **carnassial teeth,** which are enlarged, lengthened, and positioned so as to form a powerful shearing mechanism for slicing tendons and skin (Figure 28-5 and figure on p. 114). The carnassials of hyenas are as effective as large tin-snips. Slight shifting of the mandibular joint, and slight motion at the ligamentous symphysis of the jaw enable the carnivore to shear on one side at a time.

Some piscivorous fishes have teeth on the bony tongue. Reptiles and birds may have analogous horny spines on the tongue. Close-set horny projections on the tongues of some carnivores—notably the big cats—transform the organ into a rasp with which it licks the flesh from bones (Figure 28-17). The horny teeth of hagfishes can also cut flesh.

Raptorial and scavenging birds, having no teeth in their ancestry, must divide their food another way. Hawks and owls have the horny upper bill curved down at the tip to make a formidable meat hook. Pushing down with its talons and pulling up with its beak, the bird tears its prey apart. Some extinct reptiles had similar beaks. The smaller shrike and kestrel, which eat grasshoppers and beetles, also have strong hooked bills and may use claws or thorns in dismembering an insect.

Figure 28-17   TONGUE OF A MOUNTAIN LION showing rasping surface.

Finally, various flesh eaters, including certain sharks and crocodilians, may grasp the prey in the powerful jaws and then violently twist, thrash, and roll to dismember the food object.

Animals that eat a variety of plant and animal foods are said to be **omnivorous.** Their relatively unspecialized teeth tend to be intermediate between those of carnivores and herbivores. The molars are moderately broad and have low cusps, the condition termed **bunodont.**

GRINDING OF PLANT FOOD   Once it is free in the mouth, plant food is swallowed with little or no chewing by some tortoises and birds, but is very thoroughly chewed by most herbivores. This means that feeding must be slow. Most artiodactyls swallow their food a first time with little chewing, but regurgitate it, one bolus at a time, to be chewed as cud at the animal's leisure. As much as 7–10 hr/day may be spent chewing. (Some marsupials regurgitate and chew an occasional bolus, but spend little time at it.) Large quantities of saliva must be secreted, particularly if the food is dry.

In order to triturate coarse vegetation, a grinding mill is needed. As explained on page 22, the requirements are most stringent for large animals, so the adaptations of ungulates and elephants will be mentioned first. Because there is one kind of job to do, the cheek teeth are similar to one another; the functional premolars become molariform in nature. The occlusal surfaces of the teeth are flat and large in area: A single mammoth tooth may present a grinding surface of more than 250 cm² (Figure 28-18). Such large and heavy teeth must be supported by multiple roots.

Furthermore, provision is made for long wear because the teeth are used so much of the time and because the material ground is itself coarse and often has an admixture of grit. To provide for long wear, the roots are deep in the jaw of the young animal and the crown (the part of the tooth above the roots, not just above the gums) is high. Such teeth are called **hypsodont** (= high + tooth). As the exposed part of the tooth wears down, the roots slowly rise higher in the jaw, and bone fills in under them. This exposes more of the crown. When most of the crown is gone and the roots are just under the gums, the animal, probably now very old, may suffer from malnutrition.

Elephants have evolved a different mechanism that assures complete wear of all of the crown. The teeth lie in a groove in the jaw rather than in sockets. Each groove is deep at the back of the jaw and slopes upward toward the front. Six teeth form in succession at the back of each groove and migrate slowly forward. Each erupts first at its forward corner and starts to wear there long before the posterior part of the tooth is in service. As forward migration continues, a tooth rises in the sloping groove to compensate for wear. Finally, the crown of each part of each tooth wears off completely as it reaches the anterior end of the tooth row. The short roots are pushed up out of the gums and fall away. Only two of these great teeth function at one time in each jaw. As the last tooth moves along, the groove is filled in from behind by spongy bone. The series of six teeth lasts the elephant the 70 or more years that it will live.

Hyposodont teeth have another adaptation. Their occlusal surfaces are broad and flat, like millstones. But in order to grind effectively, millstones must not be

MAMMOTH, *Mammuthus*
Grinding surface

Cement

Enamel forming
cutting ridges

Dentine

Anterior

Gum line

Crown

Root

HORSE, *Equus*
Lateral surface

Cutting ridges

Cement

Enamel

Dentine

Pulp cavity

Anterior

ELEPHANT, *Loxodonta*
Longitudinal section

Anterior

Lateral

Cement

Pulp cavity

Enamel

Dentine

WATER BUFFALO, *Bubalus*
Cross–section

Figure 28-18   CHEEK TEETH ADAPTED FOR GRINDING.

smooth. The enamel of these teeth folds up and down, in and out of the substance of the tooth, thus forming a succession of hills and valleys. The pattern of folding is precise for each species, but varies widely among species. The valleys may remain open (many artiodactyls) or may be filled with cement shortly before the tooth erupts (horses, elephants). In the latter case, as the tooth wears there will be a horizontal succession of enamel–dentine–enamel–cement (Figure 28-18). Since enamel is harder than dentine or cement, it is a bit more resistant to wear and projects a little above the other materials to form **cutting ridges.** This provides and maintains the needed roughness.

The cheek teeth of small herbivores are less specialized in that they are smaller and usually have less complicated patterns of infolded enamel, yet they may be more specialized in being open-rooted, evenly curved in their sockets,

Figure 28-19   DISSECTION OF THE SKULL OF A GOPHER, *Thomomys,* SHOWING OPEN-ROOTED, EVER-GROWING TEETH.

and ever-growing (rabbits, some rodents) (Figure 28-19). Individual teeth of herbivorous reptiles never achieve the complexity of those of the large mammalian grazers and browsers, but nature provided the duck-billed dinosaurs with an effective analog. Several hundred relatively simple teeth became pressed together to form a large rough grinding plate. Some vegetarian fishes have crowded thousands of small teeth together in a comparable way. Others have no teeth in the mouth, but shred plant food with pharyngeal teeth.

CRUSHING AND CRACKING NONEDIBLE MATERIAL   Some animals must crush or crack shells, hulls, woody seeds, or other hard materials to make digestible food available. These animals are termed **durophagus.** Examples of fishes that crush shells to secure the soft meat within are the stringray, chimaeras, dipnoans (see figure on p. 112), some cichlid fishes, and the porcupine fish. Parrot fishes crush quantities of sand and rock as they eat algae; the grit shreds the plant food. These fishes have powerful jaws that are usually autostylic. Most have few teeth that are pavementlike and strong. Stingrays have numerous teeth that fit together like bathroom tiles to form crushing plates (Figure 28-20).

As their name implies, the extinct euryapsid reptiles called placodonts had large flat teeth for crushing mollusks. Several lizards and several turtles crush snails. The molars of the sea otter and walrus are flattened for the same purpose. The platypus has horny plates for crushing snails and other food.

Grain- and nut-eating birds have particularly muscular gizzards for grinding their food. Grit is eaten to make the mill more effective. Similarly, sturgeons, the gizzard shad, and the mullet, which are detritus-eaters, have muscular stomachs with tough linings for crushing miscellaneous invertebrates.

Some other vertebrates crack rather than crush their hard foods. The grackle (a bird) cuts acorns open against a keel on the inside of the upper bill. The hawfinch cracks cherry and olive pits. The beak is very stout and deep at the base, and the jaw adductors are remarkably extensive. It has been claimed that this bluebird-sized finch can bite a pit with a force of more than 45 kg. The dusky shark has a biting pressure of 3000 kg/cm^2. Hyenas crack open the bones of large ungulates to secure the marrow. Their adaptations include extra heavy jaws and teeth, enormous jaw muscles and sagittal crest, and early closure of cranial sutures. Since the food is touched only by the points of the carnassials and several other teeth, great pressure is generated.

STINGRAY, *Holorhinus*

PORCUPINE FISH, *Diodon*

CHIMAERA, *Hydrolagus*

SEA OTTER, *Enhydra*

Figure 28-20 MANDIBLES AND LOWER TEETH OF DUROPHAGUS VERTE-BRATES.

CONCLUSION

29

Vertebrate Morphology: Diversity, Unity, and Challenge

Vertebrate morphology is too vast a subject to be "covered" by one book, one course, one student, or one professor. This book has merely summarized descriptive anatomy, and even in areas selected for emphasis it has omitted much. Some additional adaptations that could be subjects for supplemental reading or special reports are the following.

Provisions for defense and escape include a diversity of structures and functions, among which are armor, teeth, claws, horns, antlers, tusks, beaks, spines, quills, camouflages, and mechanisms for stinging, poisoning, crushing, inflating the body, dropping the tail, and producing slime, irritants, or noxious odors.

There are structural as well as physiological and behavioral adaptations for living at high elevations, deep in the sea, and in caves. Some vertebrates are modified for walking on snow, mud, or floating vegetation. Adaptations for prenatal development and parturition include placental morphology, length and structure of the umbilical cord, and shape of the uterine cavity in relation to the shape, flexibility, and densities of different parts of the fetus. The larynx, syrinx, or other sound-producing mechanism exhibits marked specializations, as does the ear.

Man has received scant attention in this book, but he also has distinctive locomotor skills. Even if other animals could have the incentive to try and the patience to learn such "artificial" activities as gymnastics, diving, skiing, figure skating, and pole vaulting, none could match his ability. Also, man is not only uniquely expert but also impressively skilled at throwing. The mechanics of throwing is complex and difficult to analyze. A combination of structural adaptations favor the achievement of high velocity at the hand, just as they favor high velocity at the feet of cursors. Behavioral modification (i.e., training) assures that velocity of the projectile (ball, javelin, discus, shot) is increased by rotating several joints in the same direction at the same time, by the application of force to the object for as long as possible, by conserving the linear momentum of a runup, and by conserving angular momentum as successively lighter parts of the body and arm are made to rotate in sequence.

It is less useful to stress how much of the accumulated knowledge about vertebrate structure has been omitted from our course of study (since it has, nonetheless, included much) than to emphasize the unity of what has been presented and the extent of the engaging material that remains to be learned.

The subject matter of Parts II and III of this book are not as distinct as may at first seem apparent. Structure that characterizes the major vertebrate taxa, and the relatively general changes in structure that relate major taxa in time were stressed in Part II, whereas structure patterns that adapt vertebrates to specific modes of locomotion and feeding were presented in Part III. This distinction was made because (*1*) the emphasis of much current research justifies greater stress on function than has been usual in textbooks of morphology, and (*2*) in my judgment, when the material here presented in Part III is mixed in with that of Part II, it may seem to constitute digressions and exceptions that cloud the story of the broad outline of evolution.

Nonetheless, this division is merely a convenience in presenting and analyzing the subject. The material of the two parts of the book is not, after all, independent and significantly different in kind. Much function *was* included in Part II; the adaptations described in Part III are *also* the product of evolution. Distinctions between structure that characterizes major taxa and structure that adapts to specfic behaviors break down if carried far: Osteichthyes is surely a major taxon, yet nearly all its members share one general mode of locomotion. One could neither study the phylogeny of fishes without reference to adaptation for swimming, nor analyze the structure of fins without reference to evolutionary relationships. A double-circuit heart and complex cerebellum (described in Part II) characterize the class Aves, yet are just as necessary for masterful flight as are airfoils and flight muscles (described in Part III).

Furthermore, and importantly, essentially the same evolutionary processes that produced such generally beneficial structures as jaws, gills, lungs, and paired appendages also produced structures of more limited benefit such as keeled sternums, suspensory ligaments, and adhesive mechanisms. Evolution provides that individual animals tend to survive, that each species tends to multiply, and that the environment tends to be used to the utmost. Specialization permits specific kinds of animals to survive without serious competition in closer proximity to other kinds of animals than would be possible if all shared the same requirements.

Unity of our subject also emerges from the recognition of concepts and relationships that cut across chapter headings. In coming to the end of the book it would be helpful, therefore, to turn again to the principles and considerations discussed in Chapter 1 to be sure that they have not been lost sight of. It was necessary at the outset to state the ideas in general terms because the vocabulary and specifics of our subject were not yet at hand. Now the student might augment that introductory statement by gleaning additional examples of each concept from the body of the book.

I enjoy vertebrate morphology because it shows me that the animals can do so many things so well. A runner can dash faster than I drive my car on freeways. A springer can leap 14 times higher than its body length. A digger can thrust with 32 times its body weight. A climber can walk upside down on polished glass. A swimmer can sprint twice as fast in water as a man can run on land. A flyer can dart backwards as well as forwards. A feeder can swallow objects three times larger than its head, and another can strain from water food particles having 1/5000 the diameter of a pencil lead.

There are living direction finders, strain gauges, force multipliers, flow equilizers, suction cups, bifocal lenses, locking devices, pressure sensors, self-lubricating bearings, electric generators, gas analyzers, echo-rangers, depth gauges, recoil mechanisms, magnetic compasses, countercurrent exchangers, and compact computers. The list could go on and on, yet we have not recognized nearly all of nature's methods, let alone understood them. It will be long before we know "very much" of what there is to learn about vertebrate structure and behavior. In the meantime, the continued observation and interpretation of form and function promise to increase man's pleasure and wonder in the complexity, diversity, and perfection of the vertebrate body.

Glossary

A-, ab-. Prefix meaing without, from, away.

Abducens *(away + to lead)*. Denoting the sixth cranial nerve.

Abductor *(away + to lead)*. A muscle that moves a part away from the sagittal plane, or separates two parts.

Acceleration. Rate of increase of velocity; force divided by mass.

Acoelous *(without + hollow)*. Said of a centrum that is more or less flat at each end; platyan.

Acetabulum *(vinegar cup)*. The cup-shaped depression in the innominate bone that holds the head of the femur.

Acousticolateralis *(listen + side)*. Denoting the lateral line system plus the inner ear.

Acrodont *(extremity + tooth)*. Having rootless teeth fused at their bases to the jawbone.

Acrosome *(tip + body)*. The structure that caps the head of a sperm cell and functions in fertilization.

Actinopterygium *(ray + fin)*. A fin having bony radials and no fleshy stalk or skeletal axis.

Ad-, af-, ag-, etc. Prefix meaning toward.

Adaptation. Ability of a feature to execute its biological role with a minimum expenditure of energy.

Adductor *(toward + to lead)*. A muscle that moves a part toward the sagittal plan, or draws two parts together.

Adenohypophysis *(gland + under + growth)*. The part of the hypophysis derived from the hypophyseal pouch.

Adhesion. The sticking together of dissimilar materials. The molecular attraction exerted between surfaces in contact.

Adrenal *(near + kidney)*. An endocrine gland adjacent to the kidneys.

Afferent *(toward + to bear)*. Conducting toward or into.

Agnath *(without + jaw)*. Any jawless vertebrate.

Alar *(wing)*. Winglike.

Allantois *(sausage + form)*. The fetal membrane of amniotes that is derived from the hindgut and is functional in respiration and excretion.

Allometry *(different + to measure).* *Study of the correlation between form and size.*

Alula *(small wing).* The first digit of the bird wing together with its feathers.

Alveolus *(small cavity).* A small lobular cavity.

Ameloblast *(enamel + germ).* A cell that forms enamel.

Amnion *(fetal membrane).* The innermost fetal membrane of amniotes.

Amniote. Those vertebrates whose embryos are surrounded by an amnion; reptiles, birds, and mammals.

Amphiarthrosis *(both + joint).* A joint allowing limited motion.

Amphicoelous *(both sides + hollow).* Said of a centrum that is concave at both ends.

Amphistyli *(both + pillar).* Suspension of the jaws from the chondrocranium in part directly and in part through the hyomandibula.

Ampulla *(flask).* A dilation of a canal, such as a semicircular canal of the inner ear.

Amygdala *(almond).* A cluster of nuclei derived from the archistriatum.

Analogy. Structural correspondence based on common function.

An-. Prefix meaning without, not.

Anamniote *(without + small lamb).* Those vertebrates whose embryos have no amniotic membrane; fishes and amphibians.

Anapsid *(not + arch).* Having no opening in the temporal region of the skull.

Anastomosis *(coming together).* A communication between two blood vessels.

Androgen *(male + produce).* Any male sex hormone.

Angle of attack. Angle between the chord of an airfoil and the oncoming airstream.

Animal pole. The region of the egg where the nucleus is located and metabolic activity is highest.

Ant-, anti-. Prefix meaning against, opposite.

Antagonist *(against + fight).* A muscle the action of which opposes that of another.

Anticlinal *(against + to lean).* Said of a thoracic vertebra having its neural spine transitional between backward-leaning and forward-leaning.

Antler *(before + eye).* The bony, deciduous outgrowth from the head of deer.

Aorta *(to lift).* The main arterial trunk.

Aponeurosis *(away from + tendon).* A tough flat sheet of connective tissue serving to distribute the tension of a muscle.

Apophysis *(away from + growth).* A process of a vertebra. Specific processes are indicated by a prefix.

Arachnoid *(spiderlike).* The middle meninx of the brain of mammals.

Archenteron *(first + gut).* The embryonic digestive tube.

Arch-. Prefix meaning first, primitive, ancestral, chief, ruler.

Archicortex *(primitive + bark).* The more medial part of the cerebral cortex.

Archinephros *(ancestral + kidney).* A hypothetical ancestral kidney that develops from all of the nephrotome; holonephros.

Archipterygium *(ancestral + fin).* A fin having a fleshy stalk and a central skeletal axis flanked by radials on both sides.

Archistriatum *(primitive + striped).* The more ventral of the basal ganglia.

Artery. A large vessel that carries blood away from the heart.

Aspect ratio. The ratio of span to average chord of a waterfoil or airfoil.

Atrium *(entrance hall).* A cavity; the division of the heart between the sinus venosus and ventricle.

Auditory *(to hear).* Relating to hearing.

Autonomic *(self + law).* Relating to the part of the peripheral nervous system that supplies visceral motor nerves to vascular, nutritive, reproductive, and some other involuntary organs.

Autostyli *(self + pillar).* Suspension of the jaws directly from the chondrocranium without participation of the hyomandibula.

Axon *(axis).* The process of a neuron that transmits impulses away from the nerve cell body.

Baleen *(whale).* Horny plates in the mouths of toothless whales that filter food from the water.

Basal membrane. The thin membrane that separates epidermis from dermis.

Basal nuclei. The complex of ganglia in the brain of mammals that corresponds to the corpus striatum.

Bi-. Prefix meaning two, twice, double, life.

Blastocyst *(germ + bladder).* The blastula of mammals, which is characterized by a large blastocoel.

Blastoderm *(germ + skin).* The blastula derived from macrolecithal eggs, which consists of a disc of cells spread on the yolk.

Blastocoel *(germ + hollow)*. The cavity of the blastula.

Blastomere *(germ + part)*. Any one of the cells into which the egg divides during cleavage.

Blastopore *(germ + opening)*. The opening into the gastrocoel.

Blastula *(small germ)*. The early embryo consisting of one tissue layer of several hundred cells.

Boundary layer. The region surrounding a swimming or flying object within which shearing forces occur; the region responsible for frictional drag.

Brachial *(arm)*. Relating to the arm.

Brachiation. Climbing by arm swinging.

Branchial *(gill)*. Relating to gills.

Brev-. Prefix meaning short.

Bronchus *(windpipe)*. An airway within the lung that is supported by cartilage.

Bunodont *(mound + tooth)*. Having low cusps on the molar teeth, as for most omnivorous mammals.

Caecum *(blind)*. A pouchlike extension from the digestive tract.

Calcar *(spur)*. Cartilaginous rod supporting a flight membrane.

Camber. The front to back curvature of an airfoil; the transverse bowing of a wing.

Cantilever. A projecting beam or similar member that is supported at only one end.

Capillary *(relating to a hair)*. A microscopic blood vessel through which diffusion takes place.

Carapace *(hard covering)*. A dorsal bony covering, as of a turtle or armadillo.

Cardiac *(heart)*. Relating to the heart.

Cardinal *(red)*. Relating to the primitive system of veins that drains the head, dorsal body wall, and kidney.

Carnassial. Denoting the teeth of carnivorous mammals that shear past one another: $P^{\underline{4}}$ and $M_{\overline{1}}$ for living forms.

Carnivorous *(flesh + to eat)*. Eating meat.

Carotid *(heavy sleep)*. One of the large arteries of the neck.

Caudal *(tail)*. Relating to the tail.

Cavernosus *(cavern)*. Having internal cavities, as the erectile tissue of the penis.

Cement. The acellular, fibrous bone that joins teeth to their sockets.

Center of buoyancy. The point in an immersed body that represents the center of gravity of the displaced water; the point through which the resultant force of buoyancy acts.

Center of gravity. The point in a body through which the resultant force of gravity acts; that point from which a body can be suspended in equilibrium in any position.

Centrifugal *(center + to flee)*. Said of a force that tends to impel a moving object outward away from a center of rotation.

Centripetal *(center + to approach)*. Said of a force that tends to impel a moving object inward toward a center of rotation.

Centrum *(center)*. The body of a vertebra.

Cephalic *(head)*. Relating to the head.

Cephalization *(head + state of)*. The tendency to concentrate neurosensory and feeding mechanisms in a head.

Ceratotrich *(horn + hair)*. Slender, horny, unsegmented fin ray.

Cerebellum *(small brain)*. The derivative of the dorsal part of the metencephalon.

Cerebrum *(brain)*. The hemispheric derivatives of the telencephalon.

Cervical *(neck)*. Pertaining to the neck.

Chiasma *(figure X)*. A crossing of fibers, as of the optic nerves.

Chondrocranium *(cartilage + skull)*. The part of the head skeleton other than the splanchnocranium that consists of cartilage or replacement bone.

Chondrocyte *(cartilage + cell)*. A cartilage cell.

Chord. The straight line distance between the leading and trailing edges of a waterfoil or airfoil in the line of travel.

Chordamesoderm. The roof of the archenteron, which induces the neural plate, and itself forms notochord and mesoderm.

Chorion *(skin)*. The outermost fetal membrane of amniotes.

Choroid *(skinlike)*. A vascular layer of the brain or eye.

Chromaffin *(color + affinity)*. Denoting endocrine tissue that is functionally related to the adrenal medulla but is diffuse.

Chromatophore *(color + to bear)*. A pigment cell.

Ciliary *(eyelash)*. Relating to a hairlike structure.

Cladistic *(branching)* **classification.** Classification that reconstructs phylogenetic sequences by deductive processes that analyze primitive and derivative character states of related organisms to generate dichotomously branched sister groups.

Cleavage. The cell divisions that convert the zygote to a blastula.

Cleidoic *(locked up).* Descriptive of the eggs of amniotes that can survive in air.

Clitoris. The sensitive female homolog of the male penis.

Cloaca *(sewer).* A common passageway for products of the digestive and urogenital systems.

Cochlea *(snail shell).* The spiraled auditory part of the inner ear of mammals.

Coelom *(hollow).* Any body cavity that is derived (in vertebrates) by splitting of the hypomere, and hence is lined by mesoderm.

Colic *(colon).* Relating to the colon.

Collagen *(glue + producer of).* The substance of collagenous fibers, which are present in all connective tissue.

Commissure *(connection).* A tract joining equivalent structures on the two sides of the central nervous system.

Compression. Stress in an elastic solid resulting from a load directed toward the object and perpendicular to its surface.

Conjunctiva *(to connect).* The membrane covering the front of the eyeball.

Conus. The most anterior of the primitive heart chambers.

Convergence. Evolutionary change in two or more lineages such that corresponding features that were formerly dissimilar became similar.

Coprodeum *(dung + way).* The dorsal part of a partially divided cloaca.

Copulation *(to join).* The act that accomplishes internal fertilization.

Cornea *(horny).* The transparent superficial part of the eyeball.

Corona radiata *(crown + radiating).* The cells derived from the ovarian follicle that surround the egg at ovulation; the branching of the pyramidal tract in the cerebral hemispheres.

Corpus *(body).* Any mass or solid part of an organ.

Cortex *(bark).* The outer part of an organ.

Cosmine *(ornament).* Dentine characterized by internal tufts of radiating canals.

Crista *(crest).* Sensory cells of the ampullae of semicircular canals.

Crossopterygium *(fringe + fin).* A fin having a fleshy stalk and a skeletal axis flanked by radials on one side.

Ctenoid *(comblike).* Denoting a fish scale having a serrated margin.

Cupula *(small cask).* The gelatinous structure in which the sensors of a neuromast are imbedded.

Cursorial. Adapted for running.

Cutaneous *(skin).* Relating to the skin.

Cuticle. A thin, noncellular, external covering of the skin of some animals.

Cycloid *(circle).* Denoting a fish scale having a smooth margin.

Cystic *(bladder).* Relating to a bladder or pouch.

De-. Prefix meaning away, down, from.

Decussation *(cross).* A tract joining unlike structures on the two sides of the central nervous system.

Deferent *(away + carrying).* A duct that carries away, as the sperm duct.

Delamination *(from + layering).* The formation of a tissue layer by the separation and subsequent aggregation of cells from a preexisting tissue layer.

Dendrite *(tree).* The processes of a neuron that receive and propagate impulses toward the nerve cell body.

Denticles *(small teeth).* Small toothlike structures that may either project from dermal armor and scales or occur independently as small scales.

Dentine. A tissue of mesectodermal origin that contributes to teeth, denticles, and some fish scales, being usually softer than enamel but harder than bone.

Derived character. A feature that departs from the ancestral condition for the taxon in question.

Dermal bone. Bone that ossifies directly without replacing cartilage.

Dermatocranium *(skin + skull).* The part of the head skeleton that consists of dermal bone.

Dermatome *(skin + to cut).* The outer division of the epimere.

Dermis *(skin).* The inner part of the skin, derived from mesoderm.

Diaphragm *(partition).* The muscular partition of mammals separating the pleural and peritoneal cavities.

Diaphragmatic *(partition).* Said of a thoracic vertebra having its prezygapophyses facing more or less dorsally, but its postzygapophyses facing more or less laterally.

Diaphysis *(through + growth).* The shaft of a long bone.

Diapsid *(two + arch).* Having two openings in the temporal region of the skull.

Diarthrosis *(two + joint).* A freely movable joint having a joint cavity.

Digitigrade *(finger + walking).* Having only the digits and distal ends of the metapodials in contact with the ground when standing or moving, as for cats and dogs.

Diencephalon *(through or between + brain)*. The posterior derivative of the embryonic prosencephalon; becomes the anterior part of the brainstem.

Dioecious *(two + house)*. Having the male gonads in one individual and the female gonads in another.

Diphycercal *(twofold + tail)*. Said of a fish tail that is about symmetrical and has the spinal axis extending to its tip.

Diphyodont *(two + to grow + tooth)*. Having two developmental sets of teeth.

Diplospondyly *(double + vertebra)*. Having two vertebrae per primary body segment, in the caudal regions of certain fishes.

Distal. Located away from the central axis of the body.

Dorsal *(back)*. Said of the back, or vertebral side of the body.

Drag. Resistance to the motion of a body through water or air.

Duodenum *(twelve)*. The first segment of the small intestine.

Durophagous *(hard + to eat)*. Eating that requires the breaking up of hard materials such as shells or nuts.

Dynamic soaring. Flight that is sustained by extracting energy from wind in a shear layer—usually over the ocean.

Ec-. Prefix meaning out, outside.

Ectoderm *(outside + skin)*. The outermost of the three embryonic germ layers.

Ectomesenchyme. Mesenchyme derived from neural crests.

Efferent *(away + to bear)*. Conducting away from or out of.

Elasmobranch *(plate + gill)*. A fish having septal gills.

Elasticity. Capacity of a strained (or deformed) elastic solid to recover its original size and shape after a load is removed.

Enamel. The exceedingly hard, acellular tissue of ectodermal origin that caps teeth, denticles, and some fish scales.

Enameloid. An enamel-like tissue derived from mesectoderm.

Endocrine *(within + separate)*. Secreting into the blood stream.

Endoderm *(within + skin)*. The innermost of the three embryonic germ layers.

Endolymph *(within + clear fluid)*. The fluid within the labyrinth of the ear.

Endometrium *(within + uterus)*. The soft glandular tissue that lines the uterus.

Endostyle *(within + pillar)*. A mucous gland lying below the pharynx of lower chordates.

Enterocoely *(gut + hollow)*. The formation of coelom by pouching from the archenteron.

Epaxial *(upon + center line)*. Said of muscles of the trunk lying dorsal to the lateral septum.

Ependymal *(outer garment)*. Relating to the cells that line the cavities of the central nervous system.

Epidermis *(upon + skin)*. The outer part of the skin, derived from ectoderm.

Epididymis *(upon + testes)*. The coiled part of the sperm duct that is adjacent to the testis.

Epiglottis *(upon + laryngeal opening)*. The valvelike closure of the glottis.

Epimere *(upon + part)*. The somite, or segmented dorsal division of the embryonic lateral mesoderm.

Epimysium *(upon + muscle)*. The membrane surrounding a muscle.

Epiphysis *(upon + growth)*. A separate ossification forming the end of a long bone.

Epithelium *(upon + nipple)*. A layer of cells covering a surface or lining a cavity.

Erythrocyte *(red + cell)*. A red blood cell.

Estrogen *(mad desire + produce)*. A female hormone responsible for secondary sexual characteristics.

Euryapsid *(broad + arch)*. Having one opening in the temporal region of the skull that is bordered below by the postorbital and squamosal.

Evolutionary *(unrolling)* **classification.** Classification that reconstructs phylogenetic sequences by judgments based on all available data, including the fossil record, and that incorporates both linear and branching evolutionary processes.

Evolutionary trend. A gradual, adaptive change in the evolution of a feature within a phyletic line.

Ex-. Prefix meaning out, beyond.

Exocrine *(out + separate)*. Denoting a gland that discharges into a duct.

Facial *(face)*. Relating to the face.

Falciform *(sickle + shape)*. Having the shape of a sickle.

Fascia *(band)*. Fibrous connective tissue.

Fascicle *(small bundle)*. A bundle of nerve or muscle fibers.

Fenestra *(window)*. Any large opening, as in the innominate bone.

Fissure *(a split)*. A cleft or groove, as on the ventral surface of the spinal cord.

Follicle *(small bag)*. A structure having a cavity.

Force. The product of mass times acceleration; a push or pull that causes, or tends to cause, motion.

Fossorial. Highly adapted for digging.

Fovea *(a pit).* Depression in the retina where the sharpest image is formed.

Free-body diagram. A drawing of an isolated mechanical system showing as vectors all external translational and rotational forces acting on the system.

Friction. Mechanical resistance to relative motion. Dry static friction is the product of the normal force and the coefficient of friction.

Frictional drag. Drag on a moving object caused by friction within the boundary layer.

Frontal. Said of planes that divide the body into dorsal and ventral parts.

Funiculus *(string).* A bundle or region of nerve cell fibers of the spinal cord.

Gait. A regularly repeating manner and sequence of moving the feet when walking or running.

Gamete *(spouse).* An egg or sperm.

Ganglion *(swelling).* An aggregate of nerve cell bodies—particularly when located outside of the central nervous system.

Ganoine *(brightness + made of).* Thick, lamellar enamel.

Gastralia *(belly).* Bony supports of the abdomen of some tetrapods.

Gastrocoel *(stomach + hollow).* The cavity of the gastrula.

Gastrula *(small stomach).* The early embryo consisting of two, and potentially three, tissue layers.

Gastrulation. The events that convert a blastula to a gastrula.

Glide. A controlled descent in air at an angle of less than 45° to the horizontal.

Glomerulus *(small ball).* The tuft of capillaries within a renal capsule; any of several aggregates of nerve fibers.

Glossal *(tongue).* Relating to the tongue.

Glossopharyngeal *(tongue + throat).* Denoting the ninth cranial nerve.

Glottis. The opening from the pharynx into the larynx.

Gnathostome *(jaw + mouth).* Any vertebrate having jaws.

Gonad *(seed).* A sex gland; the ovary or testis.

Gonopodium *(seed + foot).* The copulatory organ on the anal fin of some male teleosts.

Graviportal *(heavy + to carry).* Adapted for supporting great body weight.

Gyrus *(circle).* An elevated convolution of the cerebrum or cerebellum.

Habenula *(strap).* A small cluster of nuclei in the epithalamus.

Hemal arch. The part of certain caudal vertebrae that arches under the caudal vessels.

Hemibranch *(half + gill).* A gill bar bearing filaments on one surface only.

Hemipenes *(half + penis).* The paired male copulatory organs of lepidosaurs.

Hemopoiesis *(blood + create).* The production of blood cells.

Hepatic *(liver).* Relating to the liver.

Herbivorous *(plant + to eat).* Eating the leaves or stems of plants.

Hermaphrodite *(Hermes + Aphrodite).* An individual having both male and female sex organs.

Heterocercal *(different + tail).* Said of a fish tail having the spinal axis extending into the dorsal, larger lobe.

Heterocoelous *(different + hollow).* Said of a centrum having saddle-shaped ends, as in birds.

Heterodont *(different + tooth).* Having several kinds of teeth.

Hippocampus *(sea horse).* A derivative of the archicortex in the shape of an arching band impinging on the lateral ventricle.

Holoblastic *(whole + germ).* Said of cleavage that is total, that is, that divides the entire egg.

Holobranch *(whole + gill).* A gill bar bearing filaments on both anterior and posterior surfaces.

Holonephros *(whole + kidney).* A hypothetical ancestral kidney that develops from all of the nephrotome.

Homeotherm *(same + heat).* An animal that maintains a nearly constant body temperature.

Homocercal *(same + tail).* Said of a fish tail having dorsal and ventral lobes of about the same size, and both extending beyond the spinal axis.

Homodont *(same + tooth).* Having one functional kind of teeth.

Homology *(same + ratio).* Structural correspondence based on common ancestry.

Hormone *(excite).* A chemical released in one part of the body, transported by the circulatory system, and causing a response in another part of the body.

Hyoid *(U-shaped)* **arch.** The second visceral arch.

Hyomandibula *(U-shaped + jaw).* The dorsal, and principal segment of the hyoid arch.

Hyostyli *(U-shaped + pillar).* Suspension of the jaws from the chondrocranium primarily through the hyomandibula.

Hypaxial *(under + center line).* Said of muscles of the trunk lying ventral to the lateral septum.

Hyper-. Prefix meaning above, beyond, over.

Hypobranchial *(under + gill).* Muscles of the throat derived phylogenetically from hypaxial musculature.

Hypocercal *(under + tail).* Said of a fish tail having the spinal axis extending into the ventral, larger lobe.

Hypoglossal *(below + tongue).* Denoting the twelfth cranial nerve.

Hypomere *(under + part).* The unsegmented ventral division of the embryonic lateral mesoderm.

Hypophysis *(under + grow).* An endocrine gland lying below the hypothalamus; the pituitary.

Hypsodont *(high + tooth).* Having cheek teeth with high crowns, as for ungulates and some rodents.

Induced drag. The drag at the ends of an airfoil that is a consequence of the difference in pressure above and below the airfoil.

Induction. The effect of one embryonic part on another through a chemical stimulus.

Infra-. Prefix meaning below, under.

Infundibulum *(funnel).* The ventral outgrowth of the diencephalon that forms the neurohypophysis.

Inner cell mass. The inner part of the mammalian blastocyst from which the embryo is derived.

Insulin *(island).* A hormone of the pancreas.

Inertia. The tendency of an object to remain at rest or in uniform motion in a straight line unless acted upon by external forces.

Insectivorous *(insect + to eat).* Eating insects and similar food.

Intercentrum *(between + center).* The more anterior unit of the centrum of certain labyrinthodonts.

Invagination *(in + sheath).* The folding of tissue from the vegetal pole of the blastula inward to establish the archenteron of the gastrula.

Involution *(to roll up).* The migration of cells into the gastrula at the blastopore.

Isometric *(equal + measure).* Muscle contraction without shortening.

Isotonic *(equal + strain).* Muscle contraction without change of tension.

Jugular *(throat).* Relating to the throat or neck.

Keratin *(horn).* A hard, nearly insoluble protein or albuminoid, present in the epidermis and some of its derivatives.

Kinetic. Relating to motion. Said of tetrapod skulls having several somewhat movable units.

Kinetic energy. Energy derived from motion.

Kinocilium *(move + eyelid).* The longest and most complex hairlike sensor of a neuromast cell.

Labyrinth *(tortuous passage).* The membranous structure of the inner ear.

Labyrinthodont *(tortuous passage + tooth).* Having teeth with complicated patterns of infolded enamel on their side walls.

Lagena *(flask).* An extension from the saccule of the inner ear.

Lamella *(thin plate).* A thin membrane or layer.

Laminar flow. Passage of water or air over a moving body without the formation of eddies or turbulence.

Larynx. The cartilaginous box at the anterior end of the trachea.

Lepidotrich *(scale + hair).* Bony, segmented fin ray.

Leucocyte *(white + cell).* A white blood cell.

Lever A rigid structure that transmits forces by turning, or tending to turn, at a pivot.

Lever arm. The perpendicular distance between the line of action of an applied force (or component of such force) and the associated pivot.

Lift. The force generated by an airfoil (or waterfoil) that is at right angles to the oncoming airstream (or waterstream).

Ligament *(band).* A cord or sheet of connective tissue serving to join two or more skeletal parts.

Lingual *(tongue).* Relating to tongue.

Load. Any burden or force applied to a solid object.

Lumbar *(of the loins).* Pertaining to the region between the ribs and the pelvis.

Lymph *(clear fluid).* The fluid of the lymphatic system and tissue spaces.

Machine. A mechanism that transmits force from one place to another, usually changing its magnitude and for a useful purpose.

Macrolecithal *(large + yolk).* Said of large eggs having much yolk.

Macula *(spot).* Sensory epithelium of the saccule or utricule.

Mammary *(breast).* Relating to the milk glands.

Mandibular *(jaw)* **arch.** The first visceral arch.

Mass. The numerical measure of an object's inertia, or resistance to being accelerated; the quantity of matter an object contains.

Mediastinum *(medial).* The septum in mammals that separates right and left pleural cavities.

Medulla *(pith).* The inner part of an organ; the posterior part of the brainstem.

Melanophore *(black + to bear).* A pigment cell that contains the black pigment melanin.

Meninx *(pl.* **meninges)** *(membrane).* Membranous envelope surrounding the central nervous system.

Meniscus *(small moon).* A pad of fibrous cartilage located within a joint capsule, as at the knee.

Meroblastic *(part + germ).* Said of partial cleavage, that is, that does not penetrate the yolk mass.

Mesectoderm. Mesenchyme derived from neural crests.

Mesencephalon *(middle + brain).* The middle primary brain vesicle; the midbrain.

Mesenchyme *(middle + infusion).* Embryonic connective tissue composed of branched, loosely organized cells, often with the capacity to migrate.

Mesentery *(middle + gut).* A membrane that supports an internal organ from the body wall.

Mesoderm *(middle + skin).* The middle one of the three embryonic germ layers.

Mesolecithal *(middle + yolk).* Said of eggs having a moderate amount of yolk.

Mesomere *(middle + part).* The small middle division of the embryonic lateral mesoderm.

Mesonephros *(middle + kidney).* The functional kidney of fetal amniotes, which develops from a middle part of the nephrotome.

Mesorchium *(middle + testis).* A mesentery that supports the testis.

Mesovarium *(middle + ovary).* A mesentery that supports the ovary.

Metamere *(after + part).* One of serially repeated structural units along the body axis.

Metanephros *(after + kidney).* The adult kidney of amniotes, which develops from a short posterior part of the nephrotome.

Metencephalon *(after + brain).* The anterior derivative of the embryonic rhombencephalon; becomes the cerebellum and pons.

Microlecithal *(small + yolk).* Said of small eggs having little yolk.

Modulus of elasticity. Stress divided by strain; a measure of the deformation caused per unit load.

Momentum. The capacity of an object moving in a straight line to overcome resistance; the product of mass times velocity.

Morphology *(form + science).* The science of relating and interpreting observed structures.

Mucosa *(juice).* A tissue that contains or secretes mucus.

Mucus *(juice).* A clear, slippery secretion.

Myelencephalon *(spinal cord + brain).* The posterior derivative of the embryonic rhombencephalon; becomes the medulla.

Myelin *(marrow).* The fatty sheath of a nerve fiber.

Myocardium *(muscle + heart).* The muscle layer of the heart.

Myocoel *(muscle + hollow).* The transitory cavity of the myotome.

Myomere *(muscle + part).* The axial muscle of one body segment.

Myometrium *(muscle + uterus).* The muscular part of the uterus.

Myoseptum *(muscle + barrier).* The partition of connective tissue separating myomeres.

Myotome *(muscle + to cut).* The middle division of the epimere.

Neocortex *(new + bark).* The medial, and in mammals the largest part of the cerebral cortex.

Neostriatum *(new + striped).* The more dorsal of the basal ganglia.

Neoteny *(recent + to stretch).* The retardation of the development of a somatic feature so that it remains juvenile in later developmental stages.

Nephric *(kidney).* Relating to the kidney.

Nephrocoel *(kidney + hollow).* The cavity of the mesomere.

Nephron *(kidney).* The functional unit of a kidney.

Nephrostome *(kidney + mouth).* A ciliated opening leading from the coelom into an excretory tubule.

Nephrotome *(kidney + to cut).* The nephrogenic part of the mesomere.

Neural arch. The part of the vertebra that arches over the spinal cord.

Neural crests. Small, paired, segmented aggregates of ectodermal cells that flank the embryonic neural tube.

Neural folds. The longitudinal folds that flank the neural plate and arch together during neurulation.

Neural plate. The thickened part of the ectoderm, lying over the chordamesoderm, which will form the central nervous system.

Neural tube. The embryonic central nervous system.

Neurilemma (*nerve + sheath*). The thin sheath that surrounds the myelin of a nerve fiber or, if the fiber is unmyelinated, the axon cylinder.

Neurocoel (*nerve + hollow*). The cavity of the neural tube.

Neurocranium (*nerve + skull*). The part of the skull derived from the chondrocranium.

Neuroglia (*nerve + glue*). The supportive tissue of the central nervous system.

Neurohypophysis (*nerve + under + to grow*). The part of the hypophysis derived from the infundibulum.

Neuromast (*nerve + hill*). One of the hair-cell organs of the inner ear or lateral line system.

Neuron (*nerve*). A nerve cell.

Neurotransmitter. A chemical, released by nerve endings, that modulates the firing of other neurons.

Neurulation. The process that converts the neural plate into a neural tube.

Notochord (*back + chord*). The fibrocellular rod that forms the skeletal axis of embryonic, and some adult, vertebrates.

Nuchal (*neck*). Relating to the neck.

Nucleus (*kernel*). The inner part, as of a cell; an aggregation of nerve cell bodies within the central nervous system.

Oculomotor (*eye + motion*). Denoting the third cranial nerve.

Odontoblast (*tooth + germ*). A cell that produces dentine.

Olfactory (*to smell*). Relating to the sense of smell.

Omasum (*paunch, tripe*). A muscular part of the ruminant stomach.

Omentum (*membrane*). A membrane that joins one internal organ to another.

Omnivorous (*all + to eat*). Eating a variety of plant and animal food.

Ontogeny (*a being + become*). The development of the individual from fertilized egg to adult.

Oogonia (*egg + generation*). The proliferating cells of the ovary that will become ova.

Operculum (*cover*). The flap covering the gills of actinopterygians and holocephalians.

Ophthalmic (*eye*). Relating to the eye.

Opisthocoelous (*behind + hollow*). Said of a centrum that is concave posteriorly and convex anteriorly.

Opisthonephros (*behind + kidney*). The adult kidney of anamniotes, which develops from all or most of the nephrotome posterior to the pronephros.

Optic (*vision*). Relating to the eye.

Oral (*mouth*). Relating to the mouth.

Osteocyte (*bone + cell*). A bone cell.

Osteoderm (*bone + skin*). A bone in the dermis of some reptiles.

Osteon. A cylinder-within-cylinder structural unit of dentine or bone.

Ostium (*mouth*). A small opening into a duct or space.

Ostracoderm (*shell + skin*). An agnathous vertebrate having bony armor or scales.

Otolith (*ear + stone*). A calcified body within the inner ear.

Ovary (*egg*). The female gonad.

Oviduct (*egg + duct*). The duct that conveys eggs to the cloaca or uterus.

Oviparous (*egg + to bear*). Animals that lay eggs.

Ovisac (*egg + sac*). A part of the reproductive tract where eggs are retained, but not nourished, prior to laying.

Ovoviviparous (*egg + alive + to bear*). Animals that retain their eggs in the body until the time of hatching, yet do not nourish their embryos while in the ovisac or uterus.

Ovum (*egg*). The egg cell.

Paedomorphosis (*child + form*). The retention of ancestral juvenile characters by later developmental stages or descendants.

Palatoquadrate (*palate + squared*). The dorsal segment of the mandibular arch.

Paleocortex (*ancient + bark*). The more lateral part of the cerebral cortex.

Paleostriatum (*ancient + striped*). The more central of the basal ganglia.

Parachute. A partially controlled fall at an angle of more than 45° to the horizontal.

Paraganglia (*beside + knot*). Chromaffin tissue that is adjacent to sympathetic nerve ganglia.

Parallism. Evolutionary change in two or more lineages such that corresponding features undergo equivalent alterations without becoming markedly more or less similar.

Paramesonephric (*beside + kidney*). Said of the primordium of the female reproductive tract.

Parasympathetic (*beside + with + suffering*). Denoting the part of the autonomic nervous system having craniosacral outflow.

Parathyroid (*near + shield +*

form). An endocrine gland near the thyroid gland that controls calcium and phosphate metabolism.

Parietal *(wall).* Relating to an outer wall, as of the chest or skull.

Parthenogenesis *(virgin + origin).* Development of the embryo without fertilization.

Patagium *(a border).* A fleshy membrane used as an airfoil in flying or gliding.

Pecten *(comb).* The nutritive organ within the eyeball of birds.

Peduncle *(small foot).* One of the tracts supporting the cerebellum; the constriction in front of a fish tail.

Penis. The male organ of copulation of amniotes.

Pericardial *(around + heart).* Surrounding the heart.

Perichondrium *(around + cartilage).* The fibrous membrane covering cartilage.

Perilymph *(around + clear fluid).* The fluid surrounding the labyrinth of the ear.

Periosteum *(around + bone).* The fibrous membrane covering bone.

Peritoneal *(around + stretch).* Relating to the cavity containing the digestive viscera.

Phagocyte *(eat + cell).* A cell capable of ingesting foreign matter.

Pharynx *(throat).* The part of the gut between the mouth and esophagus.

Photophore *(light + to bear).* A light-producing cell.

Phyletic line. A lineage that is relatively continuous and complete in the fossil record.

Phylogeny *(a race + become).* Evolutionary history of a lineage.

Physoclistous *(bladder + closed).* Having a gas bladder not joined to the gut by a duct.

Physostomous *(bladder + mouth).* Having a lung or gas bladder joined to the gut by a duct.

Pineal *(pine cone).* A glandular outgrowth of the epithalamus.

Pinnate *(feather).* Said of muscles having their fibers sloping in toward one or more central tendons.

Piscivorous *(fish + to eat).* Eating fish.

Pitch. Rotation of a swimmer or flyer around its transverse axis.

Pituitary *(slime).* An endocrine gland lying below the hypothalamus; the hypophysis.

Placenta *(small flat cake).* An organ of fetal and maternal tissues that are associated for physiological exchange between the respective blood streams.

Plantigrade *(sole + walking).* Having the sole of the foot in contact with the ground when standing or moving, as for man and bears.

Plasma *(something formed).* The acellular component of blood or lymph.

Plastron *(breastplate).* The ventral part of a turtle shell.

Platyan *(flat).* Said of a centrum that is more or less flat at each end; acoelous.

Pleural *(side).* Relating to the wall of the thoracic cavity.

Pleurocentrum *(side + center).* The paired, more posterior units of the centrum of certain labyrinthodonts, and any centrum derived therefrom.

Pleurodont *(side + tooth).* Having rootless teeth that are connected to the jawbone on their outer surfaces.

Plexus *(a braid).* An interweaving of nerves or vessels.

Poikilotherm *(changeful + heat).* An animal that does not maintain a constant body temperature.

Polyphyodont *(many + to grow + tooth).* Having many successive developmental sets of teeth.

Pons *(bridge).* The ventral derivative of the metencephalon.

Pontine *(bridge).* Relating to the pons of the brain.

Portal *(gate).* Said of a venous circuit that joins two capillary beds.

Potential energy. Energy derived from position, either in relation to gravity or elastic loading.

Preadaptation. Evolutionary change of function for a given structure with minimal change of form.

Prehensile. Adapted for grasping by wrapping around.

Pressure. Force per unit area.

Pressure drag. Drag on a moving object caused by displacement of the fluid medium, backflow, and the formation of pressure gradients.

Primitive character. A feature that resembles the ancestral condition for the taxon in question.

Primitive streak. The structure of amniotes that forms most of the embryonic dorsolateral mesoderm by the migration and involution of cells.

Procoelous *(in front + hollow).* Said of a centrum that is concave anteriorly and convex posteriorly.

Proctodeum *(anus + way).* The posterior part of a partially divided cloaca, derived from ectoderm.

Profile drag. All the drag on a flying or gliding object other than that associated with the ends of the wings.

Progenesis *(before + be born).* Paedomorphosis produced by precocious sexual maturation of an organism that is still in a morphologically juvenile stage.

Progesterone *(before + to bear).*

A female hormone that maintains pregnancy.

Pronator *(to bend forward).* A muscle that rotates the palms or soles downward.

Pronephros *(in front + kidney).* The most anterior of vertebrate kidneys and the first to develop in ontogeny.

Proprioceptor *(one's own + take).* An organ in muscle, tendon, or joint that senses tension.

Prosencephalon *(in front + brain).* The most anterior of the three primary brain vesicles; the forebrain.

Proximal. Located toward the central axis of the body.

Pterygiophore *(fin + bearer of).* The skeletal supports of dorsal and anal fins.

Putamen *(shell).* A lateral part of the mammalian neostriatum.

Pulmonary *(lung).* Relating to the lungs.

Pygostyle *(rump + pillar).* The bladelike bone forming the posterior end of the spine of birds.

Pyloric *(gatekeeper).* Relating to the part of the stomach adjacent to the intestine.

Recapitulation. The repetition of ancestral adult stages in embryonic or juvenile stages of descendants.

Rectum *(straight).* The terminal segment of the large intestine.

Renal *(kidney).* Relating to the kidney.

Replacement bone. Bone that replaces cartilage as it ossifies.

Resultant. The single vector that is equivalent to a given set of vectors.

Rete *(network).* A network of small vessels or fibers.

Reticular *(net).* Netlike in structure.

Reynolds number. The ratio of the inertia of a fluid medium to its viscosity. For fast-swimming vertebrates this dimensionless number ranges from $10^{5.5} - 10^{8.5}$, depending on length and velocity.

Rhinal *(nose).* Relating to the nose.

Rhombencephalon *(rhomboid + brain).* The most posterior of the three primary brain vesicles; the hindbrain.

Ricochet. A bipedal hopping gait using the hind legs in unison, as for the kangaroo.

Roll. Rotation of a swimmer or flyer around its horizontal (anteroposterior) axis.

Rumen *(throat).* The largest chamber of the ruminant stomach.

Ruminate. To chew the cud.

Saccule *(small sack).* The more ventral chamber of the inner ear.

Sacral *(sacred).* Pertaining to the region where the spine articulates with the pelvic girdle.

Sagittal. Said of planes that divide the body into right and left parts.

Saltatorial. Adapted for jumping or hopping.

Sarcolemma *(flesh + husk).* The membrane enclosing a muscle fiber.

Sarcopterygium *(flesh + fin).* A fin having a fleshy stalk.

Scansorial. Adapted for climbing.

Schizocoely *(split + hollow).* The formation of coelom by splitting of the hypomere.

Sclera *(hard).* The tough outer envelope of the eye.

Sclerotome *(hard + to cut).* The inner division of the epimere.

Scute *(shield).* Any large, flat, horny plate on the surface of a reptile.

Sebaceous *(tallow).* Relating to oil or fat.

Secondary palate. A palate that separates the respired airstream from the mouth cavity.

Serial homology. The correspondence of structures that occupy different spatial positions in a series of like structures.

Sesamoid bone *(resembling a sesame seed).* A bone imbedded in, or interrupting, a tendon.

Sexual homology. The correspondence of male and female structures that develop from identical embryonic primordia.

Shear. Stress in an elastic solid resulting from loads directed in opposite directions along parallel, closely adjacent lines.

Sinus *(cavity or hollow).* A cavity in an organ or tissue.

Sinus venosus. The most posterior of the primitive heart chambers.

Sinusoid *(cavity or hollow).* An expanded capillary, as found in the liver and certain glands.

Somatic *(body).* Relating to parts of the body other than the viscera.

Somatopleure *(body + side).* A membrane derived from ectoderm and the outer sheet of the hypomere.

Somite *(body + part).* The epimere, or segmented dorsal division of the embryonic lateral mesoderm.

Specialized character. A feature that has become modified to perform a restricted function, often with great effectiveness.

Species *(kind or sort).* A group of actually or potentially interbreeding natural populations that is reproductively isolated from other such groups.

Sperm *(seed).* The male gamete.

Spermatogonia *(seed + generation)*. The proliferating cells of the testis that will become sperm.

Sphincter *(to bind)*. A flat, washer-shaped muscle that restricts an orifice.

Spiracle *(breathing hole)*. An opening into the pharynx derived from the first gill cleft.

Splanchnocranium *(viscera + skull)*. The skeleton of the visceral arches.

Splanchnopleure *(viscera + side)*. A membrane derived from endoderm and the inner sheet of the hypomere.

Stall. Loss of lift by an airfoil.

Static soaring. Flight that is sustained by updrafts.

Statoacoustic *(standing + hearing)*. Denoting the eighth cranial nerve.

Stereocilia *(solid + eyelid)*. The numerous hairlike sensors of a neuromast cell.

Sternum *(chest)*. The breastbone.

Stomodeum *(mouth + road)*. Anterior invagination of the ectoderm that comes to line the mouth.

Strain. Deformation of an elastic solid that is caused by a load.

Stratum. A tissue layer.

Strength. Capacity of supportive material to resist force without breakage or permanent deformation; capacity of muscle to produce much force.

Stress. Pressure within an elastic solid that results from strain.

Stroma *(bed)*. The connective tissue framework of an organ, as of the ovary.

Subclavian *(under + collar bone)*. Located in the shoulder.

Sulcus *(furrow)*. A groove, as on the dorsal surface of the spinal cord or cerebrum.

Super-, supra-, sur-. Prefix meaning over, above.

Supinator *(to bend backward)*. A muscle that rotates the palms or soles upward.

Suture *(to sew)*. The union, or seam, between the bones at an immovable joint.

Sympathetic *(with + suffering)*. Denoting the part of the autonomic nervous system having thoracolumbar outflow.

Symphysis *(together + growth)*. An amphiarthrosis having a pad of collagenous fibers or fibrous cartilage separating the bones.

Synapse *(junction)*. The junction of nerve with nerve or nerve with muscle for the transmission of an impulse.

Synapsid *(together + arch)*. Having one opening in the temporal region of the skull that is bordered above by the postorbital and squamosal.

Synarthrosis *(together + joint)*. An immovable joint.

Syndesmosis *(together + tie)*. An amphiarthrosis having a band of collagenous fibers between the bones that allows moderate motion.

Synergist *(together + work)*. A muscle having about the same action as another.

Synovial. Relating to the viscid lubricating fluid occurring within joint capsules and tendon sheaths.

Synsacrum *(together + sacred)*. The unit of the avian skeleton formed by the fusion of the pelvis and numerous vertebrae.

Syrinx *(tube)*. The vocal organ of birds, located near the bifurcation of the primary bonchi.

Systemic. Relating to the parts of the circulatory system that are not associated with lungs or gills.

Taxon *(pl. **taxa**) (to arrange)*. A group of organisms recognized as a unit in classification.

Tectum *(roof)*. The roof of the midbrain.

Telencephalon *(end + brain)*. The anterior derivative of the embryonic prosencephalon; becomes the cerebrum.

Telolecithal *(end + yolk)*. Said of eggs having the yolk massed toward the vegetal hemisphere.

Tendon *(cord)*. A tough cord or band of connective tissue serving to join a muscle to a bone.

Tension. Stress in an elastic solid resulting from a load directed away from the object and perpendicular to its surface.

Testis *(testicle)*. The male gonad.

Thalamus *(bed)*. A lateral wall of the diencephalon.

Thecodont *(sheath + tooth)*. Having teeth rooted in sockets.

Thermal. A doughnut-shaped bubble of circulating warm air.

Thoracic *(chest)*. Pertaining to the chest, or region of the ribs.

Thrombocyte *(clot + cell)*. A kind of blood cell associated with the clotting process.

Thymus *(sweetbread)*. A gland in or near the neck that relates to the immune system.

Thyroid *(shield + form)*. An endocrine gland in the neck; a cartilage of the larynx.

Torque. Moment or turning force; the product of a force times the perpendicular distance between its line of action and the pivot around which an object turns, or tends to turn, as a consequence of its application.

Trabecula *(small beam)*. A small rod, bar, or support; a certain part of the chondrocranium; a spicule of bone.

Trachea *(windpipe).* The airway from the larynx to the bronchi.

Transverse. Said of planes that divide the body into anterior (head end) and posterior (tail end) parts.

Trigeminal *(three originating together).* Denoting the fifth cranial nerve.

Trochlear *(pulley).* Relating to the fourth cranial nerve, or to the bony grooves within certain joints.

Trophoblast *(nourish + germ).* The outer wall of the mammalian blastocyst.

Tunica *(coat).* A surrounding layer, as of a testis or blood vessel.

Truss. An assemblage of solid parts arranged to form a rigid framework.

Tympanum *(drum).* The eardrum.

Unguis *(hoof).* The tough, outer or lateral (not under) material of a hoof, claw, or nail.

Unguligrade *(hoof + walking).* Having only hoofs in contact with the ground when standing or moving, as for horses and deer.

Ureter *(urinary canal).* The duct of a metanephric kidney.

Urethra *(canal).* The duct that discharges urine from the bladder.

Urophysis *(tail + grow).* Ventral swelling on the caudal part of the spinal cord of many fishes.

Urodeum *(urine + way).* The ventral part of a partially divided cloaca.

Urostyle *(tail + pillar).* The rodlike bone forming the posterior end of the spine in Anura.

Uterus *(womb).* The organ in which the egg or fetus develops.

Utricule *(small bag).* The more dorsal chamber of the inner ear.

Vagina *(sheath).* The canal that receives the penis during copulation.

Vagus *(wandering).* Denoting the tenth cranial nerve.

Vector. A quantity that has both magnitude and direction, such as force or velocity; a line that represents such a quantity by its length and orientation.

Vegetal pole. The region of the egg where yolk is concentrated and metabolic activity is lowest.

Vein. A large vessel that carries blood toward the heart.

Velocity. Rate of change of position in a given direction.

Ventral. Said of the belly, or underside of the body.

Ventricle *(small cavity).* A cavity of the heart or brain.

Vesicle *(small bladder).* A small sac or space.

Vestigial *(a trace).* The state of a structure that was functional in an ancestor but is no longer useful and is reduced in size or complexity.

Visceral *(entrail)* **arch.** One of the bars separating adjacent gill slits or pharyngeal pouches; pertaining to viscera.

Viscus *(pl.* **viscera).** An internal organ, particularly of the abdominal cavity.

Vitelline *(yolk).* Relating to the yolk of an egg.

Viviparous *(alive + to bear).* Animals that give birth to "living" young.

Volant. Capable of sustained flight.

Wave drag. Drag on a boat or swimmer caused by the formation of surface waves.

Wing loading. The weight of a flyer divided by the area of its airfoils.

Work. The product of the tension of a muscle times the distance through which it acts.

Yaw. Rotation of a swimmer or flyer around its vertical (dorsoventral) axis.

Yolk sac. The fetal membrane that surrounds and absorbs yolk.

Zona pellucida *(girdle + clear).* A prominent egg membrane of mammals.

Zygapophysis *(yolk + away from + growth).* A process that joins one vertebra to the next.

Zygote *(yolked).* The fertilized egg.

References

PARTS I & II GENERAL REFERENCES

Clark, W. E. L. 1965. The tissues of the body; an introduction to the study of anatomy. Clarendon Press, Oxford. 423 p. Excellent for supportive, vascular, and nervous tissues and skin.

Colbert, E. H. 1969. Evolution of the vertebrates. 2nd ed. John Wiley & Sons, New York. 535 p. Good survey of vertebrate paleontology.

Cole, F. J. 1949. A history of comparative anatomy. Macmillan, London. 524 p.

DeBeer, G. R. 1958. Embryos and ancestors. 3rd ed. Oxford University Press, London. 197 p.

Dellman, H. D. 1971. Veterinary histology; an outline atlas. Lea & Febiger, Philadelphia. 305 p.

Dullemeijer, P. 1974. Concepts and approaches in animal morphology. Van Gorcum and Comp. B. V., Assen, Netherlands. 264 p. A somewhat esoteric consideration of method in relating form and function. Examples provide historical and philosophical perspective.

Gorden, M. S., G. A. Barthalomew, A. D., Grinnell, C. B. Jørgensen, and F. N. White. 1977. Animal physiology; principles and adaptations. 3rd ed. Macmillan, New York. 699 p.

Gould, S. J. 1966. Allometry and size in ontogeny and phylogeny. Biol. Rev. 41:587–640.

Gould, S. J. 1977. Ontogeny and phylogeny. Harvard University Press, Cambridge. 501 p. A significant and scholarly reexamination of a classical but ever important subject.

Grassé, P. P. (ed.). 1950–1970. Traité de zoologie; anatomie, systématique, biologie. Masson et Cie, Paris. Vertebrates, v. 13–17. A monumental work.

Hailman, J. P. 1976. Homology: logic, information, and efficiency, p. 181–198. *In* R. B. Masterson, W. Hodos, and H. Jerison (eds.), Evolution, brain, and behavior: persistent problems. Erebaum, Hillsdale, N.J.

Hildebrand, M. 1968. Anatomical preparations. University of California Press, Berkeley. 100 p.

Irvine, W. 1955. Apes, angels, and victorians; the story of Darwin, Huxley, and evolution. McGraw-Hill, New York. 399 p.

Jaeger, E. C. 1955. A source-book of biological names and terms. 3rd ed. Thomas, Springfield, Ill. 319 p.

Leake, L. D. 1975. Comparative histology: an introduction to the microscopic structure of animals. Academic Press, New York. 738 p. Vertebrates are covered in 290 p.

Løvtrup, S. 1977. The phylogeny of vertebrata. John Wiley & Sons, New York. 330 p. Cladistics is used to classify vertebrates and relate them more closely to mollusks and arthropods than to echinoderms.

Mayr, E. 1963. Animal species and evolution. Harvard University Press, Cambridge. 797 p.

Mayr, E. 1969. Principles of systematic zoology. McGraw-Hill, New York. 428 p.

Patt, D. I., and G. R. Patt. 1969. Comparative vertebrate histology. Harper & Row, New York. 428 p.

Pielou, E. C. 1979. Biogeography. John Wiley & Sons, New York. 351 p.

Piveteau, J. (ed.). 1952–1958. Traité de paléontologie. Masson et Cie, Paris. Vertebrates, v. 4–7.

Raup, D. M., and S. M. Stanley. 1971. Principles of paleontology. Freeman, San Francisco. 388 p.

Rensch, B. 1960. Evolution above the species level. Columbia University Press, New York. 419 p. (Originally in German, 1954.)

Rixon, A. E. 1976. Fossil animal remains: their preparation and conservation. Athlone Press, University of London. 304 p. Also a useful guide to the repair and mounting of skeletons.

Romer, A. S. 1959. The vertebrate story. 4th ed. University of Chicago Press, Chicago. 437 p. Structure of vertebrates, living and extinct.

Romer, A. S. 1966. Vertebrate paleontology. 3rd ed. University of Chicago Press, Chicago. 468 p. An advanced textbook with extensive summary of classification and list of references.

Simpson, G. G. 1949. The meaning of evolution. Yale University Press, Newhaven, Conn. 364 p. Excellent general reference, readable and nontechnical. An abridged paperback edition available from The New American Library.

Simpson, G. G. 1953. The major features of evolution. Columbia University Press, New York. 434 p.

Simpson, G. G. 1961. Principles of animal taxonomy. Columbia University Press, New York. 247 p.

Simpson, G. G., A. Roe, and R. C. Lewontin. 1960. Quantitative zoology. Rev. ed. Harcourt, Brace, Jovanovich, New York. 440 p.

Sneath, P. H. and R. R. Sokal. 1973. Numerical taxonomy; the principles and practice of numerical classification. Freeman, San Francisco. 573 p.

**CHAPTERS 3 & 4
VERTEBRATE
ANIMALS**

Brodal, A., and R. Fänge. 1963. The biology of Myxine. Universitets-forlaget, Oslo. 588 p.

Gans, C. A., d'A. Bellairs, and T. S. Parsons. 1969–. Biology of the reptilia. Academic Press, New York. Many volumes published; others projected.

Greenwood, P. H., R. S. Miles, and C. Patterson (eds.). 1973. Interrelationships of fishes. Zool. J. Linnean Soc. 53, Suppl. 1. 536 p.

Hardesty, M. W., and I. C. Potter. 1971–1972. The biology of lampreys. Academic Press, New York. 2 v.

Marshall, N. B. 1965. The life of fishes. Weidenfeld and Nicolson, London. 402 p.

McFarland, W. N., F. H. Pough, T. J. Cade, and J. B. Heiser. 1979. Vertebrate life. Macmillan, New York. 875 p.

Moy-Thomas, J. A. 1939. Revised by R. S. Miles. 1971. Palaeozoic fishes. 2nd ed. Saunders, Philadelphia. 259 p.

Noble, G. K. 1931. The biology of the amphibia. McGraw-Hill, New York. 577 p. (Republished by Dover, New York. 1954.)

Parsons, T. S., and E. E. Williams. 1963. The relationships of the modern amphibia: a re-examination. Quart. Rev. Biol. 38:26–53.

Simpson, G. G. 1945. The principles of classification and a classification of mammals. Am. Mus. Nat. Hist., Bull, 85, 350 p. Still an authoritative classification of living and extinct mammals.

Stahl, B. J. 1974. Vertebrate history: problems in evolution. McGraw-Hill, New York. 594 p. A paleontology text oriented to unresolved problems.

Thompson, K. S. 1969. The biology of lobe-finned fishes. Biol. Rev. 44:91–154. Evaluation of theories on evolution and function.

Thomson, K. S. 1971. The adaptation and evolution of early fishes. Quart. Rev. Biol. 46:139–166.

Van Tyne, J., and A. J. Berger. 1976. Fundamentals of ornithology. 2nd ed. John Wiley & Sons, New York. 808 p. Includes a chapter on anatomy and an account of each family.

Welty, J. C. 1975. The life of birds. 2nd ed. Saunders, Philadelphia. 623 p.

Young, J. Z. 1957. The life of mammals. Oxford University Press, New York and Oxford. 820 p. Broad interpretation of mammalian morphology.

Hamilton, W. J., and H. W. Mossman. 1972. Hamilton, Boyd and Mossman's human embryology; prenatal development of form and function. 4th ed. Heffer, Cambridge. 646 p. Outstanding illustrations. Includes comparative placentation.

Hörstadius, S. 1950. The neural crest. Oxford University Press, London. 111 p.

Luckett, W. P. 1977. Ontogeny of amniote fetal membranes and their application to phylogeny, p. 439–516. *In* M. K. Hecht, P. C. Goody, and B. M. Hecht (eds), Major patterns in vertebrate evolution. Plenum Press, New York.

Nelson, O. E. 1953. Comparative embryology of the vertebrates. Blakiston, New York. 982 p. A broad and comprehensive book; profusely illustrated.

Patten, B. M. and B. M. Carlson. 1974. Foundations of embryology. 3rd ed. McGraw-Hill, New York. 650 p.

Torrey, T. W. and A. Feduccia. 1979. Morphogenesis of the vertebrates. 4th ed. John Wiley & Sons, New York. 570 p. Combines anatomy and embryology.

**CHAPTER 5
EARLY
DEVELOPMENT**

Halstead, L. B. 1974. Vertebrate hard tissues. Wykeham Pub., London. 179 p.

Koller, E. J. 1972. The development of the integument: spatial, temporal, and phylogenetic factors. Am. Zool. 12:125–136.

Maderson, P. F. A. 1965. The structure and development of the squamate epidermis, p. 129–153. *In* A. G. Lyne and B. F. Short (eds.), The biology of the skin and hair growth. American Elsevier, New York.

**CHAPTER 6
INTEGUMENT
AND ITS
DERIVATIVES**

Maderson, P. F. A. 1967. A comment on the evolutionary origin of vertebrate appendages. Am. Naturalist 101:71–78.

Maderson, P. F. A. (ed.). 1972. The vertebrate integument. Symposium *in* Am. Zool. 12:12–171. Excellent collection of articles.

Montagna, W. 1956. The structure and function of skin. Academic Press, New York. 356 p. Only human skin is described.

Moss, M. L. 1968. Bone, dentin, and enamel and the evolution of vertebrates, p. 37–65. *In* P. Person (ed.), Biology of the mouth. Am. Assoc. for the Advancement of Sci., Washington.

Ørvig, T. 1951. Histologic studies of placoderms and fossil elasmobranchs. I. The endoskeleton, with remarks on the hard tissues of lower vertebrates in general. Arch. för Zool., Ser. 2, 2:321–454. An important work on the classification of hard tissues with a theory on the origin of dermal armor.

Parakkal, P. F., and N. J. Alexander. 1972. Keratinization; a survey of vertebrate epithelia. Academic Press, New York. 59 p. Excellent illustrations.

Quay, W. B. 1972. Integument and the environment: glandular composition, function, and evolution. Am. Zool. 12:95–108.

Regal, P. J. 1975. The evolutionary origin of feathers. Quart. Rev. Biol. 50:35–66.

Sengel, P. 1976. Morphogenesis of skin. Cambridge University Press, London. 277 p. Covers amniotes.

CHAPTER 7
TEETH

Butler, P. M. 1956. The ontogeny of molar pattern. Biol. Rev. Cambridge Philosophical Soc. 31:30–70.

Edmund, A. G. 1960. Tooth replacement phenomena in the lower vertebrates. Roy. Ontario Museum, Toronto. Life Sciences Division, Contribution 52. 190 p.

Gregory, W. K. 1934. A half century of trituberculy. The Cope-Osborn theory of dental evolution, with a revised summary of molar evolution from fish to man. Am. Philosophical Soc., Philadelphia, Proc. 73:169–317.

Kurten, B. 1954. Observations on allometry in mammalian dentitions; its interpretation and evolutionary significance. Acta Zool. Fennica 85:1–13.

Moss, M. L. 1970. Enamel and bone in shark teeth: with a note on fibrous enamel in fishes. Acta Anat. 77:161–187.

Osborn, J. W. 1974. On the control of tooth replacement in reptiles and its relationship to growth. J. Theor. Biol. 46:509–527.

Osborn, J. W., and A. W. Crompton. 1973. The evolution of mammalian from reptilian dentitions. Brevoria 399:1–18.

Peyer, B. (translated and edited by R. Zangerl). 1968. Comparative odontology. University of Chicago Press, Chicago. 347 p. Includes discussion of the nature of hard tissues.

CHAPTER 8
HEAD
SKELETON

Crompton, A. W., and P. Parker. 1978. Evolution of the mammalian masticatory apparatus. Am. Sci. 66:192–201. Evolution of the jaw and ear, and of their functions.

De Beer, G. R. 1971. The development of the vertebrate skull. Oxford University Press, London and New York. 554 p. (Originally published in 1937.)

Edwards, J. L. 1977. The evolution of terrestrial locomotion, p. 553–557. *In* M. K. Hecht, P. C. Goody, and B. M. Hecht (eds.), Major patterns in vertebrate evolution. Plenum Press, New York.

Frazzetta, T. H. 1968. Adaptive problems and possibilities in the temporal fenestration of tetrapod skulls. J. Morphol. 125:145–158.

Hall, B. K. 1975. Evolutionary consequences of skeletal differentiation. Am. Zool. 15:329–350.

Hall, B. K. 1978. Developmental and cellular skeletal biology. Academic Press, New York. 304 p.

Halstead, L. B. 1974. Vertebrate hard tissues. Wykeham Pub., London. 179 p.

Hunt, R. M., Jr. 1974. The auditory bulla in Carnivora: an anatomical basis for reappraisal of carnivore evolution. J. Morph. 143:21–76.

Jollie, M. T. 1977. Segmentation of the vertebrate head. Am. Zool. 17:323–333.

Miles, R. S. 1968. Jaw articulation and suspension in *Acanthodes* and their significance, p. 109–127. *In* T. Ørvig, (ed.), Current problems of lower vertebrate phylogeny. Wiley-Interscience, New York.

Moss, M. L. 1968. The origin of vertebrate calcified tissues, p. 360–371. *In* T. Ørvig (ed.), Current problems of lower vertebrate phylogeny. Wiley-Interscience, New York.

Parrington, F. R. 1949. A theory of the relations of lateral lines to dermal bones. Zool. Soc. London, Proc. 119:65–78.

Parrington, F. R. 1956. The patterns of dermal bones in primitive vertebrates. Zool. Soc. London, Proc. 127:389–411. Interpretive, not descriptive.

Parrington, F. R., and T. S. Westoll. 1941. On the evolution of the mammalian palate. Roy. Soc. London, Philosophical Trans. Ser. B 230:305–355.

Romer, A. S. 1956. Osteology of the reptiles. University of Chicago Press, Chicago. 772 p.

**CHAPTER 9
BODY SKELETON**

Burt, W. H. 1960. Bacula of North American mammals. University of Michigan Museum Zool., Misc. Publ. 113:1–76.

Eaton, T. H. 1951. Origin of tetrapod limbs. Am. Midland Naturalist 46:245–251.

Enlow, D. H. 1962. A study of the post-natal growth and remodeling of bone. Am. J. Anat. 110:79–102.

Enlow, D. H., and D. B. Harris. 1964. A study of the postnatal growth of the human mandible. Am. J. Orthodontics 50:25–50. Analysis of remodeling during growth.

Gunter, G. 1956. Origin of the tetrapod limb. Science 123:495–496.

Haines, R. W. 1942. The evolution of epiphyses and of endochondral bone. Biology Rev. 17:267–292.

Jenkins, F. A., Jr. 1971. Limb posture and locomotion in the Virginia opossum (*Didelphis marsupialis)* and in other non-cursorial mammals. J. Zool. 165:303–305.

Lacroix, P. 1951. The organization of bones. Blakiston, Philadelphia. 235 p. Excellent account of histology, development, and growth of long bones.

Montagna, W. 1945. A reinvestigation of the wing of the fowl. J. Morphol. 76:87–113.

Mookerjee, H. K. 1936. The development of the vertebral column and its bearing on the study of organic evolution. Indian Sci. Cong., Proc. 23:307–343. Review article by a major contributor to the field.

Panchen, A. L. 1977. The origin and early evolution of tetrapod vertebrae. Linnean Soc. Symp. Ser. 4:289–318.

Ridewood, W. G. 1921. On the calcification of the vertebral centra in sharks and rays. Roy. Soc. London, Philosophical Trans., Ser. B, 210:311–407.

Schaeffer, B. 1941. The morphological and functional evolution of the tarsus in amphibians and reptiles. Am. Mus. of Nat. Hist., Bull. 78:395–472.

Schaeffer, B. 1967. Osteichthyan vertebrae. Linnean Soc. London, Zool. J. 47:185–195.

Verbout, A. J. 1976. A critical review of the "Neugliederung" concept in relation to the development of the vertebral column. Acta Biotheoretica 25:219–258. The concept of the recombination of sclerotomal segments in the ontogeny of vertebrae is analyzed and rejected.

Wake, D. B. 1970. Aspects of vertebral evolution in the modern Amphibia. Forma et Functio 3:33–60.

Webb, J. E. 1973. The role of the notochord in forward and reverse swimming and burrowing in the amphioxus *Branchiostoma lanceolatum*. J. Zool., London 170:325–338.

Westoll, T. S. 1958. The lateral fin-fold theory and the pectoral fins of ostracoderms and early fishes, p. 180–211. *In* T. S. Westoll (ed.), Studies on fossil vertebrates. University of London Press, London.

Williams, E. E. 1959. Gadow's arcualia and the development of tetrapod vertebrae. Quart. Rev. Biol. 34:1–32. Important reevaluation of a long-held theory.

CHAPTER 10 MUSCLES AND ELECTRIC ORGANS

Cheng, Cze-Ching. 1955. The development of the shoulder region of the opossum *Didelphis virginiana*, with special reference to the musculature. J. Morphol. 97:415–472. Illustrates use of embryology for establishing homologies of muscles.

Goslow, G. E., R. M. Reinking, and D. G. Stuart. 1973. The cat step cycle: Hind limb joint angles and muscle lengths during unrestrained locomotion. J. Morph. 141:1–42.

Grundfest, H. 1960. Electric fishes. Sci. Am. 203(4); 115–124.

Howell, A. B. 1937. Morphogenesis of the shoulder architecture: Part VI, Therian mammalia. Quart. Rev. Biol. 12:440–463.

Huddart, H. 1975. The comparative structure and function of muscle. Pergamon Press, New York. 397 p.

Huxley, H. E. 1958. The contraction of muscle. Sci. Am. 199 (5):67–82.

Lissmann, H. W. 1958. On the function and evolution of electric organs in fish. J. Exp. Biol. 35:156–191.

Murray, J. M., and A. Weber. 1974. The cooperative action of muscle proteins. Sci. Am. 230(2):58–71.

Porter, K. R., and C. Franzini-Armstrong. 1965. The sarcoplasmic reticulum. Sci. Am. 212(3):73–80.

Straus, W. L., Jr. 1946. The concept of nerve-muscle specificity. Biol. Rev. 21:75–91.

Sullivan, G. E. 1962. Anatomy and embryology of the wing musculature of the domestic fowl (*Gallus*). Australian J. Zool. 10:458–518.

CHAPTERS 11 & 12 DIGESTIVE SYSTEM AND COELOM

Chivers, D. J., and C. M. Hladik. 1980. Morphology of the gastrointestinal tract in primates; comparisons with other mammals in relation to diet. J. Morph. 166:337–386.

Clark, R. B. 1964. Dynamics in metazoan evolution; the origin of the coelom and its segments. Clarendon Press, Oxford. 313 p.

Doran, G. A. 1975. Review of the evolution and phylogeny of the mammalian tongue. Acta Anat. 91: 118–129.

Elias, H. 1955. Liver morphology. Biol. Rev. 30:263–310. Review article by a leading student of the liver.

Goodrich, E. S. 1930. Studies on the structure and development of vertebrates. Macmillan, London. 837 p. (Republished by Dover, New York, 1958.) Coelom receives 44 p.

Gorham, F. W., and A. C. Ivy. 1938. General function of the gall bladder from the evolutionary standpoint. Field Museum Nat. Hist., Chicago, Zool. Ser. 22:159–213.

Hill, W. C. O. 1926. A comparative study of the pancreas. Zool. Soc. London, Proc. 1926:581–631.

Mitchell, P. C. 1901. On the intestinal tract of birds. Linnaean Soc. London, Trans. (Zool.), Ser. 2, 8:173–275.

Mitchell, P. C. 1916. Further observations on the intestinal tract of mammals. Zool. Soc. London, Proc. 1916:183–251. Broadly comparative study of gross structure.

Pernkopf, E. 1930. Beiträge zur vergleichende Anatomie des vertebraten Magens. Z. Anat. 91:329–390. Reference on comparative structure of the stomach.

Slijper, E. J. 1946. Die physiologische Anatomie der Verdauungsorgane bei den Vertebraten. Tabulae Biologicae 21, Pars 1:1–81. Tabulated data based on 280 references.

Tucker, R. 1958. Taxonomy of vertebrate salivary glands. Systematic Zool. 7:74–83.

Warner, E. D. 1958. The organogenesis and early histogenesis of the bovine stomach. Am. J. Anat. 102:33–63. General comments on ontogenetic and phylogenetic origins of the parts of mammalian stomachs.

Atz, J. W. 1952. Narial breathing in fishes and the evolution of internal nares. Quart. Rev. Biol. 27:366–367.

Demski, L. S., J. W. Gerald, and A. N. Popper. 1973. Central and peripheral mechanisms of teleost sound production. Am. Zool. 13:1141–1167.

Duncker, H. R. 1971. The lung air sac system of birds: a contribution to the functional anatomy of the respiratory apparatus. Ergebnisse der Anatomie und Entwicklungsgeschichte 45(6). 171 p. Excellent coverage and superb illustrations.

Duncker, H. R. 1974. Structure of the avian respiratory tract. Respiration Physiol. 22:1–19.

Duncker, H. R. 1978. General morphological principles of amniote lungs, p. 2–15. In J. Piiper (ed.), Respiratory function in birds, adult and embryonic. Springer-Verlag, New York.

Gans, C. 1970. Respiration in early tetrapods—the frog as a red herring. Evolution 24:723–734.

Gans, C. 1970. Strategy and sequence in the evolution of the external gas exchangers of ectothermal vertebrates. Forma et Functio 3:61–104.

Hughes, G. M. 1963. The comparative physiology of vertebrate respiration. Harvard University Press, Cambridge. 145 p.

Hughes, G. M. and M. Morgan. 1973. The structure of fish gills in relation to their respiratory function. Biol. Rev. 48:419–475.

Jones, F. R. H., and N. B. Marshall. 1953. The structure and functions of the teleostean swim bladder. Biol. Rev. 28:16–83. Comprehensive.

McLaughlin, R. F., W. S. Tyler, and R. O. Canada. 1961. A study of the subgross pulmonary anatomy in various mammals. Am. J. Anat. 108:149–158.

Schmidt-Nielsen, K. 1972. How animals work. Cambridge University Press, New York. 114 p. Concerns respiration, thermoregulation, and countercurrent exchange.

Staub, N. C. 1963. The interdependence of pulmonary structure and function. Anesthesiology 24:831–854.

Tenney, S. M., and J. E. Remmers. 1963. Comparative quantitative morphology of the mammalian lung: diffusing area. Nature 197:54–56, Jan.

**CHAPTER 13
RESPIRATORY
SYSTEMS AND
GAS BLADDER**

Adolph, E. F. 1967. The heart's pacemaker. Sci. Am. 216(3):32–37.

Baker, M. A. 1979. A brain-cooling system in mammals. Sci. Am. 240(5):130–139. The relation between nasal veins, carotid rete, and brain temperature.

Barnett, C. H., R. J. Harrison, and J. D. W. Tomlinson. 1958. Variations in the venous systems of mammals. Biol. Rev. 33:442–487. A review, with extensive bibliography.

**CHAPTER 14
CIRCULATORY
SYSTEM**

Carey, F. G. 1973. Fishes with warm bodies. Sci. Am. 228(2):36–44. The countercurrent exchange system of dark muscle.

Foxon, G. E. H. 1955. Problems of the double circulation in vertebrates. Biol. Rev. 30:196–228. Review of heart structure in relation to function.

Holmes, E. B. 1975. A reconsideration of the phylogeny of the tetrapod heart. J. Morphol. 147:209–228.

Jordan, H. E. 1933. The evolution of blood-forming tissues. Quart. Rev. Biol. 8:58–76.

Kampmeier, O. F. 1969. Evolution and comparative morphology of the lymphatic system. Thomas, Springfield, Ill. 620 p.

Mossman, H. W. 1948. Circulatory cycles in the vertebrates. Biol. Rev. 23:237–255. Analysis of general structure and function of elements in the system.

Quiring, D. P. 1949. Collateral circulation. Lea & Febiger, Philadelphia. 142 p. Describes functional importance of an overlooked aspect of the circulatory system.

Rennick, B. R., and H. Gandia. 1954. Pharmacology of smooth muscle valve in renal portal circulation of birds. Soc. Exptl. Biol. and Med., Proc. 85:234–236.

Satchell, G. H. 1971. Circulation in fishes. Cambridge University Press. 131 p. Summary of current knowledge of anatomy and physiology.

White, F. N. 1968. Functional anatomy of the heart of reptiles. Am. Zool. 8:211–219.

Yoffey, J. M., and F. C. Courtice. 1956. Lymphatics, lymph and lymphoid tissue. Edward Arnold, London. 510 p.

**CHAPTER 15
EXCRETORY
SYSTEM
AND OSMO-
REGULATION**

Fraser, E. A. 1950. The development of the vertebrate excretory system. Bio. Rev. 25:159–187.

Fox, H. 1963. The amphibian pronephros. Quart. Rev. Biol. 38:1–25.

Hickman, C. P., Jr., and B. F. Trump. 1969. The kidney, v. 1:91–239. *In* Fish physiology, W. S. Hoar and D. J. Randall (eds.), Academic Press, New York.

Hicks, R. M. 1975. The mammalian urinary bladder: an accommodating organ. Biol. Rev. 50:215–246.

Marshall, E. K., Jr. 1934. The comparative physiology of the kidney in relation to theories of renal secretion. Physiol. Rev. 14:133–159.

Pang, P. K. T., R. W. Griffith, and J. W. Atz. 1977. Osmoregulation in elasmobranchs. Am. Sci. 17:365–377.

Peaker, M., and J. L. Linzell. 1975. Salt glands in birds and reptiles. Cambridge University Press, London. 307 p.

Smith, H. W. 1932. Water regulation and its evolution in the fishes. Quart. Rev. Biol. 7:1–26. Explains physiological basis of kidney structure and function.

Sperber, I. 1944. Studies on the mammalian kidney. Zool. Bidrag från Uppsala 22:249–431. Excellent interpretation of variation in gross and subgross structure.

**CHAPTER 16
REPRO-
DUCTIVE
SYSTEM**

Asdell, S. A. 1964. Patterns of mammalian reproduction. 2nd ed. Cornell University Press, Ithaca, N. Y. 670 p. Reference work of wide coverage.

Cowles, R. B. 1958. The evolutionary significance of the scrotum. Evolution 12:417–418.

Fox, H. 1977. The urogenital system of reptiles, v. 6:1–157. *In* C. Gans (ed.), Biology of the Reptilia, Academic Press, New York.

Hoar, W. S. 1969. Reproduction, v. 3:1–72. *In* Fish physiology, W. S. Hoar and D. J. Randall (eds.), Academic Press, New York.

Jones, R. E. (ed.). 1978. The vertebrate ovary: comparative biology and evolution. Plenum Press, New York. 853 p.

Lofts, B. 1968. Patterns of testicular activity, p. 239–304. *In* E. J. W. Barrington, and C. B. Jorgensen (eds.), Perspectives in endocrinology. Academic Press, New York.

Lofts, B. 1974. Reproduction, p. 107–218. *In* B. Lofts (ed.), Physiology of the amphibia. Academic Press, New York.

Packard, G. C. 1977. The physiological ecology of reptilian eggs and embryos, and the evolution of viviparity within the class Reptilia. Biol. Rev. 52:71–105.

Sadleir, R. M. F. S. 1973. The reproduction of vertebrates. Academic Press, New York. 227 p.

Bernstein, J. J. 1970. Anatomy and physiology of the central nervous system, p. 1–90. *In* W. S. Hoar, and D. J. Randall (eds.), Fish physiology. Academic Press, New York.

Gardner, E. 1975. Fundamentals of neurology. 6th ed. Saunders, Philadelphia. 460 p. A good text on human neuroanatomy.

Igarashi, S., and T. Kamiya. 1972. Atlas of the vertebrate brain: morphological evolution from cyclostomes to mammals. University Park Press, Baltimore. 126 p.

Isaacson, R. L. 1974. The limbic system. Plenum Press, New York. 292 p.

Llinás, R. R. 1975. The cortex of the cerebellum. Sci. Am. 232(1):56–71.

Luria, A. R. 1970. The functional organization of the brain. Sci. Am. 222(3):66–78.

Morell, P., and W. T. Norton. 1980. Myelin. Sci. Am. 242(5):88–118.

Nieuwenhuys, R., J. Voogd, and C. van Huijzen. 1978. The human central nervous system. Springer-Verlag, New York. 253 p. Superb illustrations of structure and pathways.

Noback, C. R., and R. J. Demarest. 1977. The nervous system: introduction and review. 2nd ed. McGraw-Hill, New York. 225 p. A clear, brief, well-illustrated account of the human nervous system emphasizing pathways.

Pearson, R. G., and L. Pearson. 1976. The vertebrate brain. Academic Press, New York. 744 p. An advanced and broadly comparative reference.

Sarnat, H. B., and M. G. Netsky. 1974. Evolution of the nervous system. Oxford University Press, New York. 318 p. An excellent source for comparative and functional material.

Shepherd, G. M. 1978. Microcircuits in the nervous system. Sci. Am. 238(2):93–103.

Stevens, C. E. 1979. The neuron. Sci. Am. 241(3):55–65. (This issue is devoted entirely to the brain.)

CHAPTERS 17 & 18 NERVOUS SYSTEM

Ali, M. A. (ed.). 1978. Sensory ecology: review and perspectives. NATO Advanced Study Institute Series 18. Plenum Press, New York. 597 p. Broadly comparative. Bibliographies.

Allison, A. C. 1953. The morphology of the olfactory system in vertebrates. Biol. Rev. 28:195–244. Excellent review article.

Am. Soc. Zool., symposium. 1966. The vertebrate ear. Am. Zool. 6:368–466.

Bakhtin, Ye. K. 1976. Morphology of the olfactory organ of some fish species and a possible functional interpretation. J. Ichthyology 16:786–804.

Crescitelli, F. 1977. The visual system in vertebrates. Springer Verlag, New York. 813 p. Includes chapters on adaptations to the deep sea, vision in turtles, the pineal system, the avian eye, and comparative optics in mammals.

Cahn, P. H. (ed.). 1967. Lateral line detectors. Indiana University Press, Bloomington. 512 p. Research reports by many authors.

Detwiler, S. R. 1956. The eye and its structural adaptations. Am. Scientist 44:45–72.

CHAPTER 19 SENSE ORGANS

Dijkgraaf, S. 1963. The functioning and significance of the lateral-line organs. Biol. Rev. 38:51–105.

Eakin, R. M. 1970. A third eye. Am. Scientist 58:73–80.

Fleischer, G. 1978. Evolutionary principles of the mammalian middle ear. Advances in anatomy, embryology and cell biology, v. 55, pt. 5. 70 p.

Gamow, I., and J. F. Harris. 1973. The infrared receptors of snakes. Sci. Am. 228(5):94–100.

Kleerekoper, H. 1969. Olfaction in fishes. Indiana University Press, Bloomington. 222 p. Covers structure and behavior.

Smith, C. A., and T. Takasaka. 1971. Auditory receptor organs of reptiles, birds, and mammals, v. 5:129–178. *In* W. D. Neff (ed.), Contributions to sensory physiology. Academic Press, New York.

Walls, G. L. 1967. The vertebrate eye and its adaptive radiation. Haffner, New York. 785 p. (Originally published 1942.) Still the best reference on the comparative morphology of the eye.

Wever, E. G. 1978. The reptile ear: its structure and function. Princeton University Press, Princeton, N.J. 1024 p. Exhaustive, but with chapters on general structure and sound transmission.

**CHAPTER 20
ENDOCRINE
GLANDS**

Barrington, E. J. W. 1975. An introduction to general and comparative endocrinology. 2nd ed. Oxford University Press, New York. 281 p.

Bentley, P. J. 1976. Comparative vertebrate endocrinology. Cambridge University Press, New York. 415 p.

Bern, H. A. 1967. Hormones and endocrine glands of fishes. Science 158:455–462.

Gorbman, A. (ed.). 1967. Recent developments in endocrinology. Am. Zool. 7:81–169.

Gorbman, A., and H. A. Bern. 1962. A textbook of comparative endocrinology. John Wiley & Sons, New York. 468 p.

Pang, P. K. T., and A. Epple (eds.) 1980. Evolution of vertebrate endocrine systems. Texas Tech Press, Lubbock, Texas. 404 p.

Quay, W. B. 1969. The significance of the pineal. Soc. Endocrinology, Mem. 18:423–445.

Turner, C. D., and J. T. Bagnara. 1976. General endocrinology. 6th ed. Saunders, Philadelphia. 596 p.

**PART III
GENERAL
REFERENCES**

Alexander, R. McN. 1967. Functional design in fishes. Hutchinson, London. 160 p. Excellent; discusses swimming, buoyancy, respiration, feeding, and sense organs.

Alexander, R. McN. 1968. Animal mechanics. University of Washington Press, Seattle. 346 p. Far-ranging application of mechanical principles to animal functions.

Dyson, G. H. G. 1977. Mechanics of athletics. 7th ed. Holmes & Meier Publishers, New York. 267 p.

Gans, C. 1974. Biomechanics: an approach to vertebrate biology. J. P. Lippincott, Philadelphia. 261 p.

Gray, J. 1968. Animal locomotion. Norton, New York. 479 p. Application of mechanics to locomotion. Excellent source on fishes, amphibians, snakes, and birds. Extensive bibliography.

Hertel, H. 1963. Structure, form, and movement. Otto Krausskopf-Verlag, Germany. English edition, 1966, Reinhold, New York. 251 p. An engineer's analysis of body mechanics. Emphasis on swimming and flying.

Oxnard, C. E. 1975. Uniqueness and diversity in human evolution: morphometric studies of Australopithecines. University of Chicago Press, Chicago. 133 p. Analyzes the relationship between form and function, giving methods of study.

Smith, J. M., and R. J. C. Savage. 1956. Some locomotory adaptations in mammals. J. Linnean Soc., Zool. 42:603–622. Study of torque and gait in analysis of structure.

Walker, E. P. 1975. Mammals of the world. 3rd ed. Johns Hopkins, Baltimore, 2 v. All genera are described and illustrated by photographs. Invaluable for visualizing body form.

Wainwright, S. A., W. D. Biggs, J. D. Currey, and J. M. Gosline. 1976. Mechanical design in organisms. John Wiley & Sons, New York. 423 p.

Bassett, C. A. 1965. Electrical effects in bone. Sci. Am. 213(4); 18–25.

Bock, W. J., and B. Kummer. 1968. The avian mandible as a structural girder. J. Biomechanics 1:89–96.

Currey, J. D. 1962. Stress concentrations in bone. Quart. J. Microscopical Sci. 103:111–133. Analysis of the affect of lacunae, pits, and projections on the strength of bone.

Elliott, D. H. 1965. Structure and function of mammalian tendon. Biol. Rev. 40:393–491. Review article with extensive bibliography.

Evans, F. G. 1957. Stress and strain in bones. Thomas, Springfield, Ill. 243 p.

Frost, H. M. 1967. An introduction to biomechanics. Thomas, Springfield, Ill. 151 p. Good summary of properties of structural tissues, but does not include bone-muscle systems.

Gardner, E. 1950. Physiology of movable joints. Physiol. Rev. 30:127–176. Review article with extensive bibliography.

Hall, M. C. 1966. The architecture of bone. Thomas, Springfield, Ill. 346 p. Illustrates bone sections showing internal structure.

Herring, S. W. 1972. Sutures—a tool in functional cranial analysis. Acta Anat. 83:222–247.

Kummer, B. 1976. Biomechanics of the mammalian skeleton. Problems of static stress. Fortschritte der Zool. 24(2/3):57–73.

Vis, J. H. 1957. Histological investigations into the attachment of tendons and ligaments to the mammalian skeleton. Koninkl. Ned. Akad. van Wetenschappen, Proc. Ser. C, 60:147–157.

CHAPTER 21 STRUCTURAL ELEMENTS OF THE BODY

Dempster, W. T. and R. A. Duddles. 1964. Tooth statics: equilibrium of a free-body. J. Am. Dental Assoc. 68:652–666. Explanation and application of the use of free-body diagrams.

Frohlich, C. 1980. The physics of somersaulting and twisting. Sci. Am. 242(3):155–164.

Gans, C., and W. J. Bock. 1965. The functional significance of muscle architecture—a theoretical analysis. Ergebnisse der Anatomie und Entwicklungsgeschichte 38, IV:116–142.

Gregory, W. K. 1912. Notes on the principles of quadrupedal locomotion and on the mechanism of the limbs of hoofed animals. New York Acad. Sci., Ann. 22:267–294. Excellent for graviportal adaptations.

Kummer, B. 1959. Bauprinzipien des Säugerskeletes. Georg Thieme Verlag, Stuttgart. 235 p. Mechanics of support of the static skeleton with emphasis on the spine and femur.

CHAPTER 22 MECHANICS OF SUPPORT AND MOVEMENT

Miller, D. I. and R. C. Nelson. 1973. Biomechanics of sport: a research approach. Lea & Febiger, Philadelphia. 265 p. Includes use of free-body diagrams.

Osborn, H. F. 1929. The titanotheres of ancient Wyoming, Dakota, and Nebraska. U.S. Geological Survey, Monograph 55, v. 2:703–945. See Chapter IX for an analysis of graviportal adaptations. Many illustrations.

Slijper, E. J. 1946. Comparative biologic-anatomical investigations on the vertebral column and spinal musculature of mammals. Akad. van Wetenschappen, Afd. Natuurkunde, Tweede sectië, 42(5), 128 p.

Tricker, R. A. R., and B. J. K. Tricker. 1967. The science of movement. American Elsevier, New York. 284 p. Simple presentation of principles of mechanics as applied to athletics.

Wentink, G. H. 1979. Dynamics of the hind limb at walk in horse and dog. Anat. and Embryol. 155:179–190. A good example of eclectic research.

Williams, M., and H. R. Lissner. 1968. Biomechanics of human motion. Saunders, Philadelphia. 147 p. Good application of mechanical principles to physical therapy.

**CHAPTER 23
RUNNING AND
JUMPING**

Camp, C. L., and N. Smith. 1942. Phylogeny and function of the digital ligaments of the horse. University of California, Mem. 13:69–124.

Dawson, T. J., and C. R. Taylor. 1973. Energetic cost of locomotion in kangaroos. Nature 246:313–314.

Eaton, T. H., Jr. 1944. Modifications of the shoulder girdle related to reach and stride in mammals. J. Morphol. 75:167–171.

Gambaryan, P. P. 1974. How mammals run: anatomical adaptations. John Wiley & Sons, New York. 367 p. (Originally published in Russian in 1972.)

Hall-Craggs, E. C. B. 1965. An analysis of the jump of the lesser galago (*Galago senegalensis*). Zool. Soc. London, Proc. 147:20–29.

Hatt, R. T. 1932. The vertebral columns of ricochetal rodents. Am. Mus. Nat. Hist., Bull. 63, Article 6:599–738. Extensive survey with functional interpretation.

Hildebrand, M. 1980. The adaptive significance of tetrapod gait selection. Am. Zool. 20:255–267.

Howell, A. B. 1965. Speed in animals. Their specializations for running and leaping. Hafner, New York. 270 p. (Originally published in 1944.)

Jenkins, F. A., and S. M. Camazine. 1977. Hip structure and locomotion in ambulatory and cursorial carnivores. J. Zool., London 181:351–370.

Schmidt-Nielsen, K. 1972. Locomotion: energy cost of swimming, flying, and running. Sci. 177:222–227.

Snyder, R. C. 1962. Adaptations for bipedal locomotion of lizards. Am. Zool. 2:191–203.

Taylor, C. R. 1978. Why change gait? Recruitment of muscles and muscle fibers as a function of speed and gait. Am. Zool. 18:153–161.

Taylor, C. R., K. Schmidt-Nielsen, and J. L. Raab. 1970. Scaling of energetic cost of running to body size in mammals. Am. J. Physiol. 219:1104–1107.

**CHAPTER 24
DIGGING; AND
CRAWLING
WITHOUT
APPENDAGES**

Agrawal, V. C. 1967. Skull adaptations in fossorial rodents. Mammalia 31:300–312.

Chapman, R. N. 1919. A study of the correlation of the pelvic structure and the habits of certain burrowing mammals. Am. J. Anat. 25:185–219.

Ellerman, J. R. 1959. The subterranean mammals of the world. Roy. Soc. of South Africa., Trans. 35:11–20.

Emerson, S. B. 1976. Burrowing in frogs. J. Morphol. 149:437–458.

Gans, C. 1968. Relative success of divergent pathways in amphisbaenian specialization. Am. Naturalist 102:345–362.

Gans, C. 1970. How snakes move. Sci. Am. 222(6):82–86, 88, 93–96.

Gans, C. 1975. Tetrapod limblessness: evolution and functional corollaries. Am. Zool. 15:455–467.

Lehmann, W. H. 1963. The forelimb architecture of some fossorial rodents. J. Morphol. 113:59–76.

McNab, B. K. 1979. The influence of body size on the energetics and distribution of fossorial and burrowing mammals. Ecology 60:1010–1021.

Reed, C. A. 1951. Locomotion and appendicular anatomy in three soricoid insectivores. Am. Midland Naturalist 45:513–671.

Reed, C. A., and W. D. Turnbull. 1965. The mammalian genera Arctoryctes and Cryptoryctes. Fieldiana: Geology 15:99–170. Description of the forelimb of extinct molelike mammals.

Yalden, D. W. 1966. The anatomy of mole locomotion. J. Zool., London, 149:55–64.

**CHAPTER 25
CLIMBING**

Bock, W. J., and H. Winkler. 1978. Mechanical analysis of the external forces on climbing mammals. Zoomorphologie 91:49–61. Free body analysis of climbing animals in equilibrium.

Cartmill, M. 1974. Pads and claws in arboreal locomotion, p. 45–83. *In* F. A. Jenkins, Jr. (ed.), Primate locomotion. Academic Press, New York.

Cartmill, M. 1979. The volar skin of primates: its frictional characteristics and their functional significance. Am. J. Phys. Anthropology 50:497–510.

Emerson, S. B., and D. Diehl. 1980. Toe pad morphology and mechanisms of sticking in frogs. Biol. J. Linnean Soc. 13:199–216.

Erikson, G. E. 1963. Brachiation in New World monkeys and in anthropoid apes. Zool. Soc. London, Symposium No. 10:135–164. Reviews morphological adaptations.

Fleagle, J. G. 1974. Dynamics of a brachiating siamang [*Hylobates (Symphalangus) syndactylus*]. Nature 248:259–260.

Hiller, U. 1968. Untersuchungen zum Feinbau und zur Funktion der Haftborsten von Reptilien. Z. Morphol. Tiere 62:307–362. Some publications in English have equally fine illustrations, but this study put dry adhesion climbing of reptiles on a sound physical basis.

Jenkins, F. A., Jr., P. J. Dombrowksi, and E. P. Gordon. 1978. Analysis of the shoulder in brachiating spider monkeys. Am. J. Phys. Anthropology 48:65–76.

Jones, F. W. 1953. Some readaptations of the mammalian pes in response to arboreal habits. Zool. Soc. London, Proc. 123:33–41. Adaptations in porcupines and anteaters.

Napier, J. R., and A. C. Walker. 1967. Vertical clinging and leaping—a newly recognized category of locomotor behavior of primates. Folia Primatologica 6:204–219.

Oxnard, C. E. 1968. The architecture of the shoulder in some mammals. J. Morphol. 126:249–290. Canonical analysis distinguishes the scapula in climbers from that in nonclimbers.

Richardson, R. 1942. Adaptive modifications for tree-trunk foraging in birds. University of California Publ. Zool. 46:317–368.

Spring, L. W. 1965. Climbing and pecking adaptations in some North American woodpeckers. Condor 67:457–488.

CHAPTER 26 SWIMMING AND DIVING

Aleev, Yu. G. 1969. Functional and gross morphology in fish. Jerusalem, Israel Program for Scientific Translations. 268 p. An important reference.

Aleyev, Yu. G.* 1977. Nekton, W. Junk Publ., The Hague. 435 p. The hydrodynamics and biology of the larger swimmers (for which the Reynolds number $> 5 \times 10^3$).

Anderson, H. T. 1966. Physiological adaptations in diving vertebrates. Physiol. Rev. 46:212–243.

Bone, Q. 1975. Muscular and energetic aspects of fish swimming, v. 2:493–528. *In* T. Y.-T. Wu, C. J. Brokaw, and C. Brennen (eds.), Swimming and flying in nature. Plenum Press, New York.

Brett, J. R. 1965. The swimming energetics of salmon. Sci. Am. 213(2):80–86.

Denison, D. M. and G. L. Kooyman. 1973. The structure and function of the small airways in pinniped and sea otter lungs. Respiration Physiol. 17:1–10.

Howell, A. B. 1970. Aquatic mammals. Dover, New York. 338 p. (Originally published in 1930.) Broad and basic; more interpretive than most anatomical monographs of the period.

Kooyman, G. L. 1972. Deep diving behavior and effects of pressure in reptiles, birds, and mammals, p. 295–311. *In* M. A. Sleigh and A. G. MacDonald (eds). The effects of pressure on organisms. Symposia of the Soc. Exptl. Biol. 26. Academic Press, New York.

Lang, T. G. 1966. Hydrodynamic analysis of cetacean performance, p. 410–432. *In* K. S. Norris (ed.), Whales, dolphins, and porpoises. University of California Press, Berkeley.

Nursall, J. R. 1956. The lateral musculature and the swimming of fish. Zool. Soc. London, Proc. 126:127–143.

Simons, J. R. 1970. The direction of the thrust produced by the heterocercal tails of two dissimilar elasmobranchs: the Port Jackson shark, *Heterodontus portusjacksoni* (Meyer), and the piked dogfish, *Squalus megalops* (Macleay). J. Exptl. Biol. 52:95–107.

Storer, R. W. 1960. Evolution in the diving birds. Proc. XIIth Intern. Ornithological Congr., Helsinki 1958:694–707.

Thomson, K. S. 1971. The adaptation and evolution of early fishes. Quart. Rev. Biol. 46:139–166. Includes functional interpretations of the shape of body and tail.

Walker, W. F., Jr. 1971. Swimming in sea turtles of the family Chelonidae. Copeia 1971:229–233.

Walters, V. 1962. Body form and swimming performance in the scombroid fishes. Am. Zool. 2:143–149.

Willemse, J. J. 1966. Functional anatomy of the myosepta in fishes. Proc. Akad. Wetenschappen, Ser. C 29:58–63.

CHAPTER 27 FLYING AND GLIDING

Brown, R. H. J. 1953. The flight of birds, II. Wing function in relation to flight speed. J. Exptl. Biol. 30:90–103.

Brown, R. H. J. 1963. The flight of birds. Biol. Rev. 38:460–489. A later paper by a leading student of bird flight.

Leen, N., and A. Novik. 1969. The world of bats. Holt, Rinehart, and Winston, New York. 171 p. Outstanding photographs.

Lighthill, Sir J. 1975. Aerodynamic aspects of animal flight, V. 2:423–491. *In:* T.Y.-T. Wu, C. J. Brokaw, and C. Brennen (eds.). Swimming and flying in nature. Plenum Press, New York.

*Aleev and Aleyev are the same person. In each of the two publications his name has been translated differently.

Lissaman, P. B. S., and C. A. Shollenberger. 1970. Formation flight of birds. Science 168:1003–1005. May 22.

Norberg, U. M. 1970. Functional osteology and myology of the wing of *Plecotus auritus* Linnaeus (Chiroptera). Arkiv Zool. (Stockholm), Ser. 2, 22:483–543.

Norberg, U. M. 1972. Bat wing structures important for aerodynamics and rigidity (Mammalia, Chiroptera). Z. Morphol. Tiere 73:45–61.

Pennycuick, C. 1972. Animal flight. William Clowes & Sons, Ltd., London. 68 p.

Pennycuick, C. J. 1973. The soaring flight of vultures. Sci. Am. 229(6):102–109.

Raspet, A. 1960. Biophysics of bird flight. Smithsonian Inst., Ann. Rept. 1960:405–425. An analysis of static soaring by a vulture.

Rayner, J. M. V. 1979. A new approach to animal flight. J. Exptl. Biol. 80:17–54. Describes momentum imparted to a chain of vortex rings.

Savile, D. B. O. 1957. Adaptive evolution of the avian wing. Evolution 11:212–224.

Urry, D., and K. Urry. 1969. Flying birds. Harper & Row, Scranton, Pa. 192 p. Excellent photographs.

Vaughan, T. A. 1959. Functional morphology of three bats: Eumops, Myotis, Macrotus. University of Kansas Publ. Nat. Hist. 12:1–153.

Wilson, J. A. 1975. Sweeping flight and soaring by albatrosses. Nature 257:307–308.

**CHAPTER 28
FEEDING**

Alexander, R. McN. 1966. The functions and mechanisms of the protrusible upper jaws of two species of cyprinid fish. J. Zool., London 149:288–296.

Crompton, A. W., and K. Hiiemae. 1969. How mammalian molar teeth work. Discovery 5:23–34.

Frazzetta, T. H. 1962. A functional consideration of cranial kinesis in lizards. J. Morphol. 111:287–319.

Frazzetta, T. H. 1966. Studies on the morphology and function of the skull in the Boidae (Serpentes). Part II. Morphology and function of the jaw apparatus in *Python sebae* and *Python molurus*. J. Morphol. 118:217–296.

Gans, C. 1952. The functional morphology of the egg-eating adaptations in the snake genus *Dasypeltis*. Zoologica 37:209–244.

Gans, C. 1961. A bullfrog and its prey. Nat. Hist. 70:26–37.

Gans, C. 1967. The chameleon. Nat. Hist. 76:53–59. Includes analysis of the unique feeding mechanism.

Herring, S. W. 1975. Adaptations for gape in the hippopotamus and its relatives. Forma et Functio 8:85–100.

Hiiemae K. 1971. The structure and function of the jaw muscles in the rat (*Rattus norvegicus* L.): III. The mechanics of the muscles. Zool. J. Linnean Soc. 50:111–132.

Jenkin, P. M. 1957. The filter-feeding and food of flamingoes (Phoenicopteri). Roy. Soc. London, Philosophical Trans. Ser. B., No. 674, Vol. 240:401–493.

Liem, K. F. 1970. Comparative functional anatomy of the Nandidae (Pisces: Teleostei). Fieldiana: Zool. 56:1–66. Analyzes mechanism for swallowing entire large prey.

Lombard, R. E., and D. B. Wake. 1976. Tongue evolution in the lungless salamanders, family Plethodontidae. I. Introduction, theory and a general model of dynamics. J. Morphol. 148:265–286.

Luling, K. H. 1963. The archer fish. Sci. Am. 209(1):100–104, 106, 108.

Smith, J. M., and R. J. G. Savage. 1959. The mechanics of mammalian jaws. School Sci. Rev. 141:289–301.

Zusi, R. 1967. The role of the depressor mandibulae muscle in kinesis of the avian skull. U.S. Natl. Museum, Proc. 123:1–28.

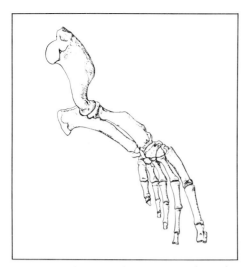

INDEX

Numbers in boldface indicate pages where the entry is illustrated.

Aardvark, 69
  brain of, 352, **353**
  feeding of, 575, 576
  fossorial adaptations of, 464,
    468, 469, 474
Abdominal pores, 300
Abdominal system of veins, 264,
  267, 269
Abdominal vein, **271**
Abducens nerve, **145,** 186, 191,
  323, 324, **334–336, 349,**
  351
Abductor muscles, 179, **230**
Aberration of lenses, 369
Abomasum, **222**
Acanthodians, 37, 38, 51, **52**
  jaws of, 560
  scales of, 95
  teeth and skeleton of, 111,
    127, 150, 160, 166
Acanthodii (*see* Acanthodians)
Acceleration (in bone-muscle
    systems), 430–432, 459
Acceleration (in ontogeny), 19
Accelerometers, 356
Accessory excretory ducts and
    glands, 286, **295,**
    300–302, 305, 306
Accessory nerve, **145,** 186, 187,
    325, **334, 335, 351**
Accommodation of eye, 369,
    370
Accuracy of data, 10
Acellular bone, 92, 95, 108

Acetabulum, 165, **166, 175, 478**
Acetylcholine, 314, 328
Acoelous (= platyan) vertebrae,
    149, 406, 519
Acousticolateralis system, 360
Acrodont tooth attachment, 109,
    110, 111, 113
Acromiodeltoid muscle, **174, 200**
Acromion process, 165, **174,**
    473, 474, 497, **499, 555**
Acromiotrapezius muscle, **522**
Acrosome, 75, **76**
Actin, **181**
Actinopterygii, 37, 46, 52, **53,**
    54, 56, 95, **112,** 261
    (*see also* Ray-finned fishes)
Actinopterygium, **162, 163,** 167
Action potential, 356
Actual forces, 418, **419**
Actual lever arms, 418, **419**
Adding of forces, 411, **412–414,**
    542
Adding of velocities, 429, **430**
Adductor femoris muscle, **195,**
    197, **199, 200**
Adductor longus muscle, **199**
Adductor mandibulae muscle,
    189–191, 195, 196, 560,
    561
Adductor muscles, 179, 189,
    190, 230
Adenohypophysis, **211,** 378, **379,**
    380, 385
Adenosine triphosphate, 182

Adhesion, 493, 502, **503**
Adrenal cortex, 382, 383
Adrenal gland, **302, 303,** 382, 383
Adrenalin, 383
Adrenal medulla, 383, 385
Adrenocorticosteroids, 383
Adrenolumbar vessels, **273–275**
Adrenotropic hormone, 379
Advanced character or taxon, 36
Afferent branchial arteries, 231,
    251, 259, **260,** 261, **262,**
    266
Afferent nerve fibers, 314, 316
    (*see also* subtypes)
Aftershaft of feather, 99
Age variation, 9
Agnatha, 37, 38, 41, **42,** 43, **44,**
    45, 111
    (*see also* Jawless vertebrates;
    specific subclasses and or-
    ders)
Air capillaries, **243**
Air sacs, **242,** 243, **244,** 245,
    299, 523
Air speed, 537, 540, 541,
    544–547, 553
Alar canal, **142**
Alar plates, 332, 333, 335, 338
Albatross, 543–545, **552, 553,**
    556, 578
Alisphenoid bone, **123, 141, 142,**
    143
Allantoic vessels, **260,** 261, 269,
    276

Allantois, 68, 83, **84,** 85, 210, **211, 260,** 261, 269, 276, 292, 305, 309
Alligator, **63, 138, 239, 319** (*see also* Crocodilians)
Allometry, **21,** 22, 184
Alula, **540,** 549, 551, 554
Alveolar duct systems, 240, **241**
Alveolar glands, 96
Alveoli of lung, 240, **241,** 526
Ambiens muscle, **195**
Ameloblasts, 108, **109**
Ammocoetes, 41, 43, **46,** 381, 382, 570
Ammonia, 288, **289**
Amnion, 83, **84,** 85
Amniotes, 85, 154, 157, 171
  excretory system of, **286, 287,** 290, **291,** 292
  nerves and spinal cord of, **318,** 320, 325, **326,** 328
  reproductive system of, **297–299, 302,** 303
  (*see also specific taxa*)
Amniotic fluid, 83, 85
Amphiarthroses, 406, 522
Amphibia (*see* Amphibians)
Amphibians, **20,** 37, 38, 58, **59,** 60
  appendicular skeleton of, 163, **164,** 165, **166,** 170**–172,** 173
  axial skeleton of, 152**–154,** 156
  brain of, 339, **343, 351,** 353
  circulatory system of, 251, **252–254, 256,** 261, **262, 266–268,** 269, 272, 278
  coelom and mesenteries of, **205,** 206
  digestive system of, 215, 219
  digging by, 466, 469
  early development of, 75, **77, 78**
  endocrine glands of, 378, 381**–383**
  excretory system of, 284, 286, 288, 292
  head skeleton of, 131, **133, 135**

integument of, **96**
muscles of, 191, **192, 193**
nerves and spinal cord of, 317, **318,** 320, 328
reproductive system of, **289, 296, 297,** 298, **299,** 301, 304**–306–**308
respiratory system of, **234,** 235, 238, **239,** 240, 244
sense organs of, 358, 359, 361, 366, 371, **374,** 375
(*see also* specific subclasses and orders)
Amphicoelous vertebrae, 148, **149,** 151, 153, 401
Amphioxus:
  circulatory system of, 249, 259, 277
  development of, 75, **77–79**
  digestive system of, 211, 212, 217, 224, 226
  general structure of, 28**–31–**34, 151, 159, 381
  nervous system of, 317, **318,** 319, 328
  and other chordates, 28, 30, **31,** 32**–**34
  urogenital system of, 282, 294
Amphisbaenians, 38, 61, 375, 405, 464, 466, 470, 480, 483, 566
  fossorial adaptations of, **465,** 469, 475, 477, 479, 481
Amphistyli, **127,** 129, 131
Ampullae (of labyrinth), **362,** 363
Ampullary organs, 360, 361
Amygdala, 342, **343, 344,** 346, **347,** 348
Anal fin, 160, 161
Analogy, 4, **5–7**
Anapophysis, 148
Anapsida (*see* Anapsids)
Anapsids, 37, 38, 60, **61,** 93
  head skeleton of, 125, **137, 138**
  (*see also specific orders*)
Anapsid skull, 136, **138, 139**
Anaspida (*see* Anaspids)
Anaspids, 38, 43, **44,** 212, 232
  fins of, 161, 166

Anastomoses, 258, 287, 298
Anatomy, 3
  (*see also* Vertebrate morphology)
Anchovy, **572**
Anconeus muscle, **174, 192, 193**
Androgens, 384
Androstenedione, 384
Angle of attack, 538, **539–**541, 544**–**546, **548,** 549, **550**
Anglerfish, **567,** 568, 573
Angular acceleration, 432
Angular bone, **126, 130, 132, 135, 137,** 142, 143
Angular process, **143**
Angular velocity, 432, 450
Animal pole, 76, **78**
Annular ligament, **396, 426**
Annular pad of lens, **374,** 375
Anoline lizards, 486, **487,** 493, 503
Antagonistic muscles, 179
Anteaters, 69, 352
  feeding adaptations of, 220, 221, 566, **575,** 576
  fossorial adaptations of, 468, 471, **475, 477, 478,** 479
  scansorial adaptations of, 486, 493, 499, 501
Antelopes, 71, 105, 438
Anterior cardinal vein, **260,** 264, **265, 267, 268, 275,** 278
Anterior chamber of eye, **368, 374**
Anterior choriod plexus, 338, **345,** 349, 354
Anterior colliculi, **335, 336,** 337, **345, 347**
Anterior commissure, **335, 344, 345,** 348
Anterior lacerate foramen, **142, 143**
Anterior peduncles of cerebellum, **335,** 337
Anterior rectus muscle, 190, **191**
Anterior vena cava (= precava), **254,** 264, **267, 268, 271, 272, 274, 275**
Antiarchi (*see* Antiarchs)

Antiarchs, 38, 47, **48,** 466
appendages of, 166
armor of, 93
lungs of, 236
spiral intestine of, 219
Anticlinal vertebra, **461,** 462
Antlers, **105**
Antrum of ovarian follicle, **297**
Anura (*see* Anurans)
Anurans, **59,** 60
circulatory system of, **252,**
255, **262,** 268, 278
feeding adaptations of, 570,
573, 577
nerves and spinal cord of, **318**
respiratory system of, **239,**
244
saltatorial adaptations of, 438,
439, 459, 462
sense organs of, 366, 371,
374, 375
skeleton of, **133,** 153, **154,**
158, **159,** 163, **164,** 171
urogenital system of, 285,
296, 298, 307
(*see also* Frogs; Toads)
Anus, **31, 46,** 82, 305, **306**
Aortic arch(es), **228,** 259, **260,**
261, **262**
Apoda, **59,** 60, 239
(*see also* Caecilians)
Aponeuroses, 178
Appearance, **5,** 6
Appendicular muscles, **186,** 187,
190, 192, 194–196, 197
Aquatic vertebrates, **506,** 507,
508, 509
adaptations of, 510–527
physiology of, 528, 529
(*see also specific kinds*)
Arachnoid meninx, 353, **354**
Arbor vitae, 339, **340**
Archaeornithes, 37, 39, 66, **67,**
155, 531, 532
Archenteron, **79**
Archerfish, 568
Archicerebellum, 340, **353**
Archicortex, 342, **343,** 345, 346,
348, 352
Archipterygium, 167, **168**

Archistriatum, 342, **343**
Archosauria (*see* Archosaurs)
Archosaurs, 37, 38, 62, **63, 64,**
112, 221
skeleton of, **138, 139,** 165
(*see also specific orders*)
Area centralis, 371, 372
Armadillos, 69, **70,** 103, 352,
576
fossorial adaptations of,
467–469, 472, **473,** 474,
479
shell of, 95, 101
Armor shields:
nature of, **42, 44, 48,** 91, 93,
94
of ostracoderms, **42,** 43, **44**
of placoderms, 45, 47, **48**
Arm swinging, 495, **496,** 497,
498, 499
Arterial duct, **262,** 263
Arterial ligament, **254**
Arteries, 247
basic pattern of, 259, **260,** 261
development of, 258
evolution of, 261, **262,** 263
structure of, 258, **259**
(*see also* Circulatory system;
specific arteries)
Arterioles, 247, 258
Arthrodira (*see* Arthrodires)
Arthrodires, 38, 47, **48**
armor of, 93, 125, 128
fins of, 166
gigantism in, 427
shearing jaws of, 111
Articular bone, **123,** 130, 134,
135, 137, 143
Artiodactyla, 71
(*see also* Artiodactyls)
Artiodactyls, **70,** 71, **103,** 280
cursorial adaptations of, 438,
441, **442,** 443, **444, 448,**
450–455, 457, **458**
digestive system of, 216, 221,
222
feeding adaptations of, **564,**
570, 580, **581**
skeleton of, **141, 172, 405**
Arytenoid cartilages, 239–241

Aspect ratio, **506,** 517, 518, 541,
542, 551, 553, 554
Aspiration ventilation, 244, 245
Association neurons, 315, 316,
332, 348
Astigmatism, 369
Astragulocalcaneum, **172**
Astragulus, **172,** 173
Asymmetrical gaits, 457, **459**
Atlas vertebra, 156, **157**
Atriopore, **29**
Atrioventricular canal, **251, 255**
Atrioventricular node, **254,** 255
Atrioventricular valve, 249, **250**
Atrium (of heart), 249, **250–255,**
268, 270
Atrium (of protochordates), **30,**
31–33
Auditory (= otic) capsules, **121,**
122, **128, 129**
Auditory (= statoacoustic) nerve,
323–325, **334–336, 349,**
351, 362
Auditory pathways of brain, **347**
Auks, 507, 520, 523, 577
Auricular lobes of cerebellum,
340, **349**
Auricular vein, **274**
Autonomic nervous system, 315,
322, **323–325, 326, 327,**
328, 339, 383
Autostyli, **127**–129, 131, 582
Aves, 37, 39, 66, **67**
(*see also* Birds)
Axial elements of fins, 167, **168**
Axial muscles, 185–194, **196**
Axis vertebra, 156, **157**
Axon cylinder, **312, 313**
Axons, **312,** 313, **314,** 316
Aye-aye, 566, **567**
Azygous vein, **268, 274**

Baculum, **176, 307,** 308
Badgers, 468, 469, 471, **473**
Balance and counterbalance,
421, **422, 433,** 424, 461
Balancing, 499, 500
Baleen plates, 101, **572**
Ball-and-socket joints, 409
Bandicoots, 68, 438, 466

Barbs of feather, **99,** 100, **101**
Barbules of feather, **99, 101**
Basal elements of fins, 160, **163,** 167
Basal nuclei of brain, 341**–343, 344**
Basal plate of chondrocranium, 122
Basal plates of central nervous system, 332, 333, 335
Basapophysis, 148, **152**
Basibranchial skeletal elements, **129**
Basilar membrane, 364, **365,** 366
Basilar papilla, 364, **365**
Basioccipital bone, **123,** 131, **138,** 141, **142**
Basisphenoid bone, **123,** 131, **137, 138, 141, 142**
Bats, 69, **70,** 176, 318, **402**
    feeding adaptations of, 220, 566, 568, 569, **574**
    scansorial adaptations of, 486, **501**
    volant adaptations of, 532, **536,** 537, 540, 545–547, 549, **550,** 551, **552,** 554, **555–557**
Beak, 98, 562, 566, **567**–571, 574, **576,** 579, 582
Bears, 69, **70, 401**
    feeding of, 566, 573
    locomotor adaptations of, **442,** 443, 490, 517
Beaver, 69, **444,** 507, 513, 520, 526, 528, 564, 568
Bending forces, **393,** 399**–401,** 402, **403**
Biceps femoris muscle, **175, 197, 200, 460**
Biceps muscle, **174, 195,** 196, **197–199, 475, 476,** 497
Bichirs, 38, **53,** 54
Bile, 221, 223, 225
Bile ducts, 224
    (*see also* Common bile duct)
Bill, 98, 562, 566, **567**–571, 574, **576,** 579, 582
Bipolar cells of retina, **370,** 371

Birds, 37, 39, 66, **67,** 428
    appendicular skeleton of, **164,** 165, **166, 171, 172,** 173
    axial skeleton of, **149, 155,** 156, 158, 159
    brain of, 331, 335, 337, 339, 341**–343,** 345, **351–353**
    circulatory system of, 254, 255, 261, 263, 264, 268, 269, 272, **275,** 276, 278, **279**
    coelom and mesenteries of, 206
    cranial kinesis in, 561, **562**
    cursorial adaptations of, 438, **439,** 442, 443
    digestive system of, 215, **217,** 219, 226
    early development of, **19,** 75, **77, 81, 84**
    endocrine glands of, 379, **380,** 382–384
    excretory system of, **272,** 287, **289–292**
    feathers of, 98, **99–101**
    head skeleton of, **134,** 136, 137, **139, 140**
    integument of, 98
    muscles of, 194, 195, **197, 198,** 555
    nerves and spinal cord of, 318, 324, 328
    reproductive system of, **272,** 295, 296, 299, **302,** 304**–306,** 307, 308
    respiratory system of, 206, **242–244,** 245
    scansorial adaptations of, 486, **488,** 498, **500, 501, 502**
    sense organs of, 358, 359, **362,** 367, 371, **372,** 375
    volant adaptations of, 531–557
    (*see also specific subclasses and kinds*)
Blastema, 293, **294,** 296, 298
Blastocoel, **77, 78**
Blastocyst, 81, 309
Blastoderm, **77,** 78, **79,** 81
Blastomeres, 77

Blastopore, 27, **78,** 79, 81
Blastula, **77,** 78
Blood, 256–258
Blood-feeding, **567,** 568, 574–576
Blood islands, 257, 258
Blood-vascular system, 247
Blood vessels:
    basic patterns of, 259, **260,** 263, 264
    development of, 258
    evolution of, 261, **262, 263,** 264, **265–268,** 269, **270–275**
    structure of, 258**–259**
    (*see also* Arteries; Veins; *specific vessels*)
Blubber, 512, 528
Body skeleton:
    of amphibians, 152–154, 156, 163, **164,** 165, **166,** 170**–172,** 173
    of birds, **149, 155,** 156, 158, 159, **164,** 165, **166, 171, 172,** 173
    of bony fishes, **149,** 152, **158, 160, 161**–163
    of cartilaginous fishes, 151, **152,** 158, **160,** 161, **162, 163**
    of jawless vertebrates, 150, 151, 158, 166
    of mammals, **154,** 156, **157,** 158, **159, 164,** 165, **166, 171, 172,** 173, **174, 175,** 176
    of placoderms, 150, 158, 160–162, 166
    of reptiles, **154**–156, **157–159, 164, 166, 171–173**
Body stalk, 82, 85
Bone, 91**–93, 94,** 95, 118, 119, 390–395, 398–410
Bone marrow, 258, 278
Bones:
    and bone-muscle systems, 417–421
    membrane (= dermal), **94,** 118, 125, **126,** 130, 150

replacement (= endochon-
dral), 118, 122, **123,** 125,
126, 130, 131, 150, 157,
169
structure of, 392–395, 398,
399–401, 402, **403**
unions with tendons and ligaments, 403, 404
(*see also specific bones*)
Bonito, **506**
Bony fishes, 37, 38, 51, 52,
53–55, 56
body skeleton of, **149,** 152,
158, 160, 161–163, 167
circulatory system of, 250,
251, 252, 256, 257, 261,
262, 264, **268, 275,** 278,
279, 280
digestive system of, **212,** 215,
216, 219
electric, 201, **202**
excretory system of, 284–286,
288, **289**
gas bladders of, 236, **237,** 238
gills of, 229, **230, 231**–233,
234, 235
head skeleton of, **127,** 129,
132
lungs of, 236
muscles of, 188, **189**–191, **519**
reproductive system of, 296,
298, 300–**307,** 308
sense organs of, 366, **374,** 375
(*see also specific subclasses
and infraclasses*)
Bony stops, 425, **473,** 479
Boundary layer, **510**–514
Boundary lubrication, 407
Bounding gaits, 457, **459**
Bow analogs of body support,
423–**424,** 556
Bowfin, 38, **53,** 518, 560
circulatory system of, 250, 277
pharynx and respiratory system
of, **212**
skeleton of, **132,** 152, **161**
Brachialis muscle, **174, 195,** 196,
198, 200, 522
Brachial plexus, 194, **319,** 320

Brachial vessels, **260,** 263, **265,
266, 270**–274
Brachiators, 495, **496,** 497, **498,
499**
Brachiocephalic artery, 263, **270,
272, 273**
Brachiocephalic vein, 264, **268,
274**
Bradycardia, 528
Brain:
anterior brainstem, **335,** 338, 339
cerebellum and pons, 339,
340, 341
cerebrum, 341–**343, 344,**
345–349
development of, 330, **331,**
332, **379**
evolution of, **46, 349,** 350,
351, 352, **353**
functional pathways of, **347**
organization of, 332, 333
posterior brainstem, 333,
334–336–338
study of, 329, 330
support and nourishment of,
353, **354,** 528
(*see also specific regions and
structures*)
Brainstem, **322,** 332–**335,
336**–339
diencephalon, **335,** 338, 339
nuclei of cranial nerves, 333,
334, 336, 347
other nuclei of medulla, 335,
336, 337
reticular formation, 335, **336,
347**
tracts and features of, **335,
336**–339
(*see also specific structures*)
Brain vesicles, 330, **331**
(*see also specific vesicles*)
Braking, 166, 523, 524, 549,
550, 551
Branchial arches, **124,** 128, 129,
131
Branchial muscles, **186,** 187,
189, **190, 192,** 194, **196,**
198, 201

Branchial rami of cranial nerves,
322
Branchiohyoid muscle, **192, 195**
Branchiomandibularis muscle, **196**
Branchiostegal rays, **132**
Bridge analogs of body support,
421, **422, 424**
Bristles of birds, 100
Broad soaring wing, **552**–554
Bronchi, 239–**241, 242,** 243,
244, 526
Bronchioles, 240, 526
Brooding of eggs, 308
Brood patch, 308
Browsing, **567**
Buccopharyngeal cavity, 244
Bulbourethral gland, **303**
Bulbus arteriosus, 250, **251**
Bunodont dentition, 580
Buoyancy, 522–**524**
Butt joints, 404, **405**

Caeca, **217, 218,** 220, 221, **223**
Caecilians, 38, **59,** 95, 239, 483,
561
excretory system of, 283, 284,
286
fossorial adaptations of, 464,
466, 469, 470, 479
reproductive system of, 294,
296, 298, 301, 304, 307,
308
sense organs of, 366, 375
teeth of, **113**
Calcaneum, **172,** 173, **494**
Calcar, 535
Calcified cartilage, 118
Calcitonin, 382
Camber, 539, **540,** 541, 549,
551, 553, 554
Camels, 71, 101, 454, 565
Canine teeth, **114**
Cannon bone, **172,** 425, **426,**
452, 453
Capillaries, **231,** 247, **258,** 259,
287
Capitulum of rib, **148,** 158
Capybara, 507, 517, **563,** 566
Carapace, **98**

Cardiac muscle, 179, **180,** 249
Cardiac region of stomach, 215
Cardiac sphincter, 215
Cardinal system of veins, 255,
    260, 264, **267, 268**
Caribou, **399,** 437, **564**
Carnassial teeth, 563–**565,** 579,
    582
Carnivora (*see* Carnivores)
Carnivores, 69, **70,** 84, 176,
    287, 299, 304
    feeding adaptations of, 220,
        563–**565,** 566, 578, **579**
    locomotor adaptations of, 438,
        441–443, **445, 447,**
        450–**453,** 454, **459,** 468,
        473, 490, 507
    (*see also specific kinds*)
Carnivorous vertebrates,
    563–**565,** 566, 568,
    578–580
Carnosauria, **64**
Carotid body, 356
Carotid duct, **262,** 263
Carotid foramen, **142**
Carotid rete, 280
Carotid system, **253, 254, 260,**
    261, **262,** 263
Carp, **5, 160, 163,** 238, 573
Carpal bones, 167, **168, 171,**
    455, 472–**474**
Carpus, 167, **168, 171,** 173, **445,**
    472–474
Cartesian coordinates, 15, **16**
Cartilage, **118,** 391, 392,
    396–398
Cartilaginous fishes, 37, 38, 47,
    49, **50, 51**
    body skeleton of, 151, **152,**
        158, **160,** 161, **162, 163,**
        167
    circulatory system of, 250,
        251, 260, 261, **262,** 264,
        265, 267–269, 276–278
    digestive system of, 212, 215,
        216, 218, 219
    electric, 201, 202
    endocrine glands of, **380,** 382,
        383

excretory system of, 284, 285,
    289, 291
gills of, 229, **230**–233
head skeleton of, **124, 127,**
    128, **129**
muscles of, 188–**190, 191**
nerves of, **318, 323,** 328
reproductive system of, **295,**
    296, 298, 301, 303, 305,
    306–308
teeth of, 109, **110,** 111, **112**
(*see also specific subclasses
    and orders*)
Cat:
    circulatory system of, **273, 274,**
        280
    digestive system of, **218**
    muscles of, **199, 200**
    urogenital system of, **303**
Catfishes, 238, 566
Cattle, 71, 105, 570
    (*see also* Cow)
Caudal artery, **260,** 263, **265,**
    266, 270, **272, 273**
Caudal fin, 160, **161,** 162, **506,**
    514, **515**–**519, 524, 525**
Caudal lymph duct, **279**
Caudal vein, **260,** 263, **265, 266,**
    268, 271, 272, 274, 275
Caudal vertebrae, 153, 155
    (*see also* Tail)
Caudate nucleus, 342, **343, 344,**
    347, 348
Caudocruralis muscle, **192**
Caudofemoralis muscle, **195,**
    197, **200, 446**
Celiac artery, **262,** 263, **265, 270,**
    272, 273
Celiac ganglion, **327**
Celiacomesenteric artery, **237,**
    238, **266**
Cement, **108,** 109, **581**
Center of buoyancy, 523, **524**
Center of gravity, 421, 523, **524,**
    525
Center of pressure, **539**
Centralia, **168,** 169, **171, 172**
Central nervous system, 314
    (*see also specific parts*)

Centrifugal force, 431, 453, **456,**
    549, **551**
Centripetal force, 431, 453, 549,
    551
Centrum, 147, **148, 149,** 151,
    152, 153, **154,** 155–**157,**
    399, 406, **518,** 519
Cephalaspids, 38, 43, **44,** 201,
    212, 232, 466, 573
    armor of, 93, 125
    fins of, 160, 161, 166
Cephalic flexure, **331,** 332
Cephalization, 28, **29, 31,** 32,
    119
Cephalochordata, 28–**31**–34,
    282
    (*see also* Amphioxus)
Ceratobranchial skeletal ele-
    ments, **124, 129**
Ceratohyal, **129, 132**
Ceratopsia, **64**
Ceratotrichs, **160,** 167, **191**
Cerebellar nuclei, **340**
Cerebellar pathways of brain,
    347
Cerebellum, **331**–333, **335,** 337,
    339, **340,** 341, **345, 347,**
    349, 351, 353, 557
Cerebral aqueduct, **331, 336, 345**
Cerebral hemispheres, 331, 332,
    341, **343, 344,** 348
Cerebral peduncles, **335, 336,**
    337, **344,** 348
Cerebrospinal fluid, 354
Cerebrospinal tract, **336**
Cerebrum:
    evolution of, **349,** 350, **351,**
        352, **353**
    general nature of, **331**–333
    structure of, 341–**345, 343,**
        346, 348, 349
Cervical enlargement of spinal
    cord, 318, **319**
Cervical flexure, **331,** 332
Cervical ganglia, **327**
Cervical plexus, **319**
Cervical ribs, **157,** 158
Cervical vertebrae, 153,
    155–**157, 461,** 462, 569

Cervical vessels, **270, 273, 274**
Cervix, 304
Cetacea, 69, **70**
  aquatic adaptations of, 507,
    512–514, **518**–523,
    526–529
Chaetognatha, 28
Chain ganglia, **326, 327**
Chameleons, 486, **487,** 494,
    497, 499, **500,** 573, 574
Check ligament, **426**
Cheek pouches, 221
Cheek teeth, **114,** 115, 563, **564,**
    580, **581**
Cheetah, **405,** 438, 440, **441,**
    444, **445, 447, 449, 450,**
    453, 457
Chelonia, **61,** 103
  (*see also* Turtles; Tortoises)
Chemical senses, 357–360, 526,
    566
Chest, 443, **444, 497, 498**
Chevron bones, 155, 156, **461**
Chimaeras (holocephalians), 38,
    47, 49, **51**
  axial skeleton of, 150, 158
  digestive system of, 219, 582
  gills of, 233
  lateral lines of, 361
  reproductive system of, 305,
    306, 308
  teeth of, 111, 582, **583**
Chimpanzee, **70, 353,** 495, 568
Chiroptera, 69, **70**
  (*see also* Bats)
Chisel-tooth digging, 464, 465,
    468–470, 475, **477**
Cholecystokinin, 216
Chondrichthyes, 37, 38, 47, 49,
    50, 51
  (*see also* Cartilaginous fishes;
    *specific subclasses and or-
    ders*)
Chondrocranium:
  bones derived from, 122, **123**
  evolution of, 126–**128, 129,**
    130
  structure of, 120, **121,** 122
Chondrocytes, **118**

Chondrosteans, 38, 52, **53,** 54
  axial skeleton of, **151**
  circulatory system of, 261,
  jaw mechanics of, **560**
    277
  respiratory system of, 233
  scales of, 95
Chondrostei (*see* Chondrosteans)
Chord (of waterfoil or airfoil),
    506, 517, 539–**542**
Chordal centrum, 151
Chordamesoderm, **78,** 79, 82
Chordata, 27–34, 119
Chorioallantoic placenta, **84,** 85
Chorion, 83, **84,** 85
Choroid layer of eye, **368, 370,**
    371, **373**
Chromaffin bodies, 383
Chromatophores, 89, **96**
  (*see also specific kinds*)
Ciliary apparatus, **368,** 370, 373,
    374, 375
Ciliary muscles, **327, 374,** 375
Cingulum, 115
Circulatory system:
  of amphibians, **252**–254, **256,**
    261, **262, 266–268,** 269,
    272, 278
  of amphioxus, 249, 259
  of aquatic vertebrates, 528
  basic pattern of, 259, **260,**
    261, 263, 264
  of birds, 254, 255, 261, 263,
    264, 268, 269, 272, **275,**
    278, **279**
  blood, 256–258
  of bony fishes, 254, 255, 263,
    264, 268, **272, 275,** 276,
    278, **279**
  of cartilaginous fishes, 250,
    251, 260, 261, **262,** 264,
    265, 267–269, 276–278
  of cyclostomes, 250, 261,
    264, 277, 278
  development of, 248, **249,**
    255, 257, **260**
  evolution of vessels, 261, **262,**
    263, 264, **265–268,** 269,
    270–274, 277–279

functional units of, 269–272,
    275–277
  general nature of, 247, 248
  heart, 28, **29, 46, 186,** 187,
    205, 206, **207,** 248,
    249–255, 528
  histology of, **256,** 257, 258,
    259
  of mammals, **254–256**–258,
    262–264, **267, 268**–272,
    273–275, 276–**279,** 280
  of reptiles, 250–**253**–255, 262,
    263, **267**–269, **270, 271,**
    275, 278, **279**
  (*see also specific structures*)
Civets, 69
Cladistic classification, 35
Cladoselachians, 38, 49, **50**
  axial skeleton of, 150
  fins of, 159
  gills of, 233
  teeth of, 111, **112**
Cladoselachii (*see* Cladoselachians)
Claspers, **50, 51, 163,** 307
Class (taxon), 34
Classification:
  of vertebrates, **37**–39
  principles of, 34–37
Claustrum, 342, **344**
Clavicle, **159, 162,** 163, **164,**
    165, 443, **444, 466, 476,**
    497, **555**
Clavobrachialis muscle, **199, 200**
Clavodeltoid muscle, **624**
Clavotrapezius muscle, **200**
Claws, 101, 464, **467,** 471,
    472–475, 501, **502, 569,**
    570, 579
Cleavage, **77,** 78
Cleidomastoid muscle, **144, 199,**
    201, **474**
Cleithrum, **130, 132, 162,** 163,
    164
Climbers, 486, **487–489,** 490
Climbing:
  adaptations for, 493–503
  advantages of, 485, 486
  basic mechanisms of, 490,
    491–493

Clinging, 501, **502**
Clitoris, 176, **306,** 307
Cloaca, **211, 217,** 219, 282, **286,**
292, **295, 296, 300, 302,**
305, **306,** 307
Cloacal bursa, 219
Cloacal loop, **263,** 266, **268,** 269
Cloacal membrane, **211**
Cloacal vein, **265, 271**
Closing plates, **228,** 364
Club glands, 90
Cochlea, **362,** 364, **365, 367**
Cochlear nerve, **365**
Cochlear nuclei, **336, 347**
Coefficient of friction, 481, 491
Coefficient of induced drag, 541,
543
Coefficient of lift, 540, 541
Coefficient of profile drag, 541
Coelacanthi (*see* Coelacanths)
Coelacanths, 38, **55,** 56, 233,
523
Coelom, 27, **79, 80,** 81, **84,** 187,
203, **204, 205,** 206, **207,**
249, 282, 283, 300, 303
Colic artery, **273**
Colic caeca, **217, 218**–220, **223**
Collagenous fibers, **88,** 95, 109,
118, 119, 396, 397, 403,
404, 406
Collar of feather, 100, **101**
Collateral circulation, 277, 354
Collateral ligaments, **396**
Collecting tubules, 286, **291**
Colloid of thyroid gland, 381
Colon, **216, 218, 223, 327**
Color change, 89
Colugo, 68, 486, 494, 501, **533,**
535
Columella, 131
(*see also* Stapes)
Commissures of brain, 333
Common bile duct, **216, 217,**
225
Common cardinal vein, **250–252,**
255, 260, 263, 264,
265–268
Common carotid artery, 254,
261, **262, 270, 273**
Common iliac artery, **270**

Common iliac vein, **266, 274,**
275
Components of forces, **414,**
415–418, **419,** 454, **456,**
480, **481, 482,** 490, **491,**
492, 515, 539, 542, 551
Components of velocities, 429,
430
Compression, 391, **392,**
393–395, 397–**399, 401,**
421, 422, 428
Concertina movement of snakes,
482, 483
Conchae (= turbinate bones),
123, 141, 358
Condor, **552**
Cones of retina, **370**–373
Conjunctiva, **368,** 371, **373**
Connective tissue (*see* Collagen-
ous fibers; Elastic fibers)
Constrictor colli muscle, **195,**
201
Constrictor muscles, 179, 189,
190, 191
Contour feathers, **99**
Conus (of eye), **374**
Conus (of heart), 249, **250–252**–
254, 262, **266**
Convergence, **17,** 18
Coprodeum, 305, **306**
Copulatory organs, 305, **306,**
307
Coracoarcualis muscle, **190**
Coracobrachialis muscle, **192,**
195, 196, **198, 199**
Coracobranchialis muscles, **190**
Coracohyoid muscle, **190**
Coracoid(s), **159, 162,** 163, **164,**
165, 443, **555**
Coracoid process, **164,** 165, **174,**
499, 555
Coracomandibularis muscle,
188, **190**
Cormorants, 507, **520,** 526, 568,
576
Cornea, **368,** 369, **373, 374, 376,**
527, 575
Corona radiata, **76, 344**
Coronary artery, **262, 270, 273,**
276

Coronary circulation, 275, 276
Coronary ligament, 206
Coronary vein, **274**
Coronoid bones, **126, 137**
Coronoid process, **143, 565,** 566
Corpora quadrigemina, **334,** 337
Corpus callosum, **344, 345,** 348,
349
Corpus cavernosum, **307**
Corpuscles of Stannius, 385
Corpus luteum, 295, **297,** 384
Corpus striatum, 341–**343,** 345
Cortex:
of adrenal gland, 382, 383
of cerebellum, 332, 333, 339,
340, 345, 347
of cerebrum, 332, 341–**343**–
345, 346, **347,** 348, 354
of gonads, 294
of hair, **102,** 104
of kidney, **291**
Cosmine, 92–94
Cosmoid scales, 93, **94**
Costocervical vessels, **273, 274**
Cotylosauria, 60, **61, 112**
Countercurrent exchange, **231,**
232, 237, 280, 290, 299,
519, 528
Cover-up digging, 464, 466, 470
Cow, **222, 335, 336, 344, 345,**
403, 426
Cranial kinesis, 137, 561, **562**
Cranial nerves:
development of, 316, 322
muscle groups innervated by,
186, 187
nature and categories of, 321,
322, 323
nuclei of, 333, **334**
relation to skull of, **145**
structure and evolution of,
323–325
Creepers, 486, 500, 567, 569
Cricoid cartilage, **134,** 239–241
Cristae of the labyrinth, **362,** 363
Crocodiles, 38, 63, **95, 157, 576**
(*see also* Crocodilians)
Crocodilians, 38, **63,** 427
aquatic adaptations of, 513,
516, 526, 528

circulatory system of, 253,
254, 255
coelom of, 206
feeding adaptations of, **576,**
578, 580
muscles of, **196**
osteoderms of, **95**
reproductive system of, 296,
307
respiration of, 245
scutes of, 97
sense organs of, 359, 375
skeleton and teeth of, 136,
138, 157, 159, 165, **576**
Crop, 215, **217,** 220, 221
Crossbill, 486, 568
Crossopterygians, 38, **55,** 56,
251
body skeleton of, 150, **154,**
163, 167
cranial kinesis in, 561
head skeleton of, **127, 130,**
131, **135,** 136
scales of, 93, **94**
teeth of, 111, **112**
Crossopterygii (see Crossopterygians)
Crossopterygium, 167, **168**
Crowns of teeth, 107–109, **113,**
580, **581**
Cruciate ligament, **426**
Ctenoid scales, 95
Cucullaris muscle, 190, **191,**
193, 196, 198
Cud-chewing, 580
Cuneate fasicles, 337
Cuneate nucleus, 337, **347**
Cupula, **360,** 361, 363
Curlew, **567**
Cursorial vertebrates,
437–**439–441**
adaptations of, 440–454
Cursors (see Cursorial verte-
brates)
Curvilinear motion, 431, 432,
453
Cusps of teeth, 109, **114,** 115
Cutaneous vessels, **266**
Cutaneus maximus muscle, **200,**
201

Cuticle (of hair), 104
Cuticle (of skin), 90
Cutting ridges of teeth, **581**
Cycloid scales, 93, 95
Cyclostomata (see Cyclostomes)
Cyclostomes, 43, **45**
brain of, 339, **349,** 353
circulatory system of, 250,
261, 264, 268, 269, 277,
278
digestive system of, 212, 215,
217, 220, 226
endocrine glands of, 379,
380–383
excretory system of, 283, 286,
290
fins of, 161
head skeleton of, **128**
integument of, 90, 93
muscles of, 188
nerves and spinal cord of,
317, **318,** 323, 325, 328
reproductive system of,
296–298, 300, 301, 305,
306, 308
respiratory system of, **230,**
232, 233, 235
sense organs of, 358, 359,
361, **362,** 366, 374
tooth analogs of, 111
Cystic duct, **218,** 225

Decussations in nervous system,
333, 339
Deep digital flexor, **426, 476**
Deep femoral vessels, **273, 274**
Deep ophthalmic nerve, **323,**
324
Deer, 71, **157, 159, 405**
antlers of, **105**
cursorial adaptations of, **442,**
444, 457, **459**
Deferent duct, **295, 296, 299,**
301, **302, 303, 306**
Deformation of supportive mate-
rials, 391
Degenerate (= vestigial) organs,
8, 19
Delamination, 81
Deltoid crest, **473,** 475, **521**

Deltoid muscles, **174, 193,** 194,
195, 196, 200, 471,
473–475, 522
Deltoid tuberosity, **174**
Dendrites, 312, **314,** 316
Density, 511, 522, 523, 524,
540, 541
Dental formula, 113–115
Dental lamina, 108, **109**
Dentary bone, **126, 130, 132,**
135, 137, 140, 142
Dentate nucleus, **340, 347**
Denticles, 90, 92–**94,** 95, 107,
111
Dentine:
of integument, 91–**94,** 95
nature of, 92
of teeth, 107, **108, 109,** 111,
581
Depressor mandibulae muscle,
193, 196, 198
Depressor muscles, 179
Depth registers, 356
Derived character or taxon, 36
Dermal (= membrane) bone(s),
94, 118, 125, **126,** 130,
150
Dermal denticles, **94,** 95, 161
Dermal papilla, 91
of feather, 100, **101**
of hair, **102,** 104
of tooth, 108, **109**
Dermatocranium, 120
Dermatome, **80,** 83, 88, 91
Dermis, **88,** 89, **90,** 91, **96, 97,**
98, 100, **101, 102,** 109
Dermoptera, 68
(see also Colugo)
Deuterostomia, 28, **31,** 82
Development:
of acousticolateralis system,
360–362
of amphibians, **77, 79**
of amphioxus, **77,** 78, **79**
of birds, **19, 77, 81**
of brain, 330, **331,** 332, **379**
of circulatory system, 248,
249, 255, 257, **260**
of coelom and mesenteries,
78–**80,** 81, **204, 205, 207**

Development (*Cont.*):
  of digestive organs, 210, **211,**
    223, 225
  early, 77–83
  of endocrine glands, 378,
    379–383
  and evolution, 18–20
  of eye, 372, **373**
  of feathers, 100, **101**
  of fishes, **77**
  of gonads, 293, **294**
  of hair, 104
  of hard tissues, 90–92
  of kidneys, 282–**284, 286**
  of long bones, **169, 170**
  of lungs and gas bladders, **211,**
    235, 236
  of mammals, **77**
  of mammary glands, 102, 103
  of muscles, 185, **186,** 187
  of nerves and spinal cord, 82,
    315, 316
  of pharynx, **211, 228,** 229
  of reptiles, **77**
  of ribs and sternum, 157–159
  of skin, 88, 89
  of teeth, 108, **109**
  of urogenital structures,
    300–307
  of vertebrae, 150
  (*see also* Embryology)
Diaphragm, 194, 206, 245, 523
Diaphragmatic vertebra, 497,
  498
Diaphysis, **169, 170**
Diapophysis, **148,** 158
Diapsid skull, 136, **138, 139**
Diarthroses, 406, **407,** 519
Diencephalon, **331,** 332, **335,**
  338, **349,** 372, **373,** 378,
  379
Digastric muscle, **144,** 198, **199**
Digestive system:
  development of, 210, **211,**
    223, 225
  evolution of, 211, **212–215,**
    216–218, 219–**222–**
    224–226
  functional adaptations of,
    220–**222, 223**

general nature of, 209, 210,
  213, **214**
  (*see also specific organs*)
Diggers, 463–**465,** 466, **467,** 468
Digging:
  adaptations for, 469–480
  advantages of, 463, 464
  different ways of, 464, 465
  general requirements for, 468,
    469
Digitigrade foot, **442,** 443
Dilator muscles, 179
Dinosaurs, 38, 63, **64**
  cursorial adaptations of, 438,
    443, 449, 451, **453**
  feeding adaptations of, 570,
    576, 582
  gigantism in, 394, 427, 428
Dioecious vertebrates, 308
Diphycercal caudal fin, **161**
Diphyodont dentition, 110, 113
Diplorhina, 37, 38, **42,** 43
Diplospondyly, **152**
Dipnoans (= lungfishes), 38, **55,**
  56
  brain of, 350
  circulatory system of,
    250–**252,** 255, 261, 268,
    269, 278
  digestive system of, 206, 219,
    220, 226
  digging by, 465, 466
  endocrine glands of, **380,** 382
  gills of, 233, **234**
  lungs of, 236
  scales of, 94
  skeleton of, **127,** 131, 150,
    163, 167
  teeth of, 111, **112,** 582
  urogenital system of, 285,
    291, 296, 300, 301,
    304–**306**
Dipnoi (*see* Dipnoans)
Distal tubule of nephron, 288,
  290, **291**
Diving, 507
  (*see also* Swimming)
Dog, 69, **114, 213, 241, 291, 400**
  cursorial adaptations of, **441,**
    442, 459

muscles of, **174, 175**
  skeleton of, **134, 138, 174,**
    175
Dolphins, **17,** 18, 69, **70, 280**
  aquatic adaptations of, 507,
    512, 514, 517, **518, 521,**
    526, 528
  skeleton and teeth of, **5,** 156,
    576
Dopamine, 314
Dormouse, 488, **489**
Dorsal aorta, **237,** 259, **260,** 261,
  262, 263, **265, 266, 286,**
  270, 272, 273
Dorsal branches of dorsal aorta,
  260, 263
Dorsal cranial nerves, 321, **322,**
  323–325
Dorsal eyes, **376**
Dorsal fin, **160,** 161
Dorsal funiculus, **317, 318,** 337
Dorsal gray columns, 316, **317,**
  332
Dorsalis scapulae muscle, 194,
  196
Dorsalis trunci muscle, 191,
  193
Dorsal median sulcus, 315–**317,**
  318, 355, 337
Dorsal mesentery, **80, 204**
Dorsal nerve ganglia, **320, 322**
Dorsal rami of cranial nerves,
  322
Dorsal rami of spinal nerves,
  186, 320, **326**
Dorsal ribs, 157, **158,** 188
Dorsal root ganglia, **320**
Dorsal roots of spinal nerves,
  315, 318, **320,** 321
Dorsal spinal nerves, **320,** 321
Dorsal thalamus, **335,** 338, **347**
Dorsobronchi, **242,** 243, 245
Dorsolateral sulcus, **317,** 318
Down feathers, **100**
Drag:
  in flying, 538, **539,** 541, **542,**
    543, 546, 549, **550,** 551,
    553
  in swimming, 510–514
Drag coefficient, 511, 541

Ducks:
  aquatic adaptations of, 507,
    514, 517, **520,** 528
  feeding of, 571
  volant adaptations of, 537,
    545, 547, 553
Duiker, **458**
Duodenum, 216, **217, 218, 223,**
  272
Dura mater, 353, **354**
Durophagous vertebrates, 582,
  583
Dynamic soaring, 544, 545

Eagles, **198,** 553, 554, 569
Ear:
  evolution of, **362,** 366, **367**
  structure and function of, 361,
    362, 363–**365**
Eardrum, 229, 364, **365, 367**
Ear ossicles, 131, **135,** 364, 527
Echidnas, 68, **69**
  feeding adaptations of, 575,
    576
  fossorial adaptations of, 465,
    466, 469, 471, **472,** 479
Echimyid rats, 488, 494, 499
Echinodermata, 28, 33, 34
Echo-location (= echo-ranging),
  367, 527, 566, 569
Ectoderm, 27, **78–81,** 82
  (*see also development of spe-*
    cific organs)
Ectodermal placodes, 316, 322,
  360, 362, 372, **373**
Ectomesenchyme, 108
Ectopterygoid bone, **126, 130,**
  137, 138, 142
Edentata (*see* Edentates)
Edentate, 113
Edentates, 69, **70,** 156, 299, 305
Eels, 464, 466, 479, 514, 516,
  517
Effective forces, 418, **419**
Effective lever arms, 418, **419**
Efferent branchial arteries, 231,
  259, **260,** 261, **262, 265,**
  266
Efferent ductules, **295, 296, 299,**
  300, 301

Efferent nerve fibers, 314, 316
  (*see also subtypes*)
Efficiency of bone-muscle systems, 434, 435
Egg-eating snakes, **578**
Eggs, 295–**297,** 303–305, 308
Egg tooth, 98
Elasmobranchii (*see* Elasmobranchs)
Elasmobranchs, 37, 38, 49, **50,**
  233
  circulatory system of, 250,
    251, 260, 261, **262, 265,**
    267, 277
  digestive system of, **216**
  endocrine glands of, **380**
  gills of, 229, **230–**233
  muscles of, **190, 191**
  nervous system of, **323, 343,**
    349, 350
  scales of, **94,** 95
  sense organs of, 358, 361,
    362, 366, **374**
  skeleton of, 118, **124, 129,**
    151, **152, 160, 162, 163**
  urogenital system of, 285,
    291, **295,** 298, 301, 303,
    305, **306–**308
  (*see also specific orders*)
Elasmoid scales, **94,** 95
Elastic cartilage, 118, 398
Elastic fibers, 89, 258, 397, 404,
  406
Elasticity, 391
Electric fishes, 201, **202,** 566, 569
Electric organs, 201, **202**
Electroplax, **202**
Elephants, 71, 299, 304, 309, 367
  feeding adaptations of, 221,
    563, 567, 570, 580, **581**
  support and locomotion of,
    428, **429,** 449
Elipsoid joints, 409, 495, **496**
Elliptical wing, 551, **552,** 553
Embolomeri, 153, **154**
Embryology:
  early development, 77–83
  and evolution, 18–20
  and vertebrate ancestry, 34
  (*see also* Development)

Embryonic area, 81
Enamel:
  of integument, 92–**94,** 95
  nature of, 91, 92
  of teeth, 107, **108, 109,** 111,
    581
Enameloid dentine, 91, 92, 107
Enamel organ, 91, 95, 108, **109**
Endocardium, **249**
Endochondral (= replacement)
  bone(s), 118, 122, **123,**
  125, 126, 130, 131, 150,
  157, 169
Endocrine glands, 377–385
  (*see also specific glands*)
Endoderm, 27, **78–81,** 82
  (*see also* Development, of
    digestive system, fetal
    membranes, and pharynx)
Endolymph, 363, **365**
Endolymphatic duct, **362,** 366
Endolymphatic sac, 362, 366
Endometrium, 304
Endomysium, 180, **181,** 403
Endostyle, **29, 30, 46,** 381
Endurance, 438, 449–453
Energy storage, 461
Enterocoely, **79,** 204
Enterogastrone, 216
Entopterygoid bone, **132**
Entotympanic bone, 143
Epaxial muscles, 157, 185, **186,**
  187, **189, 191**
Ependymal layer, **315,** 332, 338,
  354, 373
Epibranchial skeletal elements,
  124, 129
Epicardium, 249
Epicondyles of humerus, **174**
Epidermis, **88–**90, **96, 97,** 100,
  101–103
Epididymis, **299,** 301, **302, 303**
Epigastric arteries, **265, 266, 270,**
  273, 277
Epigastric vein, **271, 274**
Epiglottis, 241, 526, 578
Epimere, **80,** 82, **249**
Epimyocardium, **249**
Epimysium, 180, **181**
Epiphyseal plate, **169, 170**

Epiphyses, 156, **157, 169, 170,** 171, 173
Epiphysis of diencephalon, 376
Epipterygoid bone, **123, 137,** 143
Epipubic bones, 165
Epithalamus, 338, **351**
Epithelium of gut, 213, **214, 218**
Epurals, **161,** 162
Eras (geologic), **37**
Erythrocytes, **256,** 257
Erythrophores, 89
Esophageal region of stomach, 215
Esophagus, **212,** 214–**216–218,** 220, **222, 223, 228, 230,** 576, **577,** 578
Estrogens, 383, 384
Ethmoid bone, **140,** 561
Ethmosphenoid unit of braincase, **130**
Euryapsida, 37, 38, 63, **65,** 136, **139**
Euryapsid skull, **139**
Eustachian tube, 364, **365, 367**
Eutheria, 39, 68, **70**
Ever-growing teeth, 570, **582**
Evolution, 11–18
    and direction, 14–18
    and embryology, 18–20
    and habitat, 11–13
    rates of, 13, 14
    trends in, 14, **15, 16,** 17
    of vertebrates, **37**
Evolutionary classification, 35
Excretion:
    by gills, 232, 288, **289**
    by kidneys, 287–**289,** 290, **291**
    by salt glands, 213, 219, 290
    by sweat glands, 102
Excretory system, 281–292
    of amphibians, 284, 286, 288, 292
    of birds, **272,** 287, **289,** 290–292
    of bony fishes, 284–286, 288, **289**
    of cartilaginous fishes, 284, 285, **289,** 291
    of cyclostomes, 283, 286, 290
    of mammals, **287, 289,** 290, **291,** 292

of reptiles, 286, 287, **289,** 291, 292
    (*see also* Kidneys; Urogenital ducts; Urinary bladders)
Exoccipital bone, **123,** 130, 131, **133, 137, 140, 141**
Extensor muscles, **174, 175,** 179, **474, 522**
External auditory canal, **367,** 469, 526
External auditory meatus, **143**
External capsule, **344**
External carotid artery, 261, **262, 266, 270, 273**
External ear, 366, 367, **467,** 479, 512
External gills, **20,** 228, **234**
External glomerulus, **282**–285, 287
External iliac vessels, **270, 271,** 273
External jugular vein, 264, **266, 268, 271, 274**
External nares, **126, 138,** 469, 526, 575
External oblique muscle, **192, 193, 196, 199, 200**
Extinction, 13
Extrapyramidal pathways, **347,** 348
Eyelids, **368,** 371, **373**
Eye muscles:
    extrinsic, 186, 188, 190, **191,** 201
    intrinsic, **327, 374,** 375
Eyes:
    development of, 372, **373**
    dorsal, **376**
    evolution of, **46, 374,** 375
    fossorial adaptations of, **465, 467,** 469, **474, 476, 477**
    other adaptations of, 371, **372,** 497, 527, 568, 577
    structure and function of, **368,** 369, **370**–372

Facial lobe of medulla, 350
Facial nerve, **145,** 186, 187, 201, **323,** 324, **334–336, 349, 351,** 359, 360

Facial veins, **274**
Falciform ligament, 207
Falciform process of eye, **374,** 375
Falcons, 547, 553, 554, 568
Family (taxon), 34
Fascia, 178
Fascia lata, **200**
Fat, 89, **102, 103,** 297, 406, **407,** 557
Feather follicle, 100, **101**
Feathers:
    development of, 100, **101**
    kinds of, 99, 100
    origin of, 98, 99
    structure of, **99, 100**
Feather sheath, 100, **101**
Feeding, 559–583
    jaw mechanics, 559–566
    processing food, 575–583
    securing food, 566–575
    storage and digestion, 220, 221
Femoral nerve, **319**
Femoral vessels, 263, **265, 266, 270–274**
Femoritibialis muscle, **197**
Femorococcygeus muscle, **460**
Femur, **168, 175, 455,** 512
Fermentation of food, 221, **222**
Fetal membranes, 83, **84,** 85
Fibrous cartilage, 118, 398, 406
Fibula, **168, 175,** 450, **452,** 495, **496**
Fibulare, **168, 172,** 173
Filter-feeding, 30, **31,** 32, 211, 215, 570–**572**
Fingerpads, **500**–502
Fin rays, **94, 160, 162, 163, 518**
Fins:
    function of, 160, 166
    median, **160, 161,** 162
    origin of, 159, 160
    paired, **162, 163,** 166, 167
Fin spines, **160**
Fishes:
    brain of, 337, 339, 342, **343,** 345, **349,** 350, 353
    digging by, 465, 577
    early development of, **77**

endocrine glands of, 379, **380,** 382, 383, 385

integument of, **90**

nerves and spinal cord of, 317, 318, 320, **323**–325, 328

sense organs of, 356, **358,** 359, **360,** 361, **362,** 371, **372, 374,** 375

(*see also* specific taxa)

Flagellum of sperm, 75, **76**

Flamingos, 520, 571

Flapping flight, 545, **546,** 547, **548,** 549, 551

Flexor muscles, 179, **475, 476**

Flexor tibialis muscles, **195**

Flexures of brain, **331,** 332

Flicker, **488, 502**

Flight:

ascending and descending, 546–**548,** 551

fast, 537, 547, 548

hovering, 545, **546**

Flocculonodular lobes, 340, **351, 353**

Flow lines, **510, 511, 539**

Fluid film lubrication, 407

Flyers, **535,** 537, 538, **548, 550, 552**

Flying:

adaptations for, 538–557

advantages of, 532

general requirements for, 538

origin of, 531, 532

"Flying" fishes, **533,** 534

Flying reptiles (= pterosaurs), 18, 38, 63, 532, **535, 536,** 537, 540, 553–555, 557

"Flying" squirrels, **455,** 488, **533,** 535

Follicle:

of feather, 100, **101**

of hair, **102,** 104

Follicle cells of ovary, 295, **297**

Follicles of thyroid gland, **378,** 381

Follicle-stimulating hormone, 379

Footpads, 491, 492, **500**–502

Foramen ovale, **142, 143**

Foramen rotundum, **142, 143**

Force(s):

adding of, 411, **412–414**

components of, **414, 415**–418, **419,** 454, **456,** 480, **481, 482,** 490, **491, 492,** 515, **539, 542, 551**

of muscle contraction, 183, 184, 390, 460

and supportive materials, 392, **393**–397, **398–403**

vectors of, 411, **412–415,** 419

Forebrain, 330, **331, 343**

(*See also specific structures*)

Foregut, **211, 249**

Form:

and function, 6, 7

and size, **21,** 22

Formation flying, 545

Fornix, **344, 345,** 349

Fossils, 10, 11

Fossorial vertebrates, 463–**465,** 466, **467,** 468

adaptations of, 469–480

Fourth ventricle of brain, **331, 335,** 339, **345,** 354

Fovea of eye, **368,** 371, 374, 375

Foxes, 396, **401,** 437, 438, 449, 464

Free-body diagram, 432, **433**

Friction, 454, **456,** 478, 480, **481, 482,** 483, 490, **491,** 492, 497, 501, 541, 544

Frictional drag, 510–512

Frogs, 38, **59,** 60

aquatic adaptations of, 517, 520, 526

circulatory system of, **252, 262**

feeding of, 359, 570, 573, 577, 578

nervous system of, **318, 351**

parachuting of, 532, **533,** 534

respiratory system of, 239, 244

saltatorial adaptations of, 438, **439,** 459, 460, 462

scansorial adaptations of, 486, **487,** 494, 498, **500**–502

sense organs of, 359, 371

skeleton of, **133, 159,** 477

Frontal bone, **126, 130, 132, 133, 137, 138, 140, 141**

Fruit bats, **552, 555**

Function, **5–7**

Fundus of stomach, 215

Funiculi of spinal cord, **317**

Furcula, **164,** 165

Gaits, 454, 457, **458, 459**

Galagos, 438, 457, 459, 460, 486, 493, **494,** 501

Gall bladder, **46,** 210, **211, 216–218,** 221, 223, 225

Gallop, 449, 457, **459**

Gametes, 75, **76,** 293

Ganglia, **30,** 314, **315,** 320, **322, 323**–328

(*see also specific ganglia*)

Ganglion cells of retina, **370**–372

Ganoid scales, **94,** 95

Ganoine, 92, **94,** 95

Gape and suck feeding, 561, 573

Gar, 38, **53,** 578

Gas bladder, 236, **237,** 238, 364, 366, 522, 524

Gas gland, **237,** 238

Gasserian ganglion, 324

Gastralia, 95

Gastric vessels, **265, 266,** 270, **272–274**

Gastrocnemius muscle, **175, 195, 197, 199, 200,** 396, **447, 460**

Gastrocoel, **78,** 79

Gastroduodenal vein, **272**

Gastrohepatic artery, **265**

Gastrosplenic vessels, **265, 266, 274**

Gastrula, **78,** 79

Gastrulation, **78,** 79

Geckos, **5,** 486, 493, **500,** 502, **503, 533,** 534

Geese, **351,** 545, 547, 570

Generator potential, 356

Geniculate ganglion **323,** 324

Genioglossus muscle, **195**

Geniohyoid muscle, **192, 195–197,** 199

Genital ducts (see Urogenital ducts)

Genital pores, 300

Genital ridges, 293, **294**

Genital system, 293–309
  (*see also* Copulatory organs;
    Ovary; Testis; Urogenital
    ducts)
Genital tubercle, 307
Genital vessels, **265, 266, 270,
  271, 273, 274**
Genus, 34
Gerenuk, **567**
Germinal epithelium, 293, **294,**
  295, **297**
Germ layers, 27, **78**
  (*see also* specific layers)
Gibbon, 486, 495, 497, 501
Gigantism:
  advantages of, 426, 427
  examples of, 427, 428
  structural adaptations for, 398,
    399, 428, **429**
  and surface to volume ratio,
    22, 427, 528
Gill bars, **46,** 124, 212, 229,
  230, 231, 233, 234
Gill chambers, 229, **230,** 233,
  235, 573
Gill clefts, **212**
Gill filaments, 229, **230, 231,**
  233, **572**
Gill lamellae, **230, 231,** 232
Gill openings, **42, 44, 46**
Gill pouches, **124, 230,** 232
Gill rakers, 229, **230, 231,** 570,
  572
Gill rays, **230**
Gills:
  function of, 229–**231,** 232,
    288, **289**
  general structure of, 229–**231,**
    232, 234, 235
  number of, 123, 233, 234
  opercular, **230,** 233, 234
  pouched, **230,** 232, 233
  septal, **230,** 233
Gill septa, 229, **230,** 233, 234
Gill slits, **230,** 233
Giraffe, 105, 156, 428, 567, 570
Girdles:
  of fishes, **162, 163**
  of tetrapods, 163, **164,** 165,
    166, 174, 175

Gizzard, 215, **217,** 220, 221,
  272, 582
Glands:
  adrenal, **302, 303,** 382, 383
  alveolar, 96, 102, 103
  bulbourethral, **303**
  club, 90
  gonads, 383, 384
  granular, 90, **96,** 106
  hypophysis, **121,** 339, **345,
    349, 351,** 378, **379–381,**
    385
  lacrimal, 290, **327,** 371, 375,
    527
  mammary, 102, **103,** 384
  Meibomian, 102
  Moll's, 102
  mucous, 88, **90, 96,** 98, 106,
    477
  parathyroid, 382
  pineal, 385
    (*see also* Pineal body)
  pituitary (*see* Hypophysis)
  placenta, 384
  poison, 90, 213
  rectal, **216,** 219
  salivary, **213,** 221, **327, 575**
  salt, 213, 219, 290
  scent, 98, 102
  sebaceous, **102, 103,** 526
  sweat (= sudoriferous), 101,
    102
  tear (*see* lacrimal)
  thymus, 257, **270,** 278
  thyroid, **190, 211, 270, 378,**
    381, 382
  uropygial, 98
  wax, 102
Glans penis, **303,** 307
Glenoid cavity, **164, 174,** 497, **499**
Gliders, **533–535**
Gliding, 531, **533–535, 542–544**
Gliding angle, **542,** 543, 553
Globus pallidus, 342, **343, 347**
Glomeruli of olfactory bulbs, 341
Glomerulus, **282–286, 287**
Glomus, 284
Glossopharyngeal nerve, 186,
  187, **323,** 325, **334–336,
  349, 351,** 359, 360

Glottis, 239, 240, 244, 526
Glucagon, 225, 384
Gluteal muscles, **175,** 196, **200,
  448, 460**
Glycogen body, 318
Gnathostomes, 45
Golden moles, 17, 18, **467–472,
  474,** 475, 479, 480
Gomphosis, 405
Gonads, **31,** 293–299
  (*see also* Ovary; Testis)
Gonocoels, 294
Gonopodium, 305
Gophers, 464, 468–471, 474,
  477, **478,** 479, 582
Graafian follicle, **297**
Gracile fascicles, **335,** 337
Gracile nucleus, 337, **347**
Gracilis muscle, 197, **199**
Gradients, 6, 281, 315
Granular glands, 90, **96,** 106
Granular layer of cerebellum,
  339, **340**
Granulocytes, **256**
Grasping, 497–499
Graviportal vertebrates, 427,
  428, **429**
Gray commissure, 316, **317**
Gray matter, 314–**317,** 332, 335,
  346
Gray rami of spinal nerves, **326**
Greater trochanter, **175, 496**
Greater tubercle, **174**
Grebes, 507, **508,** 513, 520
Grinding ridges of teeth, **564**
Grinding teeth, 580, **581,** 582
Ground squirrels, 464, **467,** 468,
  471, 474
Growth hormone, 379
Growth of supportive materials,
  169, 170, 390, 404
Guanine, 89, 372, 375
Guard hairs, 104
Guinea pig, **563**
Gular pouch, 577
Gulls, 543, 547, **548, 550,** 553
Gyri, 339, **340,** 346

Habenula, 338, **345,** 346, 348,
  349, 376

Habenular commissure, 338, **345,** 348

Habenular nucleus, 338, **347**

Hagfishes, 37, 38, 43, **45, 205,** 379, 579

 circulatory system of, 257, 266, 269, 278

 excretory system of, 283–285, 290

 reproductive system of, 294, 296, 305, **306,** 308

 respiratory system of, 232

 spinal nerves of, 320

Hair, **102**–104

Hair cells (sensory), **359, 360,** 363–**365**

Hair follicle, **102,** 104

Hair root, **102,** 104

Hair shaft, **102,** 104

Haversian systems, 92, **93**

Hawks:

 feeding adaptations of, 568, 578, 579

 volant adaptations of, 543, 547, 553

Head fold of amnion, 83, **84**

Head fold of embryo, **81,** 82

Head kidney, 284

Head skeleton, 117–145

 of amphibians, 131, **133, 135**

 of birds, **134,** 136, 137, **139, 140**

 of bony fishes, **127,** 129, **132**

 of cartilaginous fishes, **124, 127,** 128, **129**

 components of, 120, **121,** 122, **123, 124,** 125, **126,** 127

 of jawless vertebrates, **42, 44,** 127, **128**

 of mammals, **134, 135, 139–145**

 of placoderms, **48, 127,**

 of reptiles **134, 135,** 136, **137–139,** 140

 segmental nature of, 119, 120, 123, 124

Hearing, 238, 363–367, 527, 566

Heart:

 of ammocoetes, **46**

 aquatic adaptations of, 528

 as chordate feature, 28

 development of, **186,** 187, 248, **249, 255**

 evolution of, 249, **250–254,** 255

 innervation of, **254,** 255, 325, **327**

 relation to coelom of, **205,** 206, **207**

 of Urochordata, **29**

Hedgehogs, 68, 468, 479

Hemal arch, 148, **149,** 150–**152,** 155, **161,** 162

Hemal canal, **151, 161**

Hemal nodes, 258

Hemal spine, 148, **149,** 152, 160, 162, **518**

Hemiazygous vein, 268

Hemibranch, 229, **230,** 233, 234, **260,** 261, **262**

Hemichordata, 28, 33, 34

Hemipenis, **195,** 307

Hemocoel, **29**

Hemolymph, 277, 278

Hemolymphatic system, 277, 278

Hemolymph propulsors, 278

Hemopoiesis, 257, 258

Hepatic artery, **265, 266,** 273

Hepatic diverticulum, **31,** 32, **211,** 223

Hepatic duct, **217,** 225

Hepatic portal vein, **224, 237,** 238, **260,** 264, **265–267,** 270, **271, 272, 274,** 528

Hepatic sinus, **266**

Hepatic unit of circulatory system, 264–267

Hepatic vein, **260, 263, 265–267, 271, 272, 274, 275**

Hepatoduodenal artery, **272**

Herbivores, 221, **222, 223, 563–565,** 566, 570, **571,** 580, **581**

Hermaphroditic vertebrates, 308

Herons, 547, 568

Heterocercal caudal fin, **161,** 524, **525**

Heterocoelous vertebrae, **149,** 155, 409

Heterodont dentition, 113, **114,** 579

Heterogeneity of supportive materials, 118, 119, 391

Heterostraci, 38, 43

 (*see also* Pteraspids)

High-speed wing, **552,** 553

Hindbrain, 330, 379

 (*see also specific structures*)

Hindgut, **211,** 221

Hinge joints, 408, 452, 564

Hippocampus, **343, 344,** 346, 349

Hippopotamus, 428, 507, 526, 529

Hoatzin, 221, 486, 570

Holoblastic (= total) cleavage, **77**

Holobranch, 229, **230,** 233–234, **260,** 261, **262**

Holocephali (see Holocephalians)

Holocephalians (chimaeras), 37, 38, 49, **50**

 digestive system of, 219, 220

 gills of, 233

 teeth and skeleton of, 111, **127,** 129, 582, **583**

 urogenital system of, 305, **306**

Holonephros, 283, 284

Holosteans, 38, 52, **53,** 54, 95, 132, **212,** 250, 278

Holostei (*see* Holosteans)

Homeotherms, 89

Homocercal caudal fin, **161,** 524, **525**

Homodont dentition, 111, 112

Homology, 4, **5**–7, 144

 serial, 6, 143, 321

 sexual, 6, 299, 301, 307

Hoofs, 101

Hooklets of feather, **99,** 100

Horizontal septum (of heart), **253**

Hormones, 377

Horn (= keratin), 88, 90, 96, **97,** 98, 101, 105, 106

Horned lizard, **467**

Horns, **105**

Horses:
  body support of, **398, 427**
  cursorial adaptations of, 438,
    440, 445, **453**, 454, **456,**
    457, **458, 459**
  feeding adaptations of, 565,
    570, **571, 581**
  skeleton, **15, 16**
Hovering flight, 545, **546**
Humeral-rotation digging, 465,
  466, 468, 470, 472, 475,
  476
Humeroantibrachialis muscle,
  192, 193
Humerus, **168, 174, 455,** 465,
  466, 472, 473, 496, 512,
  521, 536
Hummingbirds, 537, 545, **546,**
  553, 567, 574
Hyaline cartilage, **118,** 171, 398,
  406, **407**
Hydrostasis, 238, 522**–525**
Hydroxyapatite, 91, 119
Hyenas, 69, 565, 578, 579,
  582
Hyoglossus muscle, **199**
Hyoid apparatus, 132, **134,**
  239–241, 573, 574
Hyoid constrictor muscle, **190**
Hyoidean arch, **124,** 126, 131,
  186, 187, 191, **228,** 324
Hyomandibula, **124,** 126, **127,**
  129, 130, 131, **132,** 135
Hyostyli, **127,** 129, 131
Hypapophysis, 148
Hypaxial muscles, 157, **158,**
  185, **186,** 187, **189,**
  190–192, 193
Hyperstriatum, 342, **343,** 345
Hypobranchial artery, **262**
Hypobranchial muscles,
  186–188, **190,** 191, **192,**
  193, **196,** 322, 325
Hypobranchial nerve, **323,** 325
Hypobranchial skeletal elements,
  124
Hypocercal caudal fin, 161, 524,
  525, 534
Hypogastric vessels, **266, 271**
Hypoglossal foramen, **142**

Hypoglossal nerve, **145,** 186,
  323, 325, **334–336, 351**
Hypomere, **80,** 82, 123, 185,
  186, 187, **204, 249**
  (*see also* Development, of gut,
    heart, muscles)
Hypophyseal pouch, **211,** 378,
  379
Hypophysis, **121,** 339, **345, 349,**
  351, 378, **379–381,** 385
Hypothalamus, 339, **344,** 346,
  347, 349, **351,** 379, **381,**
  385
Hyposodont teeth, 580, **581**
Hypurals, **161,** 162
Hyracoidea (*see* Hyraxes)
Hyraxes, 71, 490, 501, **563,**
  565, 570

Ichthyosauria (*see* Ichthyosaurs)
Ichthyosaurs, **17,** 18, 38, 63, **65**
  aquatic adaptations of, 507,
    513, 514, 520, 528
  skull and teeth of, **139,** 576
Ichthyostegalia, 153, **154**
Iguana (iguanids), **134, 159, 195,**
  239, 486, 493, 570
Iliac artery, **260,** 263, **265, 266**
Iliac vein, **260,** 264, **265, 266,**
  272, 278
Iliocostalis muscle, **196, 200**
Iliofemoralis muscle, **197**
Ilio-ischiadic foramen, **166**
Iliolumbar vessels, **273, 274**
Iliopsoas muscle, **199**
Iliotibialis muscle, **197**
Ilium, **155,** 165, **166, 478**
Impala, 438, **441**
Incisors, 113, **114**
  feeding specializations of, **563,**
    564, 570, **571,** 573, 579
  fossorial adaptations of, 464,
    467, 469, 471, **477**
Inclinator muscles, **189**
Incus, 134, **135,** 143, 364, **367**
Indifferent stage of genital devel-
  opment, 294, 300, 301,
  307
Indris (indrisoid lemurs), 486,
  489, 494, **500,** 501

Induced drag, 541, **542,** 551,
  554
Induction, 82
  and integument, **88,** 89, 91,
    100
  and nervous system, 82
  and sense organs, 360, 362,
    372
  and teeth, 109, 111
  and urogenital system, 283
Inertia, 431, 432, 440, 511,
  515, 560, 577
Inferior jugular vein, 264, **265,**
  268
Inferior lobes of hypothalamus,
  349, 350
Inferior oblique muscle, 190, **191**
Inferior olive, 335, 337, 340,
  347
Inferior rectus muscle, 190, **191**
In-force, **417,** 418, **419, 420,**
  471, 474, 475
Infraorbital canal, **143**
Infraorbital nerve, **145, 323**
Infraspinatus muscle, **174,** 196,
  200, 522
Infraspinous fossa, **174**
Infundibulum, 378, **379**
In-lever, **417,** 418, **419, 420,**
  430, 446, **447, 448,** 471,
  472, 474, 495
Inner cell mass, **77,** 78
Inner ear, 238, **365, 367**
Inner epidermal generation, **97**
Innominate bone, 165, **175,** 477,
  478
Insectivora (*see* Insectivores)
Insectivores, 68, **70,** 176, **353**
Insectivorous vertebrates, 578
Insulin, 225, 384
Integument, 87–106
  functions of, 87, 88
  of amphibians, **96**
  of birds, 98, **99–101**
  of cyclostomes, 90
  of jawed fishes, **90–93, 94,** 95
  of mammals, 100–**102**–106
  phylogeny of, 106
  of reptiles, **95,** 96, **97, 98**
Interatrial foramen, **255**

Interatrial septum, 252, **255**
Intercalary plate, 151, **152**
Intercalated discs, **180**
Intercentrum, 147, 153, **154, 157,** 158
Interclavicle, **159,** 163, **164,** 165
Intercostal muscles, 193, **200,** 245
Intercostal vessels, **271, 273**
Interhyoideus muscle, **190, 192**
Intermandibularis muscle, 189, **190–193, 195, 196,** 198, **199**
Intermediate mass of thalamus, 338, **344, 345**
Intermediate segment of nephron, 288, 290
Intermedium, **168, 171, 172,** 173
Internal capsule, **335, 343, 344,** 348
Internal carotid artery, **260,** 261, **262,** 263, **266, 270, 272, 273,** 277
Internal gills, 228–231–234
Internal glomerulus, 282-286, **287**
Internal iliac vessels, **270, 272–274**
Internal jugular vein, 264, **266, 268, 271, 274**
Internal mammary vessels, **273, 274,** 277
Internal nares, **55,** 56, 136, **138, 140, 142,** 244
Internal oblique muscle, 193, **199**
Interneural cartilage, 151
Interneurons (= association neurons), 315, 316, 332, 348
Interrenal organ, 382, 383
Intersegmental vessels, 150, **266**
Interstitial cells of testis, **298,** 384
Intertemporal bone, **126, 130, 133**
Interventricular foramen, 255
Interventricular septum, **252, 255**
Intervertebral discs, 398, 406
Intestinal vessels, **273, 274**
Intestines, **30, 46, 211, 216–218–221,** 223

Invagination, **78,** 79
In-velocity, 430
Involution, 79
Iridophores, 89
Iris of eye, **327, 368,** 372, **373**
Ischiadic vein, **271, 272**
Ischiocaudalis muscle, **192**
Ischioflexorius muscle, **192**
Ischium, **155,** 165, **166, 478**
Islets of Langerhans, **225,** 226
Isopedine, 95

Jawless vertebrates, 37, 38, 41, **42,** 43, **44, 45**
  body skeleton of, 150, 151, 158, 166
  digestive system of, 212, 215, 217, 218
  head skeleton of, **42, 44,** 127, **128**
  respiratory system of, **230,** 232
  (*see also specific subclasses and orders*)
Jaw mechanics, 559, **560,** 561, **562**
Jaws, 45, 124, 126, **127,** 128, 559, **560,** 561, **562,** **563**–566
  (*see also* Mandible)
Jaw suspension, 126, **127**–131
Jerboa, **17,** 438, 451, **455, 460,** 468, 469
Joint capsule, 406, **407**
Joint cavity, 406, **407**
Joint receptors, 356
Joints, 404, **405,** 406, **407, 408**–410
Jugal bone, **126, 130, 133, 137–142**
Jugular ganglion, 325
Jugular lymph duct, **279**
Jugular sinus, **266**
Jugular vein, **267, 272,** 278, 528
Jumping, 437, 438, **439–441,** 457, 459–462
Juxtaglomerular cells, 385

Kangaroo rat, **17,** 438, **440,** 459, 462, 468, 469

Kangaroos, 68
  feeding adaptations of, 221, **222,** 570
  locomotor adaptations of, 438, **440, 447,** 451, 457, 486
Keratin, 88, 90, 96, **97**–99, **101,** 105, 106
  α keratin, 88, 96, **97,** 99
  β keratin, 88, **97,** 99
Kidneys:
  of cephalochordates, 30, 32, 33
  circulation in, **260, 267, 268,** 271, 272, **275**
  development of, 282–**284, 286**
  evolution of, 283–287
  function of, 271, 272, **275,** 287–**289,** 290, **291**
  general nature of, 281, **282**
  structure of, 284–288
Kinetic energy, 434, 543–545
Kinetic skulls, 137, 561, **562**
Kinkajou, **451,** 455, 499, 500
Kinocilium, **360,** 363
Klipspringer, 490, 492
Knee joint, 406, 409

Labial cartilage, **129, 190, 191**
Labia of genital system, 299
Labyrinthodont dentition, 111, **112**
Labyrinthodontia (*see* Labyrinthodonts)
Labyrinthodonts, 37, 38, 58, **59,** 427
  body skeleton of, 153, **154,** 157, 158, 163, **164,** 165, **166**
  head skeleton of, 131, **133,** 136, 561
  integument of, 95, 96
  respiratory system of, 239, 245
  teeth of, 111, **112**
Labyrinth of ear, 362–365
Lacrimal bone, **126, 130, 132, 137, 138, 140, 141**
Lacrimal gland, 290, **327,** 371, 375, 527
Lacteals, 248
Lactogenic hormone, 379

Lacunae, of bone, 92, **93,** 119, 395
  of cartilage, 118
Lagena, **362,** 364
Lagomorpha, 69
  (*see also* Rabbits)
Lamellar bone, 92, **93,** 170
Lamina propria, 213, **214**
Laminar flow, **510,** 513, 541
Lamina terminalis, **345**
Lampreys, 37, 38, 43, **45,** 46, 205, 206, 257, 516
  brain of, **349,** 350
  head skeleton of, **128**
  hypophysis of, **380**
  nerves and spinal cord of, 317, **318,** 319, **320**
  respiratory system of, **230,** 232
  sense organs of, **362,** 374, 376
  urogenital system of, 290, 291, 296, 305, **306**
Langur, 221, **222,** 486, 493, 495, 570
Lap joints, 404, **405**
Large intestine, **211, 216,** **217**–221, **223**
Laryngo-tracheal groove, **211**
Larynx, 132, **134,** 239–241, 243, **327,** 526
Lateral abdominal vein, **260,** 264, **265, 275**
Lateral branches of dorsal aorta, 261, 263
Lateral circumflex vessels, **273, 274**
Lateral digital extensor tendon, **396, 426**
Lateral epicondyle, **174, 473**
Lateral funiculus, **317,** 337
Lateral geniculate body, **335,** 338, 339, **347**
Lateral line canal, **360**
Lateral line system, **51,** 202, 325, 340, **360,** 361
Lateral septum, 185, **186,** 188, **189**
Lateral undulation of snakes, 480, **481**
Lateral ventricles of brain, **331,** 338, 341, **343, 344**

Laterobronchi, **242,** 243
Latimeria, 94, 233, 289, 523
Latissimus dorsi muscle, **193,** 194, **196–200,** 472, **474–476,** 497
Lead of feet in running, 457, **459**
Leaping, 493, 494
Lemurs, 69, 457, 486, 494
Lens muscles, **327, 374,** 375
Lens of eye, **368, 369, 472–474, 476,** 527
Lens placode, 372, **373**
Lens vesicle, **373**
Lentiform nucleus, 342, **344**
Lepidomorium, 91
Lepidosauria (*see* Lepidosaurs)
Lepidosaurs, 37, 38, 61, **62**
  body skeleton of, **159, 164, 172**
  circulatory system of, **262**
  epidermis of, 96, **97**
  head skeleton of, **137, 139**
  teeth of, **112**
  urogenital system of, **302,** 307
Lepidosteoid scales, 95
Lepidotrichs, **160,** 161, **162, 163,** 167
Lepospondyli, 37, 38, 58
Leucocytes, 256
Levator mandibulae muscle, **193**
Levator muscles, 179, 189
Levator scapulae muscle, 194, **196, 199, 200, 474**
Lever arm, 185, 418, **419, 420,** 430
Levers, 418, **419, 420,** 430, 446, **447**
Lift, 538, **539**–541, **542,** 546, 547, 549, 551, 553
Ligament(s), 178, 391, 392, 394, **396,** 397, **398, 400,** 403, 404, 406, **408,** 425, **426, 427,** 561
Limbic system, 346
Limbs:
  evolution of, 170, **171–175**
  origin of, 167, **168**
Lingual vessels, **266, 273, 274**
Lions, 437, **441, 447, 579**
Lips, 213, **467,** 469, 566

Lissamphibia, 37, 38, 58, **59,** 60
  body skeleton of, 153, **154, 172**
  circulatory system of, **252,** 253, **256, 262**
  head skeleton of, 131, **133**
  teeth of, 111, **113**
  (*see also specific orders*)
Liver:
  circulation of, 223, **224, 260,** 265–**267,** 269, 270
  development of, **207,** 210, **211,** 223
  evolution of, **46, 216–218,** 225
  functions of, 221, 522
  structure of, 223, **224,** 225
Lizards, 38, 61, **62,** 95, 576
  aquatic adaptations of, 507, 513, 516
  chondrocranium of, **121**
  circulatory system of, 253, **262, 279**
  coelom of, **205,** 206
  cranial kinesis in, 561, **562**
  cursorial adaptations of, 438, **439,** 443, **446,** 457, 461
  digestive system of, **217**
  fossorial adaptations of, 464, **465,** 466, 469, 479
  gliding of, **533,** 534
  integument of, **97,** 98
  muscles of, **195**
  scansorial adaptations of, 486, 493, 500, 501
  (*see also* Anoline lizards, Chameleons, and Geckos)
  sense organs of, 358, **362, 365, 374,** 375, **376**
  skeleton of, **5, 134, 137, 139, 159, 164, 172**
  urogenital system of, 292, **302,** 307, 308
Load, 390, 391, 416
Lobes:
  of kidney, **287**
  of liver, 225
  of lung, 240, **241**
  of testis, **299**
Lobules of liver, **224**
Locking devices, 424–**427,** 556

Long bones, **169, 170**
Long digital extensor muscle, **175, 522**
Long digital extensor tendon, **396**
Long digital flexor muscle, **195**
Long fibularis tendon, **396**
Longissimus dorsi muscle, **196, 200**
Long soaring wing, **552**, 553
Loons, 507, 520, 523, 568, 576
Loop of Henle, 290, **291**
Lorises, **372, 455,** 486, 494, 495, 497, 498
Lubrication of joints, 406, **407**
Lumbar enlargement of spinal cord, 318, **319**
Lumbar plexus, **319**
Lumbar vertebrae, **155,** 156, **450, 461,** 462, 497, **498, 555**
Lumbodorsal fascia, 200
Lumbosacral plexus, **319,** 320
Lung buds, **207, 211,** 236
Lungfishes, 38, **55,** 56, 236, **380**
(*see also* Dipnoans)
Lungs:
  development of, **207, 211,** 235, 236
  of fishes, **212,** 236, 237
  innervation of, **327**
  relation to coelom of, **205, 207**
  of tetrapods, 238, **239,** 240, **241,** 244, 245
Lures, **567,** 568
Luteinizing hormone, 379
Lymphatic system, 247, 248, 277–**279**
Lymphatic vessels, 248, **257,** 258, 278, **279,** 354
Lymph hearts, 248, 278, **279**
Lymph nodes and nodules, 214, 217, **257,** 278, **279**
Lymphocytes, **256**–258, 278
Lymphoid cells, 257
Lymph propulsors, 278, **279**
Lymph sinuses, 248, **257,** 278, **279,** 296, **297**

Machines, **417**
Mackerel, 506, **519,** 523

Macrolecithal eggs, 75
Maculae, **362,** 363, 366
Malleoli of tibia, **175**
Malleus, 134, **135,** 142, 143, 364, **367**
Mammalia (*see* Mammals)
Mammal-like reptiles (= synapsids), 38, 63, **65,** 104, 366
  skeleton of, **135,** 136, **139, 164**
Mammals, 37, 39, 68, **69, 70,** 71
  appendicular skeleton of, **164,** 165, **166, 171, 172,** 173, **174, 175,** 176
  axial skeleton of, **134, 135, 139–145, 154,** 156, **157**–159
  brain of, **331, 334–336, 337–340,** 341–**343–345,** 346, **347,** 352, **353, 354**
  circulatory system of, **254–256**–258, **262**-264, **267, 268**–272, **273–275,** 276–**279,** 280
  coelom and mesenteries of, 206, **207**
  digestive system of, **213,** 215, 216, **218,** 219–**222–224**
  early development of, **77,** 78, 81
  endocrine glands of, 378, **379, 380,** 382–384
  excretory system of, **287, 289,** 290, **291,** 292
  hair of, **102**–104
  integument of, 100–**102, 103**–106
  jaw mechanics of, **563–565,** 566
  mammary glands of, 102, **103,** 384
  nerves and spinal cord of, **315, 317–320,** 323, 324, 326, **327**
  reproductive system of, 295–**297–299,** 302, **303, 304,** 305, **306**–309
  respiratory system of, 204, **241,** 245

sense organs of, 358–360, **362, 365, 367, 368, 370,** 371, **372,** 375
  teeth of, **108,** 113, **114,** 115
Mammary glands, 102, **103,** 384
Mammillary body, **335,** 339, **345,** 346, **347,** 348, **351, 381**
Mammoth, 580, **581**
Man:
  brain of, **334**–336, 338, **340,** 346, 348
  physical abilities of, 438, 459, 461, 494, 587
  spinal cord and nerves of, **318, 319**
Manatee, 71, 507, **509**
(*see also* Sirenians)
Mandible, **123, 124, 126, 129, 401, 560, 561–565, 576, 577, 583**
(*see also* Jaws)
Mandibular arch, **123, 124, 126, 127,** 130, **186,** 187, 191, **228,** 324
Mandibular branch of trigeminal nerve, **145, 323,** 324, 334
Mandibular cartilage, 124
Mandibular condyle, **143, 563,** 564
Mandibular fossa, **143,** 564, 565
Mandibular salivary gland, **213**
Mandibular symphysis, 406, 565, 579
Maneuverability:
  in air, 538, 548–**550,** 551
  in water, 523–**525**
  on land, 453, 454, **456,** 469, 470
Mantle layer of nervous system, **315,** 316, 332
Manubrium, **159, 476, 555**
Manus, **168,** 169, **500**
Marginal layer of nervous system, **315,** 332
Marginocostal vessels, **270, 271**
Marlin, **506,** 519
Marmot, **70**
Marrow cavity, **170**
Marsupialia (*see* Marsupials)

Marsupial mole, 18, 466, **467,**
    469–471, 474, 475, 480
Marsupials, 39, 68, **69,** 84, **141,**
    309
  genital tract of, **304,** 308
  locomotor and feeding adapta-
      tions of, **440, 447,** 466,
      467, 486, **489,** 507, 535
Mass, 431, 432, 450, 460
Masseter muscle, **144,** 198, **199,**
    565
Mastiff bat, 537, 547, **552**–554
Mastoid process, **142, 143**
Mastoid sinuses, 367
Mauthner cells of medulla, 350
Maxilla bone, **54, 126, 130, 132,**
    133, 136, **137, 138,**
    140–142, 560, 561, **578**
Maxillary branch of trigeminal
    nerve, **145, 323,** 324
Maxillary vessels, **273, 274**
Medial geniculate body, **335,**
    336, 338, 347
Median eminence of neurohy-
    pophysis, 378–**380, 381,**
    385
Median epicondyle, **472, 474,**
    475
Median longitudinal fissure, **344**
Mediastinum, 206, **207**
Medulla:
  of adrenal gland, 382, 383
  of brain, 331–**334, 335, 345,**
      349
  of gonad, 294, 297
  of hair, **102,** 104
  of kidney, **291**
Meibomian glands, 102
Melanin, 89
Melanocyte-stimulating hormone,
    379
Melanophores, **88,** 89
Melanosomes, 89
Melatonin, 385
Membrane (= dermal) bone(s),
    94, 118, 125, **126,** 130,
    150
Meninges, 317, 353, **354,** 373
Menisci, 406
Merganser, **244, 502, 576**

Meroblastic (= partial) cleavage,
    77, 78
Mesectoderm, 108
Mesencephalon, 330, **331,** 332,
    336
Mesenchyme, 81, 83, **88**
Mesenteric ganglia, **327**
Mesenteric vessels, 263, **265,**
    266, 270, 272–274
Mesenteries, 203, **204,** 205–**207**
Mesethmoid bone, 122, **123**
Mesocoel, **331**
Mesoderm, 27, **78–81,** 82
  (see also development of spe-
    cific organs)
Mesolecithal eggs, 75
Mesomere, **80,** 82, 83, 282–286,
    293
Mesonephric duct, 285, **286,** 302
Mesonephros, **284,** 285, **286**
Mesopterygium, **162,** 167
Mesorchium, 293, **296,** 301, **302**
Mesos layer of epidermis, **97**
Mesovarium, 293, **295–297, 302**
Metacarpal bones, **168,** 169,
    171, 472, 473, 536
Metanephridia, 281, 282
Metanephros, **284,** 285, **286,**
    287, 291, 302, 303
Metapodial bones, **168,** 169,
    171, 172, 173, 451, **453,**
    520, **521, 536**
Metapophysis, 148
Metapterygium, **162,** 167
Metapterygoid bone, **132**
Metatarsal bones, **168,** 169, **172,**
    453
Metatheria, 39, 68, **69**
Metencephalon, **331,** 332
Mice, 69, 457, 468, 488
Microlecithal eggs, 75
Midbrain, 330, **331**–333, **349,**
    353, 379
  (see also specific structures)
Middle commissure, 338
Middle ear, **228,** 229, **365, 367,**
    527
Middle peduncles of cerebellum,
    335, 336, 337, 341
Midgut, **211**

Milk, 102, 103, 529
Milk lines, 103
Milk teeth, 113
Modulus of elasticity, 391
Molars, 113, **114,** 115, 579
Molecular layer of cerebellum,
    339, **340**
Mole-rats, 464, 465, 468–470,
    474, 475, **477,** 479, 480
Moles, 68, 465, **466, 467,**–472,
    475, **476, 477, 478,** 479,
    566
Moll's glands, 102
Moment (of force), 418
Moment of inertia, 432
Momentum:
  angular, 432
  linear, 431
Monkeys, 69, **172, 222, 433, 458**
  scansorial adaptations of, 486,
      493–495, **496, 498, 499**
Monocytes, 257
Monorhina, 37, 38, 43, **44,** 45
Monotremata (see Monotremes)
Monotremes, 39, 68, **69,** 367
  fossorial adaptations of, 466,
      469, **472,** 479
  pectoral girdle of, **164**
  reproductive system of, 299,
      304, 305, **306**–308
  (see also Echidnas; Platypus)
Morphology, 3
  (see also Vertebrate morphology)
Mosasaurs, 427, 507
Motion mechanics, 431, 432
Motor fibers, 314, 316, 332
  (see also subtypes)
Motor units, 182
Mountain beaver, 8, 468, 471,
    479, 563
Mouth, **211**–213, **228, 230, 456,**
    469, **560**–562, 572–574,
    576
Mucosa of gut, 213, **214**
Mucous glands:
  of gut, 213, 215, 217
  of integument, 88, **90, 96,** 98,
      106
  of olfactory epithelium, **357,**
      358

Mucus, 213, 215, **357,** 570
Mudhen, **520**
Multifidus muscle, **200**
Murre, 507, **552**
Muscle:
cardiac, 179, **180,** 249
kinds of, 179, **180**
skeletal, **180, 181**
smooth, 179, **180,** 204, **214,**
240
Muscle fibers, **180, 181**
Muscles:
actions of, 179
of amphibians, 191, **192, 193**
of birds, 194, 195, **197, 198**
and bone-muscle systems,
417–421
categories of, 185, **186,** 187
contraction of, 181, **182, 183**
of cyclostomes, 188
functions of, 177, 178
homologies of, 187, 188
innervation of, 185, **186,** 187,
191
of jawed fishes, 188, **189–192**
of mammals, **144, 174, 175,**
193–**199, 200,** 201
names of, 179
origins and insertions of, **144,**
174, 175, 178, 179
physiology of, 182, 183
of reptiles, 193–**195, 196**–198,
201
shapes of, 178
union to tendons and bones,
403, 404
Muscle spindle, **356**
Muscularis externa, **214**
Muscularis mucosae, 213, **214**
Muskrat, **223, 447,** 516, 517
Myelencephalon, **331,** 332
Myelin, **312, 313,** 316, 317
Myelinated fibers, 312, **313,** 314,
326, 332
Mylohyoid muscle, **195,** 198,
199
Myocardium, 249
Myocoel, **80,** 204
Myofibrils, 180, **181**
Myoglobin, 519, 528,

Myomeres, 188, **190,** 519
Myometrium, 304
Myosepta, 188, **189, 190,** 191,
192, 519
Myosin, **181**
Myotome, **80,** 83, 185, **186**

Nasal bone, **126, 132, 133, 137,**
138, 140
Nasal capsules, **121,** 122,
128–130
Nasal pit, 357, **358**
Nasolabial grooves, 358
Nasolacrimal canal, **143**
Navicular bone, **172, 494**
Neck, 57, 154, 155, **398, 465,**
467, 471, 475, 512, 555
Neck segment of nephron, 288,
291
Nectar and pollen eating, 220,
574
Necturus, **20,** 193, 262, **266, 296**
Neocerebellum, 340, **351, 353**
Neocortex, 342, **343,** 345, 346,
353
Neognathae, 39, **67**
Neornithes, 37, 39, 66, **67,** 155
Neostriatum, 342, **343**
Neoteny, 20, 144, 234
Nephric ducts, **282, 283, 286,**
291, **296, 300,** 301, 305,
306
Nephric ridge, 283, 293, 300
Nephric tubules, **275,** 281, **282,**
283, 286, 288, 290, **300,**
301
Nephrocoel, **80,** 204, 283
Nephrogenic cord, 283, **284, 286**
Nephron, 282–**287, 289, 291**
Nephrostome, **282,** 283–287,
303
Nephrotome, 283, **284,** 287
Nerve cell body, 311, **312,** 314,
316, 332
Nerve impulse, 313, **314**
Nerves, 314
Nervous system:
divisions of, 314, 315
elements of, 311–313

nerve impulse, 313, **314**
relation to endocrine glands,
377
synapse, 313, **314**
(*see also specific divisions of*)
Neural arches:
of chondrocranium, **121,** 122
of spine, **128, 148, 149,**
150–**154, 157,** 162
Neural canal, **148, 151,** 317, **318**
Neural crests, 82
and acousticolateralis system,
360
and endocrine glands, 383
and integument, 88, 89, 91
and nervous system, 316, 321,
353
origin of, **80,** 82
and teeth and skeleton, 108,
118, 123
Neural folds, **80, 81,** 82
Neural groove, **80, 81,** 317, **318**
Neural plate (of embryo), 79–**81,**
82
Neural plate (of selachian spine),
151, **152**
Neural spine, **148, 149, 151,**
152, 156, 160, **161,** 162,
400, **401, 450, 461,** 462,
518
Neural tube, **79, 80,** 82
Neurilemma, **312,** 313
Neurocoel, **80,** 82, **315,** 331,
353, 354
Neuroglia, 313, 315
Neurohypophysis, 378, **379,**
380, 385
Neuromasts, **360,** 361
Neurons, 311, **312,** 315, 316
Neurosecretion, **381,** 385
Neurotransmitter, 314
Neurula, 82
Neurulation, **80, 81,** 82
Nictitating membrane, 371, 375
Nighthawk, 566, **569**
Nipple, **103,** 512
Nissl granules, **312**
Nodes of Ranvier, **312,** 356
Nodose ganglion, 325
Nomenclature, 35

Noradrenalin, 314, 327, 328, 383
Nose pad, 465, **467, 477,** 480
Nostrils, 469, 526, 568, 575
Notochord, 150
  of ammocoetes, **46**
  as chordate feature, 28
  of cyclostomes, **46, 128,** 150
  development of, **79–81**
  of protochordates, 28, **29, 31,** 32
  relation to chondrocranium of, **121**
  sheaths of, 150, **151**
  of vertebrates, 150, **151,** 153, **154, 161,** 330
Nuchal crest, **143**
Nuchal ligament, 397, **398,** 424
Nuclei of brain, 332
  (see also *specific nuclei*)
Nuclei of cranial nerves, 333, **334, 336, 347**
Nuthatches, 486, **488**
Nutria, 507, **509, 563**

Oberhautchen, **97**
Obturator foramen or fenestrum, 165, **166, 175**
Obturator muscles, **175,** 197, **200**
Obturator nerve, **319**
Occipital arch, 122
Occipital condyles, 122, 131, 134, 141, **143,** 156
Occipital nerves, **323**
Oculomotor nerve, **145,** 186, 190, **323,** 324, **334, 335, 349, 351**
Odontoblasts, **109**
Odontognathae, 39, 66
Olecranon fossa, **174**
Olecranon process, **174, 429, 473, 474, 476,** 495, **496,** 497
Olfaction, 357–359, 526
Olfactory bulbs, 341, **345, 347, 349, 351, 353,** 357
Olfactory epithelium, 323, 341, **357,** 358
Olfactory nerve, **323,** 341, **351**
Olfactory nuclei, **347,** 348

Olfactory pathways of brain, 338, 342, 346, **347,** 348
Olfactory tracts, 341, 346, **347, 349, 353**
Oligodendroglia, 313
Olivary nuclei, **335, 336, 347,** 497
Omasum, **222**
Omenta, 203, **204,** 207
Omental bursa, 207
Omnivorous diet, 580
Ontogeny, 18
  (see also Development; Embryology)
Oogonia, 295, **297**
Open rooted teeth, **582**
Opercular bones, **130, 131, 132**
Opercular gills, **230,** 233, 234
Operculum, 49, **51–53, 130,** 131, **132,** 133, **230,** 233, 514
Ophthalmic branch of trigeminal nerve, **145, 323,** 324
Opisthocoelous vertebrae, **149,** 153, 406, 409
Opisthonephric duct, 285, **295, 296,** 302
Opisthonephric tubules, 285, 300
Opisthonephros, **46,** 284–286, **295, 296**
Opisthotic bone, **123,** 141
Opossums, 68, **69, 141,** 443, 501, 507, 526, 574
  scansorial adaptations of, 486, 494, 495, 498, 499
Optic canal, **143**
Optic chiasma, **335,** 339, **345, 379–381**
Optic cup, 372, **373**
Optic lobes, 337, **349, 351, 376**
Optic nerve, **145, 323,** 324, **335,** 339, **349, 351, 368, 370,** 372, **373**
Optic stalk, 372, **373**
Optic tracts, **335,** 339, **344, 347, 351,** 497
Optic vesicle, 372, **373**
Oral cavity, 211–213, **228,** 244
Oral glands, **213**

Oral hood, **46**
Orangutan, 486, 494, 495, 497
Orbit of skull, **126, 129**
Orbitosphenoid bone, 122, **123, 140, 141**
Order (taxon), 34
Organ of Corti, 364, **365,** 366
Ornithischia, 63, **64**
Orthodentine, 92, 107
Osmoregulation, 87, 232, 271, 272, 275, 287–**289,** 290, **291**
Osteichthyes, 37, 38, 51, **53–55,** 56, 127
  (see also Bony fishes)
Osteocytes, 92, **93**
Osteodentine, 92, 107
Osteoderms, 94, **95**
Osteons, 92, **93,** 119, 170
Osteostraci, 38, 43, **44**
  (see also Cephalaspids)
Ostium, **295, 296,** 303, 304
Ostracoderms, **42,** 43, **44,** 235, 324
  armor shields of, 93, **94**
  head skeleton of, **42, 44,** 127, 128
  sense organs of, 358, 361, 366, 376
  (see also *specific orders*)
Ostrich, 67, 318, 428, 570
  cursorial adaptations of, 438, **439,** 453
Otic capsules, **121,** 122, **123, 128, 129**
Otic notch, **126**
Oticoccipital unit of braincase, **130**
Otic vesicle, **46,** 362, 366
Otoliths, **29,** 356, **362,** 363, 366
Otters, 507, **509,** 513, 516
  (see also Sea otter)
Outer epidermal generation, **97**
Out-force, **417,** 418, **419, 420,** 421, 448, 471, 474, 475
Out-lever, **417,** 418, **419, 420,** 430, 446, **447, 448,** 471, 495
Out-velocity, 430, 446
Oval body of gas bladder, **237**

Oval window, 364, **365, 367**
Ovarian follicle, 295, **297, 302**
Ovarian funnel, **296, 302**–304
Ovarian vein, 263, **275**
Ovary:
    comparative anatomy of, **30,**
        294, **295–297, 302, 303**
    development of, 293, **294**
    endocrine function of, 383,
        384
    structure of, 294**–297**
Oviduct, **295, 296, 302–304**
Oviparous vertebrates, 68, 308
Ovisacs, 304
Ovoviviparous vertebrates, 308
Ovum, 75, **76,** 295**–297**
Owls, **372,** 457, 568, 578, 579
Oxytocic hormone, 379

Paca, **563,** 565
Pace, 454, **458**
Pacemaker, **254,** 255
Pacinian corpuscle, **356**
Paddlefish, 38, 54, 570, **572**
Paedomorphosis, 20, 34
Pain sense, 356
Paired appendages:
    development of, **169, 170**
    evolution of, 166**–168,** 170,
        171, 172, 173
    origin of, 45, 159, 160, 167,
        168, 169
Palaeoniscoid fishes, 38, 54
Palaeoniscoid scales, 95
Palate, 136, **140,** 526
Palatine bone, **126,** 130, 136,
        137, 138, 140–142
Palatine foramena, **138, 142**
Palatine nerve, **145**
Palatoquadrate bone, **124,** 126,
        127, 128, **129**
Paleocerebellum, 340, **353**
Paleocortex, 342, **343,** 345, **353**
Paleognathae, 39, 66, **67**
Paleontology, 10, 11, 117
Paleostriatum, 342, **343**
Pancreas, 210, **211, 216–218,**
        225, 226, 272, 384
Pancreaticoduodenal vessels,
        266, 270, 272–274

Pancreaticomesenteric vessels,
        265
Pancreozymin, 216
Pangolins, 69, 101, 103
    feeding adaptations of, 221,
        575, 576
    fossorial adaptations of,
        467–469, 471, **473,** 474,
        477, **478,** 479
    scansorial adaptations of, 486,
        499, 501
Parabronchi, **242, 243,** 245
Parachordal cartilages, **121,** 122,
        130
Parachuting, 531**–533,** 534
Parafollicular cells, 382
Paraganglia, 383
Parahormones, 385
Parallelism, **17,** 18
Parallelogram of forces, **412–414**
Parallelogram of velocities, 429,
        430
Paramesonephric ducts, **300,**
        301, 303**–306**
Paraphysis, 338, **376**
Parapineal organ, 376
Parapophysis, 148
Parasite drag (= profile drag),
        541
Parasphenoid bone, **126, 132,**
        136, **137, 140,** 142
Parasympathetic ganglia, **327,**
        328
Parasympathetic nervous system,
        326, **327,** 328, 339
Parathormone, 382
Parathyroid gland, 382
Parietal bone, **126, 130, 132,**
        133, 137–141
Parietal eye, **376**
Parietal foramen, **133,** 376
Parietal organ, 338, **349, 376**
Paroccipital process, **142, 143**
Parotid salivary gland, **213**
Parrots, 486, **488,** 499, **500,** 570,
        571
Pars distalis of adenohypophysis,
        379, **380, 381**
Pars intermedia of adenohy-
        pophysis, 379, **380**

Pars nervosa of neurohypophysis,
        379, **380, 381,** 385
Parthenogenesis, 308
Patagium, **536,** 537
Patella, 176, 196, **427**
Patellar groove, **175, 455, 496**
Patellar ligaments, **175, 403, 427**
Pecten of eye, **372,** 375
Pectineus muscle, **199**
Pectoantibrachialis muscle, **199**
Pectoral fins, **50, 55**
    (*see also* Fins; Paired appen-
        dages)
Pectoral girdle, **159, 162,** 163,
        164, 165, 555
Pectoralis muscle, **192, 193,**
        195–200, 473, 474, 475,
        476, 497
Pectoral vessels, **271, 272**
Pedicel of eye, **374**
Pedicels (of podocytes), **287**
Pedicels (of teeth), 111, **113**
Peduncle of tail, **506,** 513, 514,
        515, 519
Peg-and-socket joints, 405, 409
Peizoelectric properties, 356,
        402
Pelicans:
    aquatic adaptations of, 507,
        520, 526
    feeding of, 568, 576, 577
    skeleton of, **149, 171, 172**
Pelvic fins, **54**
    (*see also* Fins; Paired appen-
        dages)
Pelvic girdle, **8, 163,** 165, **166,**
        175, 477, **478**
Pelvic symphysis, 165, **166, 175,**
        384, 398, 406, 477, **478**
Pelvic vein, **266, 271**
Pelycosauria, **5,** 38, 63, **65**
Penguins, **5**
    aquatic adaptations of, 507,
        508, 513, 520, **521,** 523,
        525–527
    feeding adaptations of, 220,
        576
Penis, **303, 306,** 307, 512
Perch, **54, 189,** 573
Perennibranchiate urodeles, **234**

Pericardial cartilage, **128**
Pericardial cavity, **205, 207, 212, 230**
Pericardium, 206, **207**
Perichondrium, 118
Perichordal centrum, 152
Perichordal tube, 150, 156
Perilymph, 363–**365, 367**
Perimysium, 180, **181,** 403
Periodontal membrane, **108**
Periods (geologic), **37**
Periosteum, **93, 169, 170, 181,** 403, **407**
Periotic bone, **123, 141**
Peripheral nervous system, 314
Perissodactyla (see Perissodactyls)
Perissodactyls, **70,** 71
  cursorial adaptations of, 438, 443, 451, 453
  feeding adaptations of, 221, 565, 570, **571**
Peritoneal canal, **282,** 283
Peritoneal cavity, **205,** 206
Peroneus longus muscle, **197**
Pes, 169, **500**
Petrels, 290, 507, 526
Petrosal ganglion, 323, 325
Phagocytosis, 256, 258
Phalangeal formula, 173, 174
Phalangers, 68, 486, **489,** 494, 498, 499, 535, 567
Phalanges, **168,** 171, 173, 174, **472, 473, 474, 476,** 479, 501, 520, **521**
Pharyngeal membrane, **211, 228**
Pharyngeal pouches, **211, 228,** 278, 364
Pharyngeal rami of cranial nerves, **322**
Pharyngeal teeth, **212,** 577, 582
Pharyngobranchials, **124, 129**
Pharynx:
  as chordate feature, 28
  development of, **211, 228,** 229
  evolution of, **212, 230,** 232–234, 573
  nature of, 122, 123, **228,** 229
  of protochordates, 28, **29–32**
Pheasant, **134,** 438, 547, 554, **555, 556**

Pholidota, 69
  (see also Pangolins)
Photophores, 88
Photoreceptor cells, **29, 370,** 376, 385
Phrenic nerve, 194
Phyletic lines, 13, 14
Phylogeny, 18, **37**
Phylum, 34
Physoclistous gas bladders, 237, 238
Physostomous gas bladders, **212,** 237, 238
Pia-arachnoid meninx, 353
Pia mater, 353, **354**
Pigeon, **217, 272,** 547, 551
Pigment cells, 89
  (see also specific kinds)
Pigment layer of eye, **368, 370,** 372, **373**
Pineal body, **335,** 338, **345, 349, 351, 376,** 385
Pinhole camera, **369**
Pinnate muscles, 178, 183, **413,** 417
Pinnipeds, 220, 507, 512, 513, 520, **521, 522,** 525–528
Pipefish, 516, 573
Piranhas, 570, **579**
Piscivorous vertebrates, 220, 578, 579
Pisiform bone, **171,** 173, 472, **473,** 520, **521**
Pitch, 166, 523–**525,** 549
Pit organs, 356
Pituitary gland (= hypophysis), **121,** 339, **345, 349, 351,** 378, **379–381,** 385
Pivot joints, 409
Placenta, 68, 83, **84,** 85, 276, 278, 304, 309, 384
Placental mammals, 68
Placodermi (see Placoderms)
Placoderms, 37, 38, 45, 47, **48,** 324
  armor shields of, **48, 93, 94**
  body skeleton of, 150, 158, 160–162, 166
  head skeleton of, **48, 127,** 128
  scales of, 93, **94**
  sense organs of, 361, 376
  (see also specific orders)

Placodonts, 582
Placoid scales, 95
Plane joints, 409
Plantigrade foot, **442,** 443
Plasma, 256
Plastron, 479
Platelets, **256,** 257
Platyan (= acoelous) vertebrae, **149,** 406, 519
Platypus, 68, **69,** 466, 507, 514, 520, 582
Platysma muscle, 201
Plesiosauria (see Plesiosaurs)
Plesiosaurs, 38, 63, **65, 139,** 427, 520, 568, 576
Pleuracanthodii (see Pleuracanths)
Pleuracanths, 38, 49, **50,** 233
  body skeleton of, 150, 158, 161, 167
  teeth of, 111, **112**
Pleural cavity, 206, **207**
Pleurocentrum, 148, 153, **154, 157,** 158
Pleurodont tooth attachment, 109, **110,** 111, 113
Pleuropericardial membranes, **205,** 206, **207**
Pleuroperitoneal cavity, **205**
Pleuroperitoneal membranes, 206, **207**
Plexiform cells of retina, **370,** 371
Plexus of nerves, 314
Pneumatic duct, **212,** 236, 244
Podial bones, 167, **168, 171, 172**
Podocytes, **287**
Pogonophora, 27
Poikilotherms, 89
Poison fangs, 568, **569**
Poison glands, 90, 213
Polyphyodont dentition, 110, 111, 113
Polypterus, **53,** 54, 95
  body skeleton of, 158, 161
  respiratory system of, 234, 236
Pons, 332, **335,** 341, **345, 349**
Pontine flexure, **331,** 332
Pontine nuclei, **340,** 341, **347**
Popliteus muscle, **175**
Porcupine fish, 582, **583**

Porcupines, 69, 488, 499, 501, **563**

Porpoises (*see* Dolphins)

Portal systems, 247, 264–267, 271, 272, 275, **381,** 385

Postcava (postcaval vein), **237, 254, 266–268,** 269, **271, 272, 274, 275,** 528

Posterior cardinal sinus, **265**

Posterior cardinal vein, **237,** 238, **260,** 264, **265–268,** 269, **271, 275**

Posterior choroid plexus, 338, **345, 349, 351,** 354

Posterior colliculi, **335, 337, 345,** 347

Posterior commissure, 338, **345, 376**

Posterior intestinal vessels, **265**

Posterior lacerate foramen, **142**

Posterior peduncles of cerebellum, **336,** 337

Posterior rectus muscle, 190, **191,** 201

Posterior vena cava (= postcava), **237, 254, 266, 267,** 269, **271, 272, 274, 275**

Postfrontal bone, **126, 130, 133, 137,** 142

Postganglionic fibers, **326**–328, 385

Postglenoid foramen, **142**

Postorbital bone, **126, 130, 132, 133, 137–139,** 142

Postorbital ligament, 424, **562**

Postorbital process, **143**

Postparietal bone, **126, 132, 133, 141**

Postsynaptic membrane, **314**

Posttemporal bone, **132, 162**

Postzygapophysis, **148, 149,** 409, **450**

Potential energy, 433, 434, 543–545

Pottos, 486, **489,** 494, 498, **500,** 501

Pouched gills, **230,** 232, 233

Preadaptation, 6, 7, 12, 144

Prearticular bone, **126,** 142, 143

Precava (precaval vein), **254,** 264, **267, 268, 271, 272, 274, 275,** 278

Prefrontal bone, **126, 130, 133, 137, 138,** 142

Preganglionic fibers, **326**–328, 383

Prehensile tails, 499, 501

Premandibular arches, 186, 190

Premaxilla bone, **126, 130, 132, 133, 137, 138, 140, 141, 560,** 561, **578**

Premolars, **114,** 579

Preopercular bone, **132, 189**

Presphenoid bone, **123,** 141, **142,** 143

Pressure, 390, **393,** 394
  and flying, 538, **539,** 541
  and swimming, **510,** 512, 516

Pressure drag, 512, 513

Pressure sense, 356

Pressure transducers, 356

Presynaptic membrane, **314**

Prezygapophysis, **148,** 409

Primary feathers, 537, 547, **548,** 553, 554

Primary sex cords, 293, **294,** 295, 297, 298

Primary spermatocytes, **298**

Primary swimmers, 505, 507, 512

Primary vesicles of brain, 330

Primates, 69, **70,** 176
  brain of, **334,** 335, 336, 338, **340,** 348, 352
  eye of, **372,** 376
  feeding adaptations of, 567, 568
  fetal membranes of, **84**
  ground locomotion of, 438, 457, **458**
  scansorial adaptations of, 486, **489, 494, 496, 498–500,** 501
  urogenital system of, 287, 299, 305, **306**

Primitive character of taxon, 36

Primitive meninx, 353

Primitive sex cells, **294,** 295

Primitive streak, **81**

Proatlas vertebra, 156, **157**

Proboscidea, 71
  (*see also* Elephants)

Procoelous vertebrae, **148, 149,** 153, 155, 406, 409

Procoracohumeralis muscle, **192, 193**

Proctodeum, 305, **306**

Profile drag, 541

Profundus nerve, 324

Progenesis, 20

Progesterones, 383, 384

Prolactin, 384

Pronation, 179, 409, 495, **496**

Pronator muscles, 179, 472, **474, 475**

Pronephric duct, **284,** 285

Pronephric tubules, **284,** 303

Pronephros, **46, 284,** 285

Pronghorn antelope, **70,** 105, **401,** 438, **451, 452, 454,** 457

Pronghorns, **105**

Pronking gait, 457, **459**

Prootic bone, **123, 133, 137,** 141

Proprioceptive sense, 336, 340, 356

Propterygium, **162,** 167

Propulsion:
  in air, 545–548
  in water, 514–523

Prosencephalon, 330, **331**

Prostate gland, 302, **303**

Proteinaceous cells, 88

Protonephridia, 282

Prototheria, 37, 39, 68, **69**

Protractor muscles, 179

Protrusible jaws, **560,** 561, 573, 576

Proventriculus, 215, **217**

Proximal tubule of nephron, **287,** 288, 290, **291**

Pseudobranch, 230, 233, 234, 261

Pteraspids, 38, **42,** 43, 93, 232, 573

Pterinosomes, 89

Pterosauria (*see* Pterosaurs)

Pterosaurs, 63, 532, **535, 536,** 537, 540, 553–557

Pterotic bone, **132**

Pterygiophores, **160**

Pterygoid bone, **126, 130, 133, 136, 137, 138, 140, 141,** 142

Pterygoid muscles, **144,** 198, 565

Pterygoid process, **143**
Pterylae, 100
Publis, **155,** 165, **166**
Pubo-ischiadic fenestrum, 165,**166**
Puboischiofemoralis muscles,
    192, 195
Puboischiotibialis muscle, **192,195**
Pubotibialis muscle, **192, 195**
Puff adder, **569**
Puffins, 507, 514
Pulmocutaneous artery, **262**
Pulmonary artery, **253, 254,** 261,
    262, 266, 270, 273
Pulmonary circulation, 250–254,
    276, 277
Pulmonary trunk, 253, 254
Pulmonary vein, **266, 271, 274**
Pulp cavity, **94, 101,** 107, **108,**
    109, **581**
Pulse pump ventilation, 244
Pupil of eye, 368, **372**
Purkinje cells of cerebellum,
    339, **340**
Putamen, 342, **343, 344, 347**
Pygostyle, 156, 400, 501, **555**
Pyloric caeca, **216,** 219
Pyloric region of stomach, 215,
    218
Pyloric sphincter, 215
Pyramidal system, 348
Pyramidal tracts, **335, 336,** 337,
    348
Pyriform lobe, **343,** 346, 348

Quadrate bone, **123, 130, 132,**
    133, 134, **135, 137, 138,**
    140, 143, **562**
Quadratojugal bone, **126, 130,**
    133, 137–140, 142
Quadratus femoris muscle, **175,**
    196
Quail, **197,** 551
Quill of feathers, **99**
Quills of mammals, 7, 104

Rabbits, 69, **279,** 353, 468
    cursorial adaptations of, 438,
        441, 452, 455, 457
    feeding adaptations of, **563,**
        564, **565,** 566, 570, 582

Raccoon, 69, 443, **452,** 568
Rachis of feather, 99
Radial carpal extensor muscle,
    174, 522
Radiale, **168, 171**
Radial nerve, **319**
Radial notch of ulna, **174,** 409,
    451, 495, **496**
Radials of fins, 160, **162, 163,**
    168
Radius, **168, 174, 451, 473,** 495,
    496, 497, 520, **521, 536,**
    556
Rathke's (= hypophyseal)
    pouch, **211,** 378, **379**
Rats, 69, 468, 488
Ray-finned fishes, 38, 52, **53, 54,**
    56
    body skeleton of, 150, 158,
        161, **162, 163**
    brain of, 342, 345, **349,** 350
    head skeleton of, 129–**132,**
        560
    scales of, **94,** 95
    teeth of, 111, **112**
    urcgenital system of, 284,
        288, **289,** 291, 296, 300,
        301, 304
    (*see also specific infraclasses*)
Rays (and skates), 38, 47, 49,
    50, 111, **129,** 201, 235,
    307, 466, 516, 522, 582,
    583
Reaching, 494, 495
Recapitulation, 18, 19, 143,
    254, 281
Rectal gland, **216,** 219, **265,** 290
Rectilinear movement of snakes,
    481, **482**
Rectrices, 99
Rectum, **216–218,** 219, **223, 286,**
    305, **306**
Rectus abdominis muscle, **175,**
    192, 193, 195, 199
Rectus cervicis muscle, **192, 193**
Rectus femoris muscle, **175, 199**
Red body of gas bladder, **237**
Red muscle, 182, 183, **519**
Red (= ruber) nucleus, **336, 347,**
    497

Relaxin, 383, 384
Reliability of data, 10
Remiges, 99
Renal capsule, **275,** 281, **282,**
    287, 288, **289,** 290, **291**
Renal corpuscle, **282,** 284,
    287–291
Renal papillae, 287, **291**
Renal pelvis, **286,** 287, **291**
Renal portal system, 267, **268,** 271
Renal portal vein, **260, 265–268,**
    269, **271, 272, 275,** 278
Renal (= nephric) tubules, **275,**
    281, **282,** 283, **300, 301**
Renal vessels, 263, **266, 270,**
    271, 273–275
Renin, 385
Renofemoral artery, **272**
Renolumbar artery, **272**
Replacement bone(s), 118, 122,
    123, 125, 126, 130, 131,
    150, 157, 169
Reproductive strategies, 308, 309
Reproductive system, 293–309
    of amphibians, **289, 296, 297,**
        298, **299,** 301,
        304–**306**–308
    of birds, **272,** 295, 296, 299,
        302, 304–**306**–308
    of bony fishes, 296, 298,
        300–**307,** 308
    of cartilaginous fishes, **295,**
        296, 298, 301, 303, 305,
        306–308
    of cyclostomes, 296–298,
        300, 301, 305, **306,** 308
    of mammals, 295–**297**–**299,**
        303, 304, 305, **306**–309
    of reptiles, 296, 297, **302,**
        304–**306**–308
    (*see also* Copulatory organs;
        Ovary; Testis; Urogenital
        ducts)
Reptiles, 37, 38, 60, **61–65**
    appendicular skeleton of, **164,**
        166, 171–173
    axial skeleton of, **154**–156,
        157–159
    brain of, 336, 337, 339, 342,
        343, 345, 346, **351**–353

circulatory system of, 250–**253**–255, 262, 263, **267–270, 271, 275,** 278, **279**

coelom and mesenteries of, **205,** 206

cursorial adaptations of, 438, **439,** 443, 461

digestive system of, 215, **217,** 231

digging by, 466, 469

early development of, 75, **77**

endocrine glands of, **380**–384

excretory system of, 286, 287, **289,** 291, 292

head skeleton of, **134, 135,** 136, **137–139,** 140, 143

muscles of, 193–**195, 196**–198, 201

nerves and spinal cord of, **318, 319**

reproductive system of, 296, 297, **302,** 304–**306**–308

respiratory system of, **239,** 240, 245

sense organs of, 358, 359, **365,** 366, 371, **374,** 375, **376**

(*see also* specific subclasses and orders)

Reptilia (*see* Reptiles)

Respiratory bronchioles, 240, **241**

Respiratory circulation, 276

Respiratory system:
of amphibians, **234,** 235, 238, **239,** 240, 244
of birds, 206, **242–244,** 245
of bony fishes, 229, **230**–236
of cartilaginous fishes, 229, **230**–233
development of, **211, 228,** 229, 235, 236
general nature of, 227
of jawless vertebrates, **230,** 232
of mammals, 240, **241,** 245
of reptiles, **239,** 240, 245
(*see also* Gills; Lungs)

Resultant forces, 411, **412–414**

Resultant velocities, 429, **430**

Rete mirabile, 237

Rete tubules of testis, 298, **299**

Reticular formation, 335, **336, 347**

Reticular tissue, **257,** 258

Reticulum, **222**

Retina, **347, 368, 370, 372, 373, 376**

Retractor bulbi muscle, 201

Retractor muscles, 179

Retroarticular bone, **132**

Reynolds number, 511, 512, 540, 541

Rhachitomi, 153, **154**

Rhinal fissure, **343, 344,** 346

Rhinoceroses, **70,** 71, 105, **171,** 428

Rhipidistia, 38, **55,** 56, 58, **130,** 153, 261, 364

Rhodopsin, 370

Rhombencephalon, 330, **331**

Rhomboideus capitis muscle, **200**

Rhomboideus muscle, **174,** 194, **200, 474–476**

Rhomboid scales, 93

Ribs, **148, 152,** 156, **157, 158,** 526, 534, **555**

Ricochet, 437, **440,** 457, 459, **460, 461,** 462

Road-runner, 438, 577, 578

Robin, 551, **552**

Rodentia (*see* Rodents)

Rodents, 69, **70,** 84, 176
feeding adaptations of, 221, **223, 563,** 568, 570, 582
locomotor specializations of, **440, 447, 460, 461,** 464, **467,** 468, 488, 507, **509, 533,** 535
reproductive system of, 299, 304–**306**
(*see also* specific kinds)

Rods of retina, **370**–373

Roll, 160, 166, 237, 523, 549

Roots of teeth, 107, **108,** 580, **581**

Rostrum, **129,** 566, 575

Rotator muscles, 179

Round window, **365, 367**

Ruber nucleus, **336, 337, 347,** 497

Rumen, **222**

Runners, 437, 438, **439–441**

Running:
adaptations for, 440–454
advantages of 437, 438
general requirements for, 440

Running gaits, 457, **458**

Running walk, 454, 457, **458**

Saccule, **362,** 363, **367**

Sacral plexus, **319**

Sacral vertebrae, 153, **154, 155,** 156, 165

Sacrotuberous ligament, 399, **400**

Sacrum, 156, **476**

Saddle joints, 409

Sagittal crest, **143,** 565, 582

Sailfish, **5,** 507, 513, 519

Salamanders (*see* Urodeles)

Salivary glands, **213,** 221, **327,** 575

Salmon, **149, 161, 251, 349,** 507, 514, 519, **560**

Saltatorial vertebrates, 437–**439, 440, 441,** 457, 459, **460, 461,** 462

Salt excretion, 213, 219, 288, **289,** 290, **291**

Salt glands, 213, 219, 290

Sarcolemma, 180, **181**

Sarcoplasmic reticulum, 180, **181**

Sarcopterygians, 37, 38, **55,** 56, 358
skeleton of, 161, 167, **168**
(*see also* specific infraclasses)

Sarcopterygii (*see* Sarcopterygians)

Sarcopterygium, 167

Sartorius muscle, **175, 197, 199, 200**

Saurischia, 63, **64**

Sauropida, **64**

Sauropterygia, 63, **65**
(*see also* Pleiosaurs)

Scala media, 364, **365, 367**

Scala tympani, 364, **365**

Scala vestibuli, 364, **365**

Scalenus muscle, **199**

Scales:
  cosmoid, 93, **94**
  ctenoid, 95
  cycloid, 93, 95
  dermal denticles, **94,** 95, 161
  elasmoid, **94,** 95
  of fishes, 92–**94,** 95
  ganoid, **94,** 95
  horny, 96, **97,** 98, 101
  lepidosteoid, 95
  and locomotion, **465,** 477,
    481, **482,** 513
  origin of, 91
  palaeoniscoid, 95
  phylogeny of, **94,** 95
  placoid, 95
  rhomboid, 93
Scaly-tailed squirrels, 488, 492,
    493, 535
Scansorial vertebrates,
    485–**487–489,** 490
  adaptations of, 493–503
Scapula, **162,** 163, **164,** 165,
    174, 443, **444, 473,** 475,
    476, 497, 499, 555
Scapulihumeralis muscle, **197,**
    198
Scapulocoracoid, **162**
Scarf joints, **405**
Scavengers, 220, 578, 579
Scent glands, 98, 102
Schizocoely, 79, **80,** 81, 204
Schwann cells, **312, 313,** 316
Sciatic artery, **270**
Sciatic nerve, **319**
Sclera, **368,** 369, **370, 373, 374,**
    527
Sclerotic bones, **374,** 375
Sclerotome, **80,** 83, 150, 158
Scombroid fishes, **506,** 507
Scratch-digging, 464–466, 468,
    472, **473–475**
Scrotum, 299, **306**
Scutes, 97, **98,** 481, **482,** 568
Sea cows, 428, 570
  (see also Sirenians)
Sea horse, 573
Sea lions, 507, 514, 516, 517,
    521, 522, 526, 576

Seals, 69, 428, 507, 514, 526,
    576
Sea otter:
  aquatic adaptations of, 507,
    516, 520, 528
  feeding adaptations of, 568,
    582, **583**
  kidney of, **287**
Sea snakes, 507, **508,** 516, 528
Sea turtles, **5,** 507, **508,** 513,
    517, 520, **521,** 576, **577**
Sebaceous glands, **102, 103,** 526
Secondary feathers, 537, 547
Secondary palate, 136, **140**
Secondary sex cords, **294,** 295,
    297
Secondary swimmers, 505, 507,
    512
Secondary vesicles of brain, 331
Secretary bird, 569, 577, 578
Secretin, 216
Segmentation:
  of central nervous system, 321
  as chordate feature, 29
  of head skeleton, 119, 120,
    123, 124
  of vertebral column, 150
Selachians, 49, **50,** 205, 427
  body skeleton of, 151, **152,**
    158, **160, 162, 163**
  digestive system of, 206, 215,
    216, 218, 226
  gills of, **230,** 233, 235
  head skeleton of, **127, 129**
  teeth of, 111, **112**
  (see also Sharks; Rays)
Selachii (see Selachians)
Semicircular canals, **362,** 363,
    367
Semiconductor properties, 402
Semilunar ganglion, 324
Semilunar valves, 249, **250, 251**
Semimembranosus muscle, **175,**
    197, 199, 200, 448, 460
Seminal vesicle, **295,** 301, 302,
    305
Seminiferous ampullae, 297–**299**
Seminiferous tubules, 297–**298,**
    299

Semitendinosus muscle, **175,**
    197, **199, 200, 460**
Sense capsules, 356
Sense organs, 355–376
  (see also specific organs)
Sensory fibers, 314, 316, 332
  (see also subtypes)
Septa (of gills), **230,** 233, 234
Septa (of testis), 298, **299**
Septal gills, **230**
Septal nuclei, **344, 347**
Septum pellucidum, **344, 345**
Serotonin, 314
Serous membranes, 203, 204,
    214, 249
Serrate joints, **405,** 478
Serratus muscles, 194, **196, 199,**
    200, 475, 497, **499**
Sesamoid bones, 176, **396,** 397,
    426, 472, 473, 476, 520,
    521
Setae of gecho toes, 502, **503**
Sexual dimorphism, 9
Shaft of feather, **99,** 100, **101**
Sharks, **17,** 18, 38, 47, 49, **50**
  aquatic adaptations of, 506,
    507, 513, 519, 522, **524**
  body skeleton of, 151, **152,**
    158, **160, 162, 163**
  circulatory system of, 250,
    251, **260, 265, 267,** 269,
    280
  digestive system of, **216**
  feeding adaptations of, 570,
    573, 578–580, 582
  gills of, **230,** 233, 235
  head skeleton of, **127, 129**
  muscles of, **190, 191**
  nervous system of, **318, 323,**
    349
  teeth of, **110,** 111
  urogenital system of, 285, **289,**
    290, **295,** 300–303
Sharpey's fibers, 92, 109
Shear, 391, **392, 393**–395, 397,
    570, 579
Shearwater, 466, 537
Shellbreaker, **19,** 98
Shell glands, 303, 304

Shore birds, 547, 553, **562,** 566
Short fibularis muscle, **175**
Short fibularis tendon, **396**
Shrews, 68, **70,** 469
Sidewinding of snakes, **483,** 484
Significance of data, 10
Sinking speed, 543, 554
Sinuatrial node, **254,** 255
Sinuatrial valve, 249, **250, 255**
Sinuses of venous system, 264,
    265, 354
Sinusoids:
  of endocrine glands, 377, **378**
  of hemopoietic tissue, 258
  of liver, 223, **224,** 265
Sinus venosus, 249, **250–253,**
    254, **255,** 265, 268, 271
Siphons of Urochordata, **29, 30**
Sirenia (*see* Sirenians)
Sirenians, 71, 156, 221, 507,
    509, 512–514, 523,
    526–529, 570
Skates (and rays), 38, 47, 49, **50,**
    129, 201, 235, 466, 516,
    522, 582, **583**
Skeletal (= striated) muscle, **180,**
    181-185
Skeleton (*see* Body skeleton;
    Head skeleton)
Skin:
  development of, 88, 89
  functions of, 87, 88
  structure of, **88,** 89, **90, 96, 97,**
    98, 100–**102**
  (*see also* Integument)
Skull, 128
  (*see also* Head skeleton)
Slime and frictional drag, 514
Sling mechanisms, 424–**426**
Sloths, 69, 221
  feeding adaptations of, 565,
    570, **571**
  scansorial adaptations of, 486,
    489, 493–**496,** 501
Small intestine, **211, 216–218**-
    221, **223,** 384
Smell, 357–359, 526
Smooth muscle, **88,** 89, 179,
    180, 204, **214**

Snakes, 38, 61, **62,** 427
  aquatic adaptations of, 507,
    508, 514, 516, 528
  cranial kinesis in, 561, 562
  feeding adaptations of, 562,
    568, **569,** 577, **578**
  fossorial adapations of, 464,
    465, 466, 469, 470, 478,
    479
  locomotion of, 480, **481–483,**
    484
  nervous system of, **318,** 352
  parachuting of, 534
  reproduction of, 308
  respiration of, 240, 245
  scansorial adaptations of, 486,
    487, 493, 494, 498, 499
  sense organs of, 356, 366, 375
Snap joints, **408,** 425
Soaring, 543, **544,** 545
Soil-crawling, 464, **465,** 466,
    470, 475, 476
Soleus muscle, **200**
Somatic fibers, 315
  (*see also subtypes*)
Somatic layer of hypomere, **80,**
    82, 83, **204**
Somatic motor fibers, 315–**317,**
    319, **320, 322,** 324, 333,
    334
Somatic sensory fibers, 315–**317,**
    319, **320,** 321, **322**-325,
    333
Somatic sensory pathways of
    brain, **347**
Somatopleure, 83, **84**
Somites, **79–81,** 82, 321
Special cranial nerves, 322, **323**-
    325
Specialized structures, 12
Species, 34, 35
Speed, 438
  of flyers, 537–541, 544, 545,
    553
  of runners, 488–492
  of swimmers, 506, 507
Sperm, 75, **76, 298,** 301, 302,
    305, 308, 309
Spermatic vein, 263, **275**

Spermatids, **298**
Spermatogonia, **298**
Sphenethmoid bone, **123,** 143
Sphenodon, 61, **62,** 466
  circulatory system of, 263
  skull of, 136, **138**
  urinary bladder of, 292
Sphenoid bone, 141
Sphincter muscles, 179
Spider monkey, 486, 495, **496,**
    498, 499, 501
Spinal cord:
  as chordate feature, 28
  development of, **315,** 316
  evolution of, **46,** 317, **318**
  function of, 316, 317
  of protochordates, **29, 31**
  structure of, 316, **317, 318**
Spinal ganglia, 150, **315,** 316
Spinalis capitis muscle, **196**
Spinalis dorsi muscle, **200**
Spinal nerves:
  development of, **315,** 316
  evolution of, **319, 320**
  and muscle innervation, **186,**
    187, 190, 191, 194, 320
Spine (*see* Vertebral column;
    Vertebrae)
Spine of scapula, **154,** 165, **174**
Spinodeltoid muscle, **174, 200**
Spinothalamic tract, **336**
Spiracle, 49, **50,** 52, 53, **124,**
    228, 229, **230,** 233, 234,
    324
Spiral fold of conus, **252**
Spiral valve of intestine, **216,**
    218, 219
Splanchnic layer of hypomere,
    80, 82, 83, 186, 187,
    204, **249**
Splanchnocoel, 204
  (*see also* Coelom)
Splanchnocranium, 120
Splanchnopleure, 83, **84**
Spleen, **216–218,** 257, 258
Splenial bones, **126**
Splenic vessels, **265, 266, 273,**
    274
Splenius muscle, **200**

Springhare, 438, **440,** 445, 459, **461,** 468

Springing, **494**

Springing ligament, 425, **426,** 452, **456**

Spurs, 98

Squamata (*see* Squamates)

Squamates, 61, **62**
  integument of, **97**
  muscles of, **195**
  reproduction of, 308
  (*see also* Lizards; Snakes)

Squamosal bone, **126, 130,** 131, **133, 135, 137–142**

Squamous joints, 405

Squamous joints, 405

Squirrels, 69, 457
  "flying," **455,** 488, **533,** 535
  ground, 464, **467,** 468, 471
  scaly-tailed, 488, 492, **493,** 535
  tree, 488, 493, 500, 501, 568

Stability, 453, 454, **457,** 499, 500, 523, 524, 548, 549

Stall in flying, 539, 549, 553

Stapedial muscle, 198

Stapes, 131, **133,** 134, **135, 137, 365,** 366, **367**

Static soaring, 543, **544**

Statistical analysis, 10

Statoacoustic ganglia, **334**

Statoacoustic nerve, **323**-325, **334–336,** 337, **349, 351,** 362

Stegosauria, **64**

Stem reptiles, 38, **61,** 111, 135, 136, **139, 140, 154, 164**

Stereocilia, **360,** 363, 365, 366

Stereoscopic vision, 371

Stereospondyli, 153, **154**

Sternal ribs, 158

Sternebrae, **159**

Sternocleidomastoid muscle, **195**

Sternohyoid muscle, **195, 196, 199,** 212

Sternomastoid muscle, **144, 196, 199, 200,** 201, **474, 476**

Sternothyroid muscle, **195,** 199

Sternum, 158, **159, 466,** 497, **498,** 526, **555**

Stomach:
  comparative anatomy of, **30, 212,** 214–**216–218,** 223
  development of, **211**
  functional adaptations of, 220, 221, 384
  structure of, 214, 215

Stomodeum, **228,** 378

Straining food, 570–**572**

Strain in supportive materials, 391, 395, 402

Stratum compactum, 89, **90,** 96

Stratum corneum, **88, 96,** 101, **102**

Stratum germinativum, **88, 96, 97, 101, 102**

Stratum granulosum (ovarian follicle), **297,** 384

Stratum granulosum (skin), 101, **102**

Stratum spongiosum, 89, **90,** 96

Strength of supportive materials, 391, 392

Stress in supportive materials, 391–**393,** 394, 395

Stress lines, 392, **393**-395, 554

Striated (= skeletal) muscle, **180, 181**-185

Stride, 440
  length of, 440–445
  rate of, 445–449

Stroma of ovary, 295–**297**

Sturgeon, 38, **53,** 54, **151, 161,** 427, 566, 573, 582

Stylohyoid muscle, **199**

Styloid process, **174,** 409, 495

Stylomastoid foramen, **142, 143**

Subarachnoid space, 353, **354**

Subcardinal vein, 264, **268,** 269, **275**

Subclavian artery, **253, 254, 262,** 263, **265, 266, 270, 272, 273**

Subclavian vein, 264, **265, 268, 274,** 278

Subcutaneous lymph ducts, 278

Subdural space, 353, **354**

Subintestinal system of veins, **263,** 264, 269

Subintestinal vein, **263**-265, **275**

Sublingual salivary gland, **213**

Submucosa of gut, **214**

Suborbital bone, **132**

Subscapularis muscle, 194, **199, 475, 522**

Subscapular vessels, **271, 273, 274**

Substantia nigra, **336, 347**

Subvertebral ganglia, **326, 327**

Subvertebral lymph duct, 278, **279**

Suction cups, **501**

Sudoriferous glands, 101

Sulci, 339, **340,** 346

Superciliary bone, **138**

Superficial digital flexor tendon, **396, 426, 427**

Superior ganglion, 325

Superior oblique muscle, 190, **191**

Superior olive, **335,** 337, **347**

Superior rectus muscle, 190, **191,** 201

Supination, 179, **409,** 495, **496,** 497, 556

Supinator muscle(s), **174,** 179, 472, **475,** 497, **522**

Supportive materials:
  properties of, 390–392
  structure of, 392–402
  unions of, 403–410

Supportive mechanisms:
  of axial part of body, 421, **422–424**
  of limbs, 424–**426, 427,** 428

Supraangular bone, **126, 130, 132, 137, 140**

Supracardinal vein, 264, 269, **275**

Supracoracoid muscle, **192, 195, 197, 198,** 554

Supraoccipital bone, **123,** 131, **132, 137, 138, 140, 141**

Suprarenal vein, **275**

Suprascapular cartilage, **162, 164**

Supraspinatus muscle, **174,** 196, **200, 522**

Supraspinous fossa, **174**

Supraspinous ligament, **450,** 462
Supratemporal bone, **126, 130, 132, 133**
Surface-to-volume ratio, 22, 427, 528, 547
Suspensory ligament (eye), **368, 374,** 375
Suspensory ligament (ungulate foot), 425, **426,** 452, **456**
Sutures, 404, **405**
Swan, **140,** 547, **550**
Sweat glands, 101, **102**
Swimmer, **506,** 507, **508, 509**
  physiology of, 528, 529
Swimming:
  adaptations for, 510–529
  advantages of, 505, 506
  general requirements for, 510
Swinging, 495, **496,** 497
Swordfish, 506, 507, 512, 513
Symmetrical gaits, 454, 457, **458**
Symmetry, 27–29, 31, 33
Sympathetic ganglia, 326, **327**
Sympathetic nerve trunks, 326, **327**
Sympathetic nervous system, 238, **326, 327,** 328
Symphysis, 406
Synapse, 313, **314**
Synapsida (*see* Synapsids)
Synapsids (= mammal-like reptiles), 37, 38, 63, **65,** 366, 579
  skeleton of, **135,** 136, **139, 164**
Synapsid skull, 136, **139**
Synaptic cleft, **314**
Synaptic knob, **312, 314**
Synarthroses, 404, **405**
Syndesmoses, 406
Synergistic muscles, 179
Synovial fluid, 406, 407
Synovial membrane, 406, **407**
Synsacrum, **155,** 165, 243, 554, **555**
Syrinx, 243, **244**
Systemic arch, **253, 254, 262,** 263, **270, 272**
Systemic circulation, 253, 254, 277
Systemic trunk, 253, 254

Tabular bone, **126, 130, 133**
Tail:
  of climbers, **487–489,** 499–501
  of cursors and saltators, **439, 440,** 461, 462
  of diggers, 469, 470
  of flyers, 549, **550,** 554, 555
  of protochordates, **29, 31**
  of swimmers, **42, 44–46, 48, 50–55, 506, 509,** 514, **515–519,** 524, **525,** 529
Tail fold of embryo, 82
Tail kidney, 285, 286
Tapetum lucidum, **372,** 375
Tarsal bones, 167, **168, 172, 453**
Tarsier, 438, 460, 486, **489,** 494, **500,** 501
Tarsometatarsus, **172,** 452, **453**
Tarsus, 167, **172,** 173, **494**
Taste, 359, 526
Taste buds, **359**
Taxon, 34
Taxonomy, 35
Tear (= lacrimal) gland, 290, **327,** 371, 375, 527
Teat, **103,** 512
Tectorial membrane, **365**
Tectum of midbrain, 336, **347**
Teeth, 107–115
  attachment of, 109, **110**
  comparative anatomy of, 111–113
  cusps of, 109, **114,** 115
  development of, 108, **109**
  fossorial adaptations of, **467,** 469, 471, **477**
  origin of, **94,** 95, 107
  replacement of, 109–111, 113
  specializations for feeding of, **563–565,** 570, **571, 576** 578, **579,** 580, **581–583**
  structure of, **94,** 107, **108**
Telencephalon, **331,** 332, 338, 341, **343**
Teleostei (*see* Teleosts)
Teleosts, 38, 52, **54,** 56
  brain of, **343, 349,** 350
  circulatory system of, 250, **251, 262,** 266, 269, 278, 279

digestive system of, **216,** 225, 226
electric organs of, 201, **202**
endocrine glands of, **380,** 382–385
excretory system of, 285, 288, **289,** 291
gills of, 229, **230**-235
integument of, **90**
jaw mechanics of, **560,** 561
muscles of, **189**
reproductive system of, 296, 301, 302, 305, 308
scales of, **90, 94,** 95
sense organs of, **358, 362,** 371, **372, 374,** 375
skeleton of, **132,** 151, **160–163**
Telolecithal eggs, 76
Temporal bone, 141
Temporalis muscle, **144,** 189, **200, 565**
Temporal openings, 60–63, 136, **138, 139**
Tendon(s), 178, 180, **181,** 391, 392, 394, **396,** 397, 403, 404
Tendon organs, 356
Tension, 391, **392, 393**-400, **401,** 421
Tensor fasciae antibrachii muscle, **199**
Tensor fasciae latae muscle, **175, 199, 200, 460**
Tensor patagialis muscles, **197, 198**
Teres major muscle, **174,** 194, **196, 199, 447,** 465, 472, **473, 475, 476, 522**
Terminal nerve, **323**
Terrestrial life:
  adaptations for, 57, 58
Testis:
  comparative anatomy of, **30, 272, 295, 296, 297, 298, 299, 302, 303, 306,** 512
  development of, 293, 294
  endocrine function of, 384
  structure of, 297, **298, 299**
Testosterone, 384
Tetanus of muscle, **182**

Tetrapods, 57, **127,** 261
  girdles of, 163, **164,** 165, **166**
  origin of limbs of, 167, **168**
  phylogeny of vertebrae of, **154**
  terrestrial adaptations of, 57,
    58
Thalamus, **331,** 338, **344,** 349
Theca of ovarian follicle, 295,
  297, 384
Thecodont, 109, **110,** 113, 579
Thecodontia, 63
Therapsida, 38, 63, **65**
Theria, 37, 39, 68
Thermals, 543, **544**
Thermoreceptors, 356, 357, 361
Thermoregulation, 104, 280,
  470, 528
Thick filaments, **181**
Thin filaments, **181**
Third ventricle of brain, **331,**
  338, **344, 345**
Thoracic duct, 278, **279**
Thoracic vertebrae, **155,** 156,
  497, **498**
Thoracic vessels, **273, 274**
Thoracodorsal vessels, **273, 274**
Thorax, 443, **444,** 497, **498,** 526,
  555
Thrombocytes, **256,** 257
Thymus gland, 257, **270,** 278,
  385
Thyrocervical artery, **273**
Thyroglobulin, 381
Thyrohyoid muscle, **199**
Thyroid cartilage, **134,** 241
Thyroid diverticulum, **211, 379,**
  381
Thyroid gland, **190, 211, 270,**
  378, 381, 382
Thyroscapular vein, **271**
Thyrotropic hormone, 379
Thyroxine, 381
Tibia, **168, 175, 403, 452, 496**
Tibiale, **168, 172,** 173
Tibialis anterior muscle, **175,**
  195, 460
Tibiotarsus, **172**
Toads, **5,** 38, 60
  fossorial adaptations of, 466,
    471, 479
  skeleton of, **149, 154**

Tongue, 113, 131, **212,** 213
  and feeding, 571, **572,** 573,
    574, 575, 577, 579
Tooth bud, **109**
Torque, 185, **417,** 418, **419,**
  421, 422, 432, **433,** 517
Tortoises, 61, **98,** 466, **472,** 479,
  570, **571**
Touch sense, 356, 566
Trabeculae (of bones), **399,** 402,
  403, 407
Trabeculae (of chondrocranium),
  121, 130
Trachea, **211,** 238–**241, 242,**
  243, **244, 327,** 526
Tracheal rings, 132, 134, 240
Tracts of nervous system, 314,
  333, 337
  (*see also specific tracts*)
Transverse foramen, **149**
Transverse ligaments, **396**
Transverse process, **148, 149,**
  156, **450**
Transverse scapular vessels, **273,**
  274
Transverse septum, **205,** 206,
  207, 212
Transverse vein, **274**
Transversus costarum muscle, **199**
Transversus muscle, 193
Trapezius muscles, **174,** 190,
  191, 196, 198, **200,** 201,
  474, 476, 497, **499**
Trapezoid body, **335, 336,** 337
Tree frogs, **5,** 486, **487,** 494,
  498, **500**-502, 532
Tree snakes, 486, **487,** 493, 494,
  498, 499, 534
Tree squirrels, 488, 493, 500,
  501, 568
Triceps muscle, **174,** 194,
  195–200, 412, 447, 472,
  473–476, 495, 497, **522**
Trigeminal nerve, 186, 187, **323,**
  324, **334–336, 347, 349,**
  351
Triiodothyronine, 381
Trituberculata, 39, 68
Trochlear nerve, **145,** 186, 191,
  323, 324, **334, 335, 349,**
  351

Trochlear notch of ulna, **174**
Trophoblast, **77,** 78
Tropomyosin, **181**
Troponin, **181**
Trot, 457, **458**
Trout, **54, 216,** 507, **516**
Truncus arteriosus, 259
Trunk vertebrae, 153, 156
Tuatara, 38, 61, **62**
  (*see also Sphenodon*)
Tuber cinerium, **335,** 339, **351**
Tuberculum of rib, **148,** 158
Tuberosity of ilium, **175**
Tubulidentata (*see* Aardvark)
Tuco-tuco, 468, 469, 471, 479
Tuna, 280, **506,** 507, 512, **518,**
  519
Tunica albuginea, 298, **299**
Tunica externa, 258, **259**
Tunica interna, 258, **259**
Tunica media, 258, **259**
Tunicates (= Urochordata), 28,
  29, 30, 33, 34
Tunic of Urochordata, 29, **30**
Tubinate bones, **123,** 141, 358
Turbulent flow, **511,** 513, 514,
  516, 518
Turning in flight, 549, **551**
  (*see also* Maneuverability)
Turtles, 38, **61,** 113
  aquatic adaptations of, 507,
    508, 513, 517, 520, **521,**
    528
  body skeleton of, **149,** 158,
    165, **171, 173**
  brain of, **351,** 352
  circulatory system of, **253, 270,**
    271
  feeding adaptations of, 568,
    570, 573, 576, **577,** 582
  fossorial adaptations of, 464,
    466, **472,** 479
  head skeleton of, **134,** 136,
    137–139
  respiratory system of, 240, 245
  scutes of, 97, **98**
  shell of, 95, **98**
  urogenital system of, 289,
    292, 296, 307
  (*see also* Tortoises)
Two-joint muscle systems, 421

Tympanic bone, **135, 141,** 143
Tympanic bulla, **142, 143, 367,** 527
Tympanum, 364, **365, 367,** 527

Ulna, **168, 174,** 450, **451, 473, 474,** 495, **496,** 520, **521, 536,** 556
Ulnar carpal flexor muscle, **522**
Ulnare, **168, 171,** 520, **521**
Ulnar nerve, **319**
Ultimobranchial bodies, 382
Umbilical cord, 85, 529
Umbilical vessels, **84, 260,** 269, **273,** 275, 276
Uncinate process, 158, 245, 554, **555**
Underfur, 104
Unguis, 101
Ungulates, 71, 258
  cursorial adaptations of, 438, 442, 443, 445, **450,** 451, 454
  feeding adaptations of, 563, 570, 580, **581**
  support mechanisms of, 425, **426, 427**
  urogenital system of, 287, 299, 304, 309
  (*see also* Artiodactyles; Perissodactyls)
Unguligrade foot, **442,** 443
Unmyelinated fibers, 313, 314, 326
Unspecialized structures, 12
Urea, 102, 288, **289,** 290
Ureter, **286, 291,** 292, 300, **302, 303,** 305, **306**
Ureteric bud, **286**
Urethra, 302, **303,** 305, **306, 307**
Uric acid, **289,** 290, 292
Urinary bladders, 291, 292, **296, 303, 306, 307, 327**
Urinary system, 281–292
  (*see also* Excretory system; Kidneys; Urinary bladders; Urogenital ducts)
Urine, 282, 287–292, 301
Urochordata, 28, **29, 30,** 33, 34
Urodela (*see* Urodeles)

Urodeles (= salamanders), **5, 20,** 38, **59,** 60, 573, 576
  circulatory system of, 253, 255, 261, 262, **266, 268,** 278
  coelom of, **205**
  external gills of, **234**
  fossorial adaptations of, 466
  lungs of, 239, 240
  muscles of, **192, 193**
  reproductive system of, 294, **296,** 297–**299**-302, 307
  scansorial adaptations of, 486, **498–500**-502
  sense organs of, 358, 366, 376
  skeleton of, **133,** 158, 159, 163, 171, **172**
Urodeum, 292, 305, **306**
Urogenital canal, **304**
Urogenital ducts, **30, 295, 296, 300,** 301, **302, 303**-305
Urogenital sinus, **286, 303,** 305, **306,** 307
Uroneurals, **161**
Urophysis, 385
Uropygial gland, 98, **197**
Urostyle, 153, **154**
Uterus, **295, 303, 304, 306**
Utricule, **362,** 363, **367**

Vagal lobes of medulla, 350
Vagina, **303, 304, 306**
Vagus nerve, **145,** 186, 187, 238, **323,** 325, **334–336, 349,** 351, 359, 360
Valves of circulatory system, 248, 277, 278, 528
Valvula of cerebellum, 350
Vampire bats, 220, **567,** 574, 575
Vane of feather, **99**
Variation, 8–10
Vascular sac of hypothalamus, **349,** 350
Vasopressin, 379
Vastus lateralis muscle, **175, 199, 200,** 460
Vastus medius muscle, **199**
Vectors:
  of forces, 411–**413–415,** 416, **417,** 418, **419**
  of velocities, 428–**430**

Vegetal pole, 76, **78**
Veins, 247
  basic pattern of, **260,** 263, 264
  development of, 258
  evolution of, 264–**268,** 269
  structure of, **259**
  (*see also* Circulatory system; *specific veins*)
Velocity, 428–**430,** 445, 446, 448, **449,** 457, 459
Venous duct, 269
Ventilation:
  of gills, 235
  of lungs, 244, 245, 557
Ventral abdominal vein, **266**
Ventral aorta, **212, 250, 251,** 259, **260,** 262, **266**
Ventral branches of dorsal aorta, **260,** 261, 263
Ventral cranial nerves, **322, 323**-325
Ventral funiculus, **317,** 337
Ventral gray columns, 316, **317,** 332
Ventral median fissure, 315–**317,** 318, **335**
Ventral mesentery, **80, 204, 207**
Ventral rami of spinal nerves, 186, 194, 320
Ventral ribs, 157, **158**
Ventral roots of spinal nerves, **315,** 318, **320, 326**
Ventral spinal nerves, 319, **320**
Ventral thalamus, 338, **347**
Ventral thoracic vessels, **272**
Ventricle of heart, 249, **250–255**
Ventricles of brain, 331
  (*see also specific ventricles*)
Ventriculus (= gizzard), 215, **217, 272,** 582
Ventrobronchi, **242,** 243, 245
Ventrolateral sulcus, **317,** 318
Venules, 247, **258**
Vertebra(e):
  development of, 150
  evolution of, 150, **151, 152,** 153, **154, 155,** 156, **157**
  structure of, 147, **148**
Vertebral artery, **266, 270, 272, 273,** 277
Vertebral column (*see* Vertebrae)

Vertebral vein, **268, 271, 272, 274**
Vertebrata (*see* Vertebrates)
Vertebrate morphology, 3
  benefits of study of, 4
  principles of, 4–22
  relation to other disciplines, 3, 4
  trends in nature of, 22, 23
Vertebrates:
  classification of, 37–39
  general structure of, 32
  and nonchordates, 27, 28, **34**
  origin of, 33, **34**
  and other chordates, 27, 28, 34
  phylogeny of, **37**
Vesicular glands, 302, 305
Vesicular vessels, **273, 274**
Vestibular membrane, **365**
Vestibular nuclei, 334, **336,** 337, **347**
Vestibulocochlear nerve, 324
Vestigial organs, **8,** 19
Vibrissae, 104
Vicuna, **448, 453, 455**
Villi of gut, 210, 214, 216, **218,** 221
Visceral arches, 123, **124,** 259
Visceral branch of vagus nerve, **323,** 325
Visceral fibers, 315
  (*see also subtypes*)
Visceral grooves, 228
Visceral motor fibers, 315–**317,** 319, **320,** 321, **322,** 324–328, 333, 334
Visceral rami of spinal nerves, **320, 326, 327**
Visceral sensory fibers, 315–**317,** 319, **320,** 321, **322,** 324, 325, 333

Visceral skeleton:
  evolution of, **127–129**–132, **134, 135**
  muscles of, 189
  structure of, 122–**124**
Viscosity, 511, 540, 541
Vision, 367–376, 527, 566
Visual pathways of brain, 339, 346, **347**
Visual pigment, 370
Vitelline membrane, 76, **84**
Vitelline vessels, **260,** 261, 263, 265, 276
Vitreous chamber, **368**
Viviparous vertebrates, 68, 308, 309
Vocal cords, 239–241, 243
Volant vertebrates, 531, **535,** 537, 538, **548, 550, 552**
  adaptations of, 538–557
  general requirements of, 538
  origin of, 531, 532
Vomer bone, **126, 133,** 136, **140, 142,** 143, 561
Vomeronasal organ, 323, 358, 359
Vultures, 537, 541, 543, 545, 553, 568, 578

Walking gaits, 454, 457, **458**
Wallabies, **69,** 570
Walrus, 428, 529, 566, 582
Wave drag, 512, 514
Wave-riding of dolphins, 517
Wax glands, 102
Weasels, 69, 457, **459**
Weberian apparatus, 238, 366
Whales, **8,** 69, 101, 220, 280, 299, 428, 527
  aquatic adaptations of, 507, 511, 523, 526–529

  feeding adaptations of, **572,** 576
  (*see also* Cetacea)
White matter, 314–**318,** 332, 335, **340**
White muscle, 183, **519**
White rami of spinal nerves, **326**
Wing loading, 544, 547, 551, 553, 554
Wing slots, **540, 552**–554
Wing structure, 551, **552, 553,** 554
Wolf, **142–145**
Wombats, 68, **171,** 447
  feeding adaptations of, 565, 570
  fossorial adaptations of, 466, 469
Woodpeckers, 486, **488,** 492, 500, **502,** 566, 574
Work of muscles, 184

Xanthophores, 89
Xiphihumeralis muscle, **199**
Xiphisternum, **159**

Yaw, 160, 523, 525, 549
Yolk, 76, **77,** 78, **80, 81, 84,** 221, **249, 297**
Yolk plug, **78**
Yolk sac, 68, 83, **84,** 210, **211,** 257, 258, **260,** 261, 263, 276, 293, **294,** 304, 309

Zona pellucida, **76**
Zygapophyses, **148,** 153, 154, 156, 494, 497, 519
Zygomatic arch, **138, 142, 143,** 400, **401,** 424, 565
Zygomatic salivary gland, **213**
Zygote, **77**